Coalbed Methane of Western North America

Guidebook for the
Rocky Mountain Association of Geologists
Fall Conference and Field Trip

September 17–20, 1991
Glenwood Springs, Colorado

Co-Sponsored by the
Gas Research Institute

Editor-In-Chief
Stephen D. Schwochow

Associate Editors
D. Keith Murray
Michael F. Fahy

Published By

Rocky Mountain Association of Geologists
Denver, Colorado
1991

President's Message

The business objectives and purposes of the RMAG, as stated in the Articles of Incorporation, are (a) to promote interest in geology and allied sciences and their practical application; (b) to foster scientific research; and (c) to encourage fellowship and cooperation among its members. One of the main ways we achieve the scientific objectives is by publishing guidebooks and holding symposia. This year's guidebook, edited by Stephen Schwochow, Keith Murray, and Michael Fahy, is our second volume dealing with coalbed methane. I want to thank the editors and authors for their hard work in putting the volume together. Special appreciation is expressed to the senior editor, Stephen Schwochow, who also did the desktop publishing of the volume.

The theme of the 1991 Fall Conference centers around the Guidebook. Special thanks go to Ron Pritchett and his committees for their many months of planning.

I also want to express my thanks to the financial contributors to this year's Guidebook and Conference. The primary support came from the Gas Research Institute in Chicago. Their generous contribution and those of other corporate supporters and individuals have helped offset the ever-increasing costs associated with publications and meetings.

Stephen A. Sonnenberg
President
Rocky Mountain Association of Geologists

Rocky Mountain Association of Geologists
1991 Officers

President .... Stephen A. Sonnenberg
President-Elect .... Duff Kerr
1st Vice President .... Edward Coalson
2d Vice President .... Brad Morrison
Secretary .... Terry Kirkpatrick
Treasurer .... Stephanie Urban
Councilor .... Mel Glerup
Councilor .... Dorman Followwill

Library of Congress Catalog Card Number 91–62433
ISBN 0–933979–14–2

Co-Sponsor's Message

The development of coalbed methane resources has expanded rapidly over the last several years. Industry has experienced both success and failures in its attempts to produce these resources. Coal seam gas cannot be produced by simply applying conventional oilfield technology. Rather, it is necessary to understand the gas storage and production mechanisms in coal. Much of this knowledge has been developed and is described in this guidebook. Yet as more and more wells are drilled, we still find surprises. Many of the characterization techniques are not entirely adequate because of a lack of standard procedures or because all the subtleties of the physics and chemistry of this unique reservoir are not fully understood.

In this guidebook you will find the current state of understanding of gas production from coal seams and results of recent coalbed methane resource exploration and evaluation efforts. Clearly the coalbed methane story is not yet complete. Controversies still exist over the importance of several mechanisms, and new issues are being brought to light as development proceeds. Future progress in coalbed mehane depends upon development and application of new technologies. But even the most successful new tool or technique will fail if its impact on the resource is not communicated to others. As a cosponsor of this conference, the Gas Research Institute is pleased to have supported the publication of this guidebook—an important communication mechanism to industry for new ideas—and hopes that the interchange of ideas will better enable you to pursue the understanding and development of this resource.

Richard A. McBane
Manager, Coal and Shales Resource
Gas Research Institute

GRI
Gas Research Institute

Conference Chairman's Message

We are fortunate to have a conference in Glenwood Springs, Colorado. The site is ideal for comradery and exchange of news between people in research and production of coalbed methane.

The valley is one of many sacred places on the western slope of Colorado in the Southern Rockies, known by Ute Indians as "unca-pah-gre" or "hot water spring." We know something of the technical side—that the waters debouch to the Colorado River at a temperature of about 50°C and a volume of over 2,000 gallons per minute, and the waters contain total dissolved solids of about 20,000 milligrams per liter after bubbling through Paleozoic salts. One dip in the pool allows us to see through the eyes of the early Americans, who treasured the calming and restorative powers of the hot springs.

People who participate in the conference and use the Guidebook will have special insight to the region, resources, and cultures of western North America.

It is an honor and pleasure to have worked with the committee leaders, authors, and RMAG officers and staff to create the 1991 Guidebook, conference, and field trip. The generous spirit of these geoscientists is the source of the conference.

Ron W. Pritchett
General Chairman

Program Chairman's Message

The 1991 RMAG Symposium on Coalbed Methane in Western North America has been an exciting one to work with from several aspects. First and foremost are the people. RMAG has some of the finest people anywhere including this year's staff for the Symposium. I have enjoyed working with them even though it seems we were always faced with a deadline to meet.

Also, was the subject of the symposium—coalbed methane is exciting. The conference (hearing the enthusiasm of the speakers to present their papers) and field trip (seeing firsthand the coals as a source rock and wells that produce coalbed methane) have disseminated current technology to those who are in exploring and developing this new and vital natural gas resource. The impact that coalbed methane will have on the natural gas industry is just beginning to be realized at the publishing of this guidebook.

Finally, Glenwood Springs is a beautiful location and an exciting place to hold a conference.

I hope each of you have enjoyed the symposium as much as we have, and find the information useful in your search for and production of coalbed methane.

Henry J. Smith
Program Chairman

Acknowledgments

The Conference Organizing Committee gratefully acknowledges the following companies who have contributed toward the publication of this Guidebook:

Amoco Production Company
Barrett Resources Corporation
Dekalb Energy Company
Fuel Resources Development Company (Fuelco)
Gas Research Institute
Marathon Oil Corporation
McKenzie Methane Corporation
McKenzie Petroleum Company
Phillips Petroleum Company
Rocky Mountain GeoEngineering Company
Schlumberger Well Services Inc.
Taurus Exploration Inc.
Terra Tek Inc.
Western Gas Supply Company (WestGas)
Williams Field Services Company

Organizing Committee

General Chairman	Ron W. Pritchett, Resource Projects
Vice Chairman	D. Keith Murray, D.K. Murray and Associates Inc.
Program Chairman	Henry Smith, Barrett Resources Corporation
Field Trip Co-Chairman	Bradford Boyce, Bowen/Edwards Associates Inc. Clarence Harr, Soccobe Energy Bruce Kelso, ICF Resources Inc.
Field Trip Bus Leaders	John Rold, Colorado Geological Survey Richard Powers, U.S. Geological Survey J. Dan Powell, Consulting Geologist Edwin Price, CER Corporation Bruce Collins, Mid-Continent Resources George Hampton III, Hampton and Associates Robert Chancellor, Consulting Geologist
Guidebook Editors	Stephen Schwochow, Colorado School of Mines D. Keith Murray, D.K. Murray and Associates Inc. Michael Fahy, M. Fahy Consulting Inc.
Sponsors & Contributions	Bruce Kelso, ICF Resources Inc. Tracy Crist, ICF Resources Inc.
Publicity	Gary Taylor, Amoco Production Company
Publications	Stephen Schwochow, Colorado School of Mines Gary Hall, Consulting Geologist
Financial	Stephanie Urban, Consulting Geologist
Housing	Marshall Crouch III, White Eagle Exploration Company
Registration	Linda Flis, Geotechnical Services Sharon Kirts, Colorado School of Mines
Photography	Jack Rathbone, Consulting Geologist
Sketches	Andrew Alpha, Andy Alpha and Associates

Contents

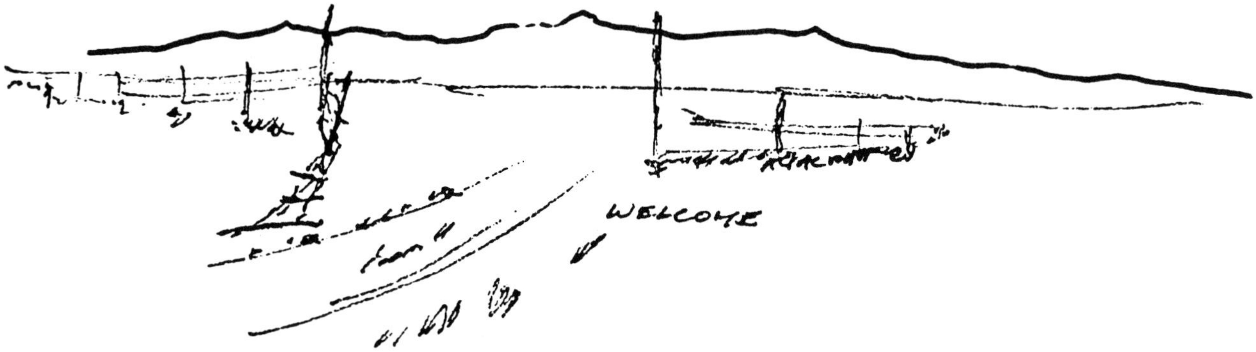
WELCOME

Depositional Environments and Stratigraphy of Mesaverde Formation, Southeastern Piceance Basin, Colorado—Implications for Coalbed Methane Exploration

Henry C. Nowak[1]

1. Shell Western Exploration and Production Inc., Houston, Texas

ABSTRACT

The Upper Cretaceous Mesaverde Formation in the southern Piceance basin of western Colorado contains, by one estimate, 36 TCF (1.02×10^{12} m^3) of natural gas in the form of coalbed methane. The intertonguing relationships of coal-bearing rocks are complex. In order to explore successfully for coalbed methane, an understanding of the depositional environments of the Mesaverde Formation is necessary. Results indicate that the lower part of the coal-bearing Mesaverde Formation, the Bowie Shale Member, was formed in a wave-dominated lower delta-plain setting. At the top of the member, areally extensive and stacked distributary channels show evidence of tidal influence in the form of herringbone cross-bedding and *Ophiomorpha* burrows.

The upper part of the coal-bearing Mesaverde Formation, the Paonia Shale Member, contains rocks formed in an upper delta-plain setting. Distributary channel sandstones are thinner and less extensive than those of the lower delta plain.

Paleogeographic reconstruction suggests that the deltaic shoreline prograded in an easterly direction. Coalbeds up to 32 ft (10 m) thick are found in the lower part of the Bowie Shale Member. These thick coalbeds trend parallel to the paleoshoreline and extend 6 to 8 miles (10 to 13 km) landward (west). Two thick coal zones within the Bowie are aligned in a north-south trend. The combination of coalbed thicknesses of up to 32 ft (10 m), and a previously determined high coalbed methane potential using additional criteria, indicates a prime area for coalbed methane within the Bowie Shale Member of the Mesaverde Formation.

INTRODUCTION

Location

The study area is located in the southeastern Piceance basin of western Colorado, in northeastern Delta County and northwestern Gunnison County, about 140 miles (225 km) southwest of Denver (Figure 1). The Mesaverde Formation crops out along the North Fork of the Gunnison River. The area covered by this study is situated between Hubbard Creek, located 5.5 miles (8.9 km) northeast of Paonia, and Paonia Reservoir 8 miles to the east. The study area covers approximately 300 mi^2 (780 km^2) extending north from the outcrop belt into the basin (Figure 2)

This paper describes the stratigraphy of the coal-bearing part of the Mesaverde Formation. Use of the term Mesaverde Formation, rather than the somewhat more familiar Mesaverde Group, will be explained below. Data from outcrops along the North Fork, from well logs, and from core descriptions were used to determine depositional environments. The stratigraphic interval covered in the subsurface encompasses a zone of high coalbed methane potential. In the southern Piceance basin, the coalbed methane in-place resource has been estimated to be 36 TCF (1.02×10^{12} m^3) (Choate and others, 1984). Later estimates made under Gas Research Institute sponsorship put the in-place resource for the entire Piceance basin at 84 TCF (2.38×10^{12} m^3) (McFall and others, 1988).

Federal coal leases to the north and adjacent to the Somerset Mine recently were relinquished, and core descriptions from holes located on those leases became available. Three of those descriptions have been incorporated into this study. The remainder of the subsurface data, including three Amoco Production Company cores from wells drilled for coalbed methane, are from oil and gas exploration activity.

Presented here is a lithostratigraphic correlation of measured outcrop and cored sections, a correlated well log section through a high coalbed methane potential area, and a diagram showing interpreted paleogeography of the area during late Campanian (Late Cretaceous) time. Finally, a model for coalbed methane exploration in southeastern Piceance basin is described, and a coal isopach map is presented, together with a map showing a possible area for future exploration.

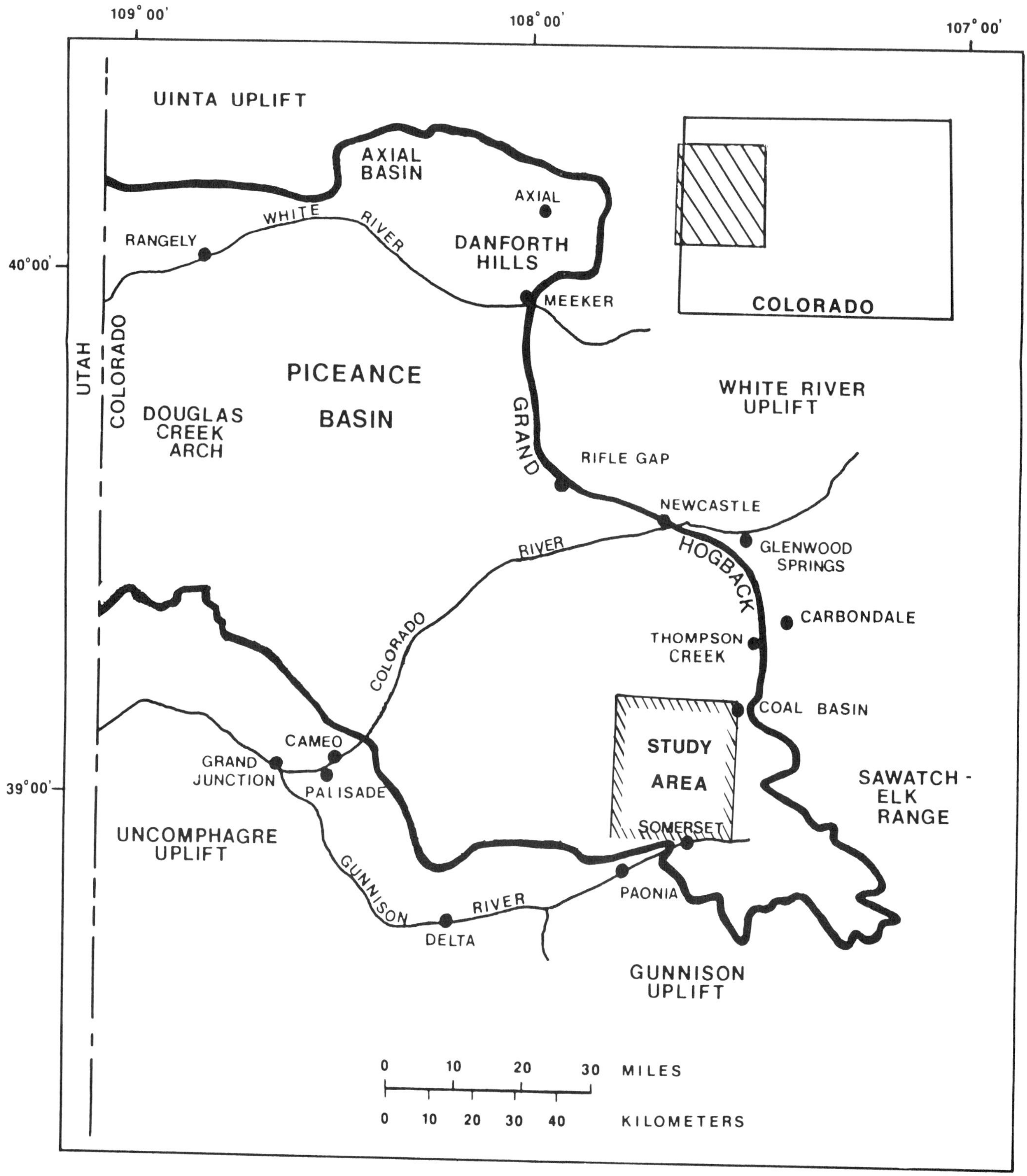

Figure 1. Location of Piceance basin study area, western Colorado, showing adjacent major tectonic elements. Modified from Collins (1976).

Previous Investigations

Within the study area, the coal-bearing part of the Mesaverde Formation is divided into two informal units (Lee, 1912). The lower unit is named the Bowie shale member. Overlying this unit is the Paonia shale member. A third unit, called the "barren member," informally named for its lack of economical coalbeds, overlies the Paonia member (Figure 3).

For the purpose of this study, the interval between the top of the Rollins Sandstone Member (basal Mesaverde)

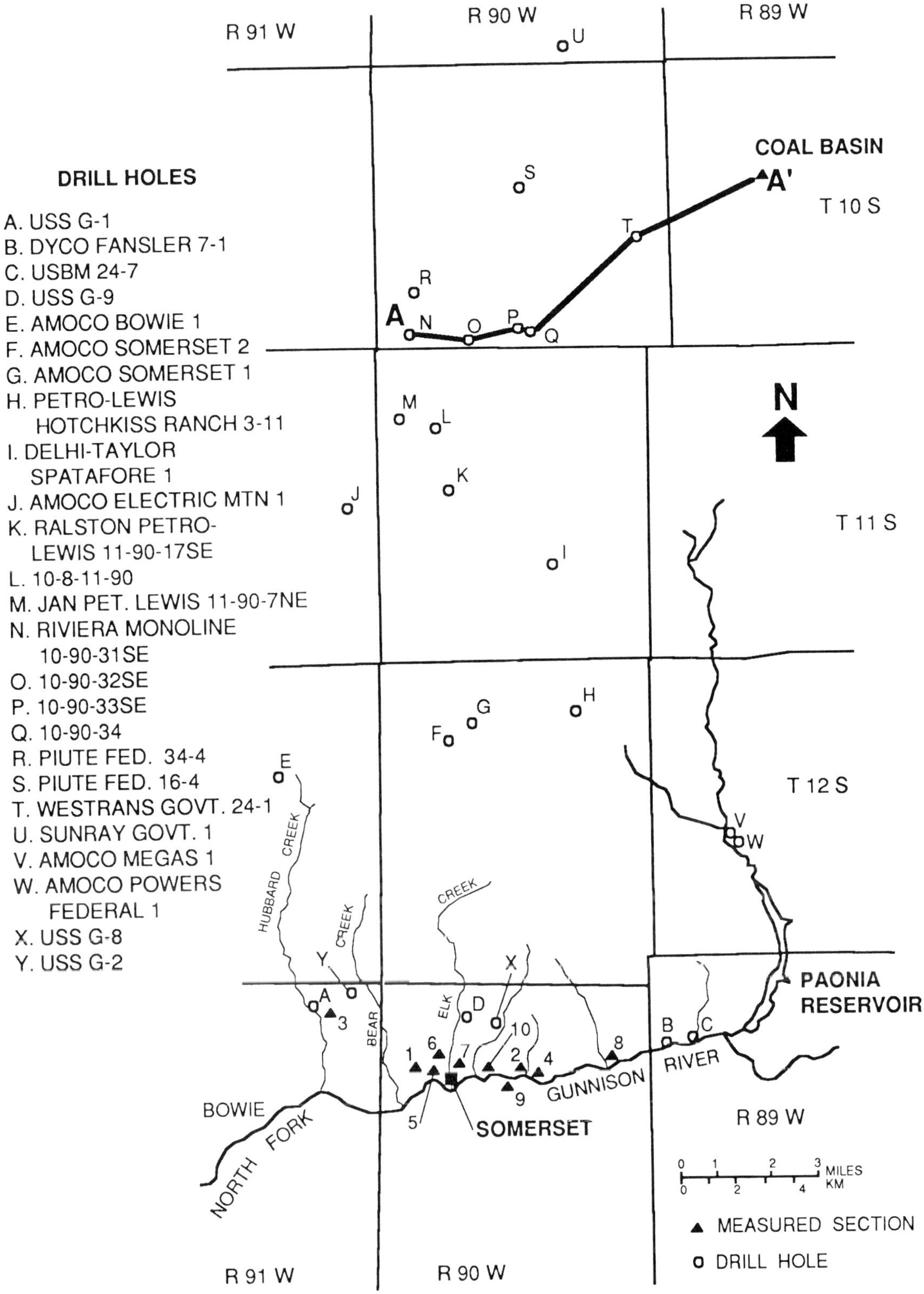

Figure 2. Locations of drill holes and measured outcrop sections in southeastern Piceance basin study area. See Figure 5 for lithostratigraphic section along North Fork of the Gunnison River between Hubbard Creek and Paonia Reservoir. See Figure 12 for log correlation between wells along line of section A–A′.

and the top of the sandstone that Hanks (1962) called the "middle sandstone" and that Wellborn (1982) termed the "Mt. Gunnison distributary channel sandstone" is named the Bowie shale member. The rocks above the middle sandstone and below the top of the F coalbed are called the Paonia shale member. The interval between the top of the

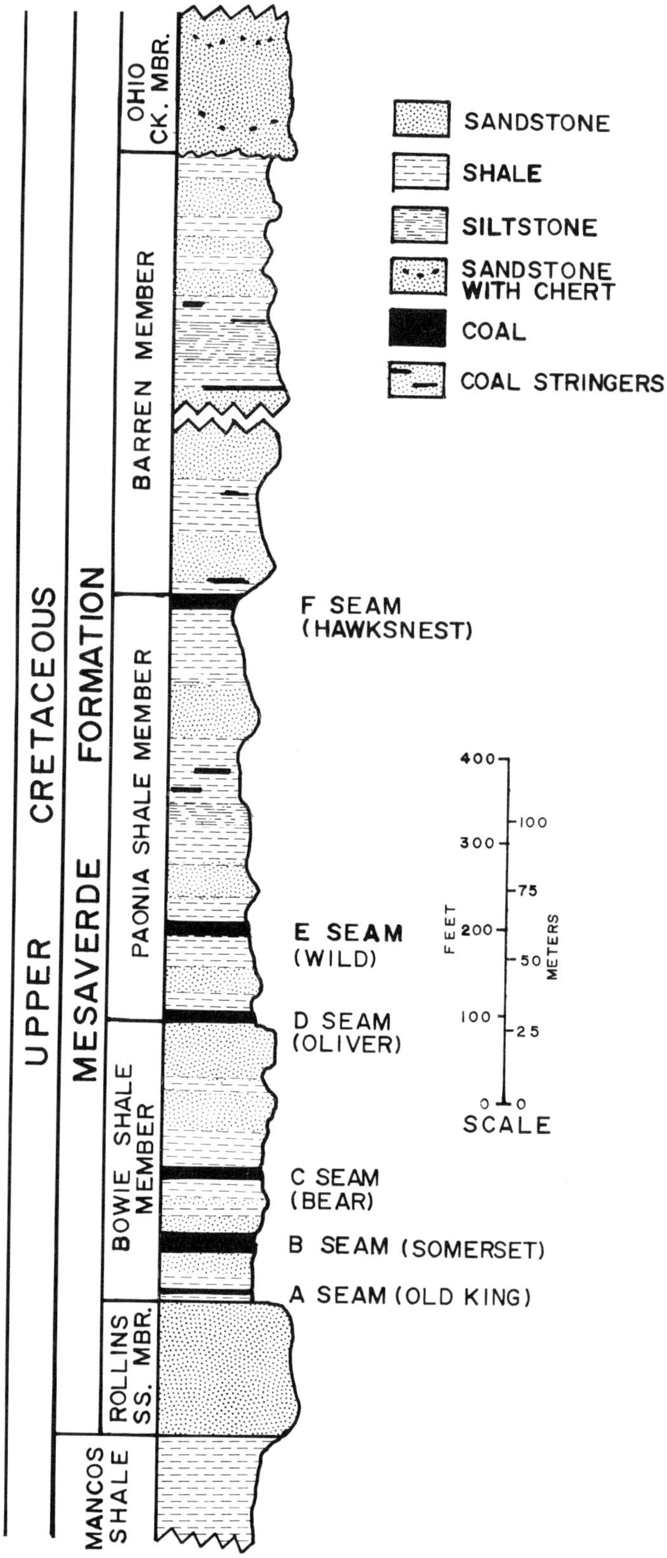

Figure 3. Generalized stratigraphic column of Mesaverde Formation in southeastern Piceance basin (modified from Daub, 1982).

F coalbed and the Ohio Creek Conglomerate is named the barren member.

Collins (1970) first proposed a deltaic origin for the Mesaverde Group in eastern Piceance basin. His study area covered Coal Basin, which is located 10 to 15 miles (16 to 24 km) north-northeast of the Mesaverde outcrop along the North Fork Valley.

Wellborn (1982) described the rocks in a stratigraphic study of the Mesaverde Formation in an area immediately south of the town of Somerset and adjacent to the southern boundary of this study area. Wellborn's interpretation of Mesaverde depositional environments supported Collins' deltaic interpretation. Part of the evidence included the occurrence within the Bowie shale member of a thick sandstone interval that Wellborn named the Mt. Gunnison distributary channel system.

While previous studies have focused on stratigraphy and coal geology, this report is the first in the southern Piceance basin to combine outcrop data with subsurface data on coal-bearing rocks that have high methane resource potential. The results are the basis for a predictive model of coalbed distribution based on lithologic and paleogeographic indicators of the depositional environments.

REGIONAL STRATIGRAPHY

Sedimentary rocks that comprise the Mesaverde Formation constitute a prograding clastic wedge that originated with erosion of the Sevier orogenic belt in central Utah (Spieker, 1949). These sediments were deposited along the western edge of, and within, the epeiric Cretaceous sea. As the depositional basin filled, the wedge prograded from west to east. Continued downwarping of the basin allowed increasing space to accommodate the accumulation of upper coastal-plain and piedmont deposits. The interfingering of marine mudstone and sandstone resulted in a stratigraphic rise of the transitional lithologic boundary of the Mesaverde Formation and Mancos Shale. The boundary becomes younger from west to east (Young, 1982).

Because of the intertonguing nature of the transitional zone between the marine Mancos Shale and the marine-nonmarine Mesaverde Formation, several discreet lithologic units have been named. In the study area, the Cozzette Sandstone Member of the Mancos Shale is the last complete cyclic pulse of transitional sandstone facies. Above the Cozzette Sandstone and separated from it by a tongue of Mancos Shale is the Rollins Sandstone, a progradational sheet sandstone that preceded nonmarine deposition.

North of the study area, the sequence of interfingering sandstone and shale, beginning in the western Piceance basin with the Sego Sandstone and ending at the top of the Rollins Sandstone, generally is referred to as the Iles Formation. As marine rocks become more dominant toward the study area, the Iles Formation becomes progressively finer grained. In the White River uplift area of the northern Piceance basin, nonmarine sandstone, shale, and commercial coal deposits are found within the Iles Formation. Southward, however, the coals become thinner, and more marine rocks appear (Collins, 1976). As Collins noted, the southernmost mineable coal in the Iles Formation occurs at Harvey Gap (Figure 1), and the last coalbed occurs just 12 miles (19 km) southeast at New Castle.

Within the study area, the southward thinning Iles Formation is finally reduced to the Rollins Sandstone, intertonguing Mancos Shale, and perhaps the Cozzette Sandstone at the southeastern basin boundary along the North Fork Valley. It is unlikely, however, that the Cozzette Sandstone is present at the eastern end of the area. This assumption is based on the overall east-southeastward thinning trend of the sandstone. The Cozzette thins from more than 35 ft (11 m) 10 miles (16 km) west of Dry Creek (Dunrud, 1989) to just several feet thick at its easternmost outcrop near Dry Creek, 6 miles (9.7 km) east of Cedaredge and 19 miles (31 km) west of Somerset. In this study the name Iles Formation therefore is not used. The Rollins Sandstone Member lies immediately below the nonmarine, coal-bearing rocks of the Mesaverde, and the Mancos Shale is below the Rollins.

The Rollins Sandstone, which Collins (1976) interpreted as a prograding bar-beach-delta front sand complex, is an important marker horizon throughout the central, southern, and eastern Piceance basin because it is overlain by the principal coal-bearing zone of the Mesaverde Formation. Warner (1964) traced the Rollins Sandstone across the southern Piceance basin on electric logs. He considered it to be the equivalent of the Trout Creek Sandstone Member of the Iles Formation on the eastern side of the basin. Warner (1964) noted that "the name Rollins is used as far north as New Castle, and the name Trout Creek is used from New Castle north along the Grand Hogback." Madden (1985, 1989) showed, through detailed mapping along Grand Hogback, that the Trout Creek Sandstone has been miscorrelated with the Rollins Sandstone at New Castle. The two units are not equivalent but actually are separated by 310 to 450 ft (94.5 to 137 m) of section; the Trout Creek at Rifle Gap lies above the Rollins Sandstone at New Castle (Madden, 1989). Above the Rollins Sandstone, the coal-bearing Mesaverde is predominantly nonmarine. In the northern Piceance basin the Mesaverde is elevated to group status, with the coal-bearing section called the Williams Fork Formation (Figure 4). Based on these relationships, the term Mesaverde Formation, rather than Mesaverde Group, is used in the study area.

Thickness of the Mesaverde Formation is variable, from 2,000 ft (610 m) in the Somerset area (Johnson, 1948) to 3,500 ft (910 m) at Coal Basin (Collins, 1976), on the north-eastern edge of the study area, to 5,155 ft (1,570 m) at Rifle Gap (Warner, 1964).

STRATIGRAPHY OF MESAVERDE FORMATION

In order to depict the data from a variety of sources in a comprehensive manner, a west-to-east lithostratigraphic cross section was constructed of the Mesaverde Formation along the North Fork Valley (Figure 5). The section

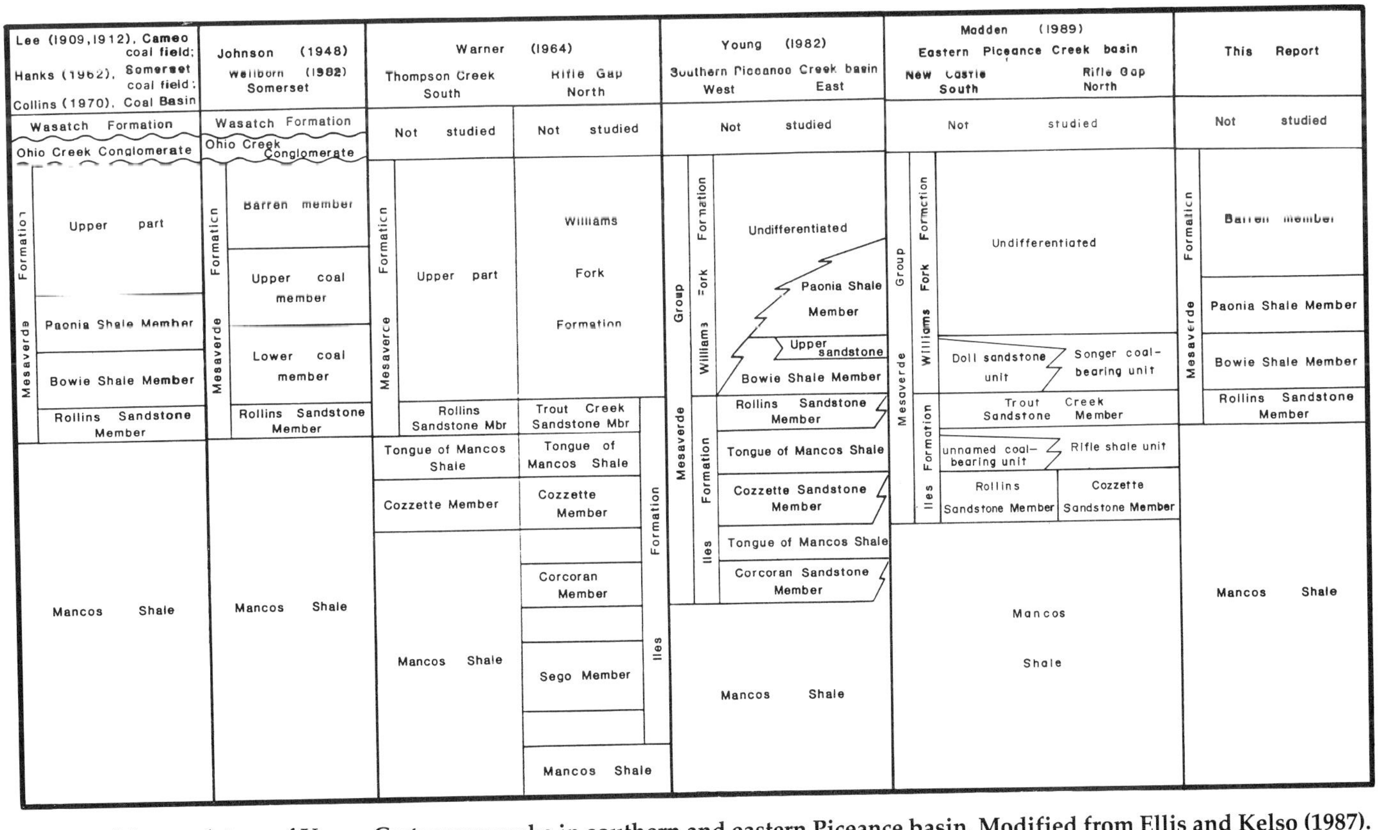

Figure 4. Nomenclature of Upper Cretaceous rocks in southern and eastern Piceance basin. Modified from Ellis and Kelso (1987).

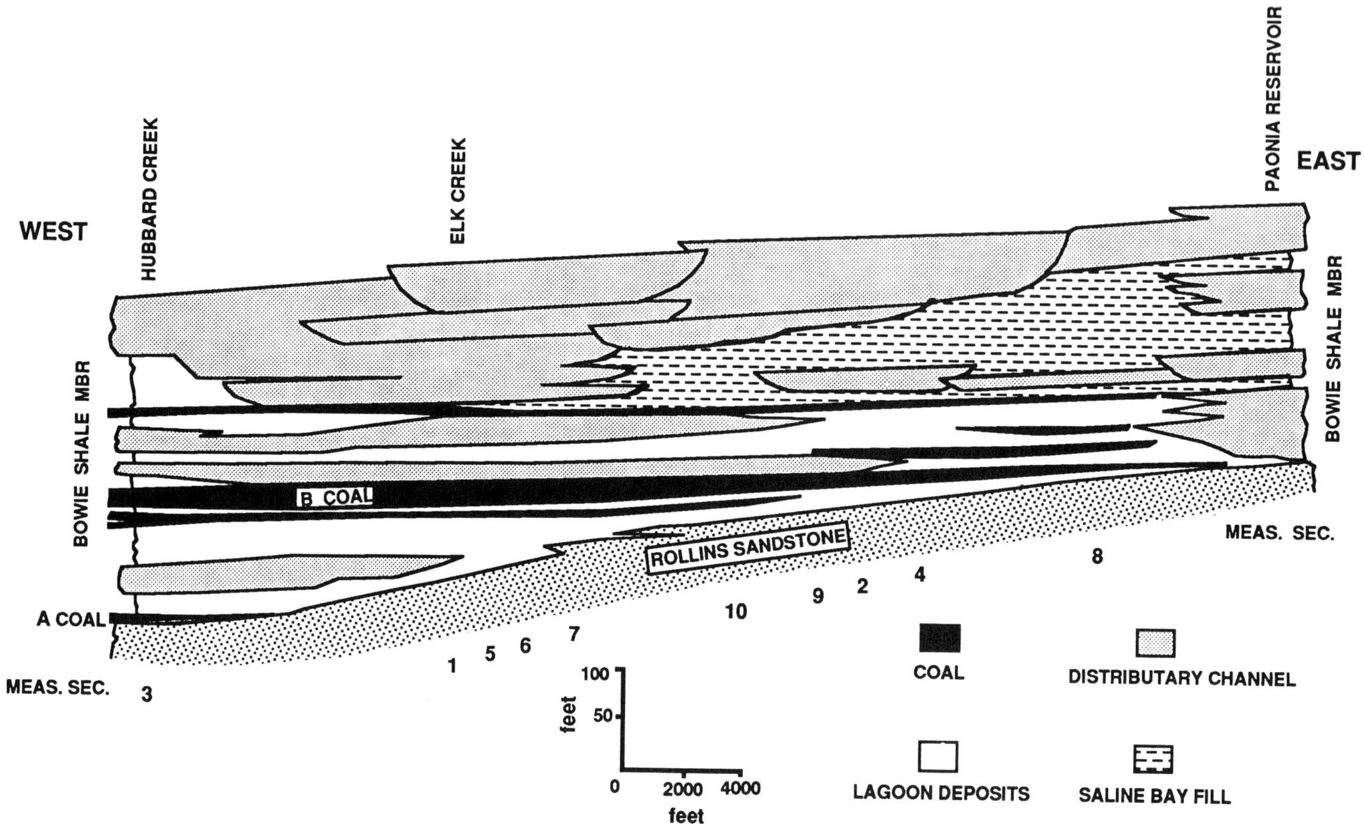

Figure 5. Lithofacies section of outcropping Bowie shale member along North Fork Valley. Several points are based on nearby core descriptions. See Figure 2 for locations of sections.

approximately parallels depositional dip of Mesaverde strata during late Campanian time.

Rollins Sandstone

The Rollins Sandstone is named for the Rollins Mine north of the town of Delta. In the study area and throughout the southern Piceance basin, it is a conspicuous cliff-forming tan sandstone 100 to 250 ft (30 to 76 m) thick (Wellborn, 1982). It usually is capped by 5 to 20 ft (1.5 to 6.1 m) of bleached white sandstone (Figure 6). The Rollins consists of medium-grained, quartzitic sand locally with lithic or feldspathic intervals. In most places the sandstone is cemented with calcite and has up to 35% matrix clay (Collins, 1970). Coalified plant remains are abundant in fresh rock samples from the uppermost part (Collins, 1970).

Bowie Shale Member

The Bowie shale member is named after the now defunct coal mining village of Bowie in Delta County at the western edge of the study area. The member lies directly above the Rollins Sandstone and ranges in thickness from 330 ft (100 m) in the western part of the study area to 260 ft (79 m) at the eastern limit. The rocks consist of coal, fine-grained sandstone, siltstone, and shale. Although the contact with the Rollins Sandstone generally is distinct, no single rock type commonly contacts the Rollins Sandstone.

Above the Rollins contact is an interval of dark-gray carbonaceous shale averaging 10 to 30 ft (3 to 9 m) thick and bearing *Corbicula* and *Ostrea* fossils (Wellborn, 1982). Core logs from Geneva Steel (USS G–1 and USS G–9, Figure 2), former operator of the Somerset Mine, also note a black, highly fossiliferous (oyster shell) carbonaceous shale 10 ft (3 m) above the Rollins Sandstone. Thin bioturbated sandstones are interbedded with carbonaceous, often mottled and convoluted, ripple-laminated siltstone.

Above the shale interval is the principal coal zone, the "B" or Somerset zone (Figure 3). At the western end of the outcrop study area the zone has an aggregate thickness of 30 ft (9 m).

Immediately above the B coal zone at the eastern end of the study area is a 75-ft (23-m) interval of medium-grained sandstone shaly at the base and containing a 7-ft-thick (2.1-m) *Ophiomorpha* zone (Toenges and others, 1952). This interval grades westward abruptly into coal, carbonaceous siltstone, and shale. The C or Bear coal zone lies above the predominantly carbonaceous, shaly, and coquina-bearing interval.

Middle Sandstone Unit

At the top of the Bowie shale member lies the middle sandstone (Hanks, 1962; Collins, 1970), which Lee (1912) defined as the top of the Bowie shale member. Wellborn

(1982) named this interval the Mt. Gunnison distributary channel system. This mostly massive sandstone is medium to fine grained, tan to buff-colored in outcrop, and cemented with calcite and authigenic clay. In the south-central part of the study area, where it crops out at Elk Creek (Figures 2 and 7), the sandstone is 175 ft (53 m) thick but thins considerably to 51 ft and 62 ft (16 to 19 m) east and west of Elk Creek, respectively.

The unit is divided into several 5- to 10-ft-thick (1.5- to 3-m) subunits consisting of thin coal, gray carbonaceous claystone, and red, silty and crumbly paleosols. Their lower parts often exhibit trough cross-stratification, locally abundant clay rip-up pebbles, and "coffee ground" coaly debris.

Ophiomorpha burrows were found in a thin interval 20 to 40 ft (6 to 12 m) above the base of the sandstone at Elk Creek and at several other locations (Figure 8). Herringbone cross-bedding also was observed above the *Ophiomorpha* zone at Elk Creek (Figure 9).

East of Elk Creek, the base of the middle sandstone is in sharp contact with thin-bedded siltstone intercalated with gray shale (Figure 5). The siltstones become thicker and coarser grained upward, and the shale partings thin and grade into sandy siltstone. Starvation ripples at the tops of the coarser grained facies exhibit bilobate feeding tracks, scattered small vertical burrows, and disarticulated remains of the bivalves, *Cardium speciosum* (Meek and Hayden) and possibly *Melania insculpta* (Meek) (Figure 10). The bases contain abundant *Thalassinoides* trace fossils and *Cylindrichnus* burrows. At measured section 6, in contrast, the middle sandstone sharply contacts coaly and carbonaceous siltstone (Figures 2 and 11) containing gray claystone and scattered reddish concretions 6 to 8 in. (15 to 20 cm) in diameter.

The upper part of the middle sandstone is devoid of fossils. Bedding structure is mostly massive with scattered planar, tabular cross-bedding. Shaly siltstone partings 6 to 24 in. (15 to 61 cm) thick become more numerous toward the top. Locally, the top of the middle sandstone is the upper stratigraphic boundary of the Bowie shale member.

Paonia Shale Member

The top of the middle sandstone marks the lower stratigraphic boundary of the Paonia shale member, which is 180 to 270 ft (55 to 82 m) thick along the North Fork Valley. In Somerset coal field the Paonia shale member contains two economically important coal zones—the "D" or Oliver coalbed, and the "F" or Hawksnest coalbed (Figure 3). The Blue Ribbon Mine at Hubbard Creek was believed to have worked the intervening E coalbed; however

Figure 6. Contact between Mancos Shale (Km) and Rollins Sandstone (Kr); Kbo, Bowie shale member. The "white cap" at the top of the Rollins is easily distinguishable. Photo courtesy of George Hampton III.

Elk Creek
Kpa
ms
Kbo
Kr
Somerset

CM
DNAG
PHOTO SCALE
IN
mm

Figure 9. Herringbone cross-bedding in the lower middle sandstone at measured section 6. Scale is 4 in. (10 cm) long.

Collins (1989, pers. comm.) stated that the mine actually worked the F coalbed equivalent at other localities. The E coaled in the study area is much thinner and does not extend east or west beyond it.

Barren Member

The barren member is the uppermost and thickest member of the Mesaverde Formation in the southern Piceance basin. Its base in the southern study area is defined as the top of the F coalbed, and the rocks consist mostly of coarse- to fine-grained sandstone. Lenticular sandstone bodies are more common than in the underlying members. The intervals between the sandstones contain abundant coaly plant material; rooted, thin, ripple-laminated silty sandstone; and interbedded concretionary shale. Palm leaf imprints also were observed in the silty sandstones.

Figure 7 (top left). Bowie shale member (Kbo) and middle sandstone (ms) at Elk Creek. Middle sandstone, at the top of the Bowie shale member, is over 120 ft (37 m) thick at this location. Top of Rollins Sandstone (Kr) is at the lower left, and Paonia shale member (Kpa) overlies the middle sandstone.

Figure 8 (bottom left). Ichnofossil, *Ophiomorpha nodosa*, near the base of the middle sandstone, Bowie shale member. Location is measured section 6.

DEPOSITIONAL ENVIRONMENTS

Stratigraphic relationships of the four generalized lithofacies within the coal-bearing Mesaverde Formation are shown in Figure 5, a west-to-east section along the North Fork of the Gunnison River between Hubbard Creek and a point just west of Paonia Reservoir. The inferred depositional environments and processes attributed to each lithofacies are based on rock descriptions obtained from outcrops and core observations and on wireline log characteristics.

Rollins Sandstone Member

In the southeastern Piceance basin the Rollins Sandstone Member represents the leading edge of a prograding clastic sediment sheet. The Rollins Sandstone is composed of offshore, littoral, tidal, and backshore deposits (Young, 1982). The offshore part consists of prodeltaic deposits. The littoral component consists of the delta front (including lower to upper shoreface and foreshore), distributary-mouth bar, and distributary channel. Backshore deposits are indicated in outcrop by the distinctive bleached zone at the top of the Rollins Sandstone (Figure 6). Similar bleached zones at the top of the Starpoint Sandstone in east-central Utah have been interpreted to represent a beach deposit (Flores and others, 1984). The bleached zone, or "white

cap," is interpreted to be an exhumed paleosol, through which humic acid derived from coalification in overlying zones has leached downward and reduced the iron in the sandstone.

Intertonguing of the Rollins Sandstone and Mancos Shale represents successive events of progradation and lateral shifting of depositional centers as the point sources of sediment—distributary channels along the barrier-delta front—were abandoned. The prodeltaic deposits at the base of the Rollins Sandstone consist of interbedded silty shale, siltstone, and silty sandstone. The silty sandstone beds, which are characterized by asymmetrical ripple laminations, increase upward in thickness and abundance. They contain numerous trace fossils, including *Thalassinoides*, *Arenicolites*, *Chrondrites*, and *Ophiomorpha*.

In many locations throughout the study area, prodeltaic deposits vary in thickness as a result of the changing location of the deltaic center of deposition. The autocyclic process of avulsion and subsequent delta abandonment changed the locus of deposition along the delta-barrier coastline complex. This style of deposition represents an interdistributary offshore sequence of delta-barrier front deposits reworked by longshore currents that moved generally south-southwest (Madden, 1989).

Thicker sections within the Rollins Sandstone represent the complete delta-front facies, which consists of distributary channel, distributary-mouth bar, shoreface, and backshore sandstones. Where the Rollins is thinner, such as in the well Riviera Monoline 10–90–31 ("N" in Figures 2 and 12), the stacking effect of one or more distributary channels or distributary-mouth bars is absent. Where present, distributary channels that are stacked on top of or scour into the Rollins are represented by a coarsening-upward or a "gamma ray response decreasing-upward" sequence on logs. Saxena (1982) described similar well log signatures from modern Mississippi delta deposits.

Channel-mouth bars have been observed in outcrops of the Starpoint Sandstone in east-central Utah (Flores and others, 1984). In this example, distributary-mouth bars are up to 60 ft (18 m) thick and are lower in elevation than distributary-channel sandstones. Figure 6 shows a possible distributary-mouth bar sandstone in the lower Rollins Sandstone west of Hubbard Creek. It appears to grade laterally into probable lower shoreface deposits.

Lithofacies of Bowie Shale Member

Four lithofacies (distributary channel, coal, lagoon deposits, saline bay-fill deposits) have been used to reconstruct the paleogeographic setting of the coal-bearing part of the Mesaverde Formation during its deposition:

Distributary-Channel Sandstone

Distributary-channel sandstone is massive to tabular, tangential and trough cross-bedded with convoluted laminations in the lower part and asymmetrical ripples in the upper part. Individual units are 20 to 50 ft (6.1 to 15 m) thick. Clay partings and scattered coalified woody debris occur along cross-bedding planes. Chert and indurated clay pebbles are present at sharp basal contacts and along large, 3- to 6-ft (1- to 2-m) trough cross-bedding scours. Herringbone cross-bedding is prevalent 25 ft (7.6 m) above the sharp basal contact. The two current directions, based on scattered readings, appear to be generally S70°E and N20°W. The sandstone is composed of fine- to medium-grained quartzose sand and 10% to 15% lithic and ferromagnesian accessory grains, with calcitic and minor, possibly authigenic, clay cement. The only fossil observed was the ichnofossil *Ophiomorpha nodosa* (Lundgren) (Figure7). The sandstone units are often stacked to form continuous intervals up to 90 ft (27 m) thick.

This lithofacies represents a lower delta-plain distributary channel. Herringbone cross-bedding indicates bidirectional, tidal currents. The presence of *Ophiomorpha* indicates that a saline wedge extended up the channel. Conditions were not ideal for *Ophiomorpha* to thrive, however, because the fossils are smaller and fewer in number than those observed in the Rollins Sandstone. The lower delta-plain setting was under the influence of tides, which allowed salt-water wedges to extend landward into the low-gradient channels.

The presence of *Ophiomorpha* in only the lower part of the middle sandstone, in most places, reflects changing environmental conditions as the delta front prograded seaward. The conditions changed from a tidal-influenced lower delta-plain to an upper delta plain with no tidal influence above the *Ophiomorpha*/herringbone cross-bed zone of the middle sandstone.

Distributary-channel levee deposits consist of interbedded shale, siltstone, and very fine-grained silty sandstone. The sandstone is ripple cross bedded to horizontally laminated in intervals 2 to 4 ft (0.6 to 1.2 m) thick, with prevalent root structures at the top. Commonly they are stacked to form thicker sequences. The deposits interfinger laterally with distributary-channel deposits.

Abandoned-channel deposits occur within and between stacked channel sandstones and consist of carbonaceous and silty shale with 0.25- to 4-in.-thick (0.6 to 10 cm) coal lenses and scattered coalified plant matter. Thickness varies from 2 to 30 ft (0.6 to 9.1 m). Unidentified burrows are present where the base of this facies contacts underlying sandstone. The top several feet of underlying sandstone is ripple laminated. Commonly these deposits locally are capped by coal up to 10 ft (3.0 m) thick.

Figure 10 (top left). ***Cardium speciosum*** **and** ***Melania insculpta*** **in calcareous siltstone from measured section 7 (Bowie shale member).**

Figure 11 (bottom left). Base of the middle sandstone (Bowie shale member) at measured section 6. Rocks below the sharp contact are thin-bedded coal and carbonaceous shale.

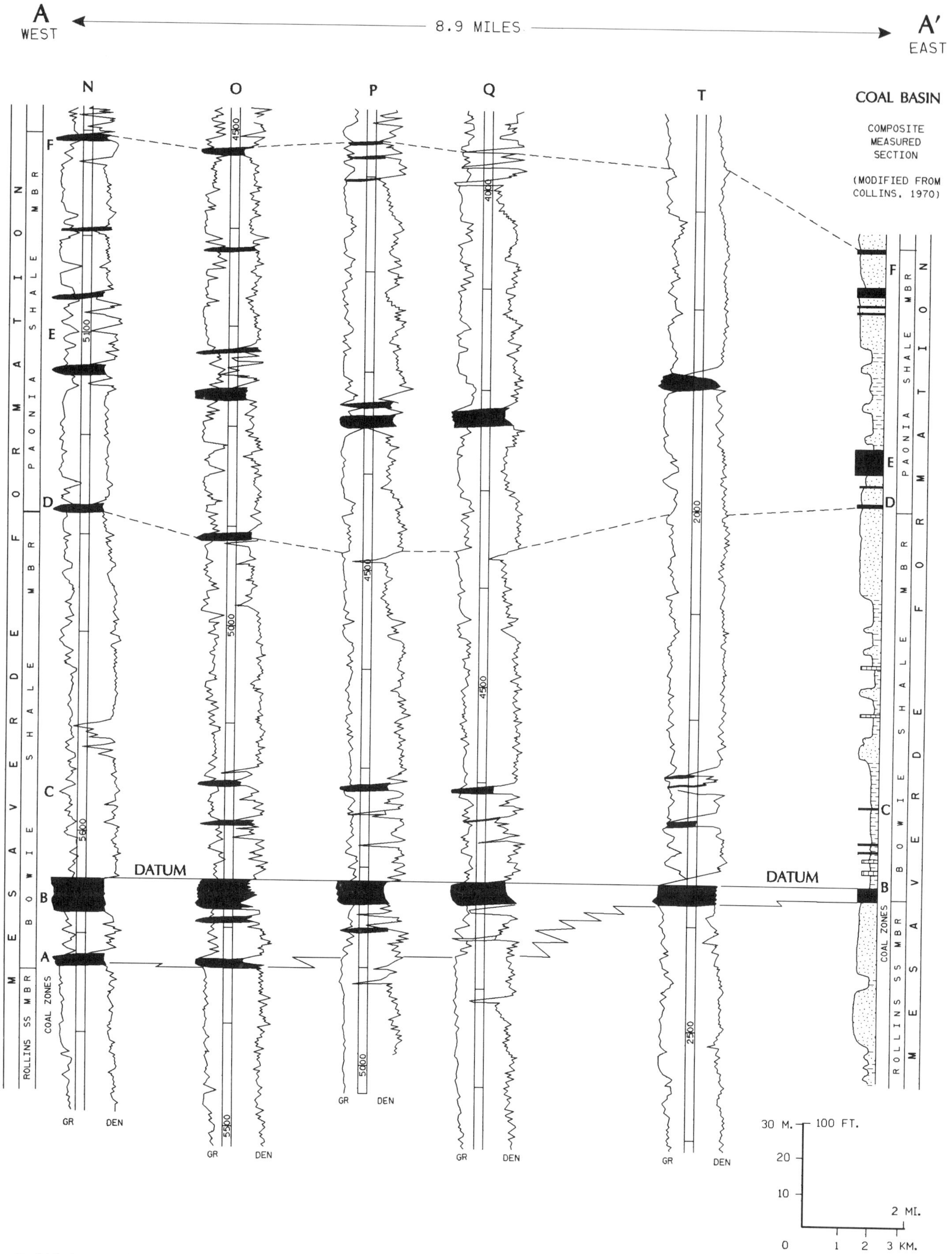
A
WEST
8.9 MILES
A′
EAST
N
O
P
Q
T
COAL BASIN
COMPOSITE MEASURED SECTION
(MODIFIED FROM COLLINS, 1970)
MESAVERDE FORMATION
PAONIA SHALE MBR
BOWIE SHALE MBR
ROLLINS SS MBR
COAL ZONES
F
E
D
C
B
A
DATUM
DATUM
GR
DEN
30 M.
100 FT.
20
10
0
1
2
3 KM.
2 MI.

Coal

The coal lithofacies contains blocky fractures and well developed face and butt cleats. Scattered, thin, resinous streaks occur on fresh surfaces. The coal is of the humic type, with an ash content generally between 5% and 15% (Toenges and others, 1952). Two coal lithotypes have been observed in outcrop—(1) black, bright-lustered, conchoidally fracturing vitrain, and (2) black, silky lustered, friable and soft fusain. Rocks below the coal layer frequently are mottled and rooted.

The depositional environment of the coal was a swamp protected from fluvial invasion and clastic deposition. McCabe (1984) suggested three possible mechanisms by which thick, low-ash peats can be preserved ultimately into economical coals. First, the acidic nature of swamp waters may have shielded the peat from clastic deposition by dissolution of minerals. Second, peat accumulated in raised or floating bogs protected by their slightly higher elevation. Third, peat deposition was not contemporaneous with clastic deposition; for example, regional clastic supply may have been cut off during swamp development. McCabe (1984) stated that the last factor probably was the most important.

Lagoon Deposits

The lagoonal facies consists of black shale with 1- to 4-in.-thick (3 to 10 cm) oyster-shell coquinas and interbedded siltstone and carbonaceous shale. Abundant brackish-water bivalves documented by Lee (1912) include *Corbula*, *Corbicula*, and *Ostrea*. Shale and interbedded carbonaceous shale are more common in the lower part, in contrast to ripple-laminated siltstone in the upper part. The lithofacies is as thick as 10 ft (3.0 m), but vertically stacked sequences can reach 30 ft (9.1 m). It often is bounded above by coal and below by either coal or the uppermost Rollins Sandstone. The depositional environment is interpreted to be a brackish-water bay or lagoon, based on lithology, fossils, lack of significant burrowing, and stratigraphic proximity to the Rollins Sandstone.

Saline Bay-Fill Deposits

The fourth lithofacies consists of ripple cross-bedded sandstone, interbedded with shale and silty sandstone 1 to 5 ft (0.3 to 1.5 m) thick. Sandstone intervals thicken upward and have sharp bases that are heavily bioturbated and slightly scoured into the underlying shale. Abundant saline trace fossils include *Thalassinoides*, *Cylindrichnus*, and other unidentified feeding tracks and burrows. Disarticulated bivalves of the species *Cardium speciosum* (Meek and Hayden) and possibly *Melania insculpta* (Meek) were found on the upper surfaces of thin calcareous siltstones (Figure 9).

Figure 12 (left). Well log sections and correlation through the northern study area. See Figure 2 for location of line of section A–A′ and names of wells.

In outcrop this lithofacies assemblage resembles the prodelta subfacies of the lower Rollins Sandstone with the exception that the degree of bioturbation in the silty sandstone is markedly less (because of brackish water) than in prodeltaic deposits below the Rollins (Figure 11). Cumulative thickness of this lithofacies averages 35 ft (11 m), in contrast to offshore prodeltaic sequences, which range up to 90 ft (27 m). This environment is interpreted to represent saline bay fill from a prograding bayhead delta or crevasse splay.

Summary of Depositional Environments

These lithofacies patterns fit the model of a deltaic system of deposition first described in this area by Collins (1970) and later detailed by Wellborn (1982). The Rollins Sandstone Member was deposited as a sheet sand at the delta front. Between channel mouths, where sediments were not being brought directly to the delta front by distributaries, sand was reworked by longshore currents into relatively thin shoreface and foreshore deposits with no accompanying channel stacking. Preservation of channels within and directly above the Rollins Sandstone resulted in thickness variations, as observed on well logs in the study area.

The Bowie shale member, which contains the thickest coalbeds, was deposited in a lower delta-plain setting influenced by tidal processes. This is indicated by brackish fauna and herringbone cross-bedding within bays and distributary channels. Thick accumulations of peat formed shoreward of the delta front at or near sea level in interdistributary areas protected from clastic input either by well developed channel levees, or as raised swamps (Figure 13).

PALEOGEOGRAPHIC RECONSTRUCTION

The rocks preserved along the North Fork Valley between Hubbard Creek and Paonia Reservoir represent the shoaling-upward sequence of an eastward-prograding barrier-delta front complex that existed along the western shore of the Late Cretaceous epicontinental sea. Marine muds and associated calcareous sediments are represented by the Mancos Shale. Collins (1970) named this delta complex the Piceance delta. It probably prograded toward the southeast from a north-northeasterly to south-southwesterly aligned paleoshoreline (Zapp and Cobban, 1960; Madden, 1989, p. 37).

A generalized paleogeographic map of the lower delta plain, distributary channel, crevasse splay, lagoon and marine bay, and barrier-delta front environments as they might have occurred in the intertonguing zones of the Rollins Sandstone and Bowie shale members is depicted in Figure 14. The map shows two delta complexes—one to the north in the early stages of abandonment as the location for delivery of sediment to the Mancos Sea migrates south-eastward to a newer active delta complex shown at the lower left. A wave-dominated system, with longshore drift

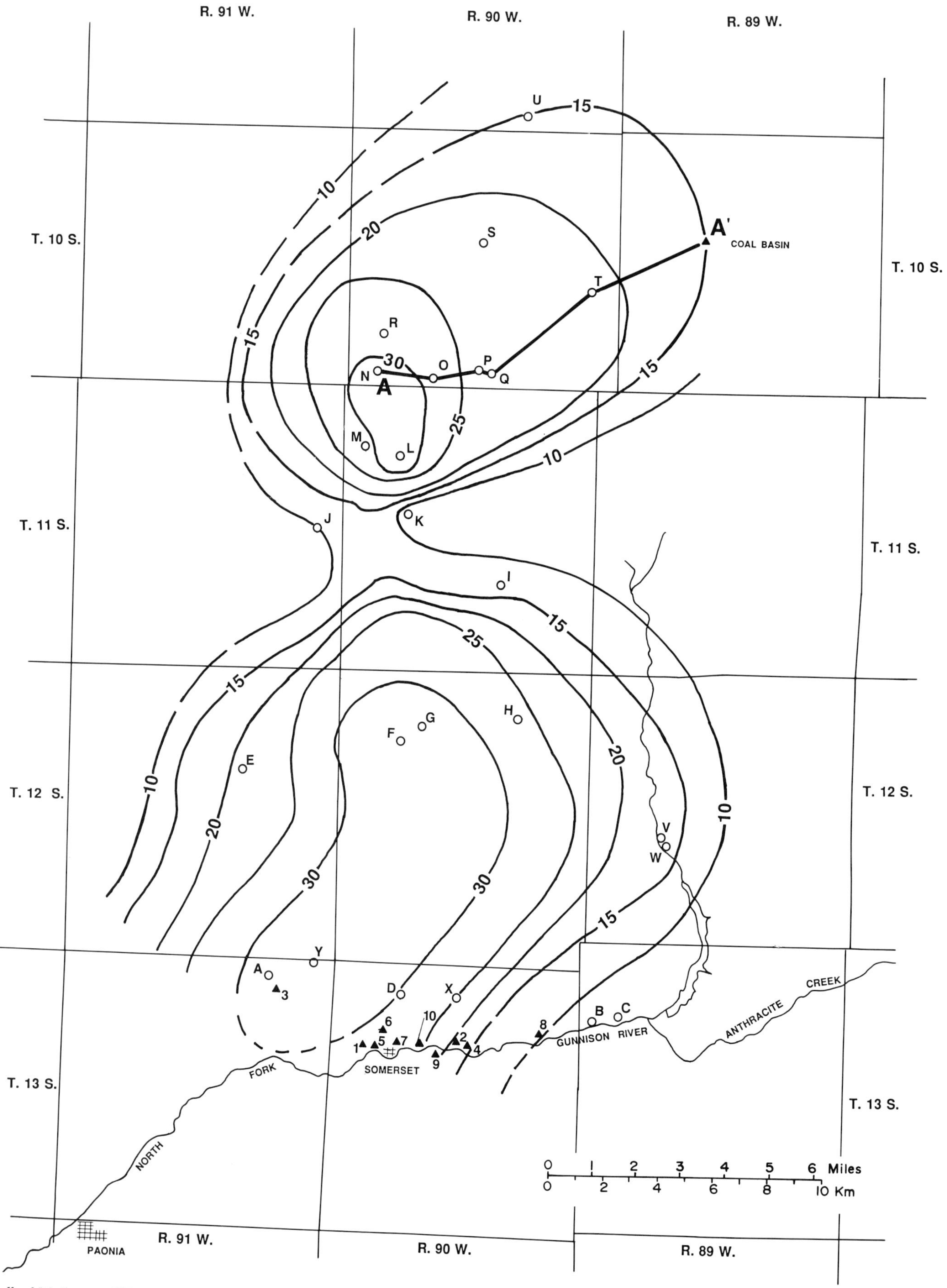
R. 91 W.
R. 90 W.
R. 89 W.
T. 10 S.
T. 11 S.
T. 12 S.
T. 13 S.
A
A'
COAL BASIN
GUNNISON RIVER
ANTHRACITE CREEK
NORTH FORK
SOMERSET
PAONIA
0 1 2 3 4 5 6 Miles
0 2 4 6 8 10 Km

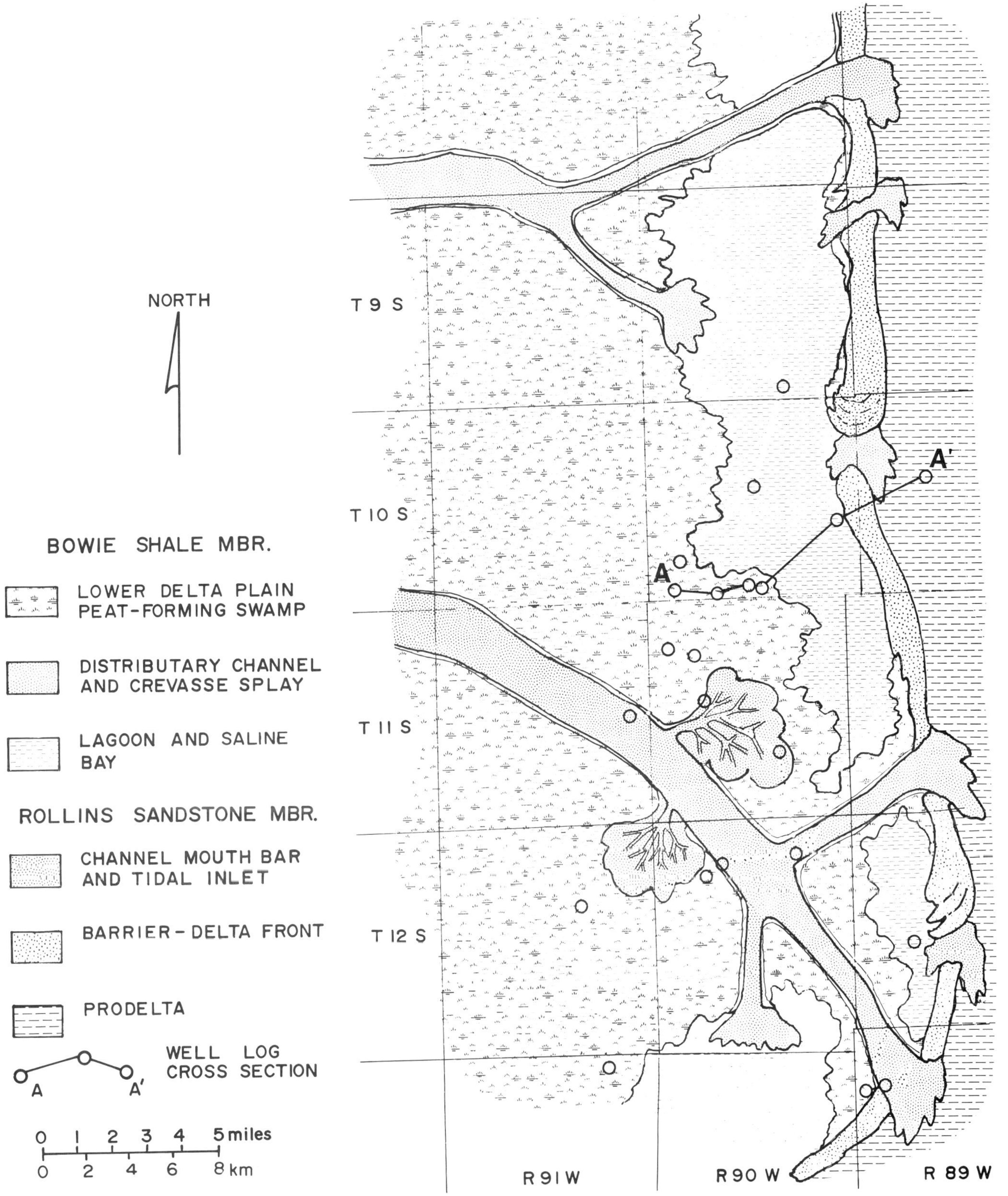

Figure 14. Paleogeographic map showing depositional environments of Bowie shale member and Rollins Sandstone Member.

Figure 13 (left). Isopach of B coal in southeastern Piceance basin. Contour interval 5 ft (1.5 m).

from north to south, is suggested by the north-to-south bending of channel-mouth bars and the spits forming on the southern ends of interdistributary barrier-delta front sand bodies.

In this reconstruction, narrow vegetated areas are beginning to form on the platform created by storm washover immediately behind the barrier sands. These sites of peat accumulation encroach landward into the lagoon environment. A similar relationship of barrier-delta front facies merging landward into coastal lagoons and swamps was interpreted for the intertonguing zone of the Starpoint-Blackhawk Formations in east-central Utah (Flores and others, 1984). North of this study area at Rifle Gap and New Castle, Madden (1989) interpreted the Trout Creek Sandstone as a north-northeast-trending strandline consisting of sandy beach ridges cut by tidal inlets. Both the Starpoint and Trout Creek examples strongly resemble the features observed in the Rollins Sandstone and Bowie shale member in the southeastern Piceance basin.

Two types of open-water bodies existed within the interdistributary/back-barrier areas of the lower delta plain. First was a brackish-water lagoon typified in core samples by dark-gray shale with oyster-shell coquinas and a paucity of other fossils or bioturbation. The other type was a saline bay indicated by preservation of a wide variety of ichnofossils, light-gray shale, and interbedded siltstone. Both environments are depicted in Figure 14. The marine bay is shown in T12&13S and R90W, with a prograding bayhead delta as interpreted from measured sections along the North Fork Valley.

The majority of coal in the Bowie shale member formed from peats that developed behind the barrier-delta front complex and in interdistributary areas. Coal preserved directly adjacent to and overlying the barrier-delta front Rollins Sandstone is thinner than that formed contemporaneously farther inland. For example, the thickest part of the B coal zone formed 6 to 8 miles (10 to 13 km) inland and is up to 32 ft (9.8 m) thick (Figures 12 and 13) but thins from west to east in the seaward direction. Where the B coal zone directly overlies the barrier-delta front facies, it is less than 2 ft (0.6 m) thick in the southern part of the study area. Underlying the B coalbeds, and separated by up to 100 ft (30.5 m) of lagoonal deposits, is the A coalbed which, in turn, directly overlies the Rollins Sandstone in the central part of the study area. Farther west, the thickest bed of the B coal zone splits and pinches out as crevasse splays and overbank flooding brought clastic sediment into the swamps.

According to Ferm and Staub (1984), sites of maximum peat accumulation appear to be delicately balanced between areas that are topographically low and easily flooded with clastics and areas that topographically are too high to permit peat accumulation. This topographic influence is controlled to the degree that underlying sediments can compact. Sediments that compact the least produce topographic highs, and sediments that compact the most produce lows. In the Bowie shale member coal is thickest where it overlies 20 to 40 ft (6.1 to 12 m) of brackish bay and lagoon deposits. It follows that compaction rates for that interval of lagoon and bay-fill sediments were optimal for thick accumulation of peat. Studies of Holocene peats indicate that the difference between "too high" and "too low" topographically is only several feet (Ferm and Staub 1984).

In summary, bodies of thickest coal, mainly the B coalbed in the study area, parallel the ancient barrier-delta front paleoshoreline. Where a coal directly overlies the barrier-delta front facies, it thickens in a landward direction for 6 to 8 miles (10 to 13 km), then begins to split and pinch out. The thickening, and landward splitting and pinching out, is directly related to interval thickness and lithology of underlying deposits between the B coal and Rollins Sandstone. The thickest part of lower delta-plain coals appears to be oriented subparallel to the northeast-southwest trend of the paleoshoreline.

IMPLICATIONS FOR COALBED METHANE EXPLORATION

The depositional models and paleogeographic reconstructions presented here can be applied to exploration for coalbed methane. A map of Choate and others' (1984) target areas for coalbed exploration in the Piceance basin is shown in Figure 15. These target areas are arranged subconcentrically from low potential, around the periphery of the basin, to high potential in the central part of the basin. The criteria used for determining potential include (1) areas of overburden greater than 2,000 ft (610 m) thick, (2) areas of thick coals, (3) areas of high heat flow, (4) areas along major synclines and anticlines, and (5) areas where igneous intrusions have caused fracturing and high heat flow. Coincidence of any two criteria constituted a "prime" target area. Further refinement of the location for thick coalbeds is introduced here for use as a model for thick coal occurrence.

As inferred from the depositional model and paleogeographic reconstruction, the thickest and most areally extensive coal formed where peat accumulation and subsidence of underlying sediments were in balance. This balance was maintained in an elongate area subparallel to the paleoshoreline and extending 6 to 8 miles (10 to 13 km) inland. Rocks representing the period of peat accumulation related to the B coalbed are termed "B coal zone," in which the average total thickness of coal is 12 ft (3.7 m) at the eastern side of the study area. To the west (landward), B coal zone thickness averages 30 ft (9.1 m), and thick individual coalbeds begin to split and pinch out 6 to 8 miles (10 to 13 km) inland (west) of the paleoshoreline. This is best shown on the isopach map of the B coal zone (Figure 13).

The Rollins Sandstone, which represents the barrier-delta front facies, is separated from the B coal zone by as much as 100 ft (30.5 m) of lagoonal and associated facies at the western end of the study area. In this landward direction an earlier cycle of peat deposition, represented by the A coal zone, directly overlies the barrier-delta front facies

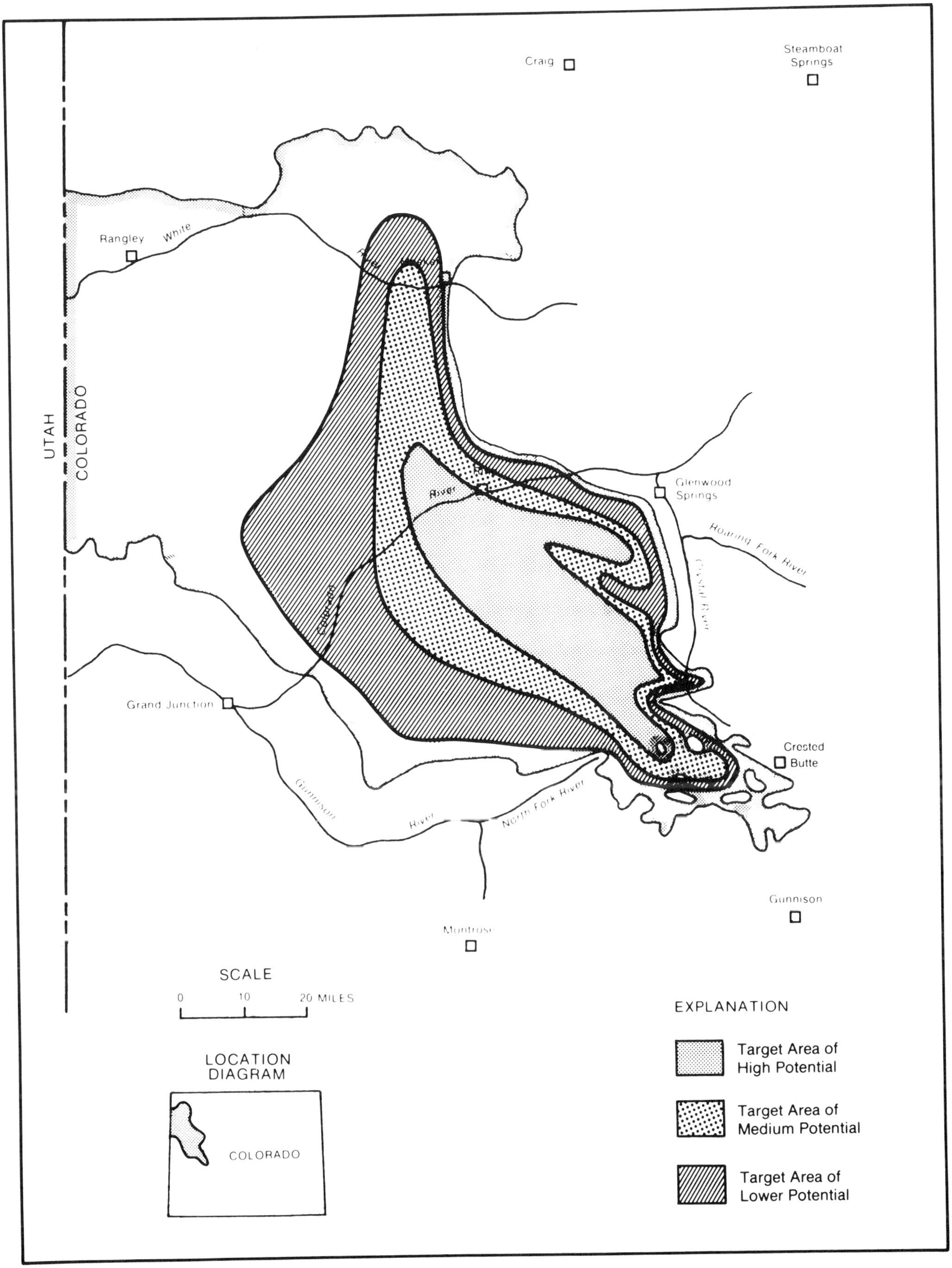

Figure 15. Target area of coalbed methane in Piceance basin. After Choate and others (1984).

and thickens in the landward direction, as does the younger B coal zone to the east.

Coalbeds representing the two cycles of peat formation in the lower delta-plain environment have geometries elongate to depositional strike of the Piceance delta. In Figure 16 the areal extents of the A and B coal zones are superimposed over the areas of high coalbed methane potential in the southeastern the Piceance basin. Boundaries of the B coal zone include the thickest coals in the lower Bowie shale member. The western extent of the A coal zone is hypothetical.

The potential for economical recovery of coalbed methane is high where these zones overlap Choate and others' (1984) target areas. In two areas the combined thickness of the A and B coalbeds exceeds 30 ft (9 m). These two thick regions are separated by an area of thinner coal. In the southern area, located in the western half of T12S R90W and the eastern part of T12S R91W, the combined thickness of A and B coals is as much as 38 ft (12 m). These thick coals trend approximately north-south. To the north, in the northwest corner of T11S R90W and southwest corner of T10S R90W, is a smaller region of combined A and B coals exceeding 40 ft (12 m) thick (Figure 13). This region also trends north-south. Similar thick coal regions may extend northward along depositional strike of the Piceance delta front. The geometry and orientation of the thickest coalbeds in the lower delta-plain depositional environment can be applied to earlier cycles of lower delta-plain coals to the west and therefore used as an aid in exploration for coalbed methane in the southwestern and other areas of the Piceance basin. Clearly, more work is needed west of the study area to determine if the A coal zone confirms this hypothesis.

CONCLUSIONS

This study was undertaken to describe the stratigraphy and to determine a depositional model for the coal-bearing part of the lower Mesaverde Formation in northeastern Delta and northwestern Gunnison Counties of western Colorado. The stratigraphy consists of, in ascending order, the Rollins Sandstone Member, which ranges between 80 and 200 ft (24 to 61 m) thick, the coal-bearing Bowie shale member, 260 to 490 ft (80 and 150 m) thick, the coal-bearing Paonia shale member with an approximate thickness of 180 ft (55 m), the barren member, which contains no economic coals and is between 1,500 and 2,000 ft thick (457 to 610 m), and the Ohio Creek Conglomerate, which was not studied in this report.

The depositional environment of the Rollins Sandstone and Bowie shale member was a wave-dominated delta system that prograded east-southeasterly. The delta consists of two major components—a lower delta plain, including delta-front deposits, and an upper delta plain. The Rollins Sandstone formed in the delta-front environment.

Rocks of the Bowie shale member, which were deposited in the lower delta plain, exhibit marine influence as a result of local subsidence and subsequent invasion by saline water. This resulted in saline and brackish facies deposited as the depocenters of fluvial sediment shifted farther inland or laterally along strike. Strata representing the saline and brackish facies are interbedded with coal in the Bowie shale member. These coalbeds formed from peat deposited in back-barrier and interdistributary platforms underlain by lagoon and barrier-delta front facies. The member is capped by a thick sequence of stacked, laterally extensive distributary-channel sandstones locally named the "middle sandstone." *Ophiomorpha* burrows and herringbone cross-bedding in the lower part of the middle sandstone are evidence of tidal influence. This suggests occasional tidally induced salt-water wedging in the lower delta-plain distributary channels. No saline indicators were found above the middle sandstone.

The thickest and most areally extensive coals, those of the B coal zone, formed where peat accumulation and subsidence of underlying sediments were in balance. The balance was maintained in an area of elongate geometry subparallel to the paleoshoreline and extending 6 to 8 miles (10 to 13 km) inland.

In the eastern part of the study area, the B coal zone is thin and directly overlies the Rollins Sandstone. In contrast, the Rollins Sandstone is separated from the B coal zone by as much as 100 ft (30.5 m) of lagoonal and associated facies at the western end of the study area. In this landward direction an earlier cycle of peat deposition, represented by the A coal zone, directly overlies the Rollins Sandstone and thickens in the landward (westward) direction. Total coal thickness of the B coal zone averages 12 ft (3.7 m), where it overlies the Rollins Sandstone in the seaward direction, to an average of 30 ft (9.1 m) 8 miles (13 km) to the west.

A model for the shape and extent of the thickest coalbeds is proposed that can be extended to earlier cycles of lower delta-plain deposition and used for exploration of coalbed methane in the lower delta-plain depositional setting. This model shows that areas of thick coalbeds trend north-south and are located adjacent to and landward of the paleoshoreline within a belt 6 to 8 miles (10 to 13 km) wide.

Two zones of A and B coal having a combined thickness of as much as 38 ft (12 m) are located in a north-south-trending belt in T10–12S R91&92W. These zones of thick coal constitute a prime target area for coalbed methane.

REFERENCES

Choate, R., Jurich, D., and Saulnier, G.J., Jr., 1984, Geologic overview, coal deposits, and potential for methane recovery from coalbeds, Piceance Basin—Colorado, *in* Rightmire, C.T., Eddy, G.E., and Kirr, J.N., eds., Coalbed methane resources of the United States: AAPG Studies in Geology, Series 17, p. 223–251.

Collins, B.A., 1970, Geology of the coal-bearing Mesaverde Formation (Cretaceous), Coal Basin area, Pitkin County, Colorado: Colorado School of Mines M.Sc. Thesis T-1284, 116 p.

Collins, B.A., 1976, Coal deposits of the Carbondale, Grand Hogback, and southern Danforth Hills coal fields, eastern

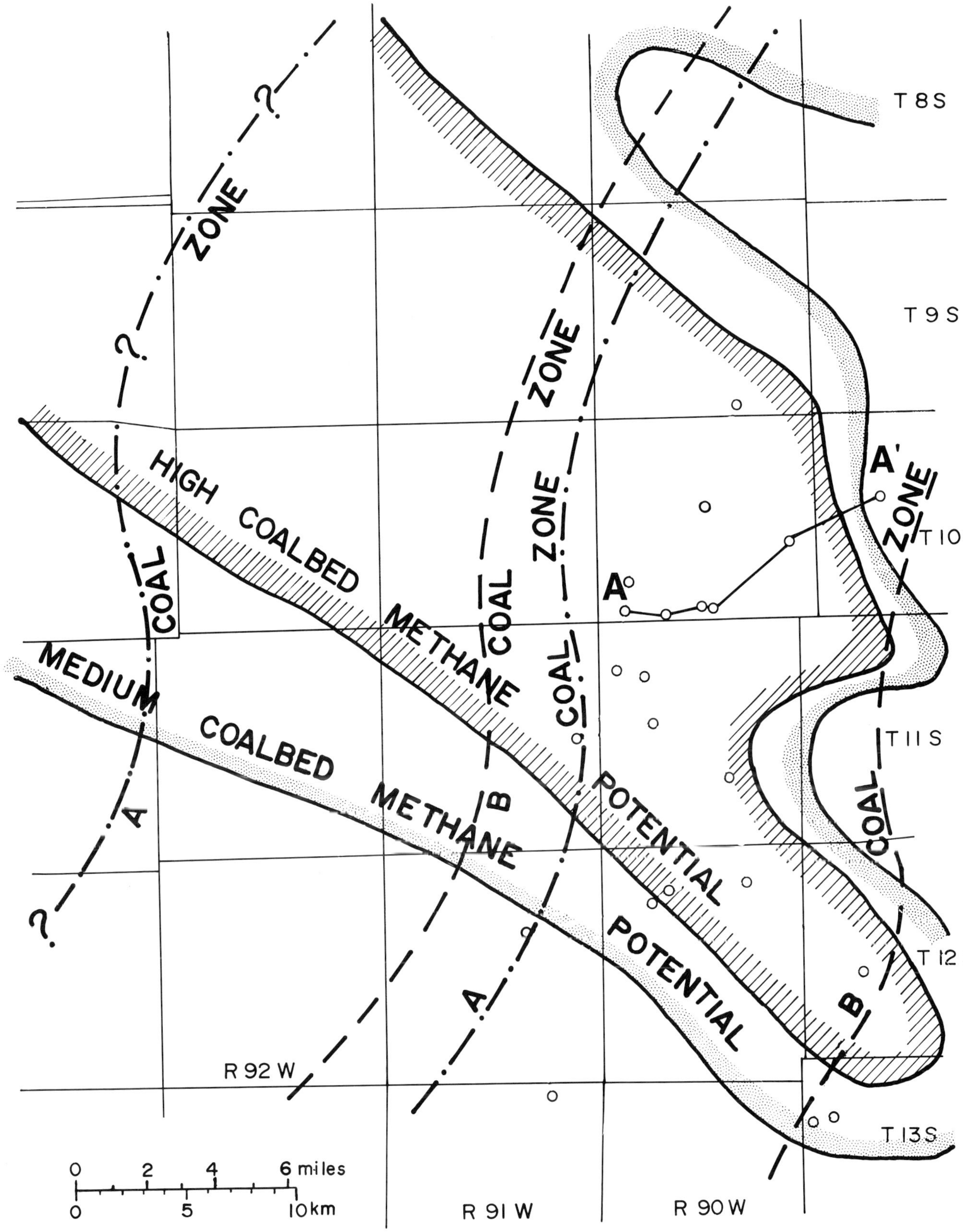

Figure 16. Coalbed methane occurrence and relationships to A and B coal zones of the Bowie shale member, southeastern Piceance basin.

Piceance Basin, Colorado: Colorado School of Mines Quarterly, v. 71, 138 p.

Daub, G.J., 1982, Stratigraphy and geology of some coal mines along the North Fork of the Gunnison River, Somerset coal field, Colorado, *in* Averett, W.R., ed., 1982 field trip guide: Grand Junction Geological Society, p. 69–77.

Dunrud, C.R., 1989, Geologic map and coal stratigraphic framework of the Cedaredge area, Delta County, Colorado: USGS Coal Investigations Map C–116, scale 1:50,000.

Ellis, M.S., and Kelso, B.S., 1987, Cross sections showing stratigraphic framework of Upper Cretaceous Dakota Sandstone, Mancos Shale, Mesaverde Group, and Mesaverde Formation, and Lower Tertiary Wasatch Formation, west-central Piceance basin, Garfield County, Colorado: USGS Miscellaneous Field Studies Map MF–2008–A.

Ferm, J.C., and Staub, J.R., 1984, Depositional controls of minable coal bodies, *in* Rahmani, R.A., and Flores, R.M., eds., Sedimentology of coal and coal-bearing sequences: International Association of Sedimentologists Special Publication 7, p. 275–289.

Flores, R.M., and others, 1984, Paleogeographic controls of coal accumulation, Cretaceous Blackhawk Formation and Star Point Sandstone, Wasatch Plateau, Utah: GSA Bulletin, v. 95, p. 540–550.

Hanks, T.L., 1962, Geology and coal deposits, Ragged-Chair Mountain area, Pitkin and Gunnison Counties, Colorado: Brigham Young University Geological Studies, v. 9, pt. 2, p. 137–160.

Johnson, R.C., and May, Fred, 1980, A study of the Cretaceous-Tertiary unconformity in the Piceance Creek basin, Colorado—The underlying Ohio Creek Formation (Upper Cretaceous) redefined as a member of the Hunter Canyon or Mesaverde Formation: USGS Bulletin 1482–B, 27 p.

Johnson, V.H., 1948, Geology of the Paonia coal field, Delta and Gunnison Counties, Colorado: USGS Preliminary Map, scale 1:48,000.

Lee, W.T., 1909, The Grand Mesa coal field, Colorado, in Contributions to Economic Geology, 1907: USGS Bulletin 341, p. 316–334.

Lee, W.T., 1912, Coal fields of Grand Mesa and the Western Elk Mountains, Colorado: USGS Bulletin 510, 237 p.

Madden, D.J., 1985, Description and origin of the lower part of the Mesaverde Group in Rifle Gap, Garfield County, Colorado: Mountain Geologist, v. 22, p. 128–138.

Madden, D.J., 1989, Stratigraphy, depositional environments, and paleogeography of coal-bearing strata in the Upper Cretaceous Mesaverde Group, central Grand Hogback, Garfield County, Colorado: USGS Professional Paper 1485, 45 p.

McCabe, P.J., 1984, Depositional environments of coal and coal-bearing strata, *in* Rahmani, R.A., and Flores, R.M., eds., Sedimentology of coal and coal-bearing sequences: International Association of Sedimentologists Special Publication 7, p 13–42.

McFall, K.S., Wicks, D.E., Kelso, B.S., and Brandenburg, C.F., 1988, An analysis of the coal-seam gas resource of the Piceance Basin, Colorado: Journal of Petroleum Technology, v. 40, no. 6, p. 740–748.

Saxena, R.S., 1982, Exploration models and recognition criteria for deltaic sand bodies, *in* Saxena, R.S., ed., Deltas—Their application to oil, gas, and coal exploration: GSA 95th Annual Meeting Short Course Notes, p. 19–70.

Spieker, E.M., 1949, Sedimentary facies and associated diastrophism in the Upper Cretaceous of central and eastern Utah, *in* Sedimentary facies in geologic history: GSA Memoir 39, p. 55–81.

Toenges, A.L., Turnbull, L.A., Davis, J.D., Reynolds, D.A., Parks, B.C., Cooper, H.M., and Abernethy, R.F., 1952, Coal deposit, Coal Creek district, Gunnison County, Colorado: U.S. Bureau of Mines Bulletin 501, 83 p.

Warner, D.L., 1964, Mancos-Mesaverde (Upper Cretaceous) intertonguing relations, southeast Piceance basin, Colorado: AAPG Bulletin, v. 48, no. 7, p. 1091–1107.

Wellborn, J.E. 1982, Stratigraphy of the Mesaverde Formation, Mt. Gunnison coal property, Gunnison County, Colorado: Colorado School of Mines M.Sc. Thesis T–2506, 91 p.

Young, R.G., 1982, Stratigraphy and petroleum geology of the Mesaverde Group, southern Piceance Creek Basin, Colorado, *in* Averett, W.R., ed., 1982 field trip guide: Grand Junction Geological Society, p 45–54.

Zapp, A.D., and Cobban, W.R., 1960, Some Late Cretaceous strand lines in northwestern Colorado and northeastern Utah, *in* Geological Survey research 1960: USGS Professional Paper 400–B, p. B246–B249.

Lineament Analysis of Northern San Juan Basin, New Mexico and Colorado—Applications to Coalbed Methane Exploration*

Robert W. Baumgardner, Jr.[1]

1. Bureau of Economic Geology, The University of Texas at Austin, Austin, Texas

ABSTRACT

This study tested the use of lineaments for identifying highly productive coalbed methane trends in the Fruitland Formation of the San Juan basin. Lineaments were mapped on 1:250,000-scale Landsat Thematic Mapper (TM) imagery. Lineament azimuth, length, and distance were compared to initial water production (IPW) data from 420 Fruitland wells in the northern half of the basin. Linear-regression analysis showed limited significant correlation only between attributes of northeast-trending lineaments and IPW.

Landsat-identified lineaments were compared with those derived from six different scales of imagery that were published in three previous studies. Little spatial correspondence exists among lineaments reported among the separate studies. Statistically significant lineament trends from these studies vary from subparallel to perpendicular. Lineament attributes from these studies and the current study were compared with IPW and maximum monthly gas production (MMG). Wells located on lineaments do not have consistently higher production values than those not on lineaments. Azimuths of lineaments nearest wells with high values of IPW and MMG vary among studies and even within one study, for lineaments derived from different imagery. Limited significant correlation exists between these production parameters and distance to the lineament nearest each well. A weak correlation exists between the number of lineaments within a given radius and either IPW or MMG. Based on available data, analyses indicate that lineament attributes cannot be used to predict production volumes.

INTRODUCTION

The purpose of this study was to map lineaments and to assess their relationships to Fruitland coalbed methane production. Depth to gas-producing intervals in the Fruitland Formation varies from 150 ft (46 m) near the basin margins to 4,000 ft (1,220 m) near the center of the basin (W.R. Kaiser, 1990, pers. comm.). Permeability in conventional reservoirs is enhanced by fractures in many areas of the San Juan basin (Gorham and others, 1979). Lineaments may be spatially correlated with fractures in the subsurface. However, unlike some empirical studies that regard spatial coincidence between lineaments and highly productive wells as proof that the lineaments or their subsurface counterparts are enhancing production, in this study strict statistical criteria were applied to determine whether lineament attributes have a statistically significant correlation with production data. If a significant correlation between lineaments and subsurface fractures can be demonstrated, then the locations and orientations of these surface features can assist in selecting drilling sites for enhanced production of coalbed methane from fractured reservoirs.

* Published by permission of the Director, Bureau of Economic Geology, The University of Texas at Austin.

REVIEW OF LINEAMENT AND FRACTURE STUDIES IN SAN JUAN BASIN

Previous lineament studies used imagery at scales different from that used in this study (Figure 1; Table 1). Kelley and Clinton (1960) mapped fractures over the entire San Juan basin from aerial photographs and photoindices. Newman and McCord (1980), who mapped 333 lineaments on Landsat imagery covering the northern San Juan basin in southern Colorado, reported localized similarity between linear features and mapped fracture patterns, which were oriented predominantly north-northwest. Knepper (1982), also using Landsat imagery, mapped more than 6,000 lineaments across the southern Colorado Plateau. He then interpreted three "derivative" lineaments in the San Juan basin that correspond spatially to gravity highs, magnetic anomalies, and dike swarms. Decker and others (1989) reported on lineaments from aerial photographs and Landsat imagery of Cedar Hill field, New Mexico. Locally, linear features from aerial photos are parallel to coal cleat directions and subparallel to open fractures in core from the Mesa Petroleum Hamilton 3 well in Cedar Hill field (Figure 2). Wandrey (1989) mapped lineaments on aerial photographs of the northwestern San Juan basin (La Plata County, Colorado) and found recurring trends that were revealed to be joints on larger-scale photos.

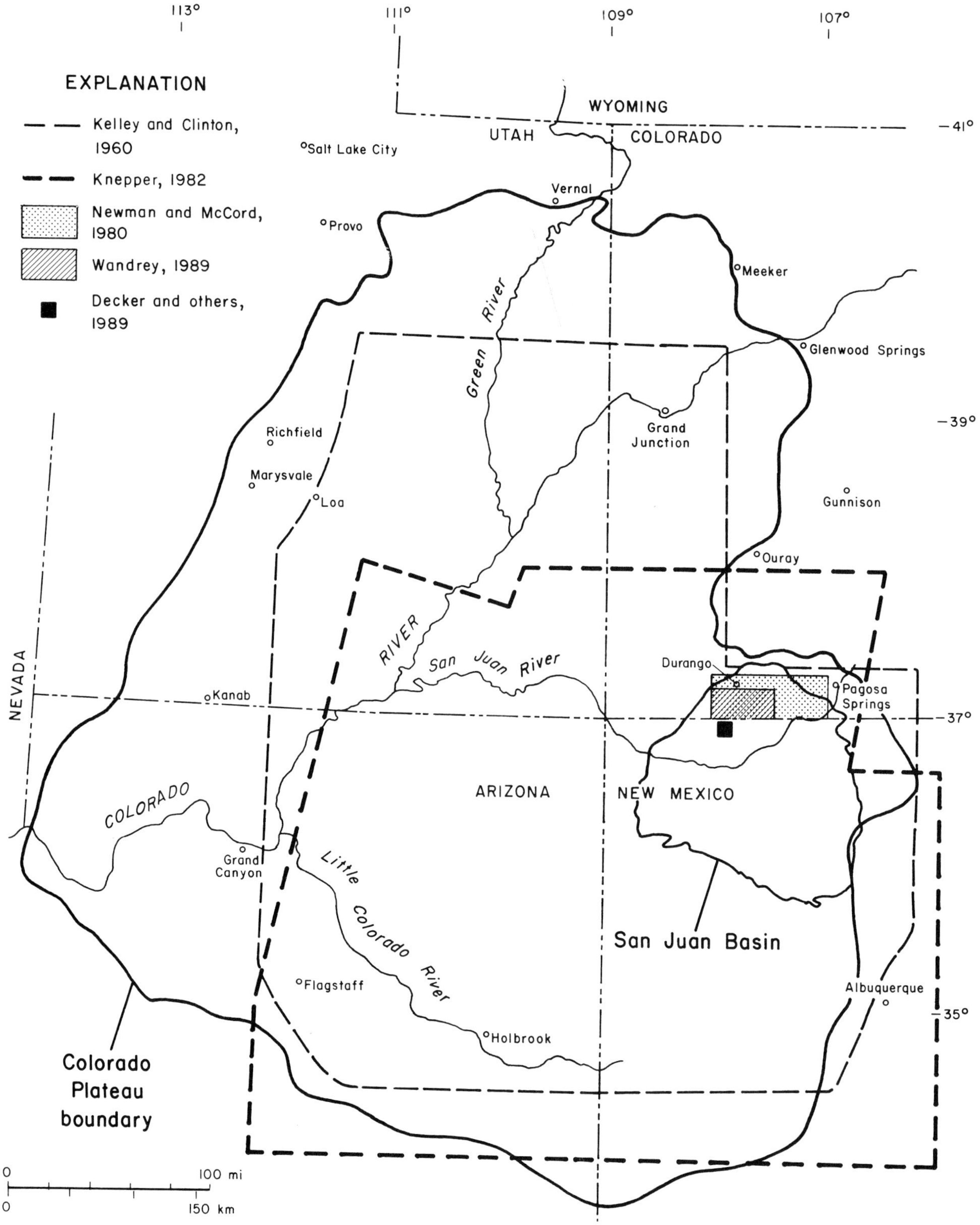

Figure 1. Lineament studies on the Colorado Plateau and in the San Juan basin vicinity (after Tremain and Whitehead, 1990). Boundary of San Juan basin is defined by the outcrop contact between Pictured Cliffs Sandstone and Fruitland Formation.

Table 1. Imagery used for lineament studies in the San Juan basin.

Authors	Scale	Imagery Type
Kelley and Clinton (1960)	Uncertain	Aerial photographs
	1:62,500	Photoindex
Newman and McCord (1980)	1:160,000	Landsat MSS [1] FCC[2] imagery
Knepper (1982)	1:800,000	Landsat MSS black and white imagery
Decker and others (1989)	1:28,000	Aerial photographs, black and white
	1:58,000	Aerial photographs, color infrared
	1:80,000	Aerial photographs, black and white
	<1:80,000	Landsat imagery
Wandrey (1989)	1:15,840	Aerial photographs, color
	1:20,000	Aerial photographs, black and white
This study	1:250,000	Landsat TM[3] FCC (bands 7, 4, 2)

1. MSS = multispectral scanner.
2. FCC = false-color composite.
3. TM = thematic mapper.

Because fracturing enhances oil and gas production in the San Juan basin, several authors have studied fractures on the surface and in the subsurface (Figure 2). From their study of fractures in the Verde, West Puerto Chiquito, and East Puerto Chiquito oil fields, Gorham and others (1979) found that competent strata are fractured in areas of maximum curvature along the Hogback monocline. Open joints trend parallel to axes of local folds. Gorham and coworkers observed that joint trends in older rocks are not necessarily manifested in overlying strata. Condon (1988) studied fractures on the northwestern side of the basin and documented four discrete sets of joints, distinguished by degree of development, abutting relationships, and orientation. Average joint-set azimuths were, in order of decreasing age, 15°, 282°, 53°, and 317° (strikes of N15°E, N78°W, N53°E, and N43°W, respectively). Relative age of joint sets was inferred from abutting relationships among joints. Younger joints terminated against older ones. The mean orientation of coal face cleats was 339° (N21°W); that of butt cleats was 48° (N48°E). Emmendorfer (1989) studied dipmeter fracture logs from 15 wells in Gavilan Mancos oil field in the eastern San Juan basin and found most fractures oriented north-northwest.

Orientations of coal cleats and joints in the San Juan basin vary from northeast to northwest. Face cleats along the southern and western basin margins predominantly trend northeastward (Figure 2) (Tremain and Whitehead, 1990). Along the northern and northwestern sides of the basin (southern Colorado), face cleats trend mostly northwestward (Condon, 1988; Tremain and Whitehead, 1990). Joints near the center of the basin also trend mostly northwestward (Tremain and Whitehead, 1990, Fig. 37), and face cleats trend northeastward and northwestward in cores from the northern half of the basin (Figure 2).

These observations suggest that if lineaments are controlled by surface fractures, lineament azimuth locally may vary considerably, as a function of local folding. Furthermore, lineaments may be perpendicular to each other, as cleats are, if these fracture trends are continuous into rocks overlying the coals. If the trends of lineaments are controlled by fracturing, then their orientations should change as cleat and joint directions change across the basin.

DATA SOURCES AND PROCEDURES

Four Landsat Thematic Mapper (TM) images at a scale of 1:250,000 were used for this study (Figure 3). False-color composite (FCC) images were generated with data from bands 7 (red), 4 (green), and 2 (blue). In bands 4 and 7 the contrast between vegetation and soil is relatively large. Some variations in soil and rock composition, such as might occur along a fault or formation contact, are more visible in FCC images generated with these three bands. The visibility of lineaments demarcated by vegetation changes, faults, or formation contacts is enhanced by these image characteristics.

Lineaments were mapped from edge to edge on each image without regard to basin boundaries. Most lineaments within the basin were checked against 1:250,000-scale U.S. Geological Survey topographic maps, 116 were compared with 1:24,000-scale topographic maps, and 23 were field checked. Man-made features mistaken for lineaments (53) accounted for 5.7% of the lineaments checked (925). Because other undetected man-made features possibly may have been mapped as lineaments, this error rate should be considered a minimum. Lineaments were compared with maps of surface geology (scale 1:250,000) to determine whether spatial correspondence existed between lineaments and mapped features, such as faults and formation contacts. None corresponded with faults; 11% coincided with formation contacts. Finally, 969 lineaments were digitized for computer-assisted analysis.

Lineaments from other studies that covered a selected area of the northwestern San Juan basin (Figure 2) were digitized directly from published maps. These lineaments were not checked against original imagery, nor was any judgment made regarding their "validity." Azimuths were analyzed just as were those from this study using the techniques described below.

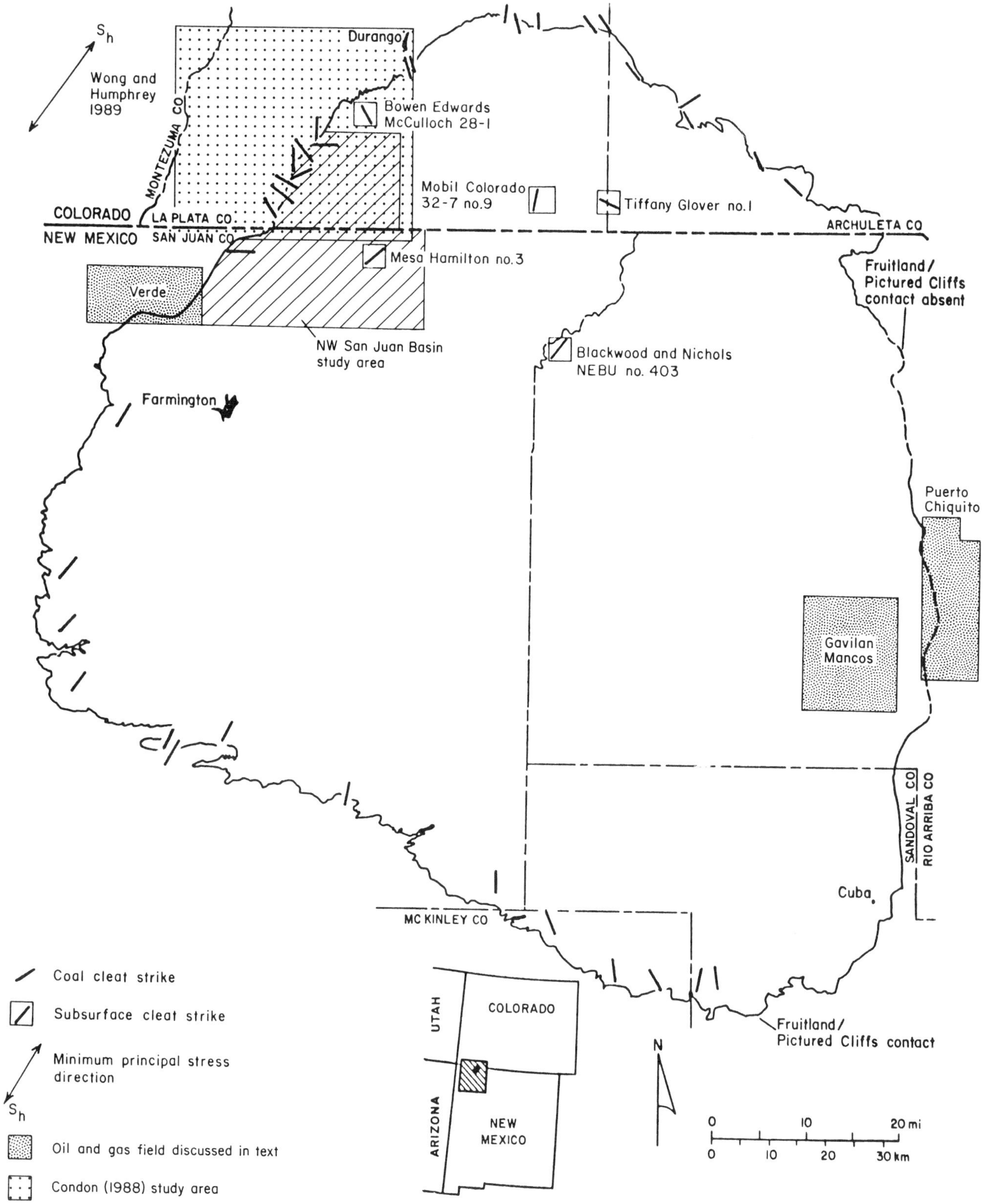

Figure 2. Coal face cleat orientations and in situ stress directions in the San Juan basin. Cleat data in Colorado are from Newman and McCord (1980), Condon (1988), and Tremain and Whitehead (1990). Cleat data in New Mexico are from Tremain and Whitehead (1990). After Tremain and others (in press).

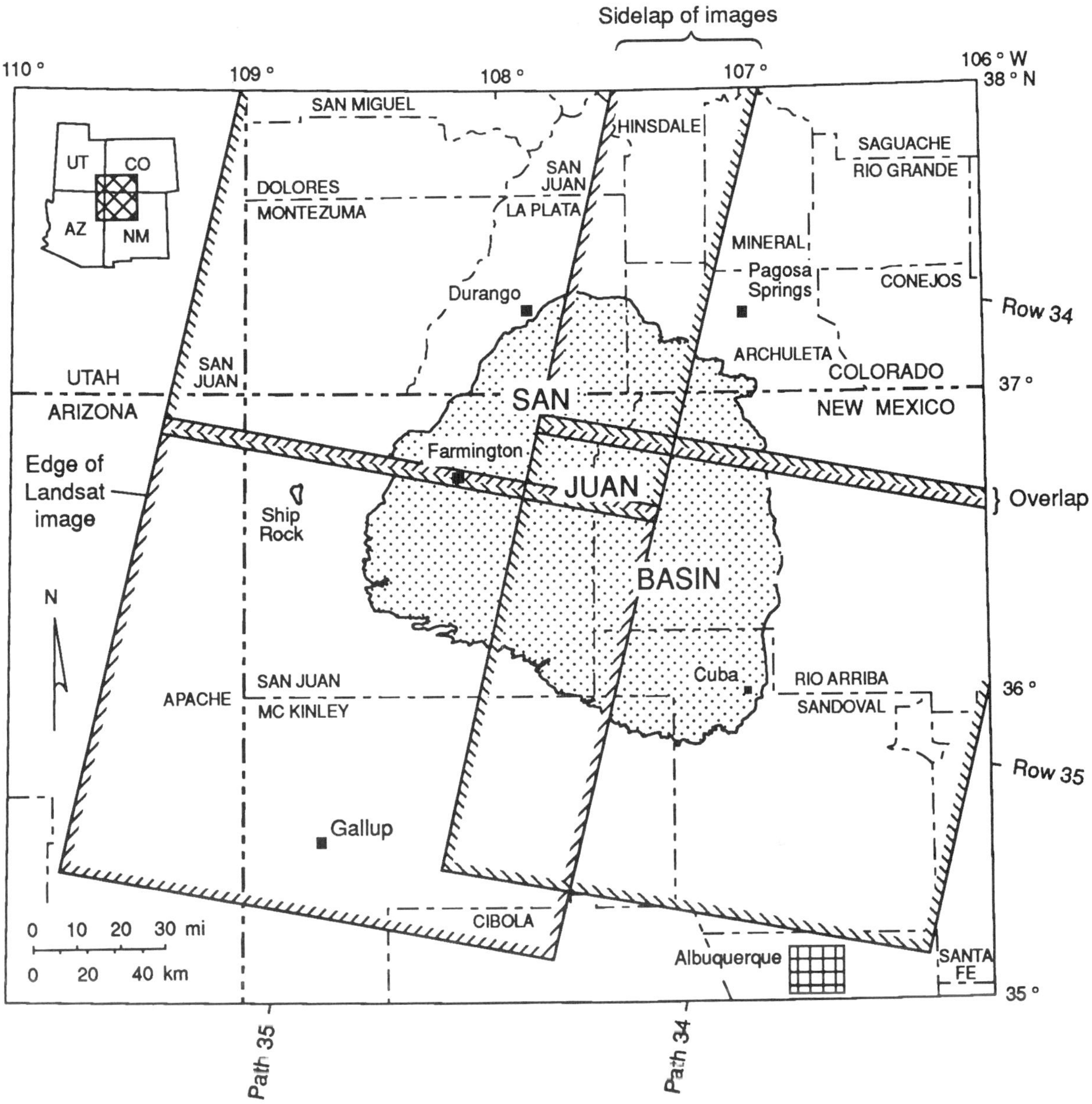

Figure 3. Coverage of Landsat Thematic Mapper imagery used to map lineaments in this study.

Values of two production parameters, initial water production (IPW) and maximum monthly gas production (MMG), were obtained from Petroleum Information Corporation, Dwight's Oil and Gas Reports (1990a, 1990b), and the Colorado Oil and Gas Conservation Commission. Wells with high IPW and MMG values are assumed to be draining more permeable fractures than wells with low values. Values of IPW and MMG were not "normalized" for different well completion techniques, completion dates, or possible interference effects from nearby wells. Nor were wells from overpressured and underpressured areas examined separately. MMG data were available for wells in the northwestern part of the basin only for 1988 and 1989. Whereas both of these parameters may be related to long-term gas production, the nature of their relationship is not well known for the study area. No statistically significant correlation ($p = .05$) occurs between MMG and IPW values in this study.

RESULTS

Length-Weighted Frequency of Lineament Azimuths

To an extent, lineament analysis is subjective, depending on vagaries of interpretation (including operator bias). To compound this problem, lineament trends often are referred to by such qualitative terms as "prominent" and "dominant." In this study lineaments were analyzed for length-weighted frequency (*F*) and Bernshtein accuracy criterion (H), as described by Dix and Jackson (1981). Numerically significant lineament trends were defined by standard statistical analysis of the orientation data. In this way, interpretation of the data is as objective as possible, although some subjectivity of the original mapping probably remains.

Lineament-azimuth data initially were displayed as rose diagrams of length-weighted frequency (Figure 4a). The parameter F expresses the total lineament length (L_s) within a 10°-wide sector of the graph, weighted in proportion to the number of lineaments (n) within the area in question:

$$F = L_s\, n / L_t$$

where L_t is total lineament length within the area.

Statistical Significance of Lineament Azimuths

Definition and Validity of Peak Values

In this study a peak on a polar graph of square root F is defined as any 10°-wide sector whose magnitude is larger than the average for that plot. To determine which greater-than-average peaks were significant, a chi-square test was used to measure the difference between each peak and the mean F value for each data set (Siegel, 1956). Dix and Jackson (1981) concluded that the 99% confidence level ($p = .01$) should be used to define geologically meaningful peaks because none of their samples with more than 100 computer-generated "lineaments" had significant peaks at that level.

Vector Sums of Greater-than-Average Peaks

Although significant azimuths of lineament orientation data can be determined using 10° sectors, significant trends that are split between two adjacent sectors may be obscured. To avoid this potential loss of important data, vector sums of adjacent greater-than-average peaks were calculated (Figure 4b).

All vector sums of greater-than-average peaks are not, however, always statistically significant. To determine which sums are significant, a chi-square test was applied to each (Figure 4c). As a further refinement, the chi-square value for each significant peak was divided by the degrees of freedom to yield the Bernshtein accuracy criterion, H (Vistelius, 1966; Dix and Jackson, 1981). In this study two vector sums, 5° and 68°, are statistically significant.

Production Data vs Lineament Attributes

To assess the effect of lineaments on coalbed methane production in the San Juan basin, IPW data from wells (Figure 5) were compared to various attributes of the nearest lineament. Lineaments are assumed to be surface manifestations of vertical fracture zones. Linear regressions were carried out between lineament attributes (independent variables) and IPW data (dependent variable). As in most statistical analyses, only correlation coefficients significant at a confidence level of $p = .05$ or better were accepted as significant. Of the more than 70 linear regressions run, only eight were statistically significant.

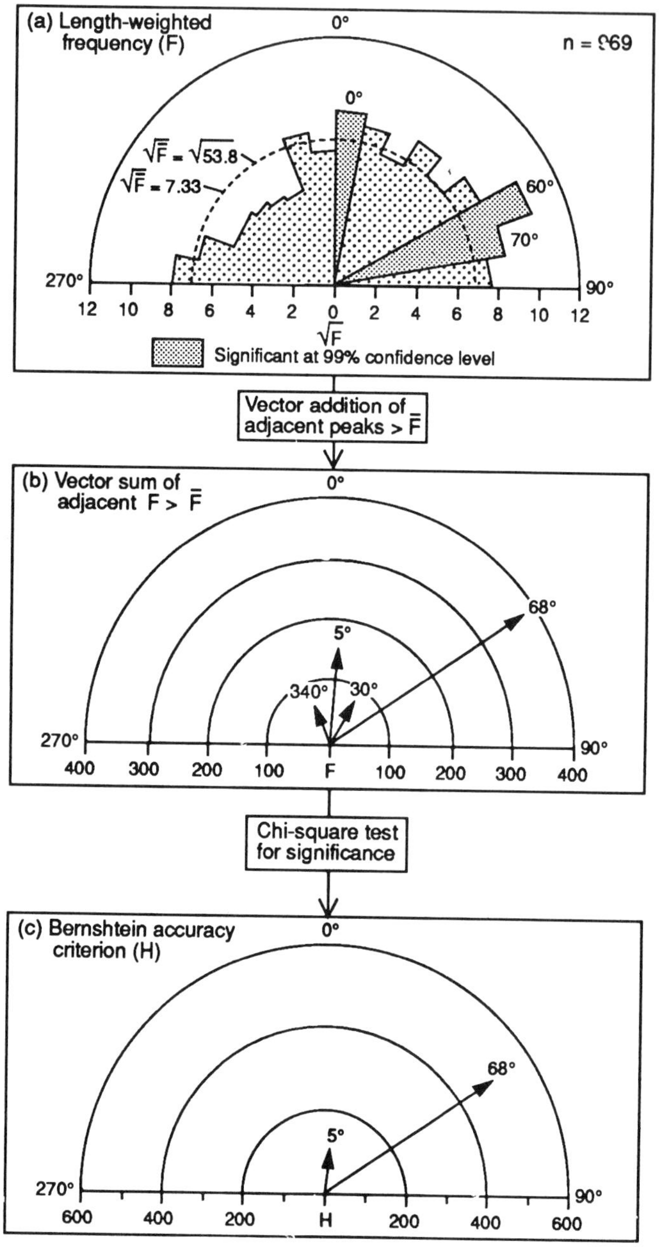

Figure 4. Polar graphs of orientation data for 969 Landsat-identified lineaments in the San Juan basin, illustrating the procedure for statistical evaluation of lineament orientation data. *a*. Length-weighted frequency (*F*) of lineaments shows three peaks significant at the 99% confidence level, as determined by a chi-square test. Low values may, in part, be due to suppression of lineaments subparallel to sun azimuth, which ranged from 133° to 148°. Square root of F is plotted to subdue areal exaggeration of large peaks. *b*. The vector sum of adjacent greater-than-average peaks of F has four peaks, but orientations are different from the F peaks. Data at 270° (90°) are adjacent to the 80° peak and are added to them in this step. *c*. Two peaks, 5° and 68°, of Bernshtein accuracy criterion (H) are significant at the 99% confidence level.

San Juan Basin

Lineament attributes (distance to the nearest lineament from a well, length of the nearest lineament, and azimuth of the nearest lineament) were correlated with initial water production (IPW) of numerous gas wells in the northern San Juan basin. Initially, lineament attributes were compared to all IPW values and to nonzero IPW values. Then, IPW values were subdivided according to natural groupings in the data (W.R. Kaiser, 1990, pers. comm.):

1 to 99 BWPD	(0.2 to 15.7 m^3/day)
100 to 249 BWPD	(15.9 to 39.6 m^3/day)
250 to 400 BWPD	(39.8 to 63.6 m^3/day)
>400 BWPD	(>63.6 m^3/day)

No lineament attributes were significantly correlated with IPW data or any subset of those data.

Then, the lineaments were separated into groups with northeast-trending (1° to 89°) and northwest-trending (273° to 358°) azimuths, and the regressions were repeated. The assumption behind this analysis was that lineaments in one azimuthal range might represent similarly oriented subsurface fractures that are more permeable than those in the other azimuthal range because of in situ stress differences. If one azimuthal set is more permeable than the other, this should be manifested in the IPW values of nearby wells.

Out of 24 linear regressions, only two significant correlations were found in these azimuthal data subsets (Figure 6). Both correlations suggest that nearby lineaments have a "negative" effect on initial water production, contrary to expected results. However, both correlation coefficients are quite small ($|r| < .28$) and account for less than 10% of the variability in IPW values ($r^2 < .10$).

In the first correlation, IPW values between 1 and 99 BWPD (0.2 to 15.7 m^3/day) increased as distance to the nearest northeast-trending lineament increases. In the second correlation, IPW values between 100 and 249 BWPD (15.9 to 39.6 m^3/day) decreased as length of the nearest

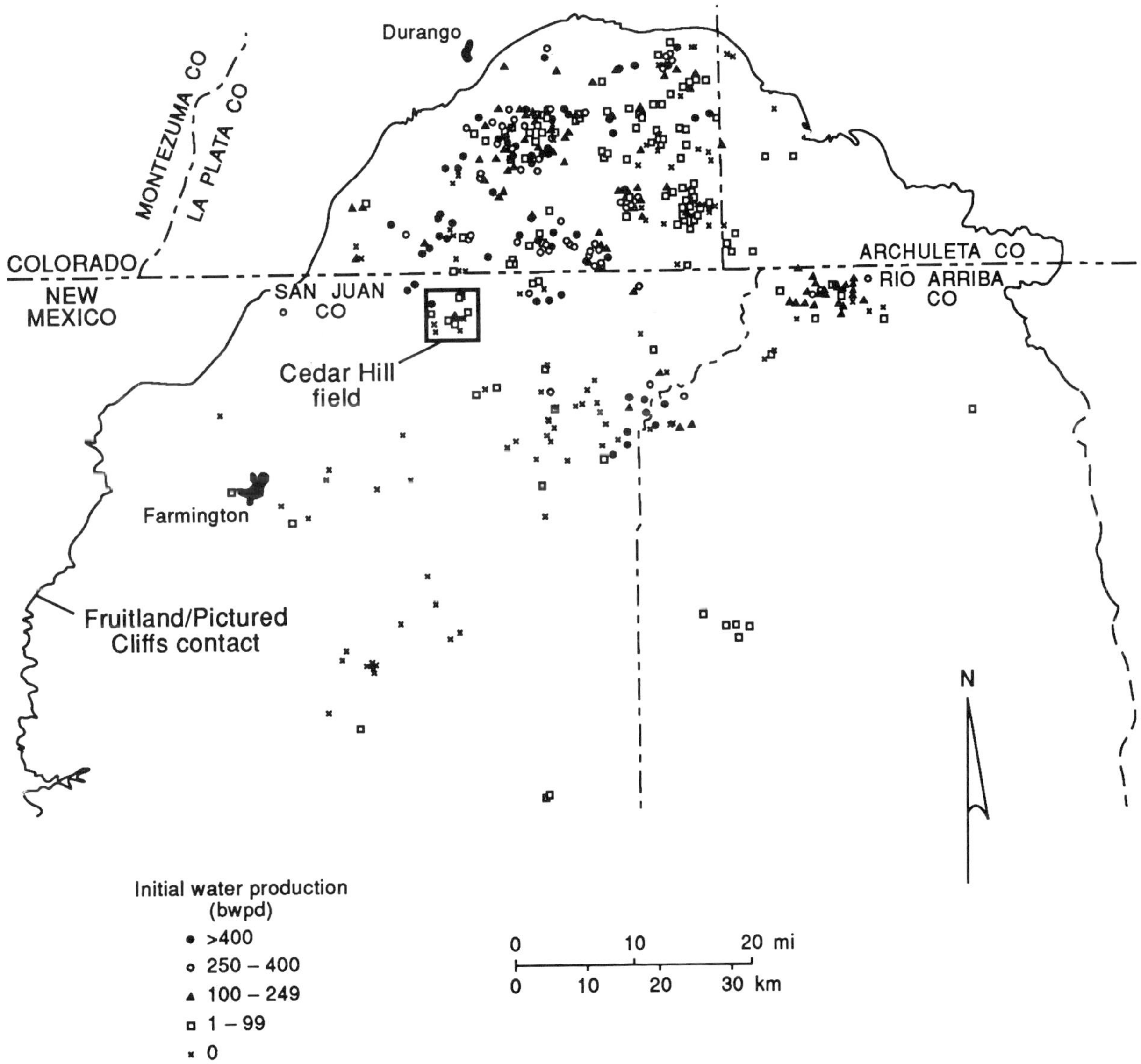

Figure 5. Locations of 420 northern San Juan basin coalbed methane wells with initial water production (IPW) data.

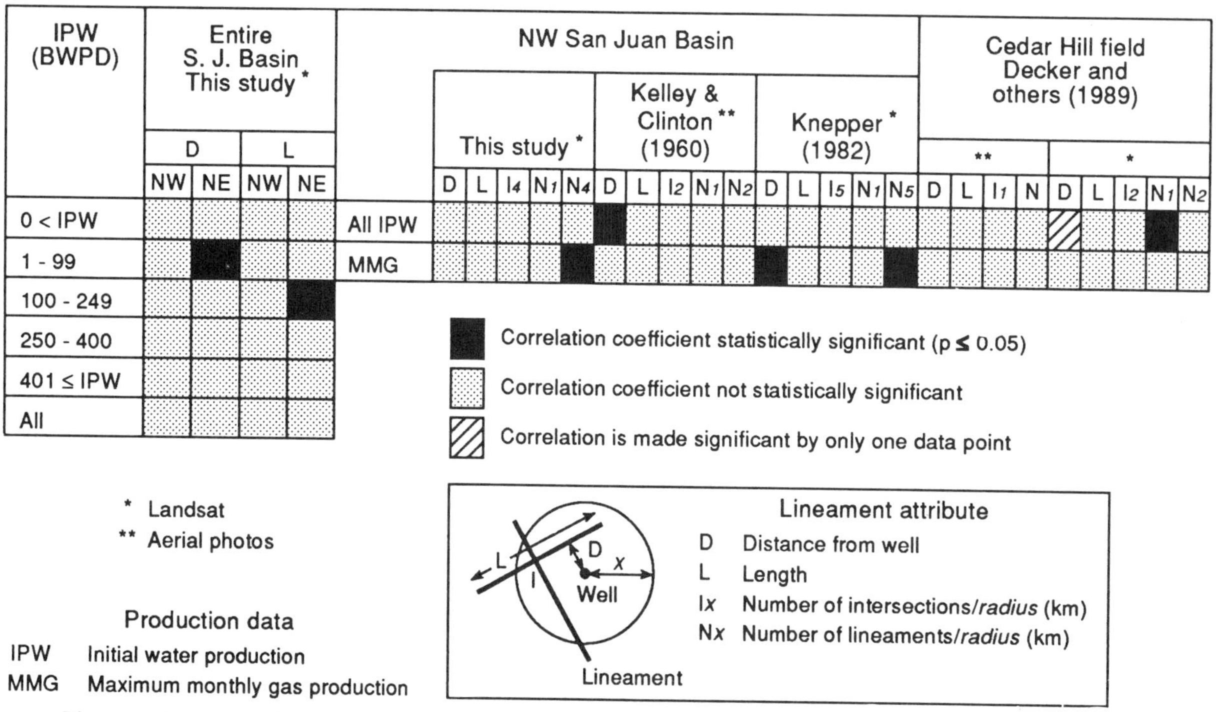

Figure 6. Results of linear regressions of two production data parameters vs four lineament attributes.

northeast-trending lineament increases. These results are opposite what would be expected if lineaments were expressions of permeable fractures in the subsurface.

More important than either of these two weak correlations is the absence of any significant correlation between lineament attributes and the highest values of IPW, ≥250 BWPD (≥39.8 m^3/day) (Figure 6). Presumably, high IPW values are those enhanced most by fracture permeability, yet they show no correlation with lineament length or distance to lineament.

Lineament azimuth cannot be used to differentiate between high and low values of IPW, either. The vector mean azimuth (27°) of lineaments nearest wells with IPW greater than 400 BWPD (63.6 m^3/day) is the same as the vector mean azimuth for those nearest wells with IPW between 1 and 99 BWPD (0.2 to 15.7 m^3/day). Together these results suggest that Landsat lineaments mapped in this study are not consistent indicators of subsurface fracture permeability.

Detailed Study Area (Northwestern San Juan Basin)

More detailed analysis of lineament attributes and production data was done for a study area in the northwestern San Juan basin (Figure 7, inset). Besides initial water production, lineament attributes were compared to maximum monthly gas (MMG) production data for 1988 and 1989 (Dwight's Oil and Gas Reports, 1990a, 1990b) (Figure 6).

Lineaments from three published studies for the northwestern San Juan basin (Table 2) were compared with one another and with lineaments mapped in this study and then correlated with production values from wells in that area. Because of the various scales of imagery used, this technique provided an opportunity to examine coincidence of lineaments mapped on different images by various observers. Of the 1,284 lineaments identified from all studies, only 52 coincided between any two studies. Of these 52 lineaments, only two coincided with more than one other study. Lineaments were considered to coincide if they (1) had the same azimuth ±5°; (2) overlapped for part of their lengths; and (3) were less than 1 mm apart (672 ft or 250 m ground distance, at a scale of 1:250,000), as measured on the original imagery.

The lack of coincidence indicates that lineament mapping is strongly dependent on the interpreter and on the type and scale of imagery used. For the same area of the basin the number of lineaments identified ranged from 51 to 914, largely the result of differences in scale (Table 2). In a smaller area near Cedar Hill field (Figure 7, inset), Decker and others (1989) showed 11 lineaments on Landsat images and 249 lineaments on aerial photographs (Table 2).

Lineament azimuth. Lineament azimuths were analyzed to determine (1) if any consistent lineament trend exists among all studies, and (2) if a consistent trend is associated with production values. For two of the three studies covering the northwestern San Juan basin (Figure 7, inset), statistically significant lineament azimuths are similar. The maximum lineament azimuth for Landsat lineaments in this study, 35° (Figure 7b), is practically the same as the significant azimuth of 38° (Figure 7c) based on aerial photographs interpreted by Kelley and Clinton (1960). However, the maximum lineament azimuth of 302° (Figure 7d) mapped on Landsat images by Knepper (1982) is nearly perpendicular to these azimuths. These results indicate that the maximum lineament azimuth can vary widely depending on the interpreter and on the scale and type of imagery used to map the lineaments.

The effect of this variability in lineament azimuth is shown in Figure 8, where maximum monthly gas production (MMG) is plotted against azimuth of the nearest lineament. The highest values of MMG, greater than 11,000 MCF (311,300 m^3), plot closest to northeast-trending lineaments in this study (Figure 8a). However, the same high values of MMG, from the same wells, plot closest to northwest-trending lineaments in Knepper's study (Figure 8b). Clearly, lineament azimuth from one study cannot be used in other lineament studies to predict MMG values.

Even different types of imagery within the same study can produce divergent azimuths. Decker and others (1989) reported lineaments mapped on aerial photographs and Landsat images. Highest values of initial water production (IPW), >100 BWPD (15.9 m^3/day), plot closest to northwest-trending Landsat lineaments (Figure 9a), but the same

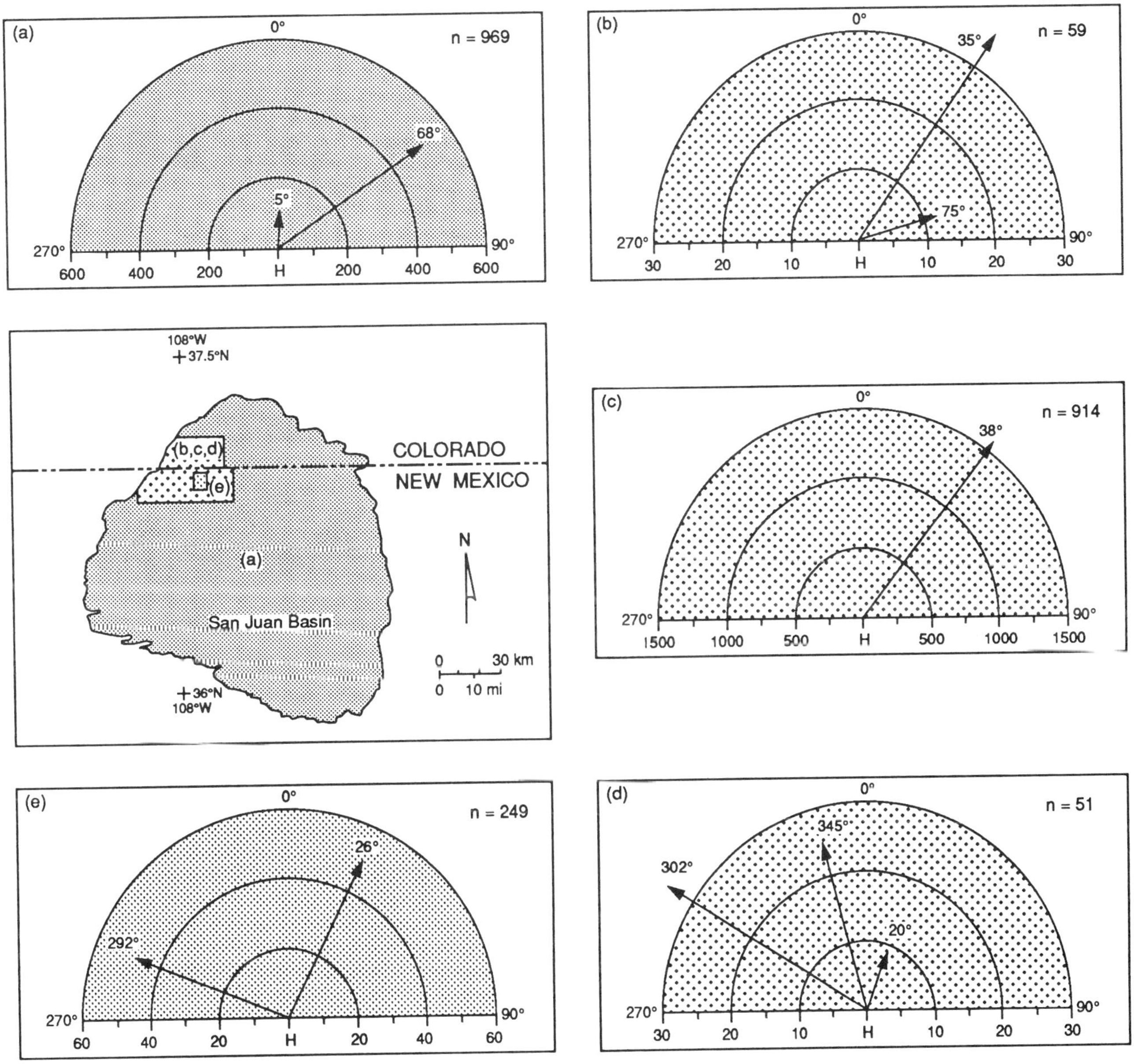

Figure 7. Polar graphs of Bernshtein accuracy criterion (H) for lineaments in this and three previous studies. Only vector sums of lineament azimuths that are significant at the p = .01 level are shown. See inset map for area covered by each study. Stippling represents areas that correspond to like data on graphs. *a*. This study, entire San Juan basin. Note that lineaments in 7b, 7c, and 7d are from different studies in the northwestern San Juan basin. *b*. Landsat lineaments, this study. *c*. Aerial photolineaments (Kelley and Clinton, 1960). *d*. Landsat lineaments (Knepper, 1982). *e*. Aerial photolineaments in Cedar Hill field area (Decker and others, 1989). Landsat lineaments mapped by Decker and others (1989) are insufficient (n = 11) to produce a meaningful value for H (Figure 4c).

Table 2. Lineaments in northwestern San Juan basin mapped in different studies. Area covered by Decker and others (1989) (Figure 7, inset) is only a small part of the area covered by the other studies. Wells were paired with the nearest lineament. Distance to nearest lineament varies with scale of imagery. Numbers of lineaments and lineament intersections were counted within a radius of a well and compared to production data. Radius increases as scale of imagery decreases to compensate for loss of resolution.

			Well-to-lineament distance					
			Minimum		Maximum		Radius	
Author	Number of Lineaments	Scale of Imagery	(ft)	(m)	(ft)	(m)	(mile)	(km)
Decker and others (1989)	249	≥1:80,000	20	6	1,200	370	0.6	1.0
	11	<1:80,000	90	27	7,100	2,170	1.2	2.0
Kelley and Clinton (1960)	914	1:62,500	70	20	13,500	4,120	1.2	2.0
This study	59	1:250,000	100	30	7,700	2,350	2.5	4.0
Knepper (1982)	51	1:800,000	430	130	25,900	7,900	3.1	5.0
Total	1,284							

wells plot closest to northeast-trending aerial photolineaments (Figure 9b).

It is clear from these comparisons that azimuths of nearest-well lineaments from different studies or even from different imagery within the same study can vary widely, even be perpendicular, and that no lineament azimuth can be described as more useful than another for predicting these production values. Furthermore, most other comparisons between lineament azimuth and production data (for example, IPW vs azimuth for Kelley and Clinton (1960), Knepper (1982), and this study) showed no preferential distribution of high values at any lineament azimuth.

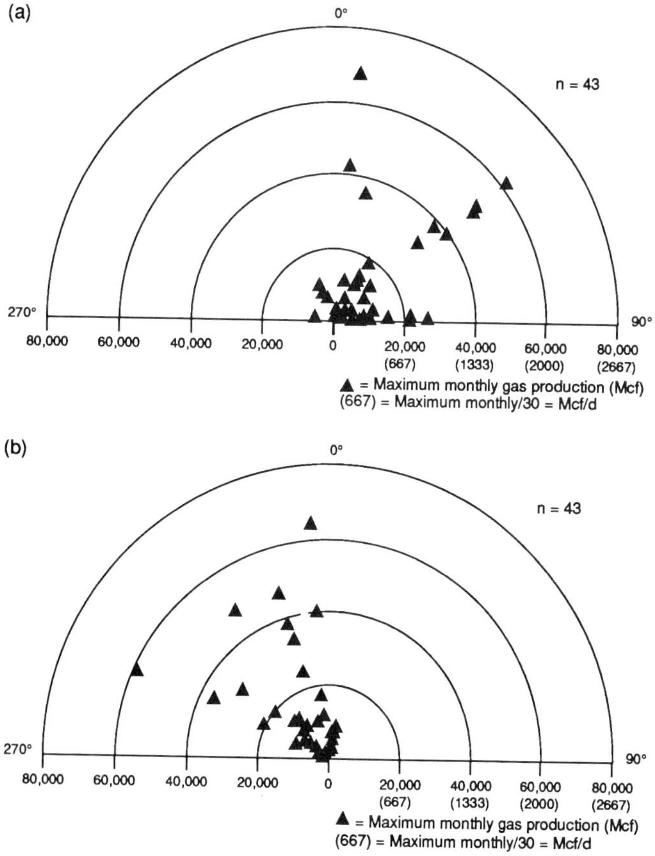

Figure 8. Maximum monthly gas production (MMG) vs azimuth of nearest lineament. MMG data on both graphs are from identical wells in northwestern San Juan basin for 1988 and 1989. *a*. This study. *b*. Knepper (1982).

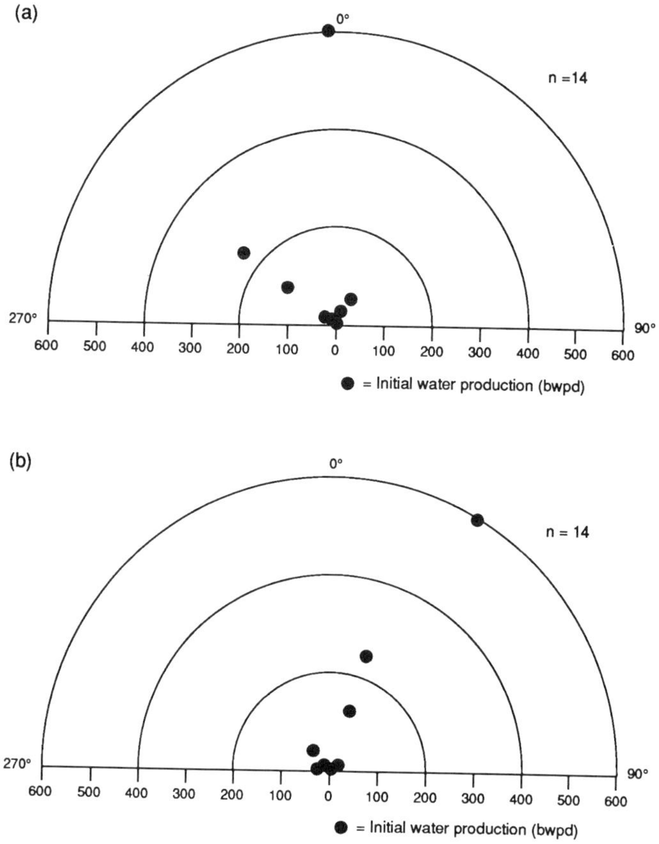

Figure 9. Initial water production (IPW) vs azimuth of nearest lineament. IPW data on both graphs are from the same wells. Lineament azimuth data are derived from different imagery in the same study (Decker and others, 1989). *a*. Landsat lineaments. *b*. Aerial photolineaments.

Wells on lineaments. Two comparisons were made to determine whether wells located on lineaments had higher production values than wells not on lineaments. In the first, mean production values from wells located less than 660 ft (200 m) from a lineament were compared with mean production values from wells more than 660 ft (200 m) from a lineament (Table 3a). (Wells situated less than this distance from a lineament were considered to be on the lineament.) This distance was selected based on the width of a pencil line at the original scale of the imagery and on possible positioning errors when digitizing the lineaments. For the second test, mean distance to the nearest lineament from wells with greater-than-average production values was compared with mean distance from wells with less-than-average production values (Table 3b). This comparison tested whether wells with greater-than-average production were significantly closer to lineaments than wells with production less than the average. No distinction was made for the relative position of well-lineament pairs. Wells along a lineament trend were treated the same as wells off the trend of, but near, a lineament.

This analysis revealed no consistent relationship between production data and a well's location on or off a lineament. If lineaments are enhancing production, MMG and IPW values both should be higher at distances less than 660 ft (200 m). However, few mean values of MMG and IPW are significantly different for wells on or off lineaments (Table 3a). For data from Knepper (1982) and Decker and others (1989), IPW values are, in fact, significantly higher in wells that are off the lineaments, opposite of what one

Table 3. Comparison between distance to nearest lineament and production data for wells in this study. Wells less than 660 ft (200 m) from a lineament are considered to be on a lineament.

***a.* Mean production data from wells located "on" lineaments and wells not on lineaments.**

	Mean MMG Production at Well-to-Lineament Distance				Mean IPW at Well-to-Lineament Distance			
	<660 ft		>660 ft		<660 ft		>660 ft	
Study	(MCF)	(m^3)	(MCF)	(m^3)	(BWPD)	(m^3/day)	(BWPD)	(m^3/day)
Decker and others (1989) [1]	39,877	1,128,519	30,071	851,009	100	15.9	4	0.6
n	9		3		11		3	
Decker and others (1989) [2]	46,997	1,330,015	34,235	968,850	0	0	*101	*16.1
n	3		9		3		11	
Kelley and Clinton (1960) [1]	15,220	430,726	17,278	488,967	210	33.4	278	44.2
n	11		32		22		54	
Knepper (1982) [2]	14,270	403,841	16,873	477,506	50	8.0	*267	*42.5
n	2		41		3		73	
This study [2]	13,712	388,050	16,980	480,534	137	21.8	265	42.1
n	3		10		4		72	

***b.* Mean distance from well to lineament for wells with production values greater than and less than the mean.**

	Mean Distance from Well to Nearest Lineament							
	for MMG				for IPW			
	> Mean		< Mean		> Mean		< Mean	
Study	(ft)	(m)	(ft)	(m)	(ft)	(m)	(ft)	(m)
Decker and others (1989) [1]	512	156	384	117	295	90	449	137
n	7		5		3		11	
Decker and others (1989) [2]	1,854	565	2,506	764	3,317	1,011	1,595	486
n	7		5		3		11	
Kelley and Clinton (1960) [1]	*3,015	*919	6,004	1,830	7,336	2,236	4,675	1,425
n	13		30		23		53	
Knepper (1982) [2]	*4,807	*1,465	13,114	3,997	12,947	3,946	8,793	2,680
n	13		30		23		53	
This study [2]	*3,767	*1,148	5,693	1,735	5,853	1,784	4,734	1,443
n	13		30		23		53	

1. Based on aerial photographs.
2. Based on Landsat images.
* Mean values are significantly different at $p = .05$ level.

might expect. Wells with above-average MMG values are closer to lineaments than those with below-average MMG values in three studies (Table 3b). This result is consistent with the hypothesis that lineaments enhance gas production, but it was not seen in Cedar Hill field area (Decker and others, 1989), where some of the highest gas production values in the basin have been observed—MMG as high as 67,840 MCF (1.9×10^6 m^3), equivalent to 2,261 MCFD (63,986 m^3/day).

Distance from well to nearest lineament. The comparison of wells on and off lineaments was followed by linear regression of production data vs distance to the nearest lineament (D, Figure 6). Distance varied according to scale of the imagery (Table 2). Only three of ten linear regressions (two types of production data vs five sets of lineaments) yielded significant correlation coefficients (Figure 6). One of those was made significant by a single data point. The correlation coefficient for Kelley and Clinton's (1960) data is positive (Figure 10a), indicating that IPW increases as distance to the lineament increases, contrary to the expected result. Knepper's (1982) data show, as expected, a negative correlation between MMG and distance to lineament (Figure 10b), but this correlation (r = –.34) explains only 12% (r^2 = .12) of the variability in MMG values.

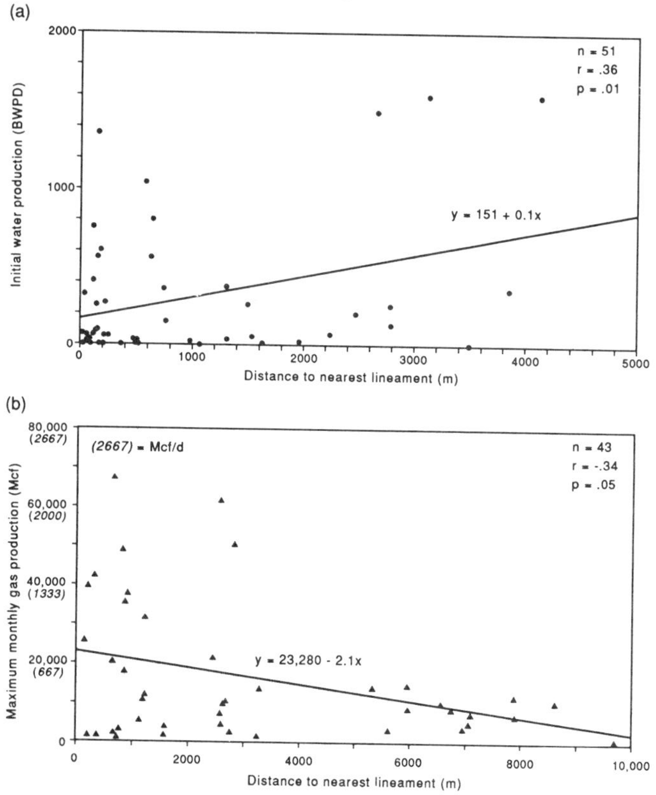

Figure 10. Production data vs distance from well to nearest lineament. *a.* IPW vs distance to nearest lineament. Data from Kelley and Clinton (1960). *b.* MMG vs distance to nearest lineament. Data from Knepper (1982).

Length of nearest lineament. Length of the nearest lineament (L, Figure 6) had no significant effect on the production data studied.

Lineament intersections within a given radius. The number of nearby lineament intersections (I_x, Figure 6) was compared with production data because if lineaments represent fractures or other subsurface zones of enhanced permeability, then their intersections should represent regions of increased permeability in the subsurface, which should be reflected in production data. However, no significant correlation was found between the number of intersections and production.

Number of lineaments within a given radius. Finally, the number of lineaments within a given radius of each well (N_x, Figure 6) was correlated with production values. To this point in the study, results showed that no lineament attribute could be used consistently to predict production values. The last comparison tests whether the number of lineaments, regardless of their length, azimuth, or relative distance from a well, correlates with IPW and/or MMG.

The number of lineaments was counted within a given radius of each well. A radius of 0.6 mi (1 km) and a "scale-determined" value were used. The latter varied with the scale of the original imagery (Table 2). Only three regressions yielded significant correlation coefficients (Figure 6 and 11). These regressions ($|r| \leq .59$) explain only about one-third of the variability in the production data ($r^2 \leq .35$). The correlations are surprising in two respects. First, all significant correlations are for Landsat-identified lineaments, rather than aerial photolineaments. In contrast, from their study of lineaments in Cedar Hill field, Decker and others (1989) concluded that aerial photolineaments are more useful in exploration programs. Second, the results show that the only lineament attribute that is significantly correlated with production values from Cedar Hill field is the number of Landsat lineaments (N_x) within a given radius of the well. However, the number of lineaments is negatively correlated with IPW (Figure 11a), indicating that as the number of lineaments within a 0.6-mile (1-km) radius of a well increases, IPW decreases. This is opposite of what usually is expected in lineament studies. Even where a positive correlation with production values was seen, as in the data from Knepper (1982) and from this study (Figure 11b and 11c, respectively), the utility of this relationship for exploration purposes is questionable because the radii used are as large as 3.1 miles (5 km).

DISCUSSION

Lineaments and Fractures

The relationship between lineaments and fractures is not always straightforward. Natural fractures in the subsurface may parallel joints mapped at the surface (Hickman and others, 1985). One study in an area of flat and gently dipping rocks reported vertical fractures in the subsurface subparallel to lineaments (Komar and others, 1971). However, the subsurface fractures were nearly perpendicular to surface fractures. Judging from such studies, the relationship between orientations of linear features at the surface and in the subsurface is not the same everywhere.

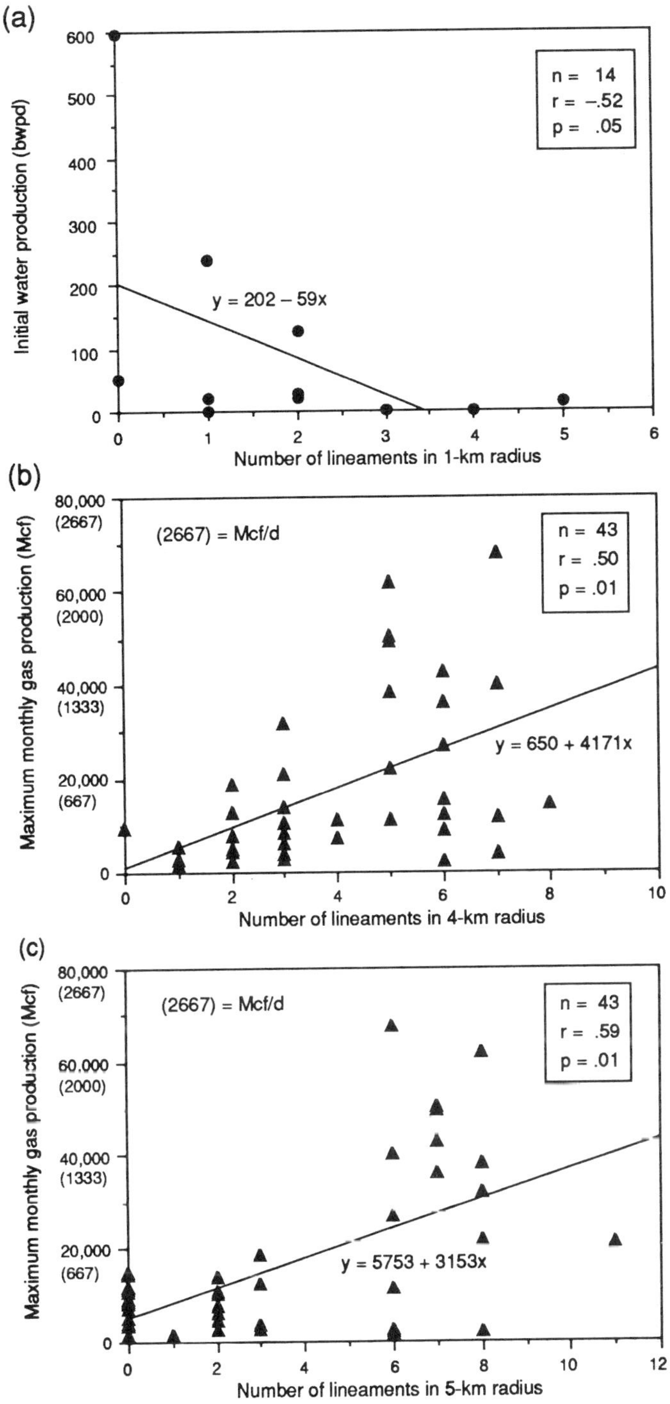

Figure 11. Production data vs number of Landsat lineaments within a given radius of a well. *a.* IPW vs lineaments within a 0.6-mile (1-km) radius (Decker and others, 1989). *b.* MMG vs Landsat lineaments within a 3.1-mile (5-km) radius (Knepper, 1982). *c.* MMG vs Landsat lineaments within a 2.5-mile (4-km) radius (this study).

Indeed, in an area of complex fracture patterns, surface and subsurface fractures can be either parallel or perpendicular in the same sedimentary basin because of variations in lithology and in burial and thermal histories of the rocks involved (Grout and Verbeek, 1985).

Lineaments and Production from Coalbeds

The results of this analysis suggest that lineament studies, on any imagery, are not very useful for locating highly productive coalbed methane wells in the San Juan basin. No consistent significant difference was found between production data from wells located on lineaments and those not on lineaments. Nor was a consistent relationship found between any lineament attribute and the entire range of values of IPW or MMG. These results do not replicate those from a study of coalbed methane in the Black Warrior basin, Alabama (Briscoe and others, 1988), where wells within 200 ft (60 m) of fracture zones had higher methane and water production. It is possible that in the San Juan basin different production parameters, such as mean annual gas production or average daily gas production, may be significantly correlated with lineament attributes, but these data were unavailable for this study.

CONCLUSIONS

This study has shown that lineaments are not reliable predictors of production data. Little agreement was found among lineaments mapped in independent studies of the San Juan basin. More than 95% of lineaments mapped in four separate studies do not coincide. The statistically significant (p = .01) vector sums of lineament azimuths in each study vary from subparallel to perpendicular to those in the other studies. Azimuths of lineaments nearest wells with high values of IPW and MMG vary between studies and even within one study, for lineaments derived from different imagery. Wells located on lineaments do not have consistently higher production values than those not on lineaments. Significant correlation was found between some production data and distance to the lineament nearest each well. Only a relatively weak correlation ($|r| \leq .59$, p = .01) exists between the number of lineaments within a given radius and either IPW or MMG.

ACKNOWLEDGMENTS

The Gas Research Institute provided funding for this work under contract no. 5087–214–1544. The author wishes to acknowledge the following persons who assisted in completion and improvement of this report. Consultation with Walter Ayers, Jr., and William Kaiser improved the analysis and presentation of results. Neil Whitehead III assisted with field-checking of lineaments. Rick Edson, Joseph S. Yeh, and Arten J. Avakian provided invaluable assistance in the use of computer-based mapping and contouring programs. Ting-Ya Hsieh assisted with integration of lineament and well data files. Insightful reviews by Walter Ayers, Jr., Tucker Hentz, William Kaiser, and Stephen Laubach improved the report and the conclusions presented herein. Figures were drafted by Wade Kolb, Joel Lardon, Kerza Prewitt, and Tari Weaver, under the direction of Richard Dillon, Chief Cartographer.

REFERENCES

Briscoe, F.H., Camp, B.S., Lottman, L.K., and Malone, P.G., 1988, A study of coal-bed methane production trends as related to geologic features, Warrior Basin, Alabama, *in* Fassett, J.E., ed., Geology and coal-bed methane resources of the northern San Juan Basin, Colorado and New Mexico: Rocky Mountain Association of Geologists Guidebook, p. 237–246.

Condon, S.M., 1988, Joint patterns on the northwest side of the San Juan Basin (Southern Ute Indian Reservation), Southwest Colorado, *in* Fassett, J.E., ed., Geology and coal-bed methane resources of the northern San Juan Basin, Colorado and New Mexico: RMAG 1988 Guidebook, p. 61–68.

Decker, A.D., Close, J.C., and McBane, R.A., 1989, The use of remote sensing, curvature analysis, and coal petrology as indicators of higher coal reservoir permeability, *in* The 1989 Coalbed Methane Symposium, Proceedings—University of Alabama, Tuscaloosa, April 17–20, 1989: School of Mines and Energy Development, Gas Research Institute, U.S. Mine Safety and Health Administration, and Geological Survey of Alabama, p. 325–340.

Dix, O.R., and Jackson, M.P.A., 1981, Statistical analysis of lineaments and their relation to fracturing, faulting, and halokinesis in the East Texas Basin: Texas Bureau of Economic Geology Report of Investigations 110, 30 p.

Dwight's Oil and Gas Reports, 1990a, Natural gas well production histories: Colorado Statewide Coal Bed Methane Report, v. 1, 373 p.

Dwight's Oil and Gas Reports, 1990b, Natural gas well production histories: New Mexico Statewide Coal Bed Methane Report, v. 1, 853 p.

Emmendorfer, Alan, 1989, Fracture orientation—Use of the dipmeter type fracture log: Mountain Geologist, v. 26, no. 2, p. 63–67.

Gorham, F.D., Jr., Woodward, L.A., Callender, J.F., and Greer, A.R., 1979, Fractures in Cretaceous rocks from selected areas of San Juan Basin, New Mexico—Exploration implications: AAPG Bulletin, v. 63, no. 4, p. 598–607.

Grout, M.A., and Verbeek, E.R., 1985, Fracture history of the Plateau Creek and adjacent Colorado River valleys, southern Piceance Basin—Implications for predicting joint patterns at depth: USGS Open-File Rept. 85–744, 17 p.

Hickman, S.H., Healy, J.H., and Zoback, M.D., 1985, In situ stress, natural fracture distribution, and borehole elongation in the Auburn geothermal well, Auburn, New York: Journal of Geophysical Research, v. 90, no. B7, p. 5497–5512.

Kelley, V.C., and Clinton, N.J., 1960, Fracture systems and tectonic elements of the Colorado Plateau: University of New Mexico, Publication in Geology 6, 104 p.

Knepper, D.J., Jr., 1982, Lineaments derived from analysis of linear features mapped from Landsat images of the Four Corners region of the southwestern United States: USGS Open-File Rept. 82–849, 79 p.

Komar, C.A., Overbey, W.K., Jr., Rough, R.L., and Lambert, W.G., 1971, Factors that predict fracture orientation in a gas storage reservoir: Journal of Petroleum Technology, v. 23, May, p. 546–550.

Newman, Karl, and McCord, John, 1980, Detailed site investigation, Northern San Juan Basin: Draft report, TRW Inc., Energy Systems Planning Division, not sequentially paginated.

Siegel, Sidney, 1956, Nonparametric statistics of the behavioral sciences: New York, McGraw-Hill, 312 p.

Tremain, C.M., Laubach, S.E., and Whitehead, N.H., III, in press, Coal fracture (cleat) patterns in the Upper Cretaceous Fruitland Formation, San Juan Basin, Colorado and New Mexico—Implications for coalbed methane exploration and development, *in* Ayers, W.B., Jr., and others, Depositional systems, hydrostratigraphy, and fracture analysis of the San Juan Basin: Gas Research Institute, Annual Report prepared for contract no. 5087–214–1544.

Tremain, C.M., Laubach, S.E., and Whitehead, N.H., III, 1990, Coal fracture (cleat) patterns in Upper Cretaceous Fruitland Formation, San Juan basin, Colorado and New Mexico—Implications for coalbed methane exploration and development, *in* Ayers, W.B., Jr., and others, Geologic and hydrologic controls on the occurrence and producibility of coalbed methane, Fruitland Formation, San Juan basin: Gas Research Institute, Topical Rept. GRI–91/0072 (August 1987–July 1990), p. 97–117.

Vistelius, A.B., 1966, Structural diagrams: New York, Pergamon Press, 178 p.

Wandrey, C.J., 1989, Lineament map of part of the Southern Ute Reservation, San Juan basin, southwestern Colorado: USGS Open-File Rept. 89–122, 1 sheet, scale 1:1,000,000.

Cleats in Coalbeds of Southern Piceance Basin, Colorado—Correlation with Regional and Local Fracture Sets in Associated Clastic Rocks

Marilyn A. Grout[1]

1. U.S. Geological Survey, Denver, Colorado

ABSTRACT

In the Piceance basin of northwestern Colorado, four of five regional sets of extension fractures (joints), and local fold-parallel sets, may be important conduits for the migration of methane from Upper Cretaceous coalbeds. The regional fracture sets are designated F_1 (oldest) through F_5 (youngest), and all are younger than the local sets that are related to late Laramide (Late Cretaceous-late Eocene) thrust faulting and folding, which created the Grand Hogback monocline and anticlines in the eastern part of the basin. The regional sets are unevenly distributed across the basin, and the local sets are found almost exclusively on the Divide Creek anticline.

Cleats occur in Upper Cretaceous and lower Tertiary coalbeds in the southern Piceance basin. Most cleats correlate in orientation, relative age, and fracture style with the intermediate-age F_3 and F_4 sets, suggesting that the coal lay unfractured for several tens of millions of years. In the central part of the basin, however, some cleats, both on outcrop and in core, correlate with the older F_1 or F_2 sets. Cleats that correlate with the youngest (F_5) set are extremely rare in the southern Piceance basin. Cleats older than F_1 joints are found in basal Tertiary coal stringers on the north-northwest-trending Divide Creek anticline (older coalbeds are not exposed). These cleats correlate both in strike and in relative age with moderately steeply dipping fractures that strike parallel to the axial trace of the anticline. It is suggested that these older cleats and fractures are fold related, having formed during extension of the outer arc of the anticline during late Laramide deformation. In general, coal cleats throughout the southern Piceance basin are similar to fractures in the enclosing clastic rocks.

INTRODUCTION

The Piceance basin is a late Laramide structural and depositional feature (Tweto, 1975, 1980) of the northeastern Colorado Plateau (Figure 1) that contains large undeveloped methane resources (Choate and others, 1984; McFall and others, 1986; ICF Resources Inc., 1990; Johnson and Rice, 1990). Most of the methane currently produced originates from lower prebasin rocks of the Mesaverde Group (Upper Cretaceous), which includes coal and interbedded sandstone, mudstone, and shale, at depths of approximately 2,500 to 7,500 ft (762 to 2,288 m) (Dunn, 1974; Donaldson and MacMillan, 1980). Matrix permeabilities of these reservoir rocks are only 0.001 to 1.3 µd (Branagan and others, 1984; Seccombe and Decker, 1986). Formation (in situ) permea-bilities are, however, several orders of magnitude greater, from 20 to 100 µd (Branagan and others, 1985; Lorenz and others, 1986), and are highly anisotropic. The large difference between laboratory- and field-tested permeabilities has been attributed to fractures that form a partly open, interconnected, three-dimensional system within the reservoir rocks through which fluid flows preferentially in certain directions.

The principal methane reservoirs so far identified in the Piceance basin are found in its southeastern part, where methane resides mostly in the matrix of coalbeds and, to a lesser extent, in coal cleats (McFall and others, 1986). Cleats are the system of natural fractures in coals. A major set (face cleat) and a minor set (end or butt cleat) commonly develop perpendicular to each other. Miners have long known of the ease of excavating coal parallel to the direction of face cleats, along which coal most easily breaks (Kendall and Briggs, 1933; McCulloch and others, 1974), but with the advent of rotary mining machines, coal can be mined easily in any direction. Mining perpendicular to the face cleats is, however, undesirable because the flow of methane and water into the mine is enhanced (McCulloch and others, 1975; Schubert, 1980).

To take advantage of methane flow along the face cleats, the Gas Research Institute in 1983 initiated a program to improve the technology for extracting methane directly from deep coalbeds through inclined and horizontal boreholes. Upper Cretaceous coalbeds in the southern Piceance basin (Figure 2) were the initial targets of this research. Knowing the directions of coal cleats and permeability

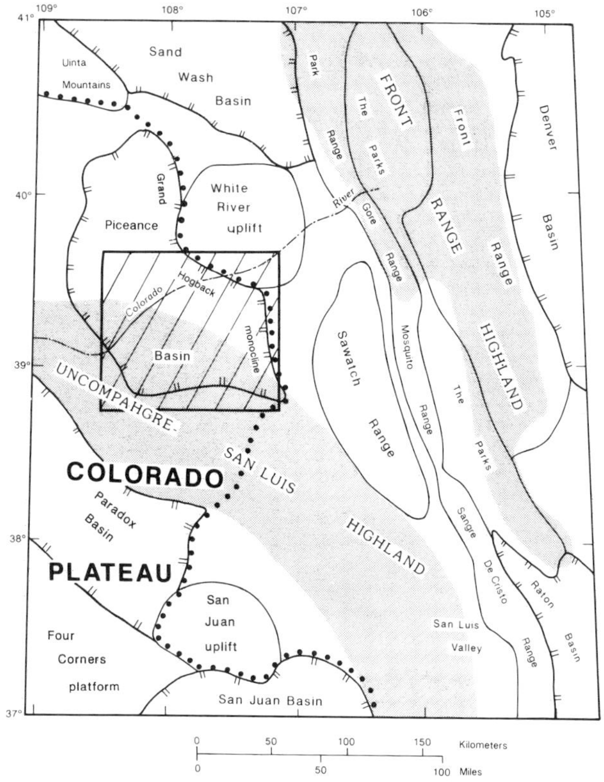

Figure 1. Location of Piceance basin study area (ruled) in relation to late Paleozoic uplifts (shaded) and Tertiary basins and uplifts. Dotted line indicates boundary of Colorado Plateau with Rocky Mountain foreland; barbed lines outline Laramide basins. Geology from Kelley and Clinton (1960), Howard and others (1972), Tweto (1980), and Gries and Dyer (1985).

anisotropy is critical in determining well spacing and in designing stimulations whereby lateral drainholes are placed directly into coalbeds for methane extraction (Seccombe and Decker, 1986).

Regional fracture studies suggest that the orientations of cleats in coalbeds and joints in other rock types are similar in some basins but not in others. For example, in the Appalachian basin of the eastern United States and in the Powder River basin of Montana, some joints in interbedded rocks have orientations similar to those of the cleats, but the joints have a greater range in strike and form more sets (Schubert, 1980). However, other studies in the Appalachian basin (Nickelsen and Hough, 1967; Kulander and others, 1980; Perry and Colton, 1981) and in the Black Warrior basin of Alabama (Murrie and others, 1976; Ward and others, 1984; Thompson and Telle, 1987) suggest little regional similarity in orientation between cleats in coal and joints in other rock types.

Cleats in Mesaverde Group coals and Wasatch Formation (Tertiary) coaly stringers in the southern Piceance basin can be correlated by relative age, orientation, and style with joint sets in the enclosing host rocks. The same techniques that were used to interpret the fracture history of Upper Cretaceous and Paleocene-Eocene rocks in the Piceance basin (Grout and Verbeek, 1983) also were used to delineate the fracture history of the interbedded coal. The data collected at each study site have been entered into a computer database developed for the Piceance basin by the U.S. Geological Survey and published in tabular format (Grout, 1988). This paper describes the fracture history of the coalbeds, compares the histories of cleats in coal and joints in other rock types, discusses cleat formation in general, and notes the influence of basin-margin thrusting and related anticline formation on orientation of cleats in Tertiary coaly stringers.

FRACTURE STUDIES IN PICEANCE BASIN

In previous studies of the Piceance basin, fracture patterns were mapped from aerial photographs by Welder (1971), Kelley and Clinton (1960), and Smith and Whitney (1979); the latter also studied fracture orientations in outcrops. Fractures and mineral fillings in oriented cores of Mesaverde Group strata from the U.S. Department of Energy Multiwell Experiment (MWX) site in the east-central basin (Figure 2) were analyzed or compared with joints in outcrops in the area and on the Grand Hogback monocline along the eastern basin margin by Clark (1983) and Lorenz (1985), Pitman and Sprunt (1986), Lorenz and Finley (1987), Finley and Lorenz (1989), and Garrett and Lorenz (1989). Lorenz and Hill (1991) also measured fracture spacings in core from slant holes and horizontal holes near the MWX site. Studies of the surface pattern of joints and slickenside lineations within the Grand Hogback monocline were made by Harper (1964), Murray (1966, 1967), and Dula (1981, 1984).

Orientations of cleats in Mesaverde coalbeds along the southern basin rim (Figures 3 and 4) were mapped by Geological Services of Tulsa (1980) and by Boreck and Strever (1980). Cleat orientations and spacings in cores from the Gas Research Institute's Red Mountain site, in the southwest-central part of the basin (Figure 2), were reported by Wiman and others (1984), Horner (1986), McFall and others (1986), and Seccombe and Decker (1986). Based on Grout and Verbeek's (1985) joint data, Seccombe and Decker (1986) suggested that a strong relationship exists between the orientations of particular joint sets in surface rocks and those of cleat sets in core at the Red Mountain site. With the exception of the fracture studies of the MWX and Red Mountain cores, generally only one aspect of fractures was measured—their orientation.

The fracture history of Mesaverde Group and overlying Paleocene-Eocene rocks in the Piceance basin has been documented in various related outcrop and underground mine studies (Verbeek and Grout, 1983, 1984a, 1984b, 1986, 1987, unpub. data; Grout and Verbeek, 1983, 1985, 1987, 1989, in press; Grout, 1988, in press, unpub. data; Perry and others, 1988; Grout and others, 1991). As determined from these studies, not only must the age relations among the fracture sets be recognized, but also the tectonic history of the basin must be known to adequately predict the joint network in correlative strata at depth. The results of these fracture and tectonic studies are summarized below.

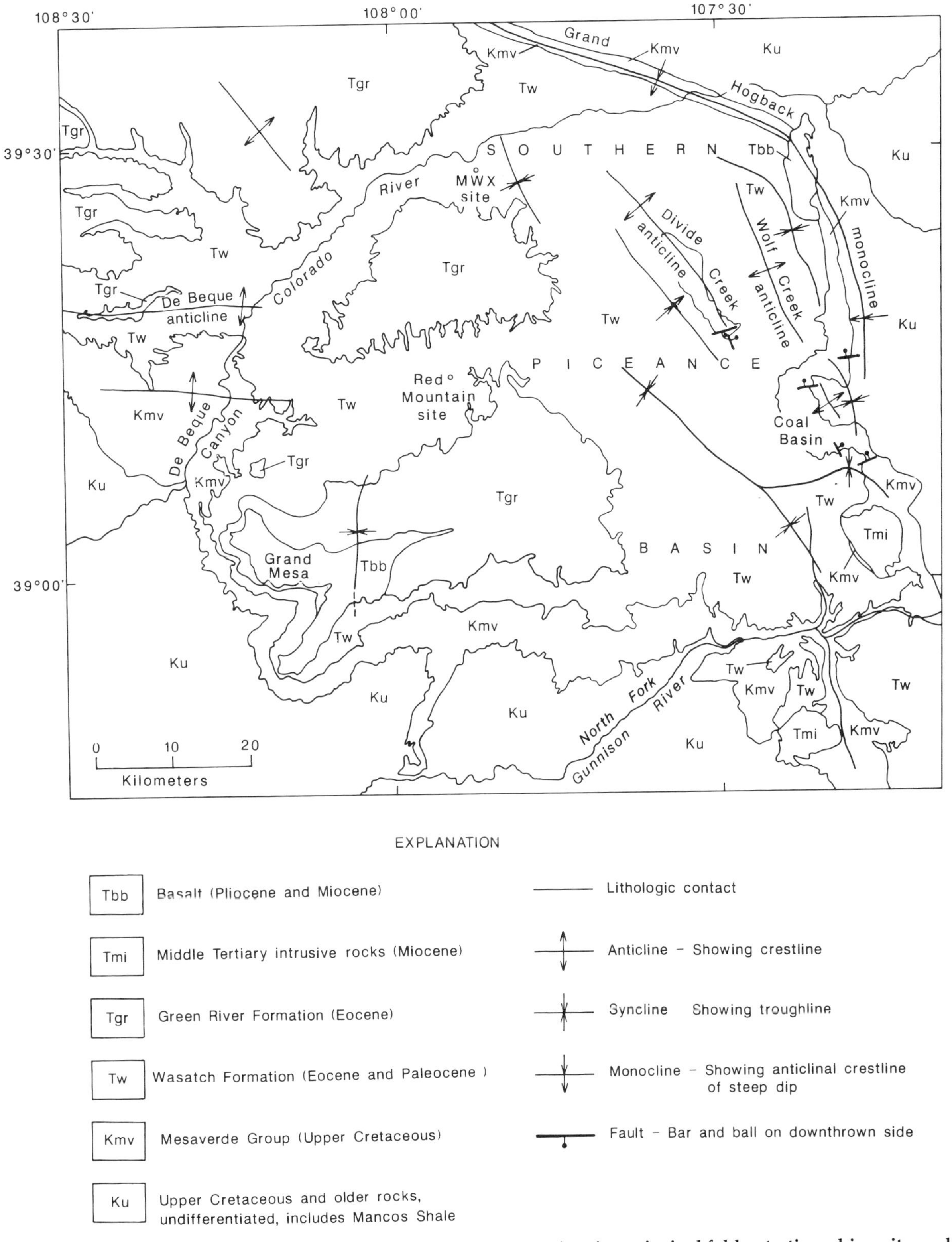

Figure 2. Generalized geologic map of southern Piceance basin showing principal folds, stratigraphic units, and oriented-core sites. Geology from Tweto and others (1978), Cashion (1973), Tweto and others (1976), Williams (1964), and Johnson (1983).

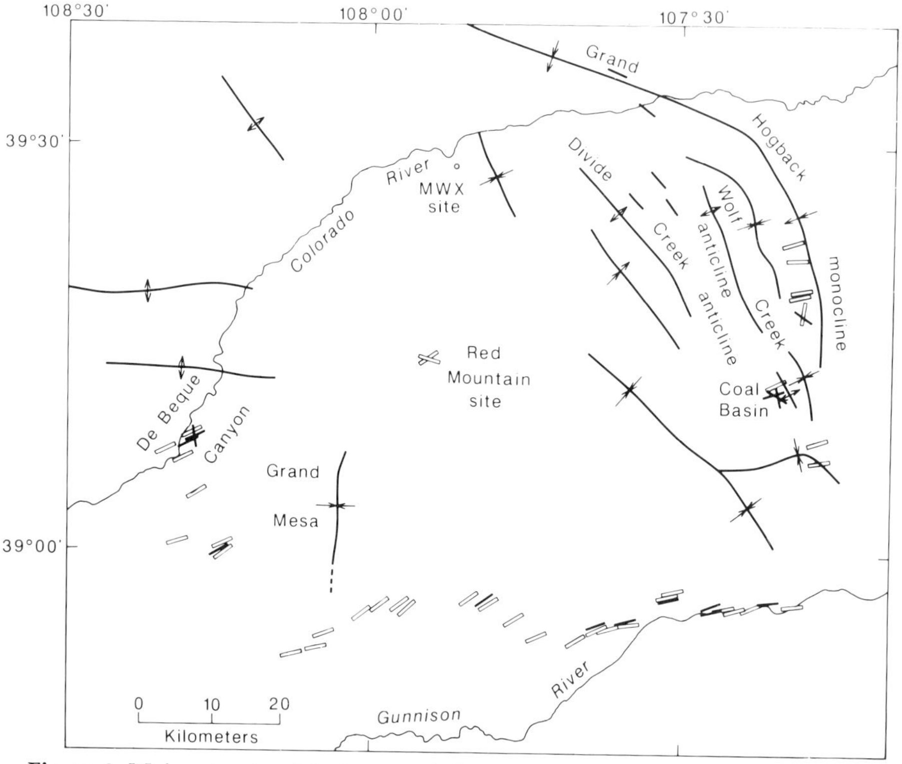

Figure 3. Major structural features and distribution and average strike of face cleats in southern Piceance basin. Filled bars represent data from this study. Data from Geological Services of Tulsa (1980), approximately located (open bars), are presented for comparison. Core data from Red Mountain site (open bars) are from Seccombe and Decker (1986) and Horner (1986).

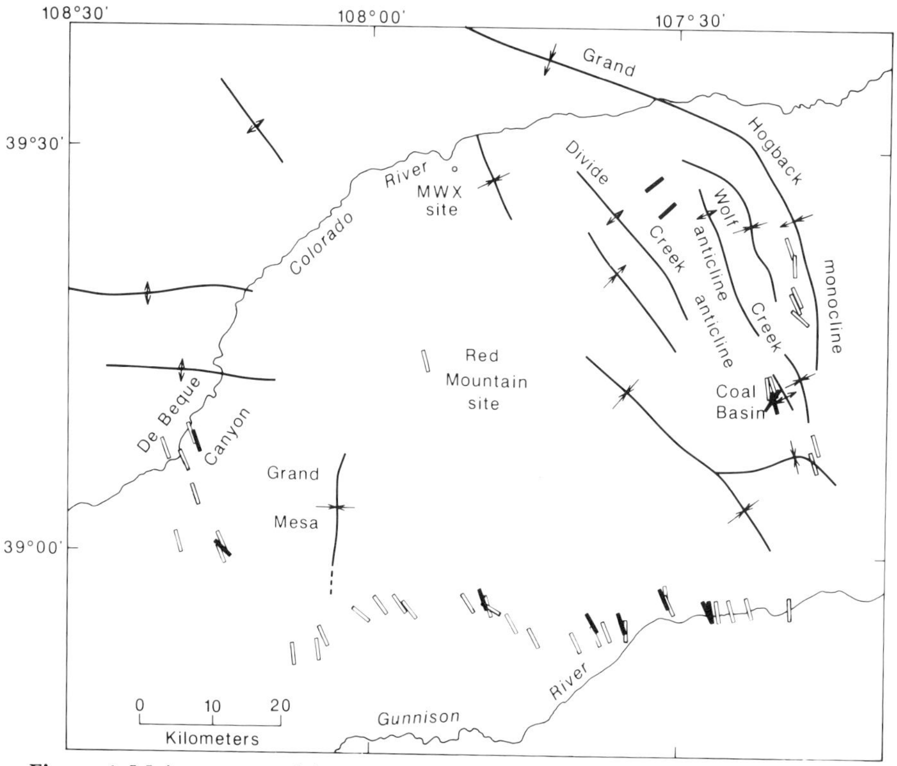

Figure 4. Major structural features and distribution and average strike of end cleats in southern Piceance basin. Filled bars represent data from this study. Data from Geological Services of Tulsa (1980), approximately located (open bars), are presented for comparison. Core data from Red Mountain site (open bars) are from Seccombe and Decker (1986) and Horner (1986).

FRACTURE HISTORY OF PICEANCE BASIN

The fracture history of Upper Cretaceous and lower Tertiary rocks of the Piceance basin encompasses the sequential formation of five regional sets of extension fractures (joints) and, locally, two older fold-related sets that are found on an intrabasin fold, the Divide Creek anticline, in the southeastern part of the basin (Figure 2). Four sets of still older joints (not discussed here), which predate both the basin sets and the final stage of formation of the Grand Hogback monocline, are found in Pennsylvanian through Paleocene rocks of the monocline; the youngest of those sets is present but weakly developed in the eastern part of the basin.

Regional Fracture Sets

Joints of the five regional sets, termed F_1 (oldest) through F_5 (youngest), show broad variations in relative abundance across the basin. For example, joints of early sets F_1 and F_2 are best developed in the northern two-thirds of the basin but are noticeably less abundant and locally absent in the south, where instead the intermediate-age F_3 set dominates the fracture network. F_4 joints are abundant throughout the basin, but F_5 joints in most places are sparse and most commonly are seen in Eocene oil shale. These differences in relative abundance of sets result in strong changes in the nature of the fracture network from place to place, although the network in most of the basin consists of the same five sets, modified only locally by one or more older sets. The same is true of the cleat system in the coal, as discussed later.

F_1 Set

Joints of the F_1 set, although present only locally in the basin, are more abundant in the north than in the south. They are widely scattered, in some places prominent and in many places absent. F_1 joints strike north-northwest (Table 1) and dip steeply to moderately steeply; some are filled with crystalline calcite. They terminate against no other joints and die out laterally as hairline cracks in the rock, except on the Divide Creek anticline, where they abut fold-parallel sets (discussed later).

F_2 Set

Joints of the F_2 set are large, both in length and commonly in height, and dominate the fracture network in the northern two-thirds of the basin, including the area of the Divide Creek and Wolf Creek anticlines. In the southern to southwestern third of the basin, however, they are much less abundant and are present mainly in Upper Cretaceous rocks. F_2 joints strike west-northwest (Table 1), dip steeply, and terminate laterally against F_1 joints, where present. Most F_2 joints in the northern part of the basin are filled with one to several layers of crystalline calcite; some in the southeast contain quartz or barite.

F_3 Set

Joints of the intermediate-age F_3 set also are large and dominate the fracture network in the southern half of the basin (Figure 5), but farther north they progressively decrease in abundance. They strike east-northeast (Table 1) and dip steeply; some are filled with crystalline calcite. Where they occur with older joint sets, F_3 joints most commonly terminate laterally against F_1 joints. F_3 joints rarely are found with F_2 joints, but both sets coexist locally along the Colorado River and commonly on the Divide Creek and Wolf Creek anticlines.

F_4 Set

Joints of the F_4 set are present throughout the basin (Figure 6). The fact that they terminate vertically against bed-parallel unloading joints and laterally against any earlier joints present strongly suggests that F_4 joints formed during or after regional uplift of the basin. In most places F_4 joints formed at high angles to whichever of the F_2 or F_3 sets was dominant; thus, they strike north-northeast in strata cut by abundant F_2 joints but strike north-northwest where the F_3 set is dominant (Table 1). Many F_4 joints in the northern part of the basin are filled with thin films of crystalline calcite; some contain microcrystalline calcite that resembles dripstone.

F_5 Set

The youngest joints in the Piceance basin, the F_5 set, are sparse but relatively most abundant in Eocene oil shale. They parallel fractures of the F_2 set (Table 1) but otherwise are unlike F_2 fractures—they are short, subplanar, and unmineralized. They terminate laterally against F_4 fractures, vertically against bed-parallel partings, and thus are of demonstrably young age relative to the other fracture sets. Their strikes parallel the present-day direction of maximum horizontal compressive stress.

Prominent Fracture Sets in Southern Piceance Basin

Regional Sets

The most prominent regional joint sets in the southern Piceance basin are the F_3 and F_4 sets. F_3 joints, which are large and strike N50°–90°E (Table 1), are present at more than 85% of the outcrops studied (Figure 5). Joints of the F_4 set, which are smaller, are even more abundant than F_3 joints and are present at nearly every outcrop examined (Figure 6). Their large dispersion in strike (N35°W to N44°E) is the result of their formation at high angles to whichever of the F_3 or F_2 sets was dominant.

Joints of the F_1, F_2, and F_5 regional sets are not abundant in the southern Piceance basin, with the partial exception of the Divide Creek and Wolf Creek anticlines, where the F_2 set is unusually well expressed relative to adjacent areas to the west. F_1 joints in the southern part of the basin generally are restricted to thick beds of sandstone and thus are of irregular distribution.

Fold-Related Sets

Two local sets of fractures are found almost exclusively on the Divide Creek anticline (Figure 2) and are restricted to thick sandstones. They strike parallel to the axial trace of the anticline (Table 1) and dip moderately steeply, intersecting bedding typically at angles of 60° to 70° (Grout, 1988; Grout and Verbeek, in press). These fractures, at each outcrop studied, form two sets that dip in opposite directions, one toward and the other away from the axial trace. The set dipping toward the fold axis consistently has the shallower dip, by 2° to 11°, apparently because of continuation of folding after joint formation. With rare exceptions, fold-parallel joints were the first to form in the Divide Creek area, and thus most simply die out as hairline cracks after having grown to appreciable size, up to 23 ft (7 m) long and 16 ft (5 m) high. Some of these joints are, however, small and terminate vertically against others of opposite dip. The fold-parallel sets predate creation of the Piceance system, having formed during anticlinal folding late in the Laramide orogeny (Grout and others, 1991; Grout and Verbeek, in press).

CLEATS IN SOUTHERN PICEANCE BASIN

Coal in the Mesaverde and Wasatch strata in the southern Piceance basin experienced stress and fracture histories similar to those of the surrounding rocks. Cleats in the coal-beds and joints in the clastic rocks exhibit similar variations in relative prominence across the basin, exhibit similar terminating relations, and similar orientations and surface structures (Tables 1 and 2). The major difference between joints in the clastic rocks and cleats in the coal is that the

Table 1. Correlation of cleat sets in coal with joint sets in clastic rocks, southern Piceance basin.

Relative Age*	Orientation	Feature and Location	Relative Abundance
OLDEST			
FP	N28°–55°W	Local fold–parallel (FP) sets in sandstone (Divide Creek anticline)	Moderate
	N38°–47°W	Face cleats (Divide Creek anticline)	Abundant
	N37°–40°W	Face cleats (Coal Basin area)	Sparse
F_1	N6°–36°W	F_1 in sandstone (Divide Creek anticline)	Sparse
	N5°–20°W	F_1 in sandstone (southern basin)	Sparse
	N14°W	Face cleats (Colorado River area)	Sparse
F_2	N46°–74°W	F_2 in sandstone (Divide Creek anticline)	Abundant
	N60°–85°W	F_2 in sandstone (southern basin)	Moderate
	N51°–70°W	Face cleats (Colorado River and Coal Basin areas)	Sparse
	N74°W	Face cleats (Red Mountain site cores) (Horner, 1986)	Sparse
F_3	N70°–90°E	Face cleats (southern basin rim) (Boreck and Strever, 1981)	Abundant
	N51°–87°E	Face cleats (southern basin rim) (Geological Services of Tulsa, 1980)	Abundant
	N62°–73°E	Face cleats (Colorado River and Coal Basin areas)	Abundant
	N50°–86°E	Face cleats (Red Mountain site cores) (Seccombe and Decker,1986; Horner, 1986)	Abundant
	N50°–90°E	F_3 in sandstone (southern basin)	Very abundant
	N54°–88°E	F_3 in sandstone (Divide Creek anticline)	Abundant
	N40°–57°E	End cleats (Divide Creek anticline)	Few measured
F_4	N35°W–N44°E	F_4 in sandstone (southern basin)	Abundant
	N10°–20°W	End cleats (Boreck and Strever, 1980)	Abundant
	N19°–31°W	End cleats (Colorado River and Coal Basin areas)	Abundant
	N11°–55°W	End cleats (southern basin rim) (Geological Services of Tulsa, 1980)	Abundant
F_5	N55°–80°W	F_5 in sandstone (southern basin)	Sparse
YOUNGEST			

* Note that even though the joint and cleat sets formed sequentially, as confirmed by consistent abutting relations across the basin, some of the sets formed in a rotational stress field (Verbeek and Grout, 1986). Some joint and cleat orientations of one set, therefore, may be nearly coincident with some from another set. The latest formed joints of the F_4 set, for example, have strikes approaching N80°–85°W and are close in orientation to the youngest joints of the F_3 set, which strike nearly east-west to N80°E and terminate against F_2 joints.

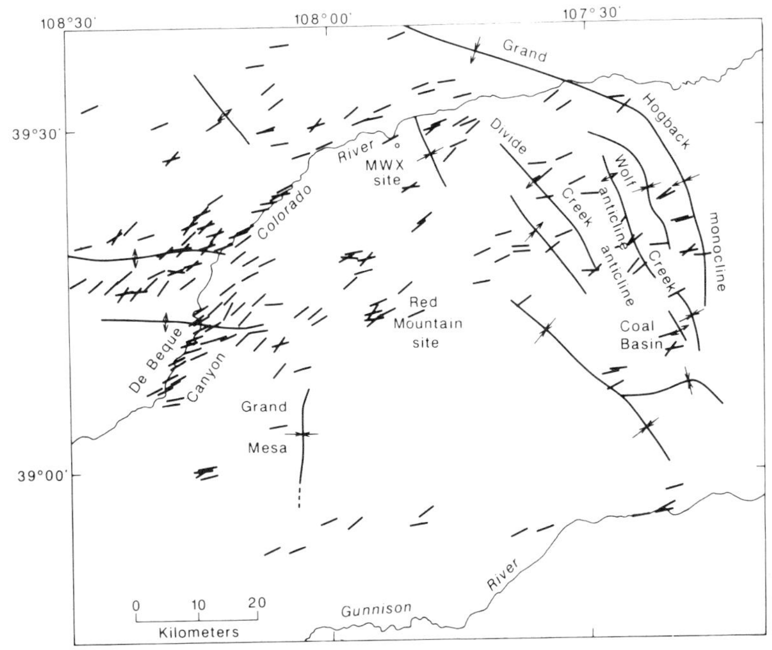

Figure 5. Distribution and average strike of joints of the F_3 set in southern Piceance basin.

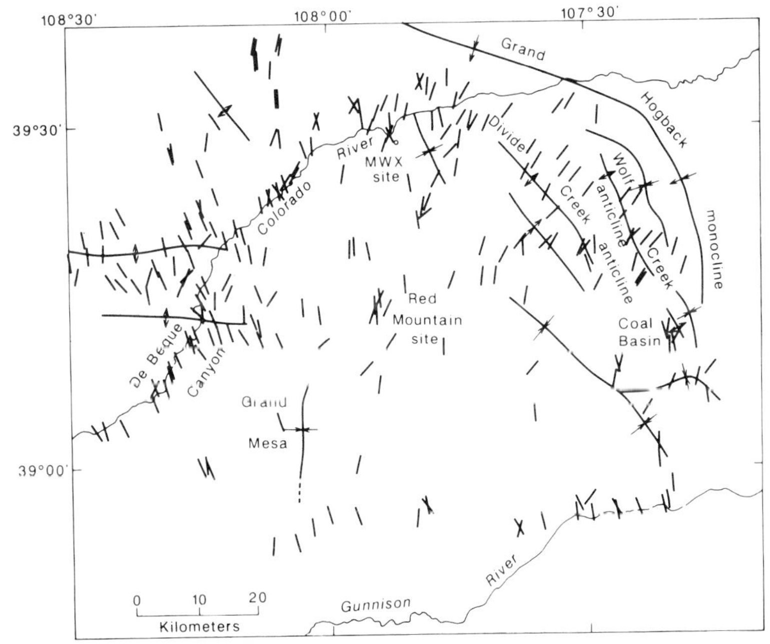

Figure 6. Distribution and average strike of joints of the F_4 set in southern Piceance basin.

cleats generally are shorter and more closely spaced because coal is much the more brittle material, and many of the coal layers or lenses where cleats were measured are thinner than the beds of enclosing host rocks.

Cleats in coalbeds of the Mesaverde Group have been studied in De Beque Canyon on the south side of the Colorado River, along the cliffs that mark the southern basin rim along the North Fork of the Gunnison River, on the Divide Creek anticline, in Coal Basin, and along the Grand Hogback monocline (Figures 3 and 4). Those along the Grand Hogback are not discussed here, but some sets are older than those found elsewhere in the basin. Cleats are nearly vertical and generally form two distinct sets—a face cleat set (Figure 3) and an end cleat set (Figure 4). Face cleats are both taller and longer than end cleats, which commonly terminate against the face cleats at nearly right angles. These characteristics indicate that face cleats formed earlier than end cleats. In some coalbeds, however, three

Table 2. Average cleat orientations in coals of the Mesaverde Group and Paleocene and Eocene basal Wasatch Formation, southern Piceance basin.

Location and Reference	Average Orientation	
	Face Cleats	End Cleats
Divide Creek anticline (Grout, 1988)	N38°–47°W	N40°–57°E
DeBeque Canyon, Colorado River area (Grout and Verbeek, 1985)	N14°W N70°W N62°–66°E	— — N19°–31°W
Colorado River area, eastern Piceance basin (Grout, 1988)	N51°–65°W	—
Red Mountain site cores (Seccombe and Decker, 1986; Horner, 1986)	N74°W N69°–86°E N50°–63°E	— — —
Coal Basin area (this study)	N37°–40°W N56°W N73°E	— N35°E N19°W
Grand Mesa and Gunnison River areas (this study)	N60°–86°E	N19°–30°W
Gunnison River area mines (Boreck and Strever, 1980)	N70°–90°E	N10°–20°W
Southern Piceance basin (Geological Services of Tulsa, 1980)	N51°–87°E	N11°–55°W

and, rarely, four or five cleat sets are present, and the traditional face cleat-end cleat terminology no longer is appropriate. The relative ages of the cleat sets coincide with the ages of regional fracture sets in clastic rocks, indicating a similar fracture history for both.

Face Cleats

Compared to end cleats, face cleats are the tallest and longest fractures in the coalbeds. They die out both laterally and vertically as hairline cracks and terminate against no other fractures; face cleats, therefore, are the first fractures to have formed in the coalbeds. With some exceptions, only one set of face cleats is prevalent. Some face cleats are taller than the thickness of the coal layer and extend upward and downward several millimeters to centimeters into beds of interlayered carbonaceous shale or siltstone. Surface structures on the face cleats are similar to those on F_1-F_3 joint surfaces in the enclosing rocks and indicate cleat formation in extension (Kulander and others, 1979). Such structures include plumes, twist-hackle fringes, and arrest lines.

Face cleats are spaced 0.04 to 9 in. (0.1 to 23 cm) apart, depending on thickness of the coal layer, cleat height, and coal rank. The thicker the coal layer and the higher the cleat (Figure 7), the wider the cleat spacing; the higher the coal rank, the closer the spacing. For example, in very thin (0.25 in. [0.6 cm]) layers of high-rank coal, a spacing of 0.06 to 0.26 in. (0.15 to 0.65 cm) is common. In coalbeds 4 to 12 in. (10 to 30 cm) thick, spacings of 0.8 to 2.8 in. (2 to 7 cm) are common. In coalbeds 12 in. (30 cm) to about 100 in. (250 cm) thick, however, spacings average 1.6 to 4 in. (4 to 10 cm), similar to spacings in beds 12 to 20 in. (30 to 50 cm) thick because average cleat height rarely exceeds 20 in. (50 cm) (Figure 7). The maximum cleat height measured was 43 in. (110 cm) in a 95-in.-thick (240-cm) coal layer. For thick coalbeds, then, an increase in bed thickness does not necessarily result in an increase in cleat spacing. Although narrow zones of overlapping cleats can extend the full

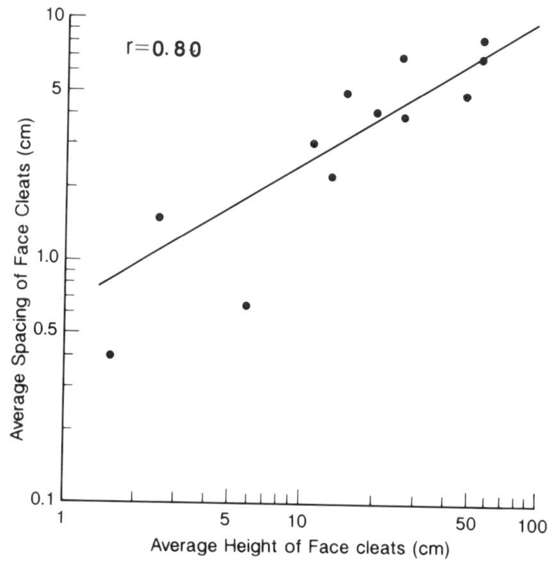

Figure 7. Relation between average spacing of face cleats and face-cleat heights for the F_3 set in subbituminous-bituminous coalbeds, southern Piceance basin.

height of thick coalbeds and resemble an individual fracture from a distance, true individual cleats within these zones are always much shorter than the zone. Cleat spacing in coalbeds, relative to layer thickness, is similar to joint spacing in clastic rocks if the cleat spans the full thickness of the coalbed and the coalbeds are of similar rank. A well defined log-log relation exists between joint spacing and layer thickness for clastic rocks (Verbeek and Grout, 1984a; 1987). Where cleats are shorter than coalbed thickness, cleat spacing increases with cleat height, given similar rank of coal (Figure 7).

Most of the face cleats strike about N60°E to N66°E in the De Beque Canyon and Grand Mesa areas and about N50°E to N86°E at Red Mountain, in Coal Basin, and along the southern basin margin (Table 2; Figure 3). These face cleats correlate in relative age and orientation with the F_3 joint set (compare Figures 3 and 5), the most prominent set in clastic rocks in the southern Piceance basin (Grout and Verbeek, 1985; Grout and Verbeek, in press; Grout, 1988; Grout, in press). Other researchers, such as Boreck and Strever (1980) and Geological Services of Tulsa (1980), found that face cleats in the southern Piceance basin range in strike from N70°E to N90°E and from N51°E to N87°E, respectively, similar to the range reported in this study (Table 2; Figure 3). Intermediate-age F_3 and F_4 sets predominate in the southern Piceance basin, except for the eastern part, suggesting that most Upper Cretaceous coalbeds (and clastic rocks) there lay unfractured until at least the late Eocene, a period of at least 30 million years.

Where more than two sets of cleats (face and end) are found in an outcrop, relative-age relations show that one or more of the older sets predate F_3. Face cleats that strike N51°W to N70°W in scattered areas of the southern Piceance basin correlate in relative age and strike with the F_2 joint set; face cleats with average strike of N14°W in De Beque Canyon correlate with F_1 (Table 1). Both sets of cleats, like their counterpart joints in clastic rocks, are subordinate to the F_3 set and are only locally prominent. Face cleats in Wasatch coaly stringers on the northern end of the Divide Creek anticline strike N38°W to N47°W, parallel to the axial trace of the fold. These cleats appear to be related to formation of the anticline and correlate in relative age with two local sets of fold-related fractures in enclosing sandstones (Table 1). Sets with similar strikes of N37°W to N40°W in Coal Basin (Figure 3) probably are related to formation of the Coal Basin anticline, which, like the Divide Creek anticline, likely owes its origin to late Laramide basin-margin thrusting (Table 1).

Most face cleats in the Piceance basin contain no mineral fillings. Several, however, contain thin seams of gypsum or an unidentified yellow powdery mineral.

End Cleats

Relative to face cleats, end cleats are both smaller and much less planar when viewed in profile. They are curviplanar, sinuous, and locally irregular in shape but nearly vertical in average dip. Extension-fracture features, commonly arrest lines, are found on most end cleats, which, like face cleats, are free of mineral filling. At many localities end cleats terminate vertically against bed-parallel parting joints and laterally, at about right angles, against any earlier cleats (F_1, F_2, or F_3) present.

In most areas of the southern Piceance basin, end cleats strike N10°W to N31°W (Table 2; Figure 4) and terminate against east-northeast-striking (F_3) face cleats (Figure 3). They thus are similar in relative age, orientation, and joint style to the regional F_4 fractures (Figure 6) in the enclosing host rocks. End cleats in Coal Basin strike north-northwest where they terminate against F_3 cleats but north-northeast where they terminate against F_2 cleats (Table 2), again similar to F_4 joints in clastic rocks. End cleats in the Divide Creek area, however, strike N40°E to N57°E (Table 2) and terminate against fold-parallel face cleats; they thus appear to correlate in orientation and relative age with the F_3 regional set. That the F_3 set corresponds to the face cleats in most of the southern Piceance basin but to the end cleats in the Divide Creek area is an additional effect of basin-margin tectonism.

Spacing of end cleats ranges from less than 0.02 in. to 6.3 in. (0.05 to 16 cm) but is commonly 0.2 to 2.4 in. (0.5 to 6 cm), depending on layer thickness and coal rank.

CLEAT FORMATION IN OTHER AREAS

According to Raistrick and Marshall (1939), the first published report on cleats appeared in 1834 by Edward Mannett, who noted extraordinary uniformity to the direction of the *slynes* (cleats) in coal. Later investigators suggested that cleats originated either through tectonic forces, such as those that deformed mountain ranges, or through nontectonic processes, such as compaction and dehydration. McCulloch and others (1974) and Schubert (1980) summarized much of the research on possible origins of cleats. Face cleats in some areas are perpendicular to major fold axes, and end cleats parallel them, leading to the common notion that face cleats are aligned with the principal compressive stress direction responsible for fold formation (for example, Ver Steeg, 1942, 1944; McCulloch and others, 1974, 1975; Diamond and others, 1975; Boreck and Strever, 1980; Geological Services of Tulsa, 1980). McCulloch and others suggested that face cleats in such areas are extension fractures that formed in a water-saturated state and under high compressive pressure during regional folding of the strata, whereas end cleats are release fractures that formed parallel to the axial trends of folds at a later time, after uplift and load removal by erosion.

On the Appalachian Plateau, Nickelsen and Hough (1967) and Nickelsen (1979) suggested that only coalbeds, to the exclusion of other sedimentary rocks, were mechanically capable of forming extension joints during early epeirogenic doming of the plateau. These cleats were likened to the planar systematic (major) joints noted by Hodgson (1961) on the Monument upwarp in Utah and Arizona. End cleats on the Appalachian Plateau are curved and terminate against face cleats; they correspond to Hodgson's nonsystematic (minor), but orthogonal, joints and are believed to have formed upon exposure and weathering of the coal. Perry and Colton (1981) concluded that coalbeds on the Appalachian Plateau were the most responsive indicators of regional tectonic stress patterns during the Alleghenian orogeny. Along the nearby Appalachian front, Alleghenian face cleats show a strikingly simple fanlike pattern, and the cleat orientations were found to be consistent with those of earlier studies of the Appalachian Plateau (Ver Steeg, 1942; Nickelsen and Hough, 1967; Kulander and others, 1980). Locally, however, coal cleats were found by Perry and Colton (1981) to show diverse orientations, and throughout the plateau no cleat orientations could be related to joints in other lithologies. Kulander and others (1980) found that face cleats on the Allegheny Plateau of West Virginia could be separated into distinct domains whose boundaries are spatially and perhaps genetically related to regional basement structure or to structural and sedimentary features in the overlying rocks (B.R. Kulander and S.L. Dean, 1990, written comm.).

In the Sydney basin of eastern Australia, Shepherd and Huntington (1981) discerned that although cleat orientations are quite variable in coal mines, cleat domains can be recognized. They concluded that the inconsistent cleat orientations across the basin probably resulted from the cumulative development of fracture systems during numerous tectonic and magmatic events. Summarizing conclusions from this and other studies, Schubert (1980) suggested that the locations, degrees of deformation, and orientations of folds and faults in coal basins appear to greatly influence the spatial distribution and orientations of fractures such as coal cleats, as well as the hydraulic conductivity of a rock mass.

In the San Juan basin, Condon (1988) concluded that the overall *average* orientation of face cleats does not correlate with joint orientations in either overlying or underlying sandstones. However, the face cleats appear to form not one but two sets, which correspond in orientation to the first- and second-formed joint sets in the sandstones, thus suggesting the opposite conclusion that a good correlation may exist between face-cleat sets and the first two joint sets in sandstone. End cleats correlate both in relative age and orientation with the third-formed of the four joint sets.

CONCLUSIONS

In the southern Piceance basin the regional fracture history documented in Upper Cretaceous and lower Tertiary sandstones and siltstones is similar to that of the interbedded coals. Cleats in coalbeds correlate with the five regional joint sets (F_1–F_5) in the host rocks, as well as with local, older fold-related sets near the thrusted eastern basin margin. Cleat sets generally are unmineralized, and some may be important conduits for migration of methane at depth. The coalbeds generally exhibit a prominent, planar set of face cleats at which terminate a smaller, curved, younger set of end cleats. The face cleats in any given area appear to have formed coevally with the earliest prominent set of regional fractures in surrounding sandstones. In most areas this corresponds to the F_3 joint set, which strikes east-northeast and is the most prominent set in the southern Piceance basin. That most of the face cleats correspond in relative age, joint style, and orientation to this joint set suggests that most Upper Cretaceous coalbeds lay unfractured until at least late Eocene time. End cleats generally correlate with a younger regional fracture set, F_4, and strike N19°W to N31°W. Where more than two sets of cleats occur, the oldest cleats correlate with the oldest (F_1 or F_2) regional sets or with the still older, local fold-related sets. On the Divide Creek anticline, for example, face cleats correlate with the fold-parallel sets and the end cleats with the regional F_3 set. The general hypothesis for the formation of cleats, that face cleats are perpendicular to the axial trace of a fold and that end cleats are parallel to it, does not hold for coal cleats in the southern Piceance basin.

ACKNOWLEDGMENTS

Funding was provided by the U.S. Geological Survey's Evolution of Sedimentary Basins Program and also partly by the U.S. Department of Energy's Western Tight Gas Sands Program. Thanks are extended to Mid-Continent Resources Inc., Powderhorn Coal Company, and Resource Enterprises Inc., for access to private or leased property; to H.D. Nowak, formerly Colorado School of Mines (now with Shell Western Exploration and Production Inc.), and E.R. Verbeek, U.S. Geological Survey, for use of unpublished coal-fracture data; and to D.B. Yager, U.S. Geological Survey, for fracture-data collection in the De Beque anticline area. I also thank D.B. Yager and C.W. Johannes for their assistance in the field and A.E. Lundberg, G.S. Hess, R.E. M'Gonigle, and E.J. McKay for volunteering their time in the field. R.E. Sabala and M.F. Duncan drafted the illustrations, and D.B. Yager, A.A. Springfield, and J.T. Springfield assisted with figure compilation and drafting. D.J. Lidke and E.R. Verbeek made many helpful suggestions that greatly improved the manuscript; D.L. Schmidt reviewed an earlier version of the abstract.

REFERENCES

Boreck, D.L., and Strever, M.T., 1980, Conservation of methane from Colorado mined/minable coal beds—A feasibility study: Colorado Geological Survey Open-File Rept. 80–5, 95 p.

Branagan, P., Cipolla, C., Lee, S.J., and Wilmer, R., 1985, Comprehensive well testing and modeling of pre- and post-frac performance of the MWX lenticular tight gas sand: Society of Petroleum Engineers Paper 13867, p. 19–22.

Branagan, P., Cotner, G., and Lee, S.J., 1984, Interference testing of naturally fractured Cozzette sandstone—A case study at the DOE MWX site, *in* Proceedings, SPE/GRI/DOE Unconventional Gas Recovery Symposium: Richardson, Texas, Society of Petroleum Engineers, p. 359–366.

Cashion, W.B., 1973, Geologic and structure map of the Grand Junction quadrangle, Colorado and Utah: USGS Miscellaneous Investigations Series Map I–736, scale 1:250,000.

Choate, R., Jurich, D., and Saulnier, G.J., Jr., 1984, Geologic overview, coal deposits, and potential for methane recovery from coalbeds, Piceance basin—Colorado, *in* Rightmire, C.T., Eddy, G.E., and Kirr, J.N., eds., Coalbed methane resources of the United States: AAPG Studies in Geology, Series 17, p. 223–251.

Clark, J.A., 1983, The prediction of hydraulic fracture azimuth through geological, core, and analytical studies: Society of Petroleum Engineers/Department of Energy Paper 11611, p. 107–114 [presented at Symposium on Low Permeability, Denver, Colorado, March 14–16].

Condon, S.M., 1988, Joint patterns on the northwest side of the San Juan Basin (Southern Ute Indian Reservation), southwest Colorado, *in* Fassett, J.E., ed., Geology and coal-bed methane resources of the northern San Juan Basin, Colorado and New Mexico: Rocky Mountain Association of Geologists, p. 61–68.

Diamond, W.P., McCulloch, C.M., and Bench, B.M., 1975, Estimation of coal-cleat orientation using surface-joint and photolinear analysis: Geology, v. 3, no. 2, p. 687–690.

Donaldson, J.C., and MacMillan, L., 1980, Oil and gas—History of development and principal fields in Colorado, *in* Kent, H.C., and Porter, K.W., eds., Colorado geology: RMAG, p. 175–189.

Dula, W.F., Jr., 1981, Correlation between deformation lamellae, microfractures, and in situ stress measurements, White River uplift, Colorado: GSA Bulletin, v. 92, pt. 1, p. 37–46.

Dula, W.F., Jr., 1984, Laramide compression and crustal segmentation, White River uplift, Colorado (abs.): GSA Abstracts with Programs, v. 16, no. 6, p. 495.

Dunn, H.L., 1974, Geology of petroleum in the Piceance Creek basin, northwestern Colorado, *in* Murray, D.K., ed., Guidebook to the energy resources of the Piceance Creek Basin, Colorado: Rocky Mountain Association of Geologists 25th Field Conference Guidebook, p. 217–223.

Finley, S.J., and Lorenz, J.C., 1989, Characterization and significance of natural fractures in Mesaverde reservoirs at the Multiwell Experiment Site: Society of Petroleum Engineers Paper 19007, p. 721–728 [presented at Rocky Mountain Regional/Low-Permeability Reservoirs Symposium and Exhibition, Denver, Colorado].

Garrett, C.H., and Lorenz, J.C., 1989, Fracturing along the Grand Hogback, Garfield County, Colorado: Sandia National Laboratories Rept. SAND88–3298, 29 p.

Geological Services of Tulsa, 1980, Geologic framework and potential structural control of methane in coal beds of southeastern Piceance Creek basin, Colorado: U.S. Department of Energy, Morgantown Energy Technology Center, 34 p. [prepared for TRW Energy Systems Group under TRW Subcontract J44432JJ0E and DOE Contract DE–AC–21–78MC08089]

Gries, R.R., and Dyer, R.C., eds., 1985, Seismic exploration in the Rocky Mountain region: Rocky Mountain Association of Geologists and Denver Geophysical Society, 300 p.

Grout, M.A., 1988, Fracture data for the Divide Creek and Wolf Creek anticlines area, southern Piceance Basin, northwestern Colorado: USGS Open-File Rept. 88–204, 66 p.

Grout, M.A, (in press), Cleat data from coal beds of the southern Piceance basin, northwestern Colorado: USGS Open-File Rept.

Grout, M.A., Abrams, G.A., Tang, R.L., Hainsworth, T.J., and Verbeek, E.R., 1991, Late Laramide thrust-related and evaporite-domed anticlines in the southern Piceance basin, northeastern Colorado Plateau: AAPG Bulletin, v. 75, no. 2, p. 205–218.

Grout, M.A., and Verbeek, E.R., 1983, Field studies of joints—Insufficiencies and solutions, with examples from the Piceance Creek Basin, Colorado, *in* Gary, J.H., ed., 16th Oil Shale Symposium, Proceedings: Colorado School of Mines Press, p. 68–80.

Grout, M.A., and Verbeek, E.R., 1985, Fracture history of the Plateau Creek and adjacent Colorado River valleys, southern Piceance Basin—Implications for predicting joint patterns at depth: USGS Open-File Rept. 85–744, 17 p.

Grout, M.A., and Verbeek, E.R., 1987, Regional joint sets unrelated to major folds—Example from the Piceance basin, northeastern Colorado Plateau (abs.): GSA Abstracts with Programs, v. 19, no. 5, p. 279.

Grout, M.A., and Verbeek, E.R., 1989, Prediction of fracture networks at depth in low-permeability reservoir rocks, Piceance and Washakie basins, western U.S. (abs.): AAPG Bulletin, v. 73, no. 9, p. 1158.

Grout, M.A., and Verbeek, E.R., (in press), Fracture history of the Divide Creek and Wolf Creek anticlines and its relation to Laramide basin-margin tectonism, southern Piceance basin, northwestern Colorado, *in* Evolution of sedimentary basins—Uinta and Piceance Basins: USGS Bulletin 1787–Z.

Harper, M.L., 1964, Mechanics of basement deformation in Glenwood Canyon, Colorado: Boulder, University of Colorado, Ph.D. dissertation, 102 p.

Hodgson, R.A., 1961, Regional study of jointing in Comb Ridge-Navajo Mountain area, Arizona and Utah: AAPG Bulletin, v. 45, no. 1, no. 8, p. 1–38.

Horner, D.M., 1986, Deep Coal Seam Project: Quarterly Review of Methane from Coal Seams Technology, v. 4, no. 2, p. 19–27.

Howard, A.D., and Williams, J.W., 1972, Physiography, *in* Mallory, W.W., ed., Geologic atlas of the Rocky Mountain region: Rocky Mountain Association of Geologists, p. 29–31.

ICF Resources Inc., compiler, 1990, The United States coalbed methane resource: Quarterly Review of Methane from Coal Seams Technology, v. 7, no. 3, p. 10–28.

Johnson, R.C., 1983, Structure contour map of the top of the Rollins Sandstone Member of the Mesaverde Formation and Trout Creek Sandstone Member of the Iles Formation, Piceance Creek basin, Colorado: USGS Miscellaneous Field Studies Map MF–1667, scale 1:253,440.

Johnson, R.C., and Rice, D.D., 1990, Occurrence and geochemistry of natural gases, Piceance basin, northwest Colorado: AAPG Bulletin, v. 74, no. 6, p. 805–829.

Kelley, V.C., and Clinton, N.J., 1960, Fracture systems and tectonic elements of the Colorado Plateau: University of New Mexico Publications in Geology 6, 104 p.

Kendall, P.F., and Briggs, H., 1933, The formation of rock joints and the cleat of coal: Proceedings, Royal Society of Edinburgh, v. 53, p. 164–187.

Kulander, B.R., Barton, C.C., and Dean, S.L., 1979, The application of fractography to core and outcrop fracture investigations: U.S. Department of Energy, Morgantown Energy Technology Center Rept. METC/SP–79/3, 174 p.

Kulander, B.R., Dean, S.L., Williams, R.E., and others, 1980, Fracture trends in the Allegheny plateau of West Virginia: West Virginia Geological and Economic Survey Map WV–11, scale 1:250,000.

Lorenz, J.C., 1985, Tectonic and stress histories of the Piceance Creek basin and the MWX site, from 75 million years ago to the present—Preliminary report: Sandia National Laboratories Rept. SAND84–2603, 50 p.

Lorenz, J.C., and Finley, S.J., 1987, Differences in fracture characteristics and related production of natural gas in different zones of the Mesaverde Formation, northwestern Colorado: Society of Petroleum Engineers Paper 16809, p. 589–593 [presented at 62d Annual Technical Conference and Exhibition, Dallas, Texas].

Lorenz, J.C., and Hill, R.E., 1991, Subsurface fracture spacing—Comparison of inferences from slant/horizontal core and vertical core in Mesaverde reservoirs: Society of Petroleum Engineers Paper 21877 [presented at Rocky Mountain Regional/Low Permeability Reservoir Symposium, Denver, Colorado, April 15–17].

Lorenz, J.C., Branagan, P., Warpinski, N.R., and Sattler, A.R., 1986, Fracture characteristics and reservoir behavior of stress-sensitive fracture systems in flat-lying lenticular formations: Society of Petroleum Engineers Paper 15244, p. 423–436 [presented at Unconventional Gas Technology Symposium, Louisville, Kentucky].

McCulloch, C.M., Deul, M., and Jeran, P.W., 1974, Cleat in bituminous coalbeds: U.S. Bureau of Mines Rept. of Investigations 7910, 25 p.

McCulloch, C.M., Diamond, W.P., Bench, B.M., and Deul, M., 1975, Selected geologic factors affecting mining of the Pittsburgh coalbed: U.S. Bureau of Mines Rept. of Investigations 8093, 72 p.

McFall, K.S., Wicks, D.E., Kuuskraa, V.A., and Sedwick, K.B., 1986, A geologic assessment of natural gas from coal seams in the Piceance Basin, Colorado: Gas Research Institute, Topical Rept. GRI–87/0060 (September 1985–September 1986), 76 p.

Murray, F.N., 1966, Stratigraphy and structural geology of the Grand Hogback monocline, Colorado: Boulder, University of Colorado, Ph.D. dissertation, 219 p.

Murray, F.N., 1967, Jointing in sedimentary rocks along the Grand Hogback monocline, Colorado: Journal of Geology, v. 75, no. 1, p. 340–350.

Murrie, G.W., Diamond, W.P., and Lambert, S.W., 1976, Geology of the Mary Lee Group of coalbeds, Black Warrior coal basin, Alabama: U.S. Bureau of Mines Rept. of Investigations 8189, 49 p.

Nickelsen, R.P., 1979, Sequence of structural stages of the Alleghany orogeny, at the Bear Valley strip mine, Shamokin, Pennsylvania: American Journal of Science, v. 279, no. 3, p. 225–271.

Nickelsen, R.P., and Hough, V.N.D., 1967, Jointing in the Appalachian Plateau of Pennsylvania: GSA Bulletin, v. 78, no. 5, p. 609–630.

Perry, W.J., Jr., and Colton, G.W., 1981, A summary of the methodology and results of regional joint-studies in the central and northern Appalachian Basin conducted by the U.S. Geological Survey as a result of Department of Energy Interagency Agreement No. EX–76–C–01–2287: USGS Open-File Rept. 81–1341, 19 p.

Perry, W.J., Jr., Grout, M.A., Tang, R.L., and Hainsworth, T.J., 1988, Wedge model for late Laramide basement-involved thrusting, Grand Hogback monocline and White River uplift, western Colorado (abs.): GSA Abstracts with Programs, v. 20, no. 7, p. A384–A385.

Pitman, J.K., and Sprunt, E.S., 1986, Origin and distribution of fractures in lower Tertiary and Upper Cretaceous rocks, Piceance basin, Colorado, and their relation to the occurrence of hydrocarbons, *in* Spencer C.W., and Mast, R.F., eds., Geology of tight gas reservoirs: AAPG Studies in Geology, Series 24, p. 221–233.

Raistrick, A., and Marshall, C.E., 1939, The nature and origin of coal and coal seams: London, The English Universities Press Ltd., 262 p.

Schubert, J.P., 1980, Fracture flow of groundwater in coal-bearing strata, *in* Symposium on Surface Mining Hydrology, Sedimentology and Reclamation: Lexington, University of Kentucky, p. 61–73.

Seccombe, J.C., and Decker, A.D., 1986, Geologic and reservoir characteristics of the Red Mountain coalbed methane test site in the Piceance basin: Gas Research Institute, Topical Rept. GRI–86/0109, 89 p.

Shepherd, J., and Huntington, J.F., 1981, Geological fracture mapping in coalfields and the stress fields of the Sydney Basin: Journal of the Geological Society of Australia, v. 28, p. 299–309.

Smith, R.S., and Whitney, J.W., 1979, Map of joint sets and airphoto lineaments of the Piceance Creek basin, northwestern Colorado: USGS Miscellaneous Field Studies Map MF–1128, scale 1:100,000.

Thompson, D.A., and Telle, W.R., 1987, Coalbed methane in the Black Warrior basin, Alabama—Geology, resources, and development: Quarterly Review of Methane from Coal Seams Technology, v. 4, no. 3, p. 2–18.

Tweto, Ogden, 1975, Laramide (Late Cretaceous-early Tertiary) orogeny in the Southern Rocky Mountains, *in* Curtis, B.F., ed., Cenozoic history of the Southern Rocky Mountains: GSA Memoir 144, p. 1–44.

Tweto, Ogden, 1980, Tectonic history of Colorado, *in* Kent, H.C., and Porter, K.W., eds., Colorado geology: Rocky Mountain Association of Geologists, p. 5–9.

Tweto, Ogden, Moench, R.H., and Reed, J.C., Jr., 1978, Geologic map of the Leadville 1°×2° quadrangle, northwestern Colorado: USGS Miscellaneous Investigations Series Map I–999, scale 1:250,000.

Tweto, Ogden, Steven, T.A., Hail, W.J., Jr., and Moench, R.H., 1976, Preliminary geologic map of the Montrose 1° × 2° quadrangle, southwestern Colorado: USGS Miscellaneous Field Studies Map MF–761, scale 1:250,000.

Verbeek, E.R., and Grout, M.A., 1983, Fracture history of the northern Piceance Creek Basin, northwestern Colorado, *in* Gary, J.H., ed., 16th Oil Shale Symposium, Proceedings: Colorado School of Mines Press, p. 26–44.

Verbeek, E.R., and Grout, M.A., 1984a, Fracture studies in Cretaceous and Paleocene strata in and around the Piceance Basin, Colorado—Preliminary results and their bearing on a fracture-controlled natural-gas reservoir at the MWX site: USGS Open-File Rept. 84–156, 30 p.

Verbeek, E.R., and Grout, M.A., 1984b, Prediction of subsurface fracture patterns from surface studies of joints—An example from the Piceance Creek basin, Colorado, *in* Spencer, C.W., and Keighin, C.W., eds., Geologic studies in support of the U.S. Department of Energy Multiwell Experiment, Garfield County, Colorado: USGS Open-File Rept. 84–757, p. 75–86.

Verbeek, E.R., and Grout, M.A., 1986, Cenozoic stress rotation, northeastern Colorado Plateau (abs.), *in* Stone, D.S., and Johnson, K.S., eds., New interpretations of northwest Colorado geology: Rocky Mountain Association of Geologists Symposium Guidebook, p. 97.

Verbeek, E.R., and Grout, M.A., 1987, Systematic joints within oil shales and associated rocks of the Green River Formation, *in* Taylor, O.J., compiler, Oil shale, water resources, and valuable minerals of the Piceance basin, Colorado—The challenge and choices of development: USGS Professional Paper 1310, p. 45–55.

Ver Steeg, K., 1942, Jointing in the coal beds of Ohio: Economic Geology, v. 37, p. 503–509.

Ver Steeg, K., 1944, Some structural features of Ohio: Journal of Geology, v. 52, p. 131–138.

Ward, W.E., III, Drahovzal, J., Jr., and Evans, F.E., 1984, Fracture analyses in a selected area of the Warrior basin, Alabama: Geological Survey of Alabama Circular 111, 78 p.

Welder, F.A., 1971, Map showing joint pattern inferred from aerial photographs, *in* Report on economics of environmental protection for a federal oil-shale leasing program: Denver, Colorado, Water Resources in Colorado Oil Shale, Advisory Committee Study, scale 1:380,000.

Williams, P.L., 1964, Geology, structure, and uranium deposits of the Moab quadrangle, Colorado and Utah: USGS Miscellaneous Investigations Series Map I–360, scale 1:250,000.

Wiman, S.K., Logan, T.L., Seccombe, J.C., and Decker, A.D., 1984, Development and evaluation of technology for methane production from a deep coal seam in the Piceance basin: Gas Research Institute, Annual Rept. GRI–85/0076, 323 p.

Mt Sopris

Coal Fracture (Cleat) Patterns in Upper Cretaceous Fruitland Formation, San Juan Basin, Colorado and New Mexico—Implications for Coalbed Methane Exploration and Development

Carol M. Tremain[1]
Stephen E. Laubach[2]
Neil H. Whitehead, III[3]

1. Colorado Geological Survey, Denver, Colorado
2. Bureau of Economic Geology, The University of Texas, Austin, Texas
3. New Mexico Bureau of Mines and Mineral Resources, Socorro, New Mexico

ABSTRACT

Producibility of coalbed methane depends in part on fracture permeability. To evaluate controls on fracture (cleat) patterns in coal seams in the Upper Cretaceous Fruitland Formation in the San Juan basin, we mapped fractures in coalbeds and adjacent sandstones in 11 cores and at 90 outcrop stations along the basin margin. Cleats are planar, perpendicular to bedding, usually uniform in strike within an outcrop or core, and arranged in closely spaced subparallel sets. Because cleats tend to terminate at laterally extensive volcanic ash interbeds (tonsteins), small faults and fracture zones that cross these layers can increase vertical connectivity within fracture networks and thus improve reservoir permeability in coal seams.

Face-cleat strikes delineate two principal domains of regional extent that are separated by a boundary of variable cleat orientation near the Colorado-New Mexico border. South of this boundary, face cleats strike predominantly northward or northeastward; north of the boundary, they strike predominantly northwestward.

The orientation, spacing, and mineralization of the cleats in the two domains create permeability anisotropy that influences methane producibility. For instance, wells may drain the coal reservoir more effectively in the face cleat direction. In the domain boundary region, the prevalence of two directions of strongly developed fractures and resulting increased coal friability may enhance the success of well completion by open-hole cavity methods.

INTRODUCTION

Natural fractures (cleats) enhance gas and water flow in coalbeds. With the increasing importance of coalbed methane as a natural gas source, information on cleat patterns is becoming critical for planning coalbed methane well placement and spacing. Moreover, fracture patterns in coal and adjacent rocks can affect success of completion and stimulation techniques, such as horizontal drilling, open-hole cavitation, and hydraulic fracture treatment.

The aim of this study is to document cleat characteristics and patterns in the Fruitland Formation. Results can help guide future coalbed methane development in the San Juan basin and provide a standard of comparison for cleat patterns in other coalbed methane-producing basins in western North America.

Previous studies of cleat on the northwestern margin of the San Juan basin documented northwest, north, and northeast fracture strikes. Newman and McCord (1980) reported joint strikes in sandstone at 70 locations in the Colorado part of the San Juan basin and cleat strikes at about a dozen stations. Condon (1988) measured cleat and/or joint strikes at 37 locations on the northwest side of the basin in Colorado. Our study builds on this previous work with observations of fracture orientations and intensity from 90 outcrop stations and from six oriented and five unoriented cores in Colorado and New Mexico. We mapped fracture patterns in coal and adjacent sandstone bedding plane (pavement) outcrops and documented cleat orientation, length, spacing, and mineral filling. Our observations are presented in the context of new stratigraphic, structural, and coal rank maps of the basin.

GEOLOGIC SETTING

The San Juan basin of the east-central Colorado Plateau, New Mexico and Colorado, is an asymmetrical structural basin of Late Cretaceous to early Tertiary age (Figure 1). As defined by the outcrop of the Upper Cretaceous Fruitland Formation, the basin is roughly circular with an area of approximately 6,700 mi^2 (17,000 km^2). It is bounded by the Chaco Slope and Zuni uplift on the south, the Hogback monocline on the west, the San Juan-Archuleta uplift on the north, and the Nacimiento uplift on the southeast. Patterns and timing of faulting along the basin margin suggest that the basin formed by both northeast- and southeast-directed shortening during the Laramide orogeny (Fassett, 1985; Ridgley and Huffman, 1990).

The Upper Cretaceous (Campanian) Fruitland Formation is the principal coalbed methane target in the basin. The Fruitland consists of coastal-plain sandstone, mudstone, coal, and carbonaceous shale. The formation is more than 400 ft (122 m) thick in the northwest, but it decreases in thickness until it disappears in the eastern part of the basin as a result of depositional thinning and erosion.

Individual Fruitland coal seams locally are as thick as 20 to 40 ft (6 to 12 m), and total net-coal thickness can reach 110 ft (34 m) (Ayers and others, 1990). Ash content typically ranges from 8% to 30% and more commonly is greater than 20% primarily because of numerous interbedded volcanic ash layers, or tonsteins, less than 5 in. (13 cm) thick. Coal rank ranges from subbituminous B in the south to high-volatile A bituminous around the northern basin margin (Scott and others, 1990). Rank increases to low-volatile bituminous in the north-central basin in T34N R8W, where coal reaches a depth of 4,200 ft (1,280 m).

CHARACTERISTICS OF FRUITLAND COAL CLEATS

Cleats are systematic natural fractures in coal that are analogous to joints in other sedimentary rocks (McCulloch and others, 1974; Ver Steeg, 1942; Raistrick and Marshall, 1939). Fruitland cleats are extension fractures that are oriented perpendicular to bedding, commonly consistent in strike within an outcrop or core, and arranged in subparallel sets. Face cleats, the major fracture set, are smooth-surfaced fractures that *usually* comprise the most prominent fracture set in terms of frequency and extent; they may be as long as several meters in plan view. Butt cleats, which exhibit predominantly rough surfaces, in most cases abut or terminate against face cleats at high angles (80° to 90°). Based on this cross-cutting relationship, face cleats are considered the earlier formed fractures. Compared to an associated set of face cleats, butt cleats are shorter, commonly less than 4 in. (10 cm) long, less continuous, and less well developed. A single face cleat set and a nearly orthogonal butt cleat set are observable in most exposures; however, butt cleats locally may be absent from a core or outcrop.

In Fruitland coal seams cleats cut across subhorizontal compaction fabrics defined by flattened objects such as coalified wood. Locally, cleats also cross concretions. Because cleats are perpendicular to bedding planes, regardless of bedding plane dip, face and butt cleat inclinations locally may be as low as ~30° to 40° in the northern San Juan basin, where coals dip as much as 50°.

Face-cleat strikes generally are consistent over wide areas of the basin (Figure 1), but in some large exposures and in several mine pit floors, curved or undulatory cleats of variable strike are evident (Figures 2 and 3). In addition to face and butt cleats, other systematic, generally less closely spaced fracture sets may occur in some coal outcrops. Among these are fractures that are indistinguishable from face cleats except for their divergent strike. Faults and associated fractures crosscut cleats in some areas; elsewhere, cleats curve as they approach faults. Striated cleats are rare. Fractures with polygonal patterns occur in some dull coals[4] where orthogonal cleat sets are absent.

Cleat Types and Cleat Spacing

Cleat spacing observed in Fruitland coals ranges from less than 0.03 in. (0.1 cm) to more than 1 ft (30 cm). To obtain comparable cleat-spacing measurements between stations, we defined a hierarchy of cleat (fracture) sizes that can exist within a given seam. These range from primary cleats, which extend across one or more coal-type (lithotype) layers, to secondary and tertiary cleats, which are vertically discontinuous within coal lithotype layers (Figure 4). These fractures give coal its blocky appearance.

Primary cleats are the most prominent cleat set observed in most Fruitland exposures. Two scales of primary cleat spacing and height are evident locally. The more closely spaced primary cleats have an average spacing of 1 to 3 in. (2.5 to 8 cm), are 1 in. to more than 3 ft (2.5 cm to more than 0.9 m) long and commonly exhibit oxidized surfaces indicative of fluid flow. Their height equals the thickness of the coal lithotype layer in which they occur, from a foot or more in thick layers to less than 0.5 in. (1.25 cm) in thin vitrain layers. They rarely cross interbedded tonsteins or thin shale partings even where interbeds are less than 0.5 in. (1.25 cm) thick. For example, in one lower Fruitland coalbed in the Pinto pit (Navajo Mine, T28N R15W), fewer than 20% of the cleats penetrate or cross interbeds 0.2 in. (0.5 cm) thick or less. On the larger scale, master cleats (Henkle and others, 1978), which are visible in large Fruitland outcrops and surface coal mines, are primary cleats that cross several coal-type layers. Their spacing ranges from 1 to 3 ft (0.3 to 0.9 m). Where coal seams are less than 6 ft (1.8 m) thick, master cleats may cross the entire seam.

Secondary cleats are fractures that are parallel to primary cleats but are *vertically discontinuous* within a lithotype.

4. Dull coal is a type of banded coal defined microscopically as consisting mainly of clarodurain and durain and of 20% or less of bright materials such as vitrain, clarain, and fusain. [AGI]

Secondary cleat spacing commonly is 1 to 2.5 in. (2.5 to 6.3 cm) in subbituminous to high-volatile C bituminous coals in the southern part of the basin and 0.25 to 0.5 in. (0.6 to 1.2 cm) in high-volatile B bituminous coals in the west. Spacing ranges from 0.12 to 0.25 in. (0.3 to 0.6 cm) in high-volatile A bituminous coals in the north. Length and height of secondary cleats generally are less than 1 in. (2.5 cm). The distinction between primary and secondary cleats is obvious in outcrop but less so in core.

Tertiary cleats are closely spaced fractures and microfractures whose spacing and height vary from 0.03 to 0.12 in. (0.1 to 0.3 cm); they are situated between secondary

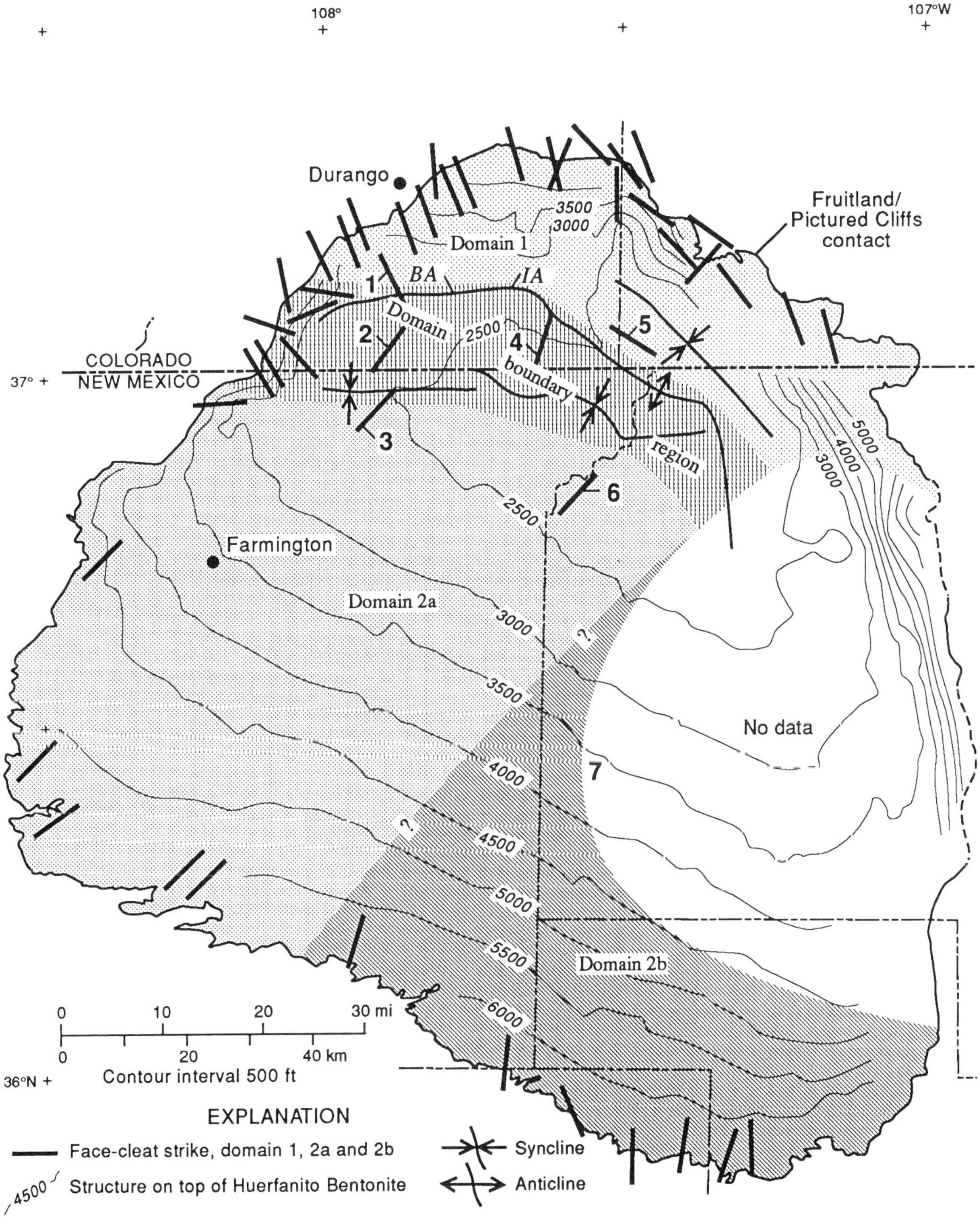

Figure 1. Structure map of San Juan basin, contoured on the Huerfanito Bentonite Bed, showing face-cleat strikes and domains in Fruitland Formation coal seams. Bold numbers indicate locations of wells from which oriented Fruitland core were examined (Tremain and others, 1990). BA, Bondad anticline; IA, Ignacio anticline.

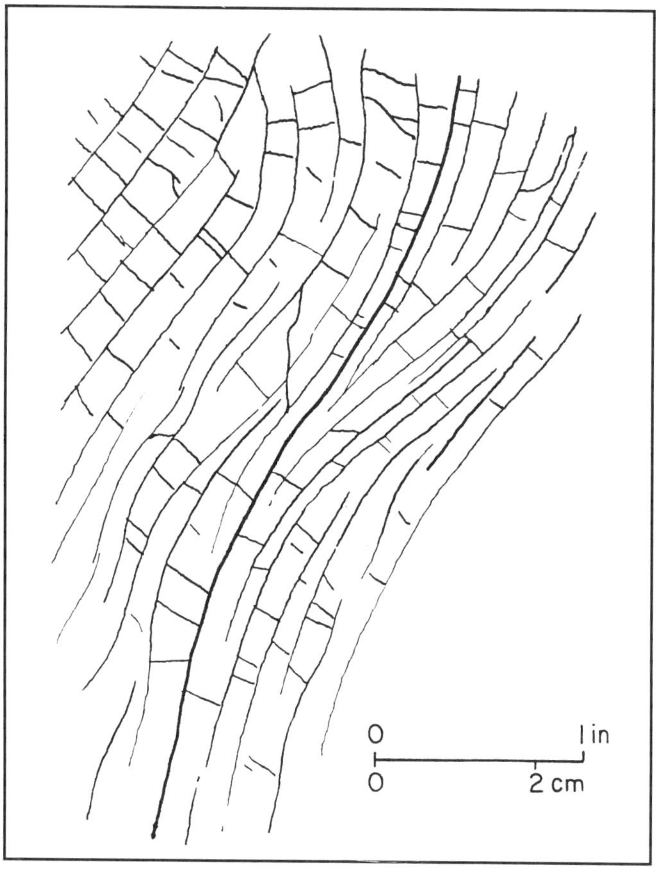

Figure 2. Map view of curved cleat traces in flat-lying coal, Fossett Gulch (T34N R5W).

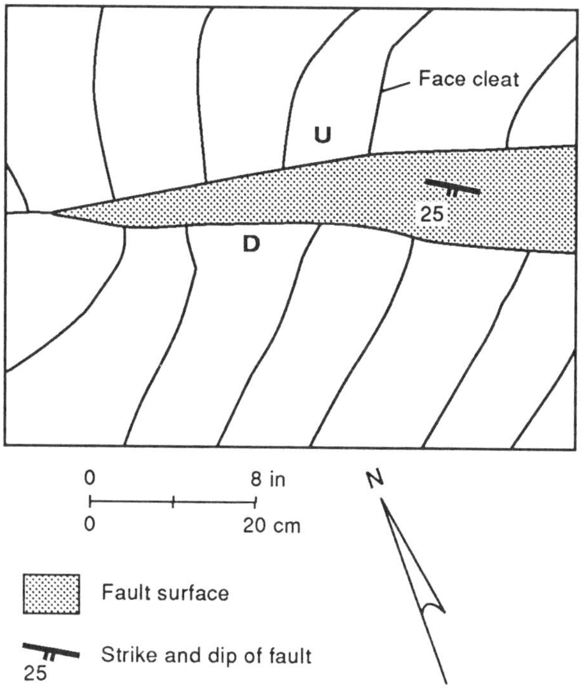

Figure 3. Cleat curvature near the termination of a small normal fault, Navajo Mine (T28N R15W).

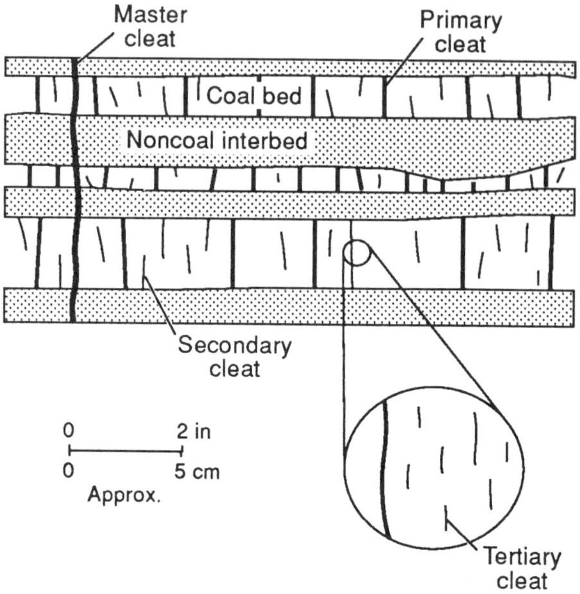

Figure 4. Hierarchy of coal cleat sizes (primary, secondary, and tertiary).

cleats. Lack of mineral infill or staining, greater abundance in weathered outcrops, and terminations at open, horizontal unloading fractures suggest that tertiary cleats formed near the ground surface and may not have been present at depth.

Spacing of coal fractures in regional cleat sets (total fracture intensity) is consistent over distances up to 1 mile (1.6 km) where coals overlie sandstones with planar upper surfaces. For example, in a 4-in.-thick (10-cm) coal seam of uniform composition lying about 15 ft (4.6 m) above the Pictured Cliffs Sandstone, as exposed over a distance of 2,300 ft (702 m) in the Navajo Mine, most cleat spacing values fall within one standard deviation of the mean for the entire data set. Where the sandstones have irregular contacts, and locally near lateral terminations or beneath thick lenticular fluvial sandstones, cleat intensity increases by as much as 25%. Such an increase could be attributed to several causes, including fracturing due to stress concentrations at irregular lithologic boundaries.

Variation in Cleat Spacing

Cleat spacing is controlled by coal rank and lithotype and by bed thickness (Ammosov and Eremin, 1960). Primary cleat spacing in the San Juan basin depends on coal rank and type and, to a lesser extent, on stratigraphic and structural position and thickness of individual coalbeds within a seam. In this study the term "coalbed" refers to divisions of a coal seam separated by noncoal layers such as shale partings or tonsteins. In Fruitland coalbeds, average primary face-cleat spacing is closer in high-rank coals in the northern one-third of the basin than in lower rank coals to the south. Mean primary face cleat spacing is 1.5 in. (4 cm) in outcrops along the northern and western basin margins. In the south, spacing ranges from to 0.75 to 6 in. (2 to 15 cm) but typically is between 2.5 and 3.5 in. (6.5 and 9 cm) in medium-brightness coal. Increased cleat spacing with decreasing bituminous rank has been reported in other basins (Ammosov and Eremin, 1960; Ting, 1977).

Primary cleat spacing depends in part on coal type (Ammosov and Eremin, 1960; Ting, 1977). In Fruitland coal, spacing is closest in bright, vitrain-rich layers and widest in dull, durain-rich layers. Some primary cleats in 0.1-in.-thick (0.25-cm) vitrain-rich layers or lenses have spacing of less than 0.03 in. (0.1 cm), whereas some thin dull (or bone) coal layers are virtually uncleated. However, these thin lenses of very bright or dull coal are volumetrically insignificant (<10%).

Fracture spacing in coal and other stratified rocks commonly is inversely proportional to bed thickness, with more closely spaced fractures in thinner beds and spacing somewhat less than bed thickness (Spears and Caswell, 1986). Bed thickness in medium-brightness coal seams in the northern and western San Juan basin ranges from less than 0.4 in. (1 cm) to more than 20 in. (50 cm), with a mean thickness of 4 in. (10 cm). Primary face cleat spacing in these beds ranges from 0.03 to 12 in. (0.1 cm to 30 cm), with a mean spacing of 1.5 in. (4 cm). For medium-brightness coalbeds, cleats generally are somewhat more closely spaced in thin beds (Figure 5). Variance of cleat spacing increases with bed thickness.

More closely spaced cleats in thin coalbeds result from both the bed-thickness effect and the tendency for thin-bedded coals to be vitrain rich. Coalbeds with numerous vitrain-rich layers and tonsteins or shale partings tend to exhibit more closely spaced cleats, but the cleat system may be poorly interconnected vertically because cleats tend to terminate at noncoal interbeds.

Cleat-Filling Minerals

Some fractures in Fruitland coalbeds have thin mineral veins. Principal cleat-filling minerals—pyrite, calcite, and gypsum—are present in outcrop and in core from depths as great as 3,900 ft (1,190 m) in the northern and central basin. Pyrite occurs in thin films and isolated patches on some cleat surfaces in mines and cores. In outcrop, primary face cleats commonly show a reddish-brown to red stain that may represent oxidation of finely disseminated sulfides formerly present in fractures. Near the Colorado-New Mexico border, isolated pyrite rosettes occur on both face and butt cleat surfaces.

Calcite occurs in cleats in several mines and cores (Tremain and others, 1990, Table 1). In Fruitland sandstone of the Sutec No. 1 Oxford core (sec. 25, T34N R8W), calcite occurs within a vertical extension fracture that is parallel to, and possibly continuous with, adjacent calcite-filled cleats. Other calcite-filled cleats and parallel calcite-filled fractures in sandstone occur in the Chimney Rock Mine (sec. 30, T34N R4W), Colorado, and vein widths are as much as 0.05 to 0.1 in. (1.3 to 2.5 mm) in some New Mexico examples. The calcite typically has a fibrous habit.

Cleat-filling fibrous gypsum is present in several outcrops and cores in New Mexico (Tremain and others, 1990, Table 2). In northern San Juan basin exposures, fine-grained clay minerals (kaolinite?) occur on some face-cleat surfaces. Where present, pyrite and calcite generally fill primary

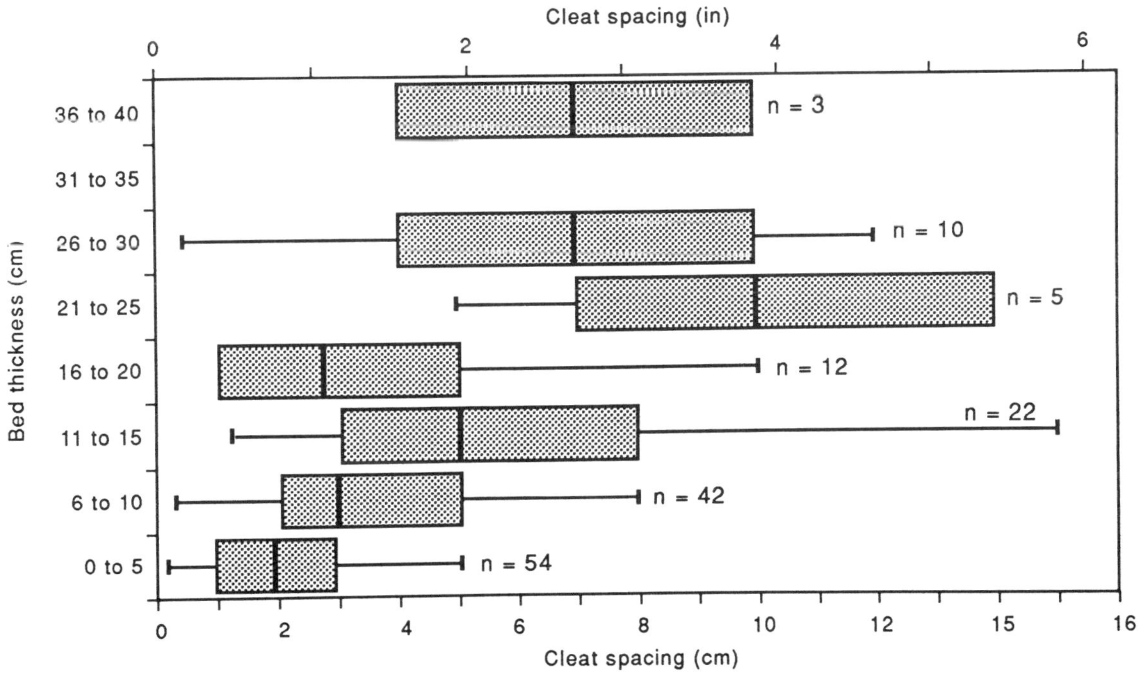

Figure 5. Cleat spacing vs bed thickness of medium-brightness coals, northern San Juan basin. Center one-half of data for the indicated range of bed thickness is represented by a rectangle (box); median value is indicated by a heavy bar.

Figure 6. Coal pavement map showing crossing and mutually abutting face-cleat (fracture) patterns in a single bedding plane at Navajo Mine (T28N R15W). Prominent, natural barren and mineral-filled fractures are shown.

and secondary cleats in a given bed, but maps of fractures in coal pavements show that mineral filling is discontinuous along fracture traces (Figure 6).

Wide mineral-filled veins and vein networks occur near some faults and in several outcrops where presence or absence of faults could not be demonstrated. Vein widths are as much as 0.39 in. (1 cm). Veins are filled with gypsum, calcite, or quartz, and subsidiary pyrite and clay minerals.

Cleat Orientation

Cleat strikes were measured in six cores and at 90 outcrop stations, including selected areas within large, active open-pit mines and abandoned mines, natural exposures, and excavations. Most stations cover areas up to 100 ft^2 (9 m^2). In selected areas cleat measurements were supplemented by maps of cleat patterns on bedding surfaces, such as that shown in Figure 6. Cleat maps were constructed either by mapping with reference to a grid marked on the outcrop or by mapping directly onto color photographs.

Face cleats strike predominantly northwest in the northern San Juan basin and predominantly north or northeast in the southern part of the basin (Figure 1). Cleat strike typically varies by less than 10° in individual cores and in most small outcrops (Tremain and others, 1990, Table 1). Larger exposures show nearly constant cleat strikes over large areas. For example, face-cleat strikes vary by less than 5° over a 2,300-ft-long (702-m) transect in the Pinto pit in the Navajo Mine. In a traverse approximately 6 miles (10 km) long in middle to lower Fruitland coals between the Animas and Florida Rivers, mean face-cleat strikes at 15 stations typically vary by less than 10°. Cleat strikes are consistent between adjacent coal seams at the old La Plata Mine (sec. 27, T35N R9W), where four lower Fruitland coalbeds have mean face cleat strikes between N30°W and N16°W, and at the south fork of Texas Creek (secs. 7 and 8, T35N R7W) and East Gap (secs. 17 and 18, T34N R10W), where cleat strikes in adjacent lower and lower to middle Fruitland coal seams vary by only 11° and 5°, respectively, from those in basal and higher coalbeds.

Great dispersion in face-cleat strike or abrupt strike variations of 20° or more occur locally. These variations result from curved cleats, gradual changes in cleat strike over wide areas, and changes in cleat strike between adjacent coalbeds. Curved cleats occur at seven stations in the Fort Lewis Mine (sec. 36, T33N R12W) and at six widely spaced stations in the northern basin. Cleats change strike by 5° to 30° within distances of several inches to several feet. Some curvature results from hooking of cleats into an adjacent fracture, probably as a result of interaction between a growing and preexisting fracture, but in several cases the entire array of face cleats curves in unison. Unlike curved cleats that define some polygonal patterns in coal, such a pattern would not be expected from brittle failure in an isotropically shrinking medium.

The greatest variation in cleat strike over short distances was observed where curved cleats occur near faults (Figure 3). At the Fort Lewis Mine, cleats change strike by 35° to 40° near a small graben. Strikes more nearly parallel the fault plane within approximately 30 ft (9 m) of the fault. In the Navajo Mine, in the hanging wall of an east-striking normal fault with 6 in. (15 cm) of net slip, northeast-trending cleats change strike by 30° over a distance of only 14 in. (35 cm) as they approach the fault plane.

Cleat strike can also vary greatly between adjacent beds. At West Gap (sec. 13, T34N R11W), Beaver Creek (sec. 13, T35N R6W), and Fort Lewis Mine, face-cleat strikes vary by 53°, 48°, and 71°, respectively, between adjacent beds. These differences show no consistent pattern with stratigraphic position. The area of greatest variability in cleat strike was observed near the Fort Lewis Mine (Figure 7). Some cleat-strike variations may reflect folds, faults, and fracture zones in this area (Newman and McCord, 1980; Laubach and others, 1990). However, this region also is part of a wider zone (described below), wherein face-cleat strike shifts by nearly 90°.

CLEAT-STRIKE DOMAINS

Face cleat strikes typically are consistent over large areas of the San Juan basin, but they are not the same throughout the entire basin (Figure 1). Face-cleat strikes delineate two principal domains of regional extent. In the northern part of the basin, northwest-striking face cleats predominate (domain 1, mean strike N16°W), but south of the New Mexico-Colorado line, north-northeast- and northeast-striking face cleats predominate (domain 2, mean strike N20°E). These domains are separated by a boundary domain that extends eastward from the state line at the western margin of the basin to T32N R7W, where it then turns southeastward into New Mexico.

Face cleats in the northern domain strike approximately 10° to 15° more northerly than butt cleats in the southern domain. Yet within each domain, face and butt cleats are orthogonal. The northern domain of northwest-striking face cleats is further characterized by gentle west- to west-northwest-trending folds, thick sandstones and coalbeds with northwesterly strikes, and higher coal rank (high-volatile A bituminous or greater) (Scott and others, 1990). In contrast, the southern domain of north- to northeast-trending face-cleat strikes is developed on a gentle, northeast-dipping homocline, the Chaco Slope. Here, fluvial channel-fill sandstone and interchannel coal deposits have predominantly northeasterly or northerly trends, and coals generally are of lower rank (high-volatile B bituminous or lower). Cleat spacing apparently is somewhat wider in domain 2.

The boundary between domain 1 and domain 2 is an area of variable face-cleat strike exposed in T33N R11&12W near the Colorado-New Mexico border. We interpret the domain boundary region to be 15 to 20 miles (24 to 32 km) wide based on the width of outcrop that shows inconsistent or opposed face-cleat strikes. In the subsurface to the east, the boundary is constrained to strike eastward or northeastward by divergent face-cleat strikes in widely spaced wells. This trend is approximately parallel to the east-trending basin hingeline and to several gentle folds (Bondad and Ignacio anticlines). Greatest cleat-strike variability was observed at the outcrop of the boundary zone. Northwest- and northeast-striking cleats commonly are equally prominent within a given bed in this area. As a result, the coal generally is friable. Near Soda Springs (sec. 1, T32N R12W), Cinder Buttes (sec. 14, T32N R12W), and Fort Lewis Mine, face-cleat strike (defined by abutting relationships) is either northeastward or northwestward in adjacent beds. Beds with mutually abutting cleat relationships and crossing cleat patterns are prominent. These patterns suggest two interfering face-cleat trends.

Based solely on face-cleat strike, the southern part of the basin can be further subdivided into a domain (2a) of northeast-striking cleats in the west, and a domain (2b) of north-striking cleats in the east (Figure 1). The location and nature of the boundary between domains 2a and 2b is difficult to determine accurately because well information on cleat strike in the central basin is lacking. The transition may be gradual in that differences in strike are slight. No indication was found that crosscutting or interfering cleat sets exist along the boundary.

FAULTS AND ASSOCIATED FRACTURES

East-, northwest-, and northeast-striking faults are present in the basin (Zapp, 1949; Wood and others, 1948; Barnes and others, 1954; Newman and McCord, 1980; Condon, 1988). Faults and associated fractures are superimposed onto cleat patterns in some areas, but elsewhere, cleats curve near the faults, suggesting that faults and cleats formed together. Faults in coal seams are poorly exposed, except in mines in the west-central part of the basin. West-northwest- to northwest-striking normal faults occur on the western margin of the basin in the La Plata and Fort Lewis mines. Northeast- and northwest-striking faults occur near Bondad, and northeast-striking faults occur south of Durango. The Cat Creek fault in Archuleta County, Colorado, strikes northeastward, and faults near the southern boundary of the basin have northeast or east-northeast strikes.

Some large faults are marked by zones of intensely fractured rock as much as several tens of feet wide. Fractures within and adjacent to these zones are vertically more continuous than is typical of primary face cleats and commonly extend across noncoal interbeds. Cleat strike at stations near some of the large faults show marked variations (Figure 7), and abrupt cleat curvature occurs near some small faults (Figure 3). Because many fault-related fractures are closed, they may be barriers, rather than conduits, to fluid flow. Faults of large enough displacement to have juxtaposed coal and noncoal layers could disrupt the continuity of regional flow paths. Such faults

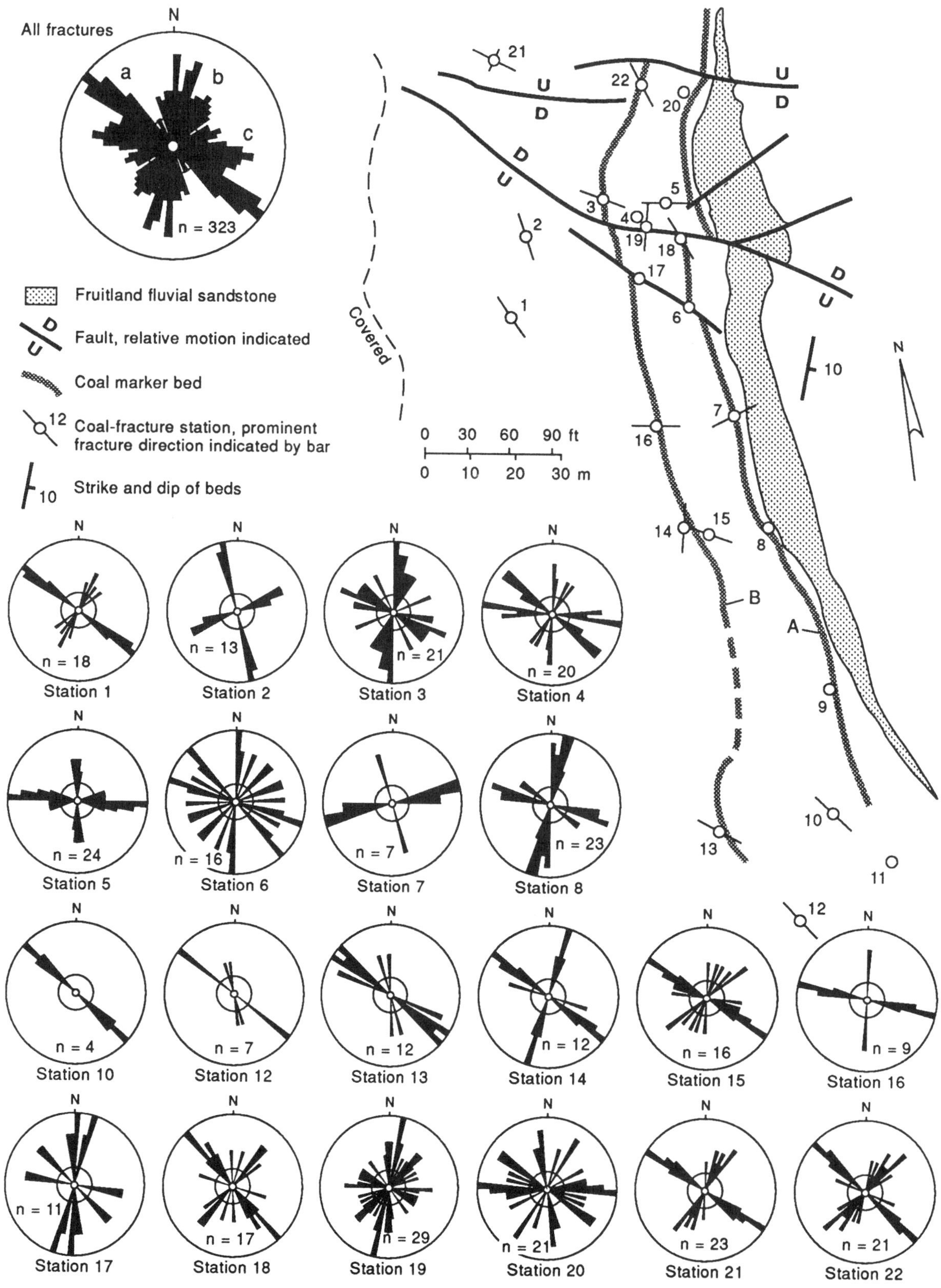

Figure 7. Faults and fracture (cleat) strikes in a part of Fort Lewis Mine area (sec. 36, T33N R12W). Cleat strikes are most variable near faults.

are visible in mines and can be mapped in the subsurface in areas of dense well control in the northwestern and north-central parts of the basin (Ayers and others, 1990).

Small faults, with net slip of less than 3 ft (0.9 m), at the Fort Lewis and Navajo mines and at Squaw Creek (sec. 9, T34.5N R5W) are isolated, polished, and striated slip surfaces with little or no variation in spacing of cleats adjacent to the faults. Such faults commonly lack or have only narrow, 4 to 8 in. wide (10 to 20 cm), associated fracture zones. As little as 0.4 to 0.8 in. (1 to 2 cm) of slip on such faults can interconnect cleat sets in adjacent coal layers that otherwise would remain isolated by unfractured tonsteins.

Another type of fault zone consists of many small striated and polished slip surfaces in anastomosing arrays. Although these zones locally are as wide as 3 to 6 ft (0.9 to 1.8 m), net slip on the microfault arrays commonly is less than 4 in. (10 cm). Because these faults are not open fractures, but instead consist of dense, comminuted coal, they may be permeability barriers. Intensity of faulting and proportions of different fault types are unknown, but observations of extensive mine highwalls suggest that faults are widely spaced in the west-central part of the basin. For example, in the Navajo Mine, only four such faults were visible on approximately 650 ft (198 m) of highwall. Faulting is more intense in the La Plata Mine.

West-northwest and northeast-striking fracture zones, where fracture intensity dramatically increases, occur in Pictured Cliffs Sandstone and overlying Fruitland coal on the western basin margin (Laubach, 1991). Zones are locally associated with and parallel to small normal faults. In such areas coal and sandstone have greater fracture frequency and connectivity.

TIMING AND CAUSE OF CLEAT FORMATION

Cleats can result from tectonic strain and compaction during coalification (McCulloch and others, 1974). Coalification converts initially ductile, fluid-rich organic material into brittle rock at shallow depths. Occurrence of cleats in shallow lignites confirms that fractures can form early in the burial history of coal (Kendall and Briggs, 1933).

In the San Juan basin the timing of Fruitland cleat development is constrained by age of the cleats relative to dated folds and burial history. Where beds have been reoriented by folding along the northern and western basin margins, systematic regional face and butt cleats remain orthogonal to bedding, indicating that cleat formation predated development of folds in early Tertiary (Paleocene to Eocene) time (Fassett, 1985). Crosscutting relationships between cleats and bedding-parallel compaction fabrics indicate that brittle failure of the coal occurred after a significant amount of compactional flattening. Lack of distorted or folded cleats shows that little flattening of coal occurred after cleat development. Compaction could have been accomplished at minimum burial depths of approximately 2,000 to 3,000 ft (610 to 915 m), which were achieved prior to Late Cretaceous to early Tertiary tilting and uplift. Therefore, cleat sets in the San Juan basin probably formed in Late Cretaceous (post-early Maastrichtian?) time prior to the folds that define the basin margin.

Cleat formation was broadly contemporaneous with orogenic shortening of the Cordilleran belt, which already was in progress during Campanian deposition of the Fruitland Formation and which continued into early Tertiary time. Other structures in Fruitland coals, such as faults, local fractures discordant to cleat, and some fracture zones, may have developed during Laramide folding and faulting or Cenozoic epeirogenic events.

Face cleats are opening-mode fractures that formed parallel to greatest horizontal compressive stress. Consistency of face-cleat strikes within large domains indicates that these fractures developed in response to a regionally coherent stress pattern probably caused by shortening or uplift. Throughout Late Cretaceous time the San Juan basin was situated in the foreland adjacent to the Cordilleran orogenic belt. Northwest- and northeast-striking cleats in domains 1 and 2 could mark maximum horizontal paleostress trajectories established during episodes of southeast- and northeast-directed compression. Coal occurrence and coal rank maps indicate that the basin hingeline, and folds such as the Bondad and Ignacio anticlines, may have been initiated by late Campanian time (Scott and others, 1990). During this time, extensive volcanism and possibly fault movement occurred in southeastern Arizona (Cumella, 1983). Orogenic uplift to the southwest may have sustained northwestward depositional strike and northeastward sediment transport in the Fruitland Formation, and stresses related to this event may have affected the Fruitland during coalification. In late Maastrichtian (or possibly Paleocene) time, depositional patterns in the McDermott Member of the Animas Formation indicate a short-lived volcanic and tectonic event northwest of the San Juan basin, and the sub-Ojo Alamo Sandstone unconformity indicates uplift and erosion of the basin prior to late early Paleocene time (Fassett, 1985). Small basinwide strains could have accompanied the early stages of any of these events, accounting for cleat development prior to folding and parallelism of shortening directions indicated by face cleats and northeast- and northwest-trending folds.

Cleat domains may represent separate deformation events or contemporaneous paleostress provinces. Abutting relationships in the domain boundary region are inconclusive, but northeast-striking face cleats abut widely spaced (3 to 6 ft [0.9 to 1.8 m]) northwest-striking fractures in the Navajo Mine, suggesting that domain 1 face cleats in the north formed before domain 2 cleats in the south. If so, it implies early southeast-directed maximum horizontal stress during domain 1 cleat development giving way to northeast-trending maximum horizontal stress during domain 2 cleat development. In the north, coalification may have advanced earlier to a stage at which cleat development could occur, as suggested by the current pattern of coal rank and deeper burial in the northern basin during Late Cretaceous time. Northwest-striking domain 1 cleats formed in highest rank coals (medium-volatile and high-volatile A and B bituminous), whereas domain 2 cleats formed primarily in high-volatile C bituminous and subbituminous coal. Multiple cleat-forming events have been inferred in other basins (Spears and Caswell, 1986).

Alternatively, crisscrossing or mutually abutting relationships and alternations in face-cleat polarity in adjacent beds in the domain 1-2 boundary region may be due to the position of this region between two distinct, contemporaneous paleostress provinces, as Kulander and Dean (1980) suggested for some cleat domains in the Allegheny Plateau. The boundary region may have experienced complex stress patterns during coalification, as the coal rank map suggests that folds such as the Ignacio anticline were developing then. Thus the overall tectonic setting and structural history of the San Juan basin may govern cleat intensity and orientation patterns.

IMPLICATIONS FOR COALBED METHANE EXPLORATION AND DEVELOPMENT

Coalbed permeability may be three to ten times greater in the face-cleat direction than in other directions (McCulloch and others, 1974). This permeability anisotropy reflects the strong preferred orientation and greater length of interconnected fractures in the face-cleat direction. Potential permeability anisotropy arising from face cleats, and differences in face-cleat strike in various cleat domains within the San Juan basin, can be taken into account when designing development well drilling programs and stimulation procedures such as hydraulic fracture treatments. Fruitland face-cleat strike, and thus greatest potential permeability anisotropy, shifts by 90° between domain 1 and 2, from northwest to northeast.

Coal reservoir simulations emphasize the importance of cleat spacing because intense fracture development can enhance the rate of gas production. Furthermore, a closely cleated coal may yield more gas than a coal with wide cleat spacing. For example, core samples of some friable coalbeds in the Appalachian basin yielded approximately 94% of total gas content, whereas the same amount of otherwise similar blocky coal from the same area yielded only 60% to 65% of total gas content (McCulloch and others, 1975).

Cleats are closely spaced in the boundary region between domains 1 and 2 because two face cleat sets overlap and interfere, and the coals have relatively high rank. Moreover, folds and faults in this area locally enhance fracture development. The highly productive Cedar Hill field (T31–32N R10W) and Meridian "400" area (T30N R6W) are developed in or near this boundary domain (Figure 1).

Observations of cleat patterns between domains 1 and 2 suggest that two interfering face-cleat sets result in highly cleated coal that could have higher overall fracture connectivity and more isotropic permeability. Furthermore, prevalence of two directions of strongly developed fractures in the domain boundary region and resulting increased coal friability may enhance the success of well completion by open-hole cavity methods in this area. In cavity completions, coal around the borehole is allowed to collapse, causing coal farther away in the seam to expand toward the cavity. This process not only opens fractures near the borehole but also induces additional fractures to form. However, potential generation of coal fines and attendant production problems may be greater here.

Great lateral continuity of tonsteins at interwell scales, together with the tendency for cleats to terminate at interbedded noncoal layers, should tend to isolate cleat systems in adjacent beds, producing a significant element of layering (heterogeneity) in permeability within Fruitland coal seams. Because tonsteins generally are thin, however, small faults with net slip of as little as 0.4 in. (1 cm) can interconnect the cleat systems in adjacent beds, even without subsidiary fracturing along fault zones. On the other hand, some small faults are closed and unlikely to enhance permeability. Local fracture intensification and greater fracture continuity in coal and adjacent sandstone are associated with some faults as well as with folds and fracture zones. Faults that are below seismic resolution could play a role in connecting otherwise isolated cleat systems.

ACKNOWLEDGMENTS

We thank Walter B. Ayers, Jr., for discussions and reviews. This research was supported by the Gas Research Institute under contract no. 5087–214–1544. We also thank the Southern Ute Indian Tribe and the owners and operators of the Carbon Junction, Chimney Rock, La Plata, and Navajo mines for access to their properties.

REFERENCES

Ammosov, I.I., and Eremin, I.V., 1960, Fracturing in coal (translated from Russian): Moscow, IZDAT Publishers, 109 p. [available from Office of Technical Services, Washington, DC]

Ayers, W.B., Jr., and Ambrose, W.A., 1990, Geologic controls on the occurrence of coalbed methane, Fruitland Formation, San Juan Basin, *in* Ayers, W.B., Jr., and others, Geologic evaluation of critical production parameters for coalbed methane resources, part 1—San Juan basin: Gas Research Institute, Annual Rept. GRI–90/0014.1 (August 1988–July 1989), p. 9–72.

Barnes, H., Baltz, E.H., Jr., and Hayes, P.T., 1954, Geology and fuel resources of the Red Mesa area, La Plata and Montezuma Counties, Colorado: USGS Oil and Gas Investigations Map OM–149, scale 1:62,500.

Choate, Raoul, and Rightmire, C.T., 1982, Influence of the San Juan Mountain geothermal anomaly and other Tertiary igneous events on the coalbed methane potential in the Piceance, San Juan and Raton Basins, Colorado and New Mexico, *in* Proceedings of the Unconventional Gas Recovery Symposium, May 16–18, 1982, Pittsburgh, Pennsylvania: SPE/DOE Paper 10805, p. 151–164.

Condon, S.M., 1988, Joint patterns on the northwest side of the San Juan Basin (Southern Ute Indian Reservation), southwest Colorado, *in* Fassett, J.E., ed., Geology and coal-bed methane resources of the northern San Juan Basin, Colorado and New Mexico: Rocky Mountain Association of Geologists Guidebook, p. 61–68.

Cumella, S.P., 1983, Relation of Upper Cretaceous regressive sandstone units of the San Juan Basin to source area tectonics, *in* Reynolds, M.W., and Dolly, E.D., eds., Mesozoic paleogeography of the west-central United States: SEPM Rocky Mountain Section, Rocky Mountain Paleogeography Symposium 2, p. 189–199.

Fassett, J.E., 1985, Early Tertiary paleogeography and paleotectonics of the San Juan Basin area, New Mexico and Colorado, *in* Flores, R.M., and Kaplan, S.S., eds., Cenozoic paleogeography of the west-central United States: SEPM Rocky Mountain Section, Rocky Mountain Paleogeography Symposium 3, p. 317–334.

Fassett, J.E., 1987, Geometry and depositional environments of Fruitland Formation coalbeds, San Juan Basin, New Mexico and Colorado—Anatomy of a giant coal-bed methane deposit, *in* The 1987 Coalbed Methane Symposium Proceedings—University of Alabama, Tuscaloosa, November 16–19, 1987: School of Mines and Energy Development, Gas Research Institute, and U.S. Mine Safety and Health Administration, p. 19–35.

Henkle, W.R., Jr., Muhm, J.R., and DeBuyl, M.H.F., 1978, Cleat orientation in some subbituminous coals of the Powder River and Hanna basins, Wyoming, *in* Hodgson, H.E., ed., Proceedings of the Second Symposium on the Geology of Rocky Mountain Coal–1977: Colorado Geological Survey Resource Series 4, p. 129–141.

Kendall, P.F., and Briggs, H., 1933, The Formation of rock joints and the cleat of coal: Proceedings, Royal Society of Edinburgh, v. 53, p. 164–187.

Kulander, B.R. Dean, S.L., Williams, R.E., and others, 1980, Fracture trends in the Allegheny Plateau of West Virginia: West Virginia Geological and Economic Survey Map WV–11, scale 1:250,000.

Laubach, S.E., 1991, Fracture patterns in low-permeability-sandstone gas reservoir rocks in the Rocky Mountain region, *in* Proceedings, Rocky Mountain Meeting and Low-Permeability Reservoir Symposium: Society of Petroleum Engineers Paper 21853, p. 501–510.

Laubach, S.E., Tremain, C.M., Whitehead, N.H., III, and Baumgardner, R.W., 1990, Fracture-trace maps of Upper Cretaceous Pictured Cliffs Sandstone pavements, San Juan Basin, Colorado—Implications for coalbed methane exploration (abs.): GSA Abstracts with Programs, v. 22, no. 7, p. A202.

McCulloch, C.M., Deul, M., and Jeran, P.W., 1974, Cleat in bituminous coalbeds: U.S. Bureau of Mines Rept. of Investigations 7910, 25 p.

McCulloch, C.M., Diamond, W.P., Bench, B.M., and Deul, M., 1975, Selected geological factors affecting mining of the Pittsburgh Coalbed: U.S. Bureau of Mines Rept. of Investigations 8093, 72 p.

Newman, K., and McCord, J., 1980, Detailed site investigation northern San Juan Basin: Unpublished draft report, available for inspection at Colorado Geological Survey, Denver, Colo.

Raistrick, A., and Marshall, C.E., 1939, The nature and origin of coal and coal seams: London, The English Universities Press Ltd., p. 42–45.

Ridgley, J.L., and Huffman, A.C., Jr., 1990, Basin analysis study of the San Juan Basin Colorado and New Mexico, *in* Carter, L.M.H., ed., USGS research on energy resources, 1990 program and abstracts—Sixth V.E. McKelvey Forum on Mineral and Energy Resources: USGS Circular 1060, p. 68–69.

Scott, A., Kaiser, W., and Ayers, W.B., Jr., 1990, Thermal maturity of Fruitland coal and composition and origin of gases in the Fruitland Formation, San Juan Basin, *in* Ayers, W.B., Jr., and others, Geologic and hydrologic controls on the occurrence and producibility of coalbed methane, Fruitland Formation, San Juan basin: Gas Research Institute, Topical Rept. GRI–91/0072 (August 1987–July 1990), p. 243–241

Tremain, C.M., Laubach, S.E., and Whitehead, N.H., III, 1990, Coal fracture (cleat) patterns in Upper Cretaceous Fruitland Formation, San Juan basin, Colorado and New Mexico—Implications for coalbed methane exploration and development, *in* Ayers, W.B., Jr., and others, Geologic and hydrologic controls on the occurrence and producibility of coalbed methane, Fruitland Formation, San Juan basin: Gas Research Institute, Topical Rept. GRI–91/0072 (August 1987–July 1990), p. 97–117.

Spears, D.A., and Caswell, S.A., 1986, Mineral matter in coals—Cleat minerals and their origin in some coals from the English Midlands: International Journal of Coal Geology, v. 6, no. 2, p. 107–125.

Ting, F.T.C., 1977, Origin and spacing of cleats in coal beds: Journal of Pressure Vessel Technology, v. 99, Series J, no. 4, p. 624–626.

Tremain, C.M., Laubach, S.E., and Whitehead, N.H., III, 1990, Coal fracture (cleat) patterns in the San Juan Basin, Colorado and New Mexico, *in* Ayers, W.B., Jr., and others, Geologic evaluation of critical production parameters for coalbed methane resources—part I, San Juan basin: Gas Research Institute Topical Report, in preparation.

Ver Steeg, K., 1942, Jointing in the coal beds of Ohio: Economic Geology, v. 37, p. 503–509.

Wood, G.H., Jr., Kelly, V.C., and MacAlpin, A.J., 1948, Geology of the southern part of Archuleta County, Colorado: USGS Oil and Gas Investigations Preliminary Map 81, scale 1:63,360.

Zapp, A.D., 1949, Geology and coal resources of the Durango area, La Plata and Montezuma Counties, Colorado: USGS Oil and Gas Investigations Preliminary Map 109, scale 1:31,680.

A G Alpha '91

Hydrogeology of the Fruitland Formation, San Juan Basin, Colorado and New Mexico

Paul R. Oldaker[1]

1. Independent Hydrologist/Hydrogeologist

ABSTRACT

The hydrogeology of the Fruitland Formation in the San Juan basin consists of reservoir characteristics, potentiometric surface, and water quality of the unit. Each will be examined and then summarized.

The intrinsic permeability of Fruitland coals and sandstones varied from 0.002 md to 811 md, with a logarithmic mean of 26 md. Permeability data showed no consistent relationship with basal coal, coal, or composite sample groups, but logarithmic permeability did have a consistent increasing relationship with logarithmic initial water production. Permeability was not related with depth. Porosity-compressibility-thickness product (ϕch) varied from 1×10^{-6} psi/ft to 270×10^{-6} psi/ft (22.6×10^{-6} to $\times 10^{-6}$ 6,107 kPa/m) with a logarithmic mean of 5.8×10^{-6} psi/ft (131×10^{-6} kPa/m). The variability of ϕch is attributed to total system compressibility because porosity and coal seam thickness do not significantly vary.

The potentiometric surface of the Fruitland Formation shows a hydrodynamic flow pattern from recharge areas in the northern and western parts of the basin to discharge into the region's perennial rivers. Overpressuring is the result of both this hydrodynamic surface and present topography. Because of the low permeabilities of the formation, steady-state flow quantity in the formation is less than 0.01% of the mean annual flows of the rivers.

Total dissolved solids concentrations increase in the direction of flow with increasing residence time for a particular flow path. With five major discharge areas, several flow paths are available, each with a different residence time.

INTRODUCTION

Data for the following paper were compiled in 1987 for Amoco Production Company. The objective was to gather and interpret all available hydrogeologic data for the Fruitland Formation. The data collected were classified into four groups—undifferentiated Fruitland lithology, basal coal, coal, and composite lithologies.

RESERVOIR CHARACTERISTICS

Intrinsic permeability and porosity-compressibility-thickness product (ϕch) are the major variables controlling flow as defined by the radial-flow equation. Fifty-five values of intrinsic permeability and thirteen ϕch values were compiled for the Fruitland Formation.

The permeability distribution (Figure 1) shows a wide range—from a minimum 0.002 md to 811 md, with a logarithmic mean of 26 md. Most of the analyses are for coal, which is a common lithology type within the formation. The wide range in permeability reflects the wide variation in coal as a lithologic type. Source organic material, depositional environment, ash lithologies, maturation process, and tectonics are several factors affecting coal lithologic type. They also appear to be factors in determining the permeability.

No patterns were found for or among the four data groups; consequently, either the variability is within one or more of the groups, or it affects all the data groups.

Initial water production is the only variable found to have a relationship to Fruitland Formation permeabilities (Figure 2). As permeability increases, initial water production increases. The relationship variability is due to other variables in the radial-flow equation, such as thickness, porosity compressibility-thickness product, and change in bottom-hole pressure. However, permeability appears to be the most sensitive variable affecting the radial-flow equation.

From the plot in Figure 3, depth and assumed increased pressure are not related to Fruitland Formation permeabilities. If identical rocks are subjected to pressure, permeability should decrease because porosity is decreasing. This does not, however, seem to be occurring in coal as it is such a heterogeneous lithologic type. Therefore depth and increased pressure do not effect consistent changes in coal permeability.

Values for porosity-compressibility-thickness product also are variable for coal in the Fruitland Formation. These values range from 1×10^{-6} psi/ft to 270×10^{-6} psi/ft (22.6×10^{-6} to $6{,}107 \times 10^{-6}$ kPa/m) with a log mean value of 5.8×10^{-6} psi/ft (131×10^{-6} kPa/m) (Figure 4). Because porosity

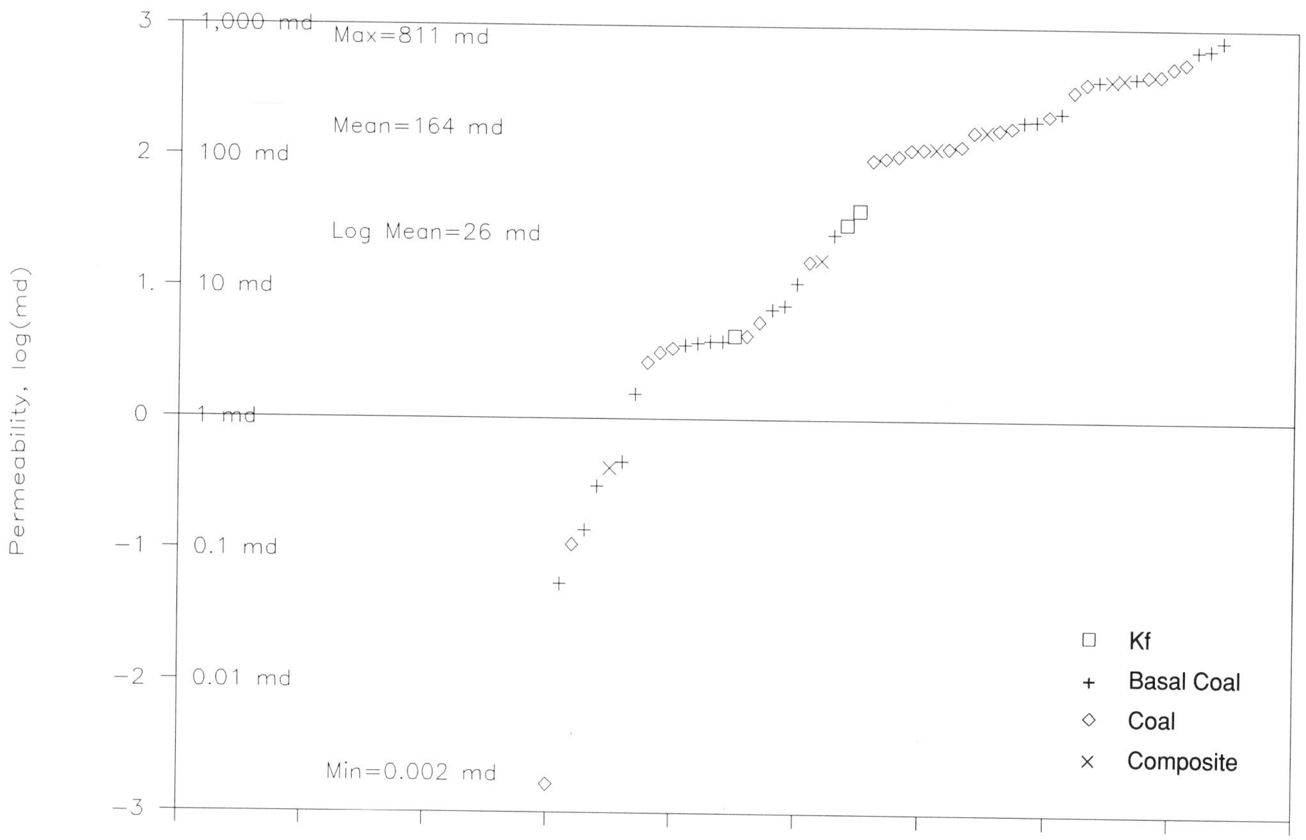
Max=811 md
Mean=164 md
Log Mean=26 md
Min=0.002 md
1,000 md
100 md
10 md
1 md
0.1 md
0.01 md
Permeability, log(md)
Kf
Basal Coal
Coal
Composite

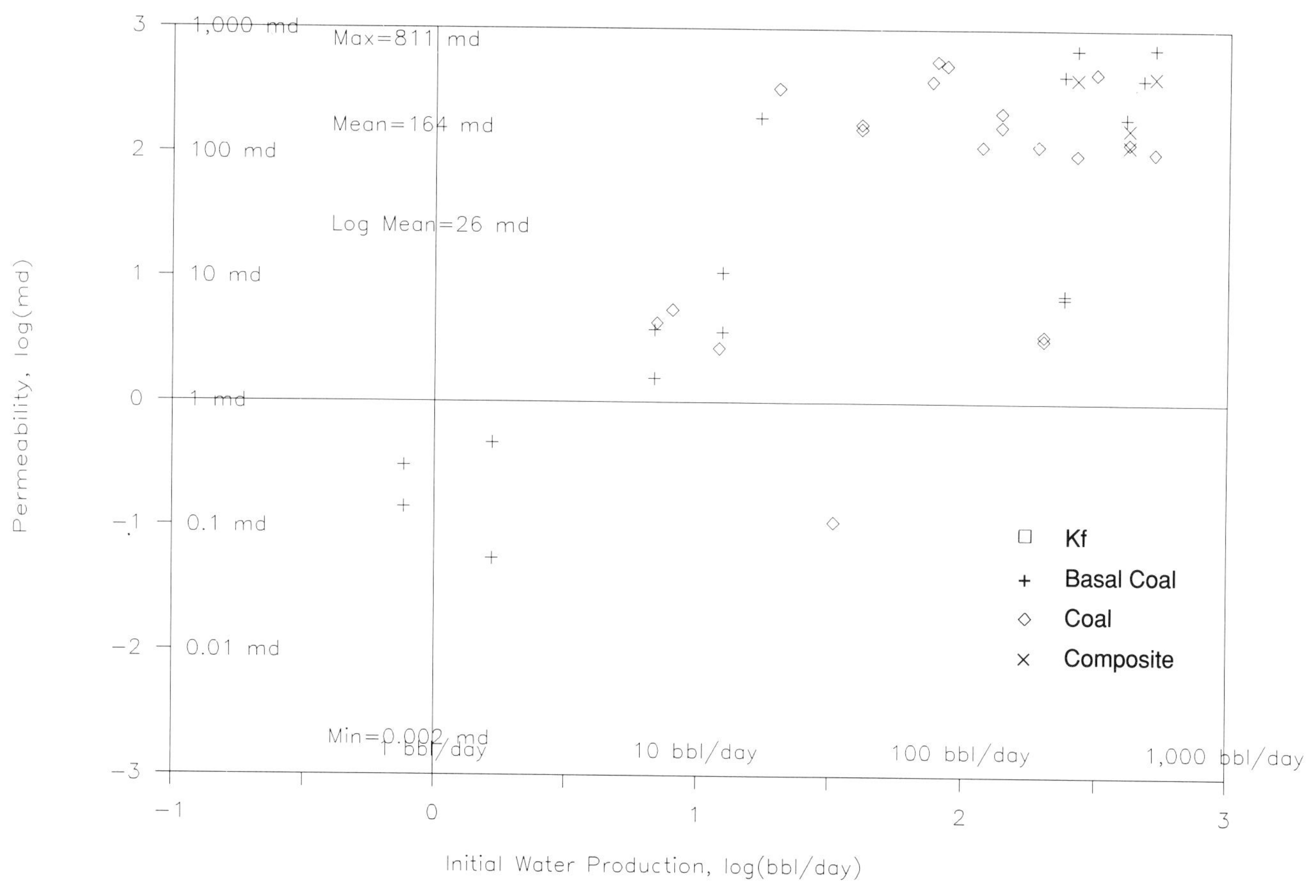
Max=811 md
Mean=164 md
Log Mean=26 md
Min=0.002 md
1,000 md
100 md
10 md
1 md
0.1 md
0.01 md
Permeability, log(md)
1 bbl/day
10 bbl/day
100 bbl/day
1,000 bbl/day
Initial Water Production, log(bbl/day)
Kf
Basal Coal
Coal
Composite

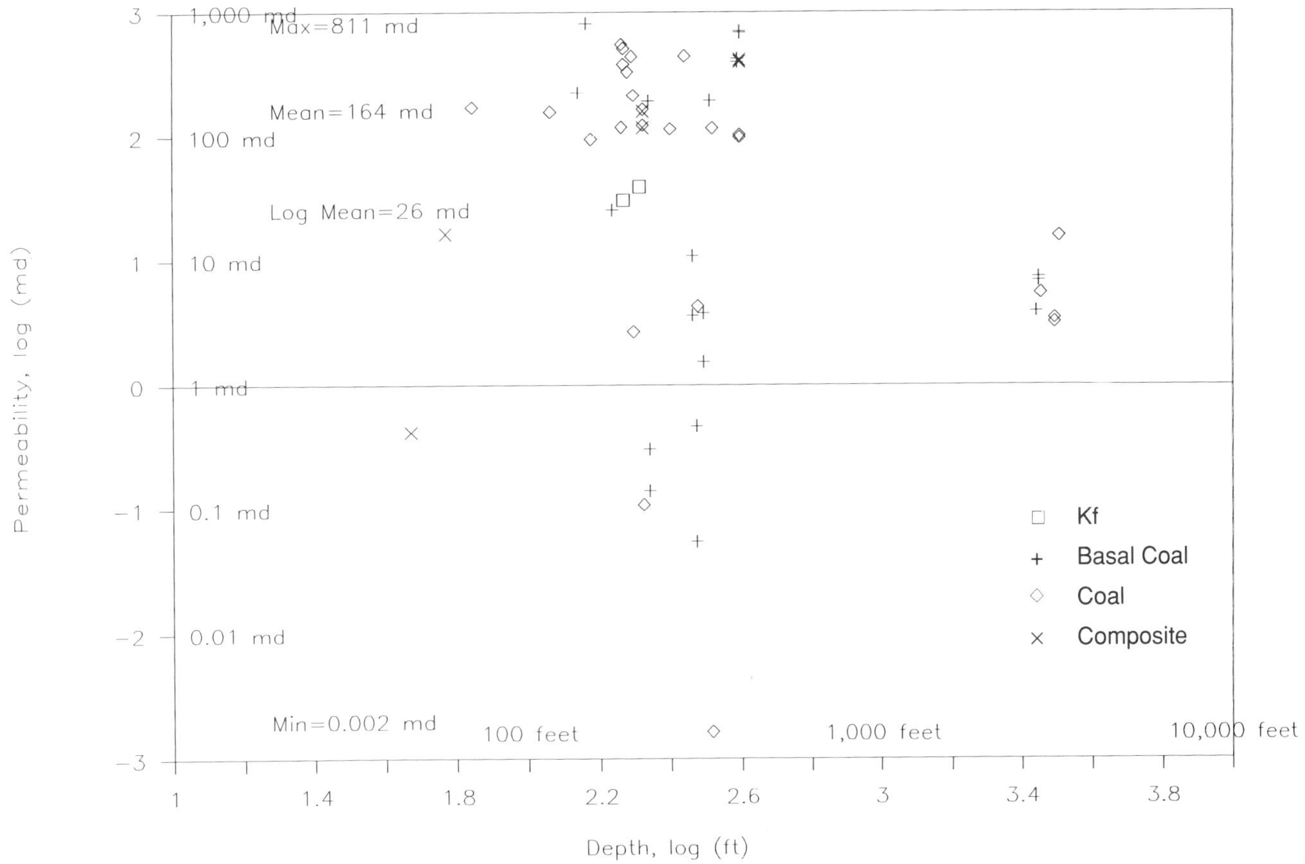

Figure 3. Permeability vs depth in the Fruitland Formation, San Juan basin.

and coal seam thickness are not widely varying, the total system compressibility must be widely varying to result in this large range of values. This again is due to the variability of lithologies in coal. The wide range in ϕch significantly affects the results calculated from the radial-flow equation generally in the radial dimension of the cone of depression and not the quantity of flow.

POTENTIOMETRIC SURFACE

The potentiometric surface map shows horizontal flow patterns within the Fruitland formation (Figure 5). Enough data are available for calculating potentiometric head along the outcrop and in the northern part of the basin, but data in the center of the basin are relatively scarce.

The potentiometric surface is controlled by discharge from the formation into five perennial rivers, from west to east—San Juan River, La Plata River, Animas River, Los Piños River, and Piedra River. Flow is from recharge areas in interfluvial outcrops to these discharge areas. This flow pattern essentially is a subdued expression of the topography probably as a result of climate-controlled water deficits and structure of the formation. Recharge occurs at high elevations with low water deficit, and discharge occurs at lower elevations with higher water deficits. Overpressuring is the result of the relative position of the hydrodynamic surface with respect to present topography.

The total steady-state ground-water flow in the formation can be calculated from the sum of the discharges. Each subcrop discharge was calculated using the Darcy equation and the variables thickness, logarithmic mean permeability, and potentiometric-head gradient. As compared in Table 1, flow at each subcrop is small, ranging from 121 to 3,872 BPD (19 to 616 m^3/day). Total flow from the Fruitland Formation is 7,750 BPD (1,232 m^3/day). The low permeability of the formation limits the total flow quantity to only a small fraction (less than 0.01%) of the combined mean annual flows of the rivers themselves.

Figure 1 (top left). Permeability distribution in the Fruitland Formation, San Juan basin.

Figure 2 (bottom left). Permeability vs initial water production, Fruitland Formation, San Juan basin.

WATER QUALITY

In a hydrodynamic flow system, water quality depends on the length of residence time in the formation and on geochemistry of the system. Total dissolved solids (TDS) concentrations increase in the direction of flow with increasing residence time for a particular flow path. On

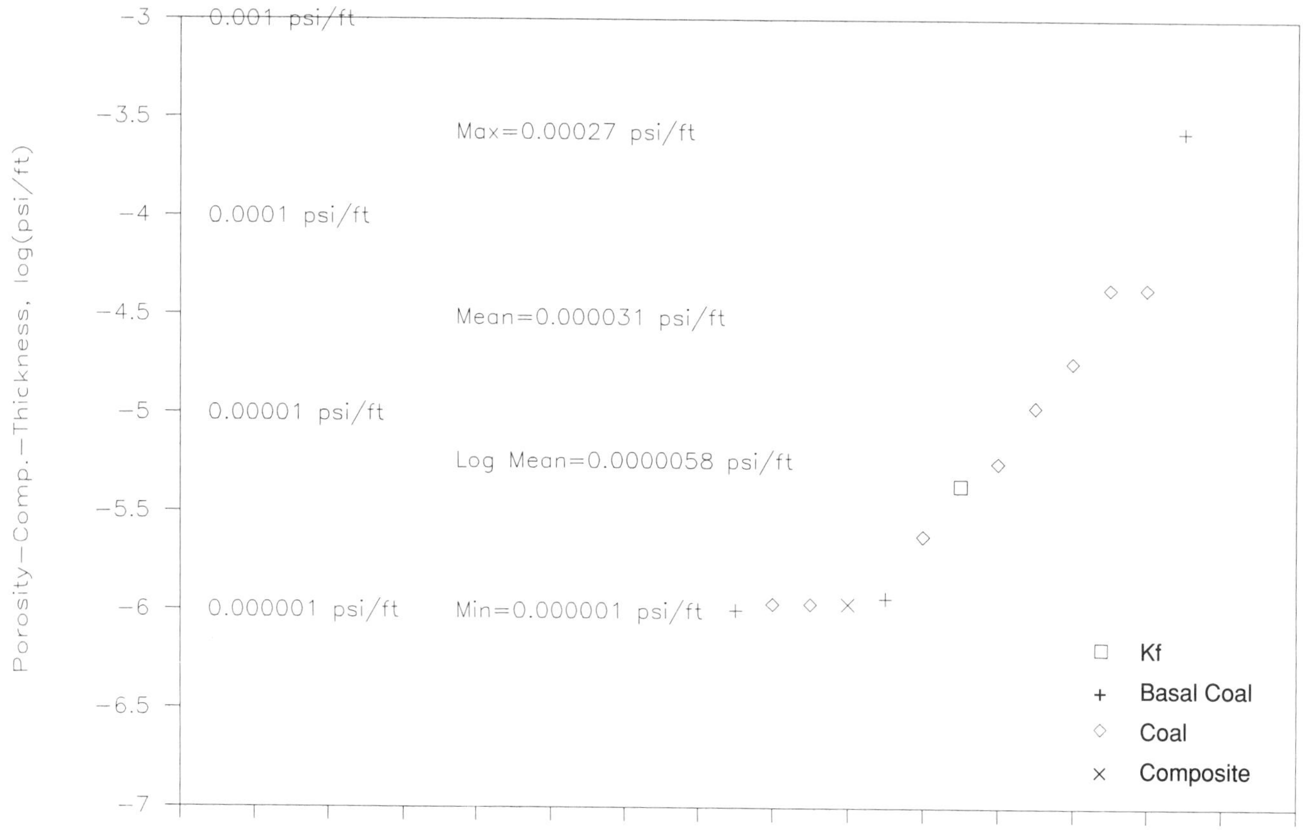

Figure 4. Porosity-compressibility-thickness product distribution, Fruitland Formation, San Juan basin.

Figure 6, flow is vertically down the y-axis from high potentiometric head to lower head. With five major discharge areas in the San Juan basin, several flow paths are available with different residence times, which results in a wide range of TDS values at any given potentiometric head. The highest TDS concentrations occur at fairly high head. This indicates that higher dissolved solids concentration should be expected in the central and southern areas of the basin where data are scarce. Otherwise, some vertical flow component is taking this water out of the formation.

Sodium and bicarbonate are the dominant ions, the result of cation exchange and sulfate reduction in the absence of chloride. In the western part of the basin chloride ion concentrations become significant, indicating that the depositional environment was more seaward. In the southeast, sulfate concentrations are significant for unknown reasons.

CONCLUSIONS

1. Fruitland Formation permeabilities and porosity-compressibility-thickness products have wide ranges because coal is a heterogeneous lithologic type.

2. The Fruitland Formation is a hydrodynamic system driven by climate and by present topography and structure of the formation. Fruitland Formation waters discharge

Table 1. Comparison of mean annual flows for five San Juan basin rivers with steady-state ground-water flow from the Fruitland Formation.

River	River Flow		Fruitland Formation Flow	
	(10^6 BPD)	(10^6 m^3/day)	(10^6 BPD)	(10^6 m^3/day)
San Juan River	19,130,171	3,041,971	622	99
La Plata River	508,269	80,815	121	19
Animas River	13,497,398	2,146,086	3,872	616
Los Piños River	8,714,856	138,566	2,822	449
Piedra River	4,986,598	792,869	313	50
Total	46,837,292	6,200,307	7,750	1,233

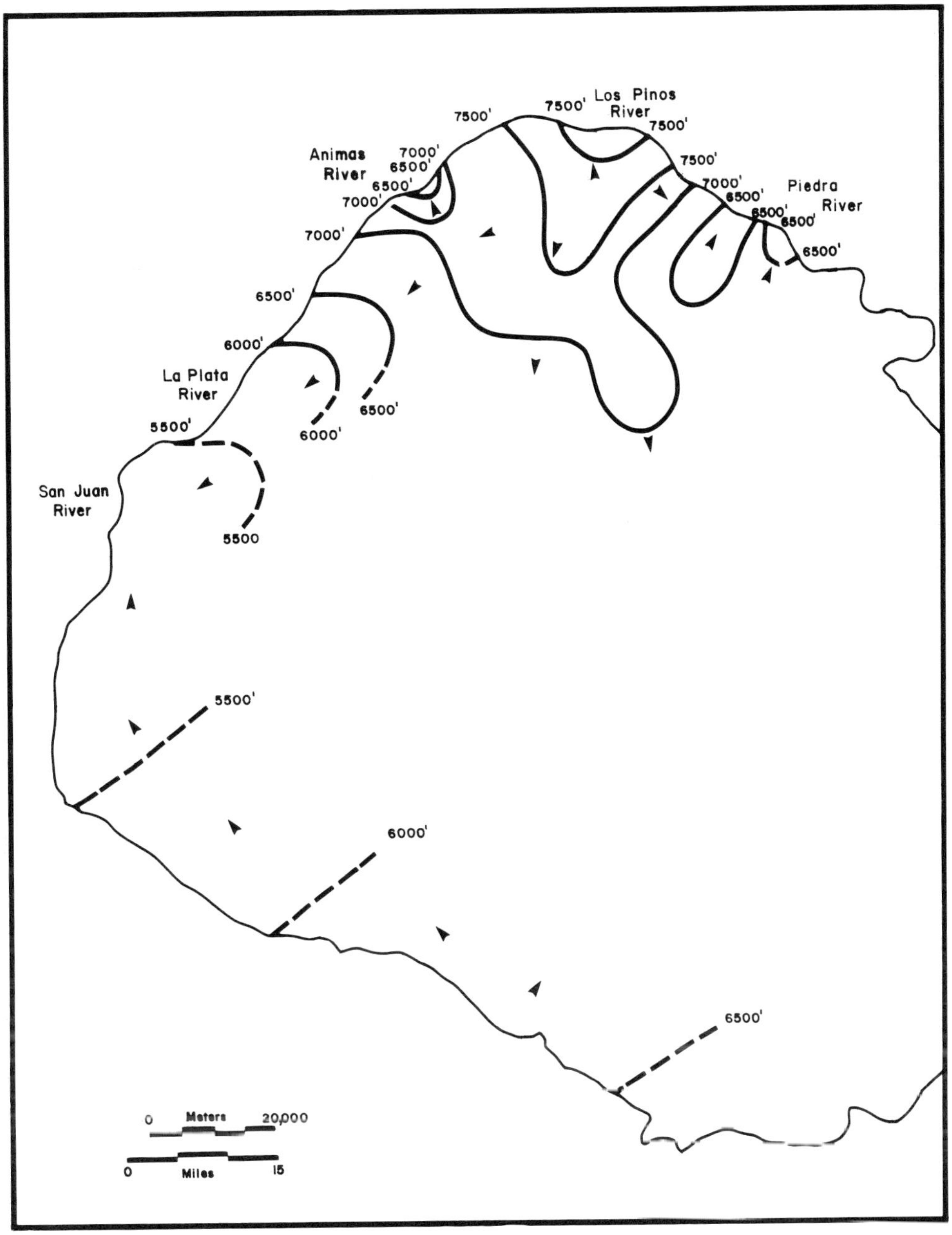

Figure 5. Potentiometric surface map of the Fruitland Formation, San Juan Basin. Contour interval 500 ft.

into five perennial rivers that cross the formation outcrop. The total formation flow is only a small fraction of mean annual river flows.

3. Total dissolved solids concentrations increase in the direction of flow, but several flow paths are available within the Fruitland Formation

ACKNOWLEDGMENTS

My thanks to Mr. Bill Hanson, Mr. Jack McAnear, and Mr. Harry Ter Best of Amoco Production Company.

REFERENCES

Berry, F.A.F., 1959, Hydrodynamics and geochemistry of the Jurassic and Cretaceous systems in the San Juan Basin, northwestern New Mexico and Southwestern Colorado: Privately printed paper.

Colorado Mined Land Reclamation Board, 1987, [Personal communication] Review of mine permits for Carbon Junction Mine and Chimney Rock Mine.

Colorado State Engineer's Office, 1987, Well permit list, April 1987.

Fassett, J.E., and Hinds, J.S. 1971, Geology and fuel resources of the Fruitland Formation and Kirtland Shale of the San Juan Basin, New Mexico and Colorado: USGS Professional Paper 676, 76 p.

Hantush, Mahdi, 1960, Theory of ground water motion: Unpublished lecture notes.

Hutchinson, E.C., and Brogden, R.E., 1976, Water-quality data for the Southern Ute Indian Reservation, southwestern Colorado: USGS Open-File Rept. 76–16, 58 p.

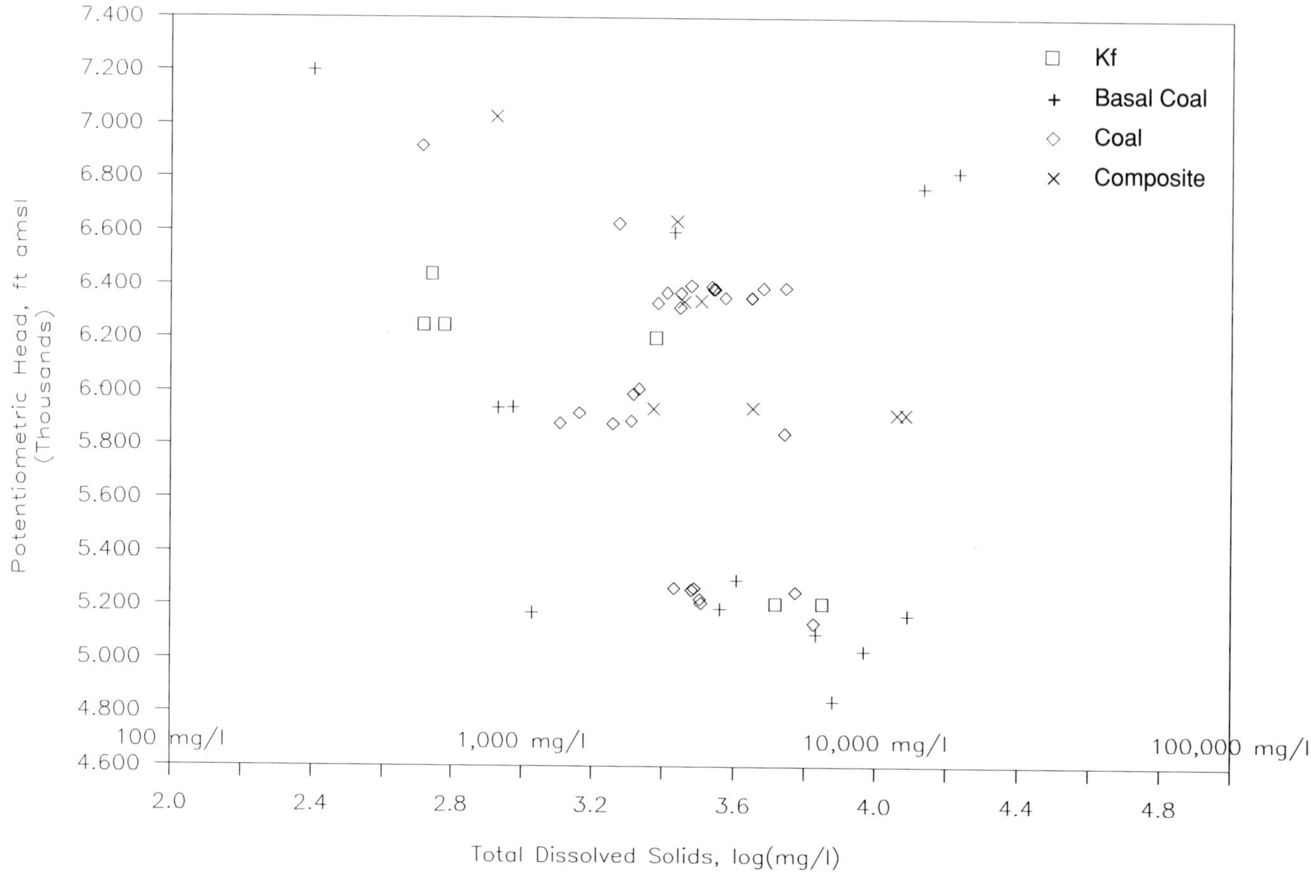

Figure 6. Potentiometric head vs total dissolved solids concentrations, Fruitland Formation, San Juan basin.

Kelso, B.S., Goolsby, S.M., and Tremain, C.M., 1980, Deep coal bed methane potential of the San Juan River coal region, southwestern Colorado: Colorado Geological Survey Open-File Rept. 80–2, 56 p.

Lyford, F.P., 1979, Ground water in the San Juan basin, New Mexico and Colorado: USGS Water-Resources Investigation 79–73, 27 p.

New Mexico Energy and Minerals Department, 1987, [Personal communication] Review of mine permits for Navajo Mine, San Juan Mine, Black Diamond Mine, La Plata Mine, Gateway Mine, De-Na-Zin Mine, Star Lake Mine, Burnham Mine, and Gallo Wash Mine.

Oldaker, P.R, 1987, Cedar Hill field, Fruitland basal coal, reservoir analyses: Notes for presentation at The Coalbed Methane Forum, Denver, Colo., March 1987.

Steven, T A., Lipman, P W., Hail, W.J., Barker, F., and Luedke, R.G., 1974, Geologic map of the Durango quadrangle, southwestern Colorado: USGS Miscellaneous Geologic Investigations Map I–764, scale 1:250,000.

Stone, W.J., Lyford, F.P., Frenzel, P.F., Mizell, N.H., and Padgett, E.T., 1983, Hydrogeology and water resources of the San Juan Basin, New Mexico: New Mexico Institute of Mining and Technology Hydrologic Report 6.

Way, S.C., Bumb, A.C., Koenig, R.A., McKee, C.R., Reverand, J.M., and Ward, W.P., 1984, Hydrologic characterization of coal seams for methane recovery: Gas Research Institute Final Rept. GRI–83/0064 (January-March 1983), 59 p.

Structural Controls on Coalbed Methane in the San Juan and Western Canada Sedimentary Basins

Jack R. Century[1]

1. J.R. Century Petroleum Consultants Ltd., Calgary, Alberta

ABSTRACT

The structural history of the San Juan basin is unique in its well documented diverse and recurring nature. Two cycles of Proterozoic crustal deformation are recorded in the Needle Mountains of southwestern Colorado, immediately bordering the northern margin of the San Juan basin. Fred Barker (1969) and Barbara Tewksbury (1989) have mapped and dated the Needle Mountains in detail. The Uncompahgre Uplift and Paradox basin regional tectonism, of Late Paleozoic age, preceded Cretaceous coal-bearing formations, followed by other distinctly different Late Cretaceous and Tertiary structural styles. The Animas Formation, straddling the Cretaceous-Tertiary boundary and the underlying McDermot Formation record a maximum 3,000 ft (915 m) of dominantly volcanic sediments, including a large amount of coarse andesite conglomerates. Major heat flow, hydrodynamic, and fracture regimes were fixed by these major tectonic events that control coal maturation and coalbed methane productivity.

The timing and duration of the Cretaceous/Tertiary and younger heat-flow pulses followed earlier crustal weakness patterns and created the thermal and related history necessary for favorable coalbed methane generation and economic producibility. Oligocene dating of the San Juan volcanic field does not fully explain Fruitland coal maturation levels.

An understanding of this complex episodic history will help evaluate the profitable development of coalbed methane and how the San Juan basin can serve as a model or scenario for other basins in the United States and western Canada.

Recent history of conventional oil and gas resource assessments indicates that the timely development of potential coalbed methane supplies could suffer if unrealistic industry or government appraisal methods are again employed. Environmental considerations and ownership issues affecting coalbed methane future production also will be discussed.

REFERENCES

Barker, Fred, 1969, Precambrian geology of the Needle Mountains, southwestern Colorado: USGS Professional Paper 644–A, 35 p.

Tewksbury, B.J., 1989, Proterozoic geology of the Needle Mountains—A summary, *in* Grambling, J.A., ed., Proterozoic geology of the Southern Rocky Mountains: GSA Special Paper 235, p. 65–73.

FLATIRONS

Factors Affecting Rate and Total Volume of Methane Desorption from Coalbeds

B.M. Das[1]
Dennis J. Nikols[2]
Z.U. Das
V. Joseph Hucka[3]

1. CANMET, Government of Canada, Devon, Alberta
2. Alberta Research Council, Edmonton, Alberta
3. University of Utah, Salt Lake City, Utah

ABSTRACT

Methane is a common product of coalification. It is released from all ranks of coal, from peat to anthracite. Traditionally, methane release during coal mining and handling has been considered a hazard because of the explosive nature of gas. However, the coal industry has come to recognize that coalbed methane is a source of clean energy that can be produced from deep-seated seams not amenable to economic exploitation in any other way.

Ultimately, any systematic evaluation of coalbed methane as a resource must address the questions of how much methane is available from a given coal seam and its rate of desorption. The processes governing methane formation, its retention within the coal seam and surrounding strata, and its release all must be understood before the full benefit of this resource can be realized. The most important factors in methane formation are coal rank, maceral composition, and the nature of rank improvement (normal or abnormal metamorphism). Retention of methane depends on the geotechnical properties of coal (strength and brittleness) and the geological factors governing its occurrence (such as stratigraphy, lithology, tectonics, and burial depth). Desorption is influenced by rank and maceral composition, amount and nature of mineral matter present, gas pressure, jointing and cleats, and strength and brittleness of the coal.

This paper analyzes these factors and relates them to estimation of the techno-economic potential of methane in coalbeds.

INTRODUCTION

Methane is a product of coal metamorphism. Although methane is formed during all stages of coalification, the rate accelerates at higher ranks (bituminous and higher). The released methane is partly retained within the coal and surrounding porous strata and partly lost to more distant traps or to the atmosphere. Gas is retained in three ways—(a) within the molecular structure of the coal matrix (although maybe not in gaseous form); (b) in an adsorbed state onto the surfaces of the coal particles themselves; and (c) in a free form within fractures and pore spaces in the coal and surrounding rock (commonly sandstone). The retained gas is desorbed when the equilibrium of the system is disturbed either by mining or by some process of gas drainage.

Desorption of methane has both deleterious and beneficial consequences. Methane emissions are considered hazardous in any confined space due to their explosive nature. In addition, release of methane into the atmosphere is believed by some to contribute to global warming. On the other hand, methane is a source of clean energy, the increased use of which could help control some environmental problems associated with global energy use. Thus, the systematic study of methane desorption is vital if this resource is to be used safely and to the fullest extent.

Three important parameters relate to methane desorption from coal—(1) composition of the desorbed gas, (2) rate of desorption, and (3) total volume of desorbed gas. These parameters combine to help determine the methane productive potential of a coal seam.

The bulk of desorbed coal gas is methane with minor amounts of higher hydrocarbons. Although in most cases only trace amounts of other gases (nitrogen, carbon dioxide, carbon monoxide, hydrogen sulfide, and some oxides of nitrogen) are present, high nitrogen and/or carbon dioxide contents sometimes result in methane proportions below

60%. Methane in coal is either thermogenic or biogenic. The former constitutes the bulk of produced gas to date and is the only form where volume can be characterized by coal rank. Thus, thermogenic methane is the primary focus of this paper.

Both the rate and total volume of desorption depend on the formation and retention of methane in the coals and their desorption characteristics. In addition, the process of desorption is affected by the physicomechanical properties of coal, nature and content of contained mineral matter, and permeability.

A significant body of work has been published on the subject of methane desorption and drainage (Liden, 1961; Airuni and Éttinger, 1966; Patching, 1970; Deul and others, 1973; Das and Hucka, 1977; Hucka, 1980; Das and others, 1982; Das, 1987). Using this existing body of work, we will analyze the factors that affect both the rate and total volume of desorption.

METHANE FORMATION AND RETENTION IN COALBEDS

Formation and retention of methane in coals are necessary precursors to desorption. Both factors are intimately interlocked, and both determine the eventual production of gas from a seam. The more methane that is formed, the higher the probability that it will be retained and, consequently, the higher the gas pressure controlling desorption.

Methane Formation

The formation of methane during coalification is governed by five factors—(a) coal rank, (b) coal type (maceral composition), (c) nature of rank metamorphism and whether it is normal or abnormal (Francis, 1961), (d) tectonic history, and (e) proximate analysis values of the coal.

Coal Rank

Little thermogenic methane is formed in the early stages of coalification. Thus, low-rank coals (peat, lignite, subbituminous) commonly are poor in methane. An exception is the Fort Union Formation of the Powder River basin, where biogenic gas is produced from subbituminous coal. It is only when coalification reaches mid-bituminous rank (>84% carbon, dry mineral-matter-free basis, dmmf) that large amounts of thermogenic methane begin to form (Figures 1 and 2).

Coal Type (Maceral Composition)

Coal is composed of three major petrological units or macerals—vitrinites, inertinites, and exinites (Table 1). Inertinites produce the least amount of gas and are termed "subhydrous." Vitrinites ("normal") produce more gas than inertinites but generally less than exinites ("perhydrous"). This relationship agrees with their respective hydrogen contents. At ranks higher than 29.5% volatile matter (VM), or 89.5% carbon (dmmf), the hydrogen content of vitrinites exceeds that of exinites. This phenomenon is known as the "coalification break" (Stach, 1953). At this rank, exinites begin to release large amounts of gas, and shrinkage cleats begin to develop.

Nature of Rank Metamorphism

During coalification, a progression in rank may occur in one of two ways. First, "normal" metamorphism may occur through gradual increase in temperature and pressure as a consequence of burial and/or minor tectonism. In this case the hydrogen content of the coal follows the normal paths for each maceral (Figure 3). Formation of methane is thus directly related to coal rank.

Second, "abnormal" metamorphism occurs when the coal is subjected to a sudden, large increase in temperature and/or pressure as, for example, in the case of igneous intrusion. The process generally is of short duration and results in a straight-line "carbonization track" (Figure 3). An important consequence of this is the rapid drop in hydrogen content compared to normal levels for any given rank of coal, resulting in a higher gas-to-rank ratio than normal.

Tectonic History

Tectonism can have a great effect on formation of methane in coals. It can affect both the rate of change in rank and the

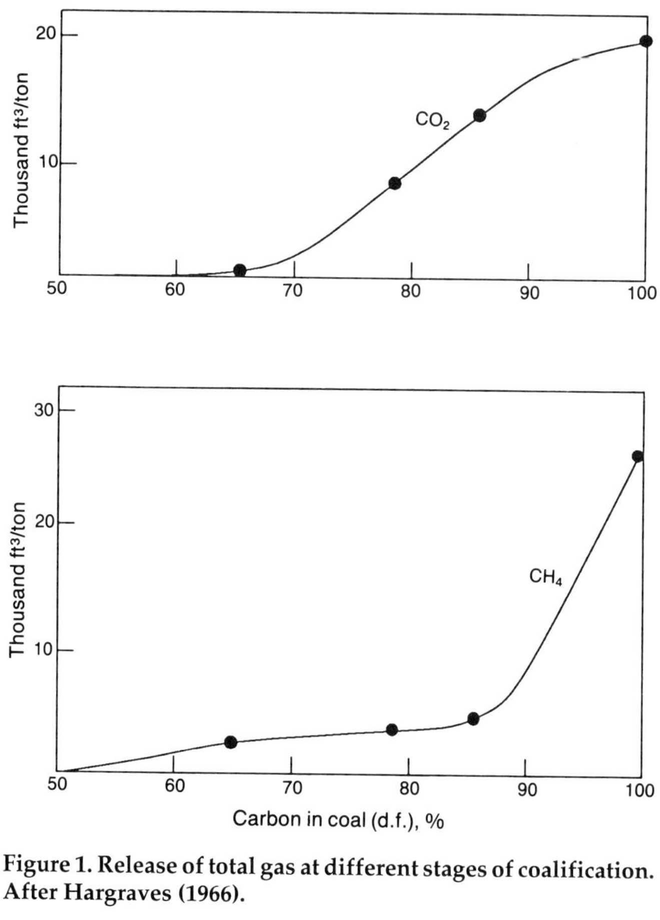

Figure 1. Release of total gas at different stages of coalification. After Hargraves (1966).

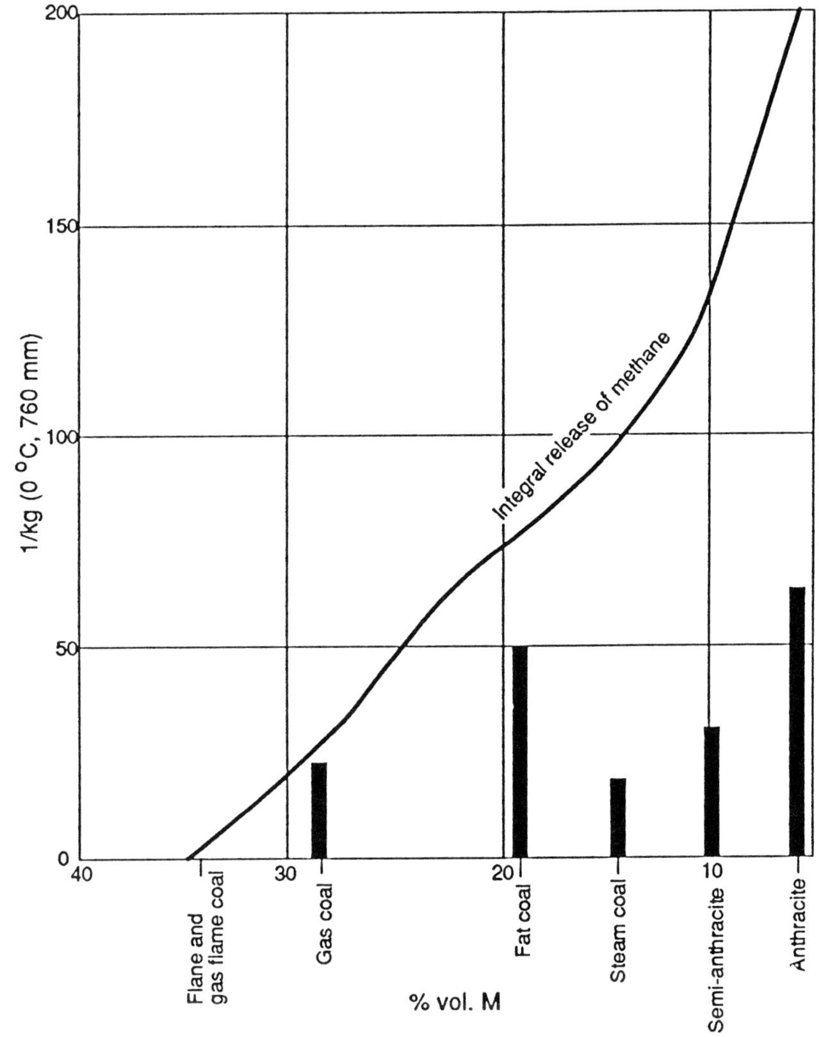

Figure 2. Formation of thermogenic methane during coalification in different rank stages of bituminous coals and anthracite. Based on Jüntgen and Karweil (1966).

final rank of coals (Francis, 1961). This is caused by an increase in temperature and/or pressure through direct influence, or by increased burial depth. As discussed above, a progression in rank leads to an increase in the amount of gas produced.

Proximate Analysis

Ash and moisture content of coal do not contribute to production of methane. Thus, the higher the proportion of these nonproductive materials, the less gas is produced per unit volume of coal.

Methane Retention

Retention of methane in coals, and their surrounding envelope of strata, is one of the main factors contributing to gas pressure and, hence, methane desorption. The primary elements governing methane retention are coal rank and type, nature of rank metamorphism, nature of seam roof and floor, tectonic history, burial depth, and proximate analysis values of the coal.

Coal Rank and Type

Coals of different ranks contain different amounts of free space. Surface area, pore volume, and cleating and fracture development control the amount of methane that can be stored in a given seam. Adsorption and fracture filling are the two major methods of storing gas within the coal.

Adsorption occurs at the interface between the coal and the gas and thus depends on surface area of the coal. Low-rank coals have large surface areas (Figure 4) but do not produce much gas. In addition, because of generally shallow burial, low-rank coals tend to lose what gas is produced to the surrounding environment. High-rank coals have a smaller surface area on which adsorption can occur, but because of the much higher levels of gas present in them, they constitute better reservoirs for adsorbed methane.

The other method of methane retention in coal depends on pore, fracture, and void volumes within and around the coalbed. Pore volume is high in low-ranking coals but decreases with increasing rank to about 91% carbon (dmmf). Thereafter, the trend reverses (Figure 5). However, coals

Table 1. Summary of nomenclature in coal petrology. After Stach (1968).

Lithotype	Maceral	Maceral Group	Microlithotype	Principal Maceral Constituents
Vitrain	Collinite Telinite	Vitrinite (V)	Vitrite	V
			Vitrinertite	V + I
			Microite	I (micrinite dominant)
Fusain	Micrinite Semifusinite` Fusinite Sclerotinite	Inertinite (I)	Fusite	I (except micrinite)
	Cutinite Resinite Sporinite Alginite	Exinite (E)	Sporite	E
Clairain			Clarite	V + E
Durain			Durite	I + E
			Duroclarite	V + E + I
			Clarodurite	I + E + V

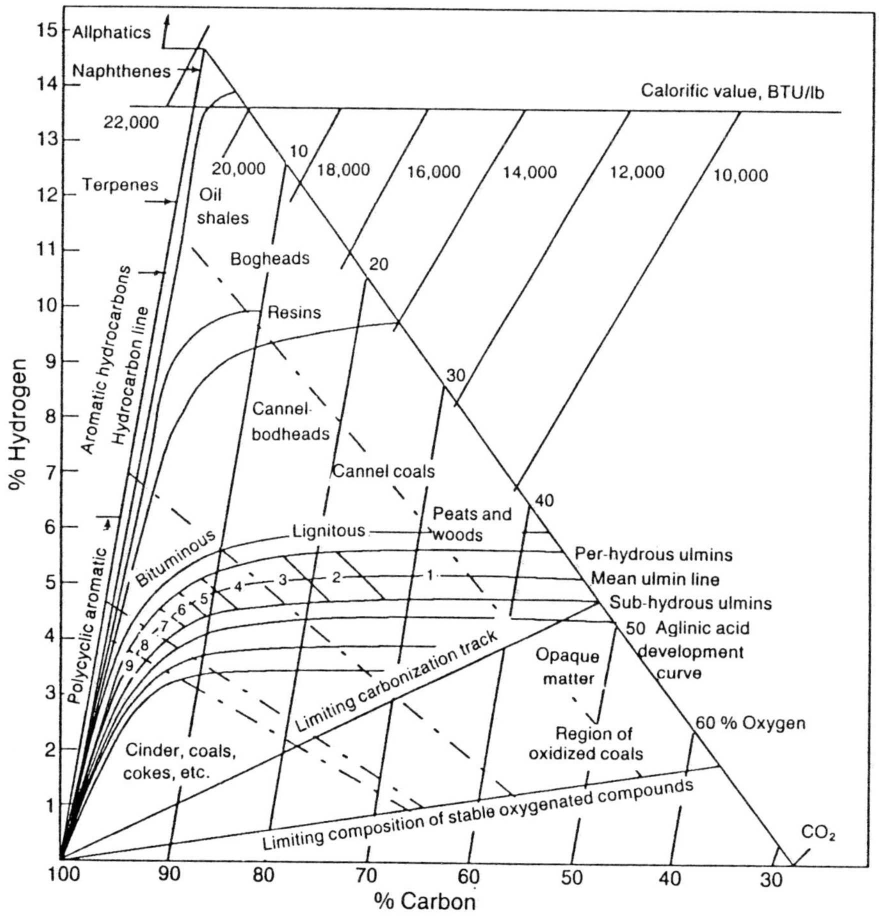

Figure 3. Basic rational classification chart. After Francis (1961).

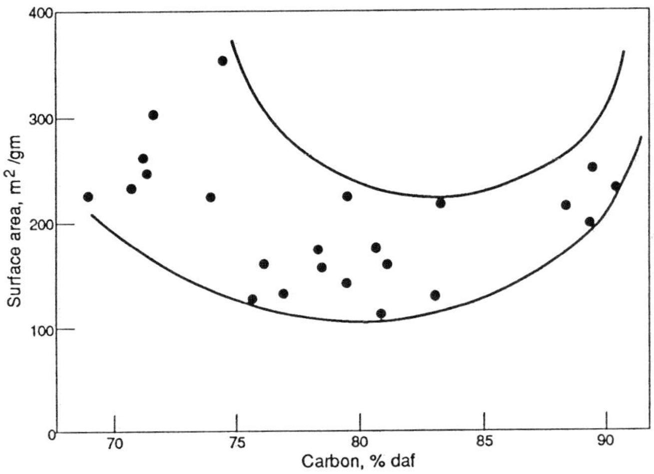

Figure 4. Rank dependence of surface area determined by adsorption of carbon dioxide at 25°C. After Gan and others (1972).

with carbon contents above 84% have low strength and high brittleness, making fracture formation more likely (Figures 5, 6, and 7). Thus, high-ranking coals (above 84% carbon dmmf) are potentially good sources of fracture- and pore-filling gas.

Nature of Rank Metamorphism

Abnormally metamorphosed coals not only produce more gas but also have greater surface area. Therefore, they are good candidates for retention of large volumes of methane.

Nature of Seam Roof and Floor Strata

During the formation of methane, only a part is retained by the coal in the adsorbed state. The rest is released and exists in a free state. What happens to this free gas is de-

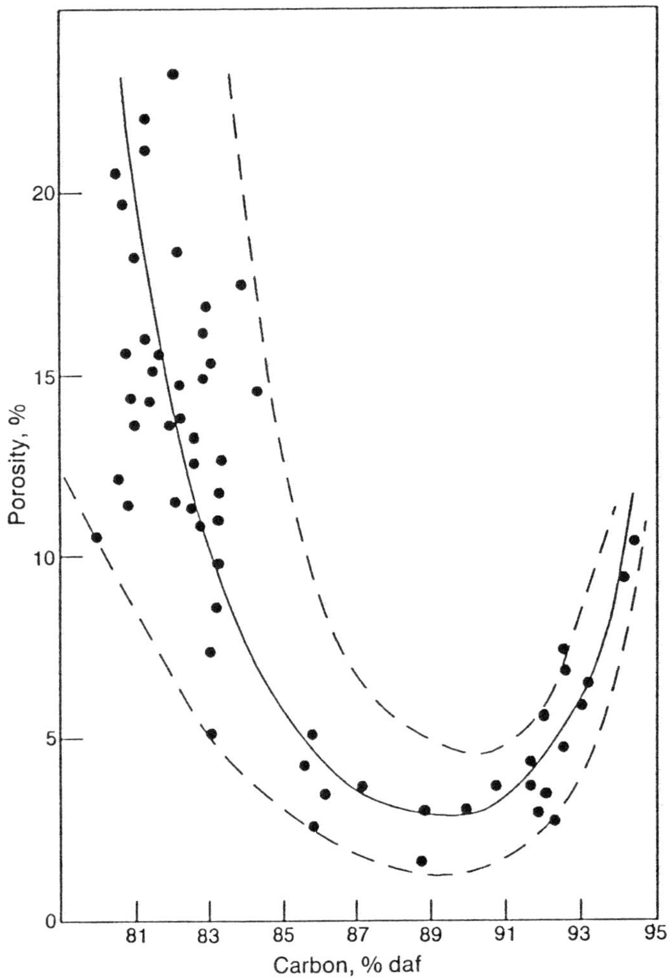

Figure 5. Variation in coal porosity with rank. After King and Wilkins (1944).

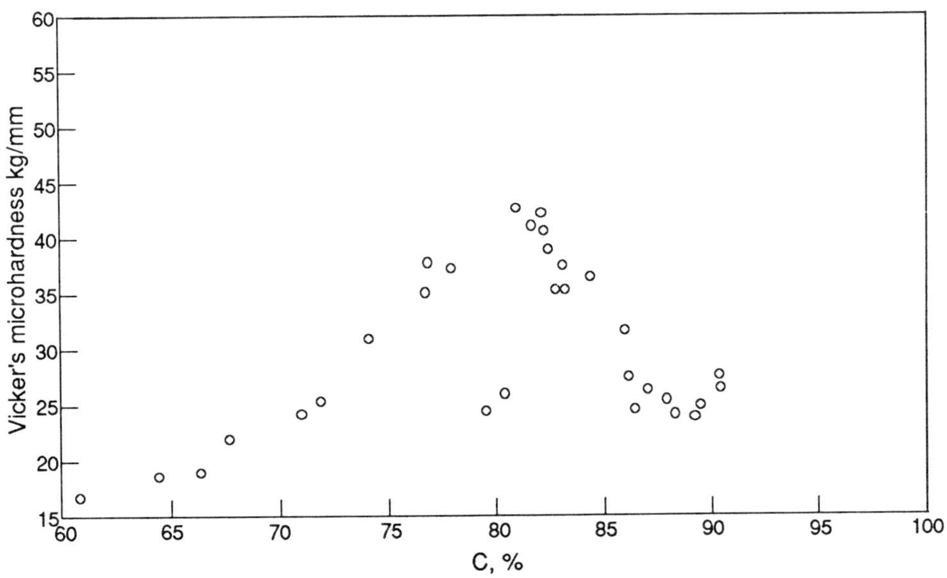

Figure 6. Vicker's microhardness vs coal rank. After Das (1987).

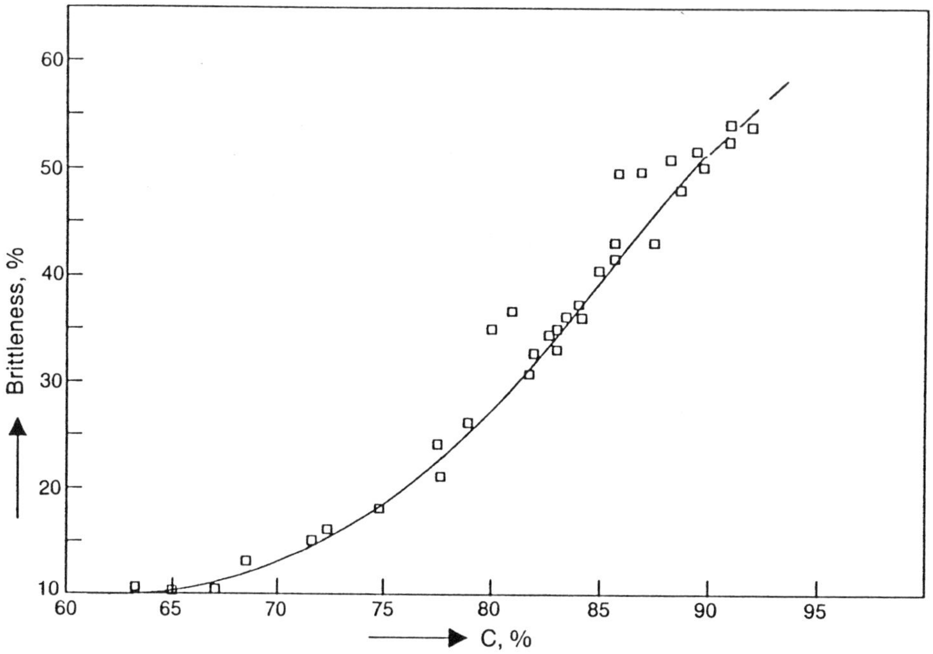

Figure 7. Brittleness vs coal rank. After Das (1987).

termined by the nature of the seam roof and floor rocks. Within a relatively impermeable rock envelope (for example, shale), gas remains within the coal void spaces, with an associated buildup in gas pressure. Permeable beds (for example, sandstone), however, allow gas to escape. Depending on the extent, depth, and nature of the permeable rocks, gas will either escape to the atmosphere or migrate to another trap elsewhere, with a resultant increase in gas pressure. If pressure buildup is sufficient and if void spaces in the coal seam increase with coalification, some gas subsequently may be forced back into the coalbed.

Tectonic History

The effect of tectonism on gas retention is primarily negative. Tectonic activity commonly increases the permeability of strata through the formation of joints, fractures, and faults. This increases the likelihood of gas migration either to a distant trap (which may be tectonically controlled) or into the atmosphere.

Burial Depth

As already seen, greater burial depths generally mean higher rank and, consequently, more gas. In addition, greater depths reduce the chance of complete escape of the gas.

Proximate Analysis

Coal composition, expressed as proximate analysis values, influences the amount of gas produced per unit volume, and thus, the probability of gas retention. In addition, higher ash content increases the strength of coal, which would reduce the incidence of fracture formation and hence inhibit the escape of the gas.

METHANE DESORPTION FROM COAL

In addition to knowledge of the mechanisms of methane formation and retention in coalbeds, it is important to understand the processes involved in its desorption. As with any fluid flow, movement of methane is governed by two fundamental parameters—*motive force* of and *resistance* to passage.

Motive Force

The motive force, in the case of methane desorption from coal, is gas pressure. This equals the difference between the gas pressures within the coal and at the receiving end of the flow. For any spontaneous flow, this is equivalent to the gradient between confined gas pressure and ambient pressure. In the case of induced flow, it depends on the pressure differential between the two ends of the flow path.

Ultimately, gas pressure depends on the formation and retention of gas within the coalbed. These depend on the factors discussed above. Systematic evaluation of these factors indicates that higher gas pressure is created in coals with high proportions of exinite, up to an equivalent rank of 89.5% carbon (dmmf), and with high proportions of vitrinite above that rank. Abnormal metamorphism also is likely to promote high gas pressure. In addition, tectonically unaffected coals at great burial depths contain gas at high pressure, as do coals of low ash content.

Resistance

In any fluid flow, as resistance increases, rate of flow decreases for any given pressure difference. Thus, as a coalbed becomes more fractured, permeability increases and, consequently, resistance to flow decreases. As discussed above, the formation of joints, cleats, and fractures in coal is determined by seam strength and brittleness.

Strength

Strength increases the resistance of a material to deformation when it is subjected to an impressed force. In Figure 6 the strength of coal is expressed in terms of *microhardness*. Low-rank coals (lignites) have low strength. As rank increases to about 81% carbon (dmmf), strength increases. Above that level (bituminous to prime coking rank), strength decreases with increasing rank. Finally, strength increases again through semianthracite and anthracite.

Brittleness

Brittleness of a material indicates the ratio of reversible energy to total energy supplied to it by a deforming force at the point of failure. The higher the proportion of reversible energy, the more violent the rupture and the higher the degree of fracturing. Materials with very low brittleness that can continuously and irreversibly dissipate deforming energy through flow are termed ductile.

Figure 7 shows the brittleness curve of coal vs rank. Low-rank coals are ductile and brittleness increases with increasing rank. Thus, higher rank coals are more prone to fracture formation.

Fracture Formation in Coal

Based on their strength and brittleness, as shown in Figures 6 and 7, coals can be divided into the following distinct groups:

1. *Soft, ductile coals*. These are low-rank coals (lignite) that deform easily without fracturing. Thus, fracture permeability generally is low. In addition, because this rank of coal is a poor producer of gas, methane desorption is of little relevance.
2. *Moderately hard, ductile coals*. This group consists of higher rank (subbituminous) coals that can be deformed by relatively large forces but without appreciable fracturing. Fracture permeability and rate of desorption thus are low. This poor performance as a gas producer is exacerbated by the low volumes of gas produced by this rank of coal.
3. *Moderately hard, brittle coals*. These include high-volatile bituminous coals that can be deformed by moderate forces, resulting in the formation of some fractures. Both the rate and total volume of desorption can be moderate because of relatively moderate permeability, gas pressure, and gas content.
4. *Weak, highly brittle coals*. The coking and prime coking coals of this group are soft but highly brittle. They contain large amounts of gas that, if trapped by an impermeable envelope, could represent ideal conditions for high rate and total volume of desorption. The total volume of gas may be enhanced by gas stored within a permeable envelope (sandstone).
5. *Extremely hard, brittle coals*. These constitute the highest ranks (semianthracite and anthracite). These coals are brittle but require large forces to deform them. Thus, their permeability will vary depending on the forces to which they were subjected during burial history. On the other hand, they are likely to contain large quantities of desorbable gas.

CONCLUSIONS

From this assessment of factors affecting the occurrence and production of coalbed methane, the following conclusions can be drawn:

1. The composition, rate, and volume of desorption of gas from a coalbed are vital factors in assessing both the hazard and the productive potential of a coal seam.
2. In most cases, the desorbed gas is predominantly methane. Other hydrocarbons and gases (nitrogen, carbon dioxide, carbon monoxide, hydrogen sulfide) are present usually in minor amounts. However, the fact that large amounts of nitrogen(<41%) have been recorded makes an analysis of any recovered gas an important precursor to further work.
3. The most important factors controlling the rate and total volume of desorbed gas are coal rank, coal type (maceral or petrological composition), nature of rank metamorphism or coalification (normal or abnormal), tectonic history, burial depth, and the strength and brittleness of the coal. These factors also control permeability development in any given coal reservoir.
4. Low-rank coals (lignite and subbituminous) are not likely to be potential sources of thermogenic methane but, in the case of subbituminous coals, are sources of biogenic methane. Higher rank coals (bituminous, semianthracite, and anthracite), which have generated both biogenic and thermogenic methane (usually more thermogenic by volume), are more likely to be better potential producers. Rates and total volume of desorption generally are highest at equivalent ranks above 84% carbon (dmmf), but prime targets exist between 87% and 91.5% carbon (dmmf). Semianthracite and anthracite will desorb large volumes of gas, but rates will depend on permeability of the coal.
5. Exinites will produce more gas than vitrinites, except at ranks higher than 89% carbon or 29.5% VM (dmmf), at which point the trend is reversed. Inertinites contribute least to gas production. These data are in agreement with hydrogen contents of the three maceral groups.
6. Abnormally metamorphosed coals follow a straight "carbonization track" on H-C graphs (Figure 3), and thus their hydrogen content is always lower than the value for normally metamorphosed coals at any rank. Abnormally metamorphosed coals will, therefore, produce more gas.
7. Tectonic activity tends to increase the permeability of the rock strata surrounding a coalbed. This might allow for the escape of gas into the surrounding sedimentary envelope.
8. In general deeper coals are more gas rich than shallow coals. This is due to rank improvement and increased

confinement with depth. Above 246 to 262 ft (75 to 80 m), gas will escape from a coal seam. Below 656 ft (200 m), unless the strata are faulted, gas will be retained.

9. High mineral-matter (ash) and moisture content will reduce the total volume of gas per unit volume of coal. High ash content can also reduce the permeability of a coal except in high-rank coal (anthracite).

10. The strength and brittleness of a coal determine its permeability. Analysis of these factors indicates that the rate of gas flow is highest for coking coal between ranks of 87% and 91.5% carbon (dmmf). However, a very high flow rate can be obtained from coals of rank above 91.5% carbon (dmmf) if fracturing is induced.

To obtain a reliable indication of either the hazard or productive potential of a coal seam, the above factors must be taken into account when carrying out desorption tests on borehole samples. The chances of arriving at a valid conclusion are greatly increased by an awareness of the complex factors that can affect the desorption of methane from coal.

REFERENCES

Airuni, A.T., and Éttinger, I.L., 1966, Gases of coal seams: Moscow, State Publishing House.

Das, B., 1987, Methane desorption in coal handling: CANMET Division Rept. ERP/CRL 87–123 (OP, TR), 29 p.

Das, B., and Hucka, V., 1977, Investigation of methane emission and drainage at the coal mines of DEVCO, Sydney, Nova Scotia: Laval University, Dept. of Mineralogy and Metallurgy, Rept. for CANMET.

Das, B., and others, 1982, Methane desorption study on borehole coal samples at Coal Valley, Alberta: CANMET Division Rept. ERP/CRL 82–06 (TR), 32 p.

Deul, M., Fields, H.H., and Elder, C.H., 1973, Degasification of coal bed—A commercial source of pipeline gas: U.S. Bureau of Mines Open-File Rept. 109–73, 9 p.

Francis, W., 1961, Coal—Its formation and composition: London, Edward Arnold Ltd., 806 p.

Gan, G., Nandi, S.P., and Walker, P.L., Jr., 1972, Fuel: London.

Hargraves, A.J., 1966, Some aspects of seam gas: West Virginia University, School of Mines Rept.

Hucka, V.J., 1980, Methane drainage in Sunnyside coal mines: Salt lake City, University of Utah Final Rept. 20381 [prepared for TLT Babcock Co., Akron, Ohio].

Jüntgen, H., and Karweil, Joachim, 1966a, Gasbildung und Gasspeicherung in Steinkohlenflözen, part I—Gasbildung: Erdol und Kohle, Erdgas, Petrochemie, v. 19, p. 251–258.

Jüntgen, H., and Karweil, Joachim, 1966b, Gasbildung und Gasspeicherung in Steinkohlenflözen, part II—Gasspeicherung: Erdol und Kohle, Erdgas, Petrochemie, v. 19, p. 339–344.

King, J.G., and Wilkins, E.T., 1944, Proceedings of the Conference on Ultrafine Structure in Coal and Cores: London, BCURA.

Lidin, G.D., 1961, Development of A.A. Skotshinskii's formulae of methane emission prognosis in coal mines: Moscow, A.A. Skotshinskii, Institute of Mines.

Patching, T.H., 1970, The retention and release of gas in coal—A review: Canadian Institute of Mining and Metallurgy [CIM] Bulletin, v. 63, no. 703, p. 1302–1308.

Stach, E., 1953, Der Inkohlensprung in Ruhrkarbon: Brennstoff-Chimie, v. 34, p. 353.

Stach, E., 1968, Basic principles of coal petrology, in Murchison, D., and Westol, T.S., eds., Coal and coal bearing strata: London, Oliver and Boyd, p. 3–17.

In Situ Determination of Desorbable Methane Content by Use of Three Decay Functions

Rex R. Plaizier[1]
V. Joseph Hucka[2]

1. Westech Engineering Inc., Salt Lake City, Utah
2. University of Utah, Salt Lake City, Utah

ABSTRACT

The desorptive properties of methane in selected western coal seams were studied through actual field measurements using a manual bubble "desorbometer" and later verified through laboratory modeling. Data were reduced using three functions—power decay, quadratic, and exponential. Methane desorption characteristics of individual coal seams were shown theoretically to be a good estimator of total methane content of a seam with a precision of ±20%.

A laboratory was equipped to properly analyze and reduce field data, in addition to providing controlled conditions for studying methane adsorption properties of coal. Field test results indicate that all three theoretical desorption functions provide a satisfactory model of the desorption process based on estimated and actual gas contents, and on high coefficients of regression.

Estimated gas contents at four western coal mines were derived through computer reduction of data obtained from field and laboratory tests. The coal seams investigated include (1) B seam, Dutch Creek Mine, Piceance basin, Colorado, (2) Rock Canyon seam, Soldier Creek Mine, Uinta basin, Utah, (3) Lower Sunnyside seam, Sunnyside Coal Mines, Uinta basin, Utah, and (4) Sub 3 seam, Castle Gate Mine, Uinta basin, Utah. Average methane contents range from 57.7 to 176.2 CF/T (1.8 to 5.5 cc/g). Linear-regression analysis of test data indicate that the quadratic decay function may provide a more accurate model of the desorption process than the other two theories discussed.

INTRODUCTION

Coalbed methane creates potential hazards in underground mines. Consequently, the methane content of a coal seam is an essential parameter for mine ventilation design and for feasibility studies of methane drainage systems. This parameter is important as well in assessing gas outburst potential. However, coalbed methane also is a valuable clean-burning fuel. Presently, the most used method for gas-content determination in United States gassy coal mines is the "direct" method developed by the U.S. Bureau of Mines (McCulloch, 1975).

Testing methods used in Germany and France for determining in situ gas content have been studied and modified for use in western United States coal seams. Results of the "bubble desorbometer" method used in this study are compared with results obtained from the direct method.

BACKGROUND AND TEST PROCEDURES FOR DETERMINING GAS CONTENT

Gas content is measured as a volume of gas within a given unit weight of coal. Methane in coal is found both as free gas in void spaces and adsorbed onto the internal surfaces of coal particles. The volume of free gas is directly related to coal porosity and in most coals makes up only 3% to 5% of the total gas content (Boxho, 1980).

Residual gas content is measured as the total gas (methane) volume per unit weight of coal retained in the coal at 0.145 psi (1 kPa) atmospheric pressure and in a pure methane environment.

French and German Influences

Researchers at France's National Coal Institute (CERCHAR) have studied methane properties in coal, gas-content testing procedures, and development of drainage systems. They have proposed a method by which either long boreholes are drilled from an entry horizontally into the seam or from a vertical borehole intersecting the seam. Coal samples are taken every 3.3 to 6.6 ft (1 to 2 m) from a horizontal borehole and tested until a relatively constant gas content is reached. Samples from vertical boreholes consist of 0.35 oz (10 g) of core or cuttings screened to include only particles larger than 0.079 in. (2 mm), which then are immediately sealed inside a transportable container.

Researchers at the German Mine Ventilation Institute in Bochum, Germany, developed a similar procedure for testing the gas content of coal seams in situ. Coal samples collected from horizontal boreholes are tested at the face with a bubble desorbometer (Figure 1) so that methane content can be estimated within minutes of sample removal. The method is similar to the French technique in the sampling procedure and quantification of residual gas and gas desorbed after the sample is sealed in the container, but it differs in the theory and determination of the lost gas that is given off before the sample is tested and sealed.

Drill cuttings samples are collected every 3.3 to 6.6 ft (1 to 2 m) in the borehole and screened to a size of 0.015 to 0.025 in. (0.4 to 0.63 mm); 0.14 to 0.17 oz (4 to 5 g) is placed inside the desorbometer cell. Samples are tested until a relatively constant gas content value is reached. The desorbometer is operated manually, but an automatic electronic desorbometer also is available.

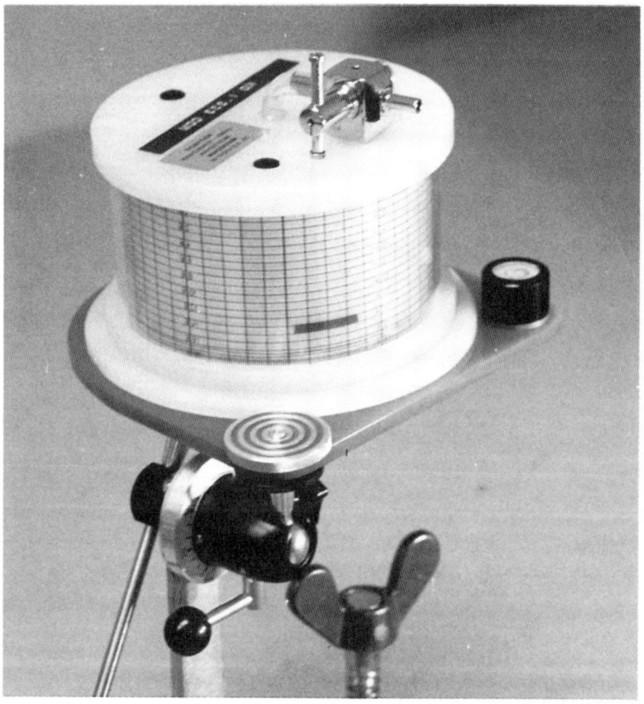

Figure 1. Manually operated bubble desorbometer.

Modified Method for Determining Gas Content

Both the French and German methods were modified into a procedure better suited for the conditions and equipment available for western coalbed evaluation. The respective theories for estimating lost gas used by the French and Germans were used for a comparative basis in this study.

Determination of Q_1

Q_1 is defined as the quantity of gas given off and lost from the time the sample is disturbed to the time it is sealed inside a transportable container. Drill cuttings samples collected from horizontal boreholes are sieved to a size of 0.0078 to 0.015 in. (0.20 to 0.40 mm) and then sealed inside the bubble desorbometer cell. Desorbed gas from the sample pushes a slug (bubble) of colored glycol through a small-diameter nylon tubing. Movement of the slug one full circumference around the desorbometer corresponds to a volume of 7.5×10^{-2} in.^{3} (1.233 cc) of gas, a value referred to as the K_d factor of the desorbometer. The tubing is graduated volumetrically, enabling the recording of elapsed time at regular desorbed-gas intervals. A maximum of 18 revolutions can be recorded on the meter, but normally only three to six are required for the test. Figure 2 shows the flow chart for the test procedure.

The theory and function describing the desorption of gas as a function of time is one of the principal concerns of this study. The French use a quadratic expression, the Germans use a decaying power function, and the modified test uses a decaying exponential function. These functions are summarized as follows:

Quadratic:
$$Q_1 = k\,(t_1)^{0.5} - k\,(t_0)^{0.5} \tag{1}$$

Power:
$$Q_1 = \frac{r_1}{(1 - K_t)} \times t_b^{(1-K_t)} \tag{2}$$

Exponential:
$$Q_1 = \frac{r_0}{-k_t} \times (e^{(-k_t t_1)} - 1) \tag{3}$$

where,

t_0 = time elapsed from sample disturbance to insertion of the sample into the desorbometer;

t_1 = elapsed time to the end of the testing period in the desorbometer (For the quadratic method, $t_1 = 2t_0$);

t_b = elapsed time to the end of the tests in the desorbometer;

r_1 = specific desorption rate at 1 minute after initial sample disturbance;

r_0 = initial desorption rate when the sample is initially disturbed.

k_q = desorption constant specific for the coal seam (quadratic method);

K_t = desorption constant specific for the coal seam (power method);

k_t = desorption constant specific for the coal seam (exponential method);

The constants r_1, r_0, k_q, K_t, and k_t are determined from the desorbometer data by plotting the volumetric desorption rate against cumulative desorption time since initial drilling. Plotting on these data on log-log paper for the German method (power decay function) and on semilog paper for the modified method (exponential decay function) yields a straight line with a slope of K_t or k_t, respectively (Figures 3a and 3b).

Determination of Q_2

The volume Q_2 is defined as gas desorbed after the sample is sealed in the transportable container and before the container is opened in the lab for further tests. This volume is determined from the gas pressure and methane content in the container. At the laboratory, the container pressure, lab temperature, ambient temperature, and methane concentration in the container are measured. The volumetric percent of methane gas in the container is calculated by the following formula in which the volume displaced by the coal sample itself is taken into account.

$$V_f = V_b \frac{p}{p_b} - \frac{M_r}{\rho} \tag{4}$$

where,

V_f = free (increased) volume in container (in.^{3} or cc);
V_b = normal volume of the container (in.^{3} or cc);
p = gas pressure in the container (psi or kPa) adjusted for temperature and pressure changes;
p_b = ambient pressure (psi or kPa);
M_r = mass of sample (oz or g);
ρ = density of coal (lb/ft^3 or g/cc).

The specific volume, q_2 (cc/g), of methane released in the container is then found by:

$$q_2 = \frac{V_f \times C}{M_{af} \times 100} \tag{5}$$

where,

C = concentration of methane in container (% vol.);
M_{af} = ash-free mass of the coal sample.

Ash content of the coal is determined immediately after milling the sample, and the ash-free mass is calculated as:

$$M_{af} = M_r \times \frac{(100 - 1.1A)}{100} \tag{6}$$

where A is ash content (% weight).

Determination of Q_3

Q_3 is defined as the volume of gas released from the sample after it was removed from the container and crushed in the testing mill. The specific volume of released gas, q_3, which is measured by the use of liquid-filled burettes (Figure 4), consists of two quantities:

$$q_3 = q'_3 + \Delta q'_3 \tag{7}$$

where,

q_3 = volume of gas remaining in coal sample after measurement of q_2 (CF/T or cc/g);
q'_3 = volume of gas released by milling the sample (CF/T or cc/g); and
$\Delta q'_3$ = volume of gas remaining in the coal after milling due to partial pressure of methane inside the mill (CF/T or cc/g).

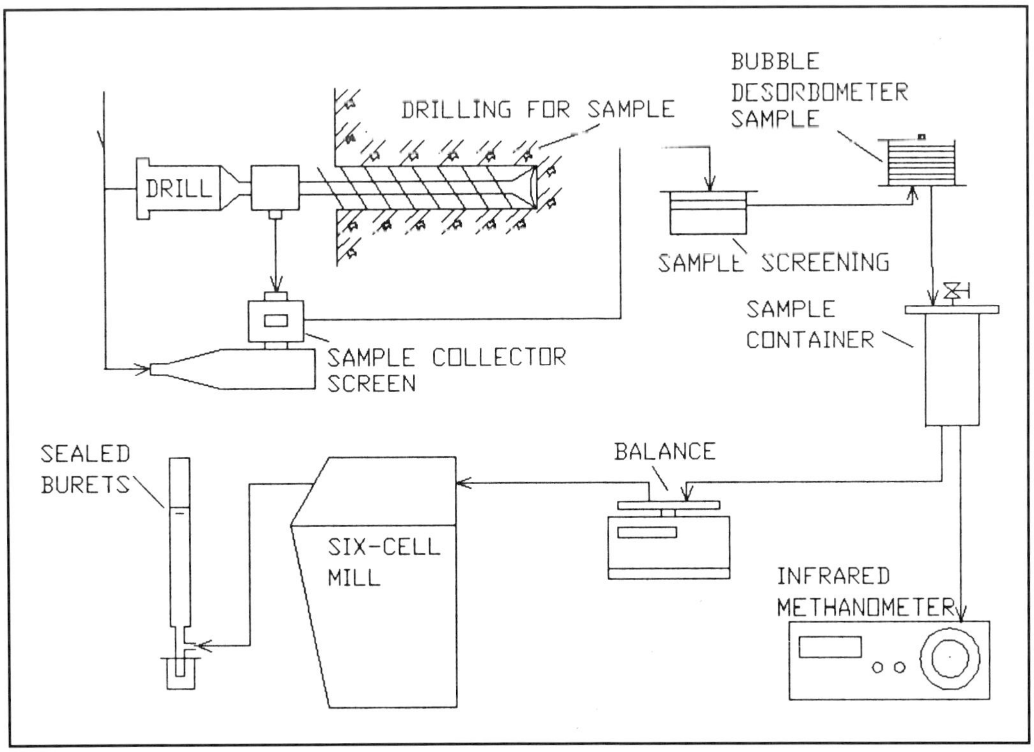

Figure 2. Flow chart of gas-content determination.

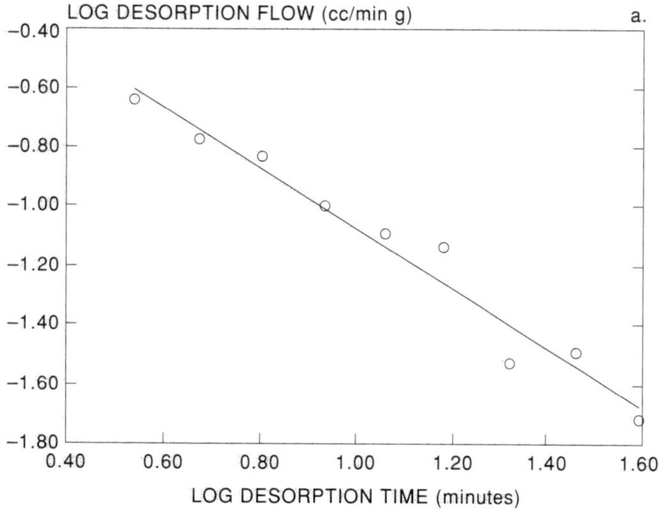

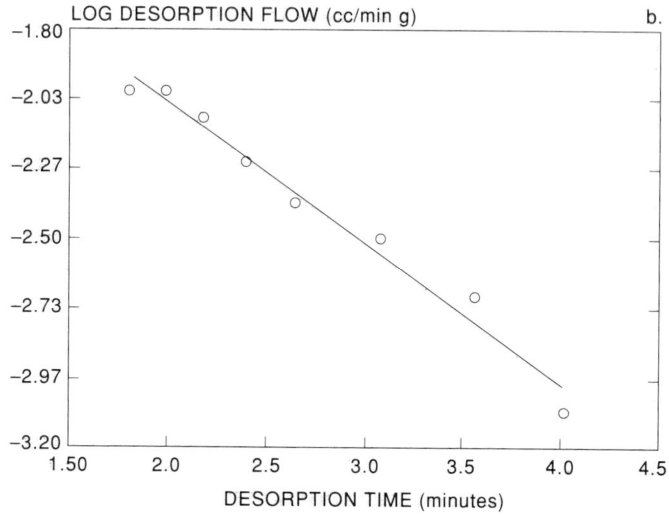

Figure 3. Desorption time vs desorption flow rate. Slope of curve determines specific desorption constant of the coal tested. *a*. Log-log plot (power decay function), slope K_t. *b*. Semilog plot (exponential decay function), slope k_t.

After the measurement of q_2 is completed, the coal sample is disc milled in 10-minute intervals. The gas released from this crushing is collected in liquid-filled burettes. Milling continues until no more bubbles of gas are released in a 2-minute interval. The value of q'_3 then is simply:

$$\Delta q'_3 = \sum \frac{Q'_3}{M_{af}} \tag{8}$$

where $\Sigma Q'_3$ is the total gas measured in the burettes.

The partial pressure of methane in the mill, which is directly proportional to the methane content in the mill, also influences the amount yet to be desorbed from the crushed coal. Consequently, the concentration of methane in the mill must be measured precisely. If the mill is purged with pure methane before milling begins, then the amount of gas remaining in the coal after milling corresponds to the value of $q_{1\,bar}$, as the partial pressure of methane in the mill will be 0.145 psi (1 kPa or 1 bar). Therefore, the methane concentration in the mill is calculated after each milling time interval according to the following equation:

$$C_n = \frac{V_f \times C_{(n-1)} + Q'_{3(n)}}{V_f + Q'_{3(n)}} \times 100 \tag{9}$$

where,

C_n = methane concentration in the milling cell after the nth milling interval (% CH_4);

$C_{(n-1)}$ = methane concentration in the milling cell after the n–1 milling interval (% CH_4);

$Q'_{3(n)}$ = gas released during the nth milling interval (cc);

V_f = free volume of the milling cell (in.^{3} or cc).

$\Delta q'_3$ then can be found by:

$$\Delta q'_3 = q_{1\,bar} \times \frac{C_n}{100} \tag{10}$$

where $q_{1\,bar}$ is residual gas content of the coal determined in the laboratory.

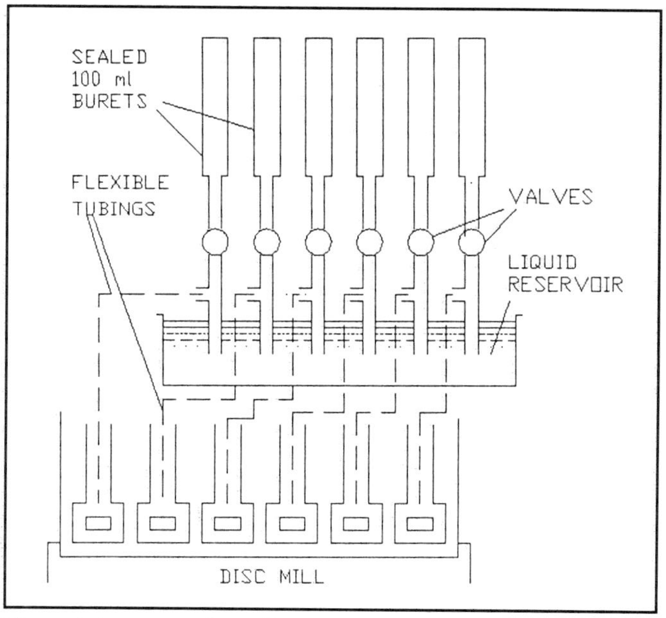

Figure 4. Apparatus for collecting and measuring specific volume of gas released from milled coal.

Laboratory Determination of Gas Content

To verify the accuracy of the field tests, six sample cells were filled with coal cuttings from the same borehole sample, screened to 0.0078 to 0.015 in. (0.20 to 0.40 mm). The cells were interconnected by copper tubing, and the total six-cell system initially was placed in a pure methane environment under increased pressures to approximate in situ adsorption conditions. The samples then were measured for their q_1 and q_3 values, which represent the total gas content adsorbed onto the sample.

Gas Content Transformed to In Situ Conditions

To relate the desorbed gas measured under laboratory conditions to those of in situ conditions, temperature and moisture content must be considered. This particular formula is described at length in Plaizier (1990).

Determination of Free Gas

Free gas is defined as the gas present within the coal in void spaces but not adsorbed onto the coal. In situ free gas content, q_f (in m^3 gas/m^3 of nondisturbed coal), normally is less than 5% of the total gas content and can be determined through the following formula (Boxho, 1980):

$$q_f = \frac{\phi_n p}{c\, p_a} \tag{11}$$

where,

ϕ_n = effective porosity for methane in coal (m^3/m^3);
C = compressibility coefficient of methane (for technical calculations, $C = 1$);
p = methane gas pressure in pores;
p_a = atmospheric pressure (psi, kPa, or bars).

The volumetric effective porosity, ϕ_n, is determined by the equation:

$$\phi_n = 3.84 \times 10^{-2} - (1.81 \times 10^{-3} \times F) + (9.38 \times 10^{-7} \times F^3 \rho) \tag{12}$$

where,

F = volatile-matter content of coal (% weight); and
ρ = coal sample density (g/cc).

TOTAL GAS CONTENT DETERMINATION

The total specific gas content of the coal then is:

$$q_{total} = q_1 + q_2 + q_3 + q_f \tag{13}$$

Desorbable gas content is, however:

$$q_d = q_{total} - q_{1\,bar} \tag{14}$$

RESULTS OF FIELD IN SITU TESTS

Field tests were performed on coal samples from several gassy mines in Colorado and Utah. In the Piceance basin, Colorado, medium-volatile bituminous coal of the B seam (Cameo Coal Group), from the Upper Cretaceous Mesaverde Group (Hucka and others, 1989), was tested in the Dutch Creek Mine, located about 6 miles (9.7 km) south of Redstone. Samples were collected from the longwall face.

In Utah tests were conducted on three coal seams of the Upper Cretaceous Mesaverde Group in the Book Cliffs coal field. The first, the Sub 3 seam in the Castle Gate Mine, located near Helper, is the basal coalbed of the Spring Canyon Member of the Blackhawk Formation. Second is the high-volatile B bituminous Rock Canyon seam at the Soldier Creek Mine. Third, the Sunnyside seam at the Sunnyside Mine (Figure 5), located south of Price, is a high-volatile A bituminous coal that overlies the Sunnyside Sandstone Member of the Blackhawk Formation. A gassy 2-ft-thick (0.6-m) rider seam occurs above the main seam.

A complete tabulation of field test results is given in Plaizier (1990). Results are categorized according to the three desorption theories discussed earlier (German power decay function, French quadratic decay function, and Plaizier's exponential decay function). A sample output sheet from the field study analysis appears in Table 1.

Upon compilation of field results and desorption characteristics of treated samples, the German desorption constant (K_t) was discovered to be greater than 1 for many samples. As equation 15 below suggests, a value of $K_t > 1$ yields a negative desorbed gas volume, which obviously is impossible.

$$q_{(0-b)} = \frac{r_1}{1 - K_t} \times t_b^{(1-K_t)} \tag{15}$$

This deviation from physical possibility using equation 15 is due to the parent (or original) integral (equation 16) becoming an improper integral because the logarithmic function is undefined at $t = 0$.

$$q_{(0-b)} = \int_{t_0}^{t_b} r_1 t^{-K_t}\, dt \tag{16}$$

The desorbed gas can be approximated using computer numerical integration algorithms as long as the lower limit of the integration is slightly more than zero. The German power decay function representing the methane desorption rate (equation 2) implies that the desorption rate at the instant the coal sample is disturbed ($t = 0$) is infinite.

Because of the difficulties in modeling methane desorption from some samples using the power function, a theory that methane desorption from coal could be modeled by an exponentially decaying function therefore was incorporated into testing procedures and calculations for this work, and tests were carried out to verify its validity and physical accuracy.

Figure 6 represents a plot of desorption rate vs time based on this exponential relationship. This curve can be represented by the general equation:

$$r = r_0 e^{-kt} \tag{17}$$

where r is the desorption rate at any time t, and r_0 is initial desorption rate at time t_0. The amount of gas desorbed between times t_0 and t_1 can be found by integrating:

$$q_1 = \int_{t_0}^{t_1} r\, dt \tag{18}$$

Using equations 17 and 18 and assuming $t_0 = 0$, the specific volume q_1 becomes:

$$q_1 = \frac{r_0}{-k_t} (e^{(-k_t t_1)} - 1) \tag{19}$$

The only unknowns in equation 19 are the curve constant k_t and initial desorption rate, r_0. The constant k_t can be

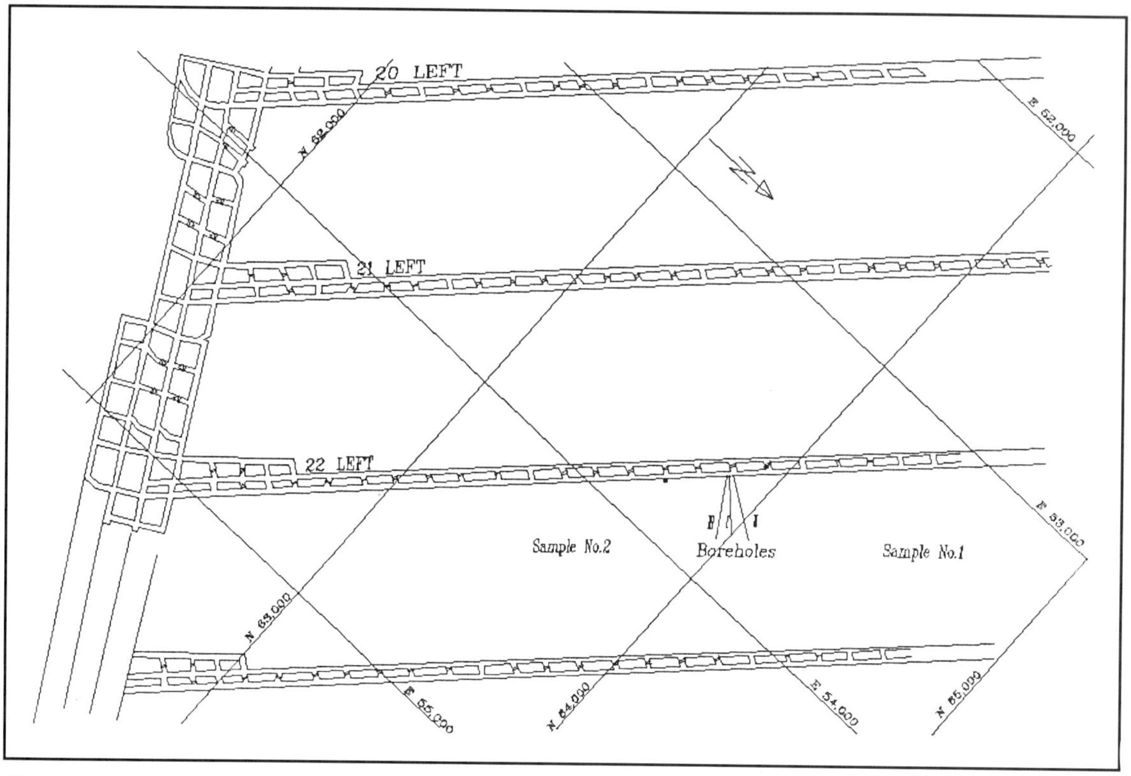

Figure 5. Map of underground workings at Sunnyside Mine, Utah, showing locations of horizontal boreholes and coal sample areas.

found with data points measured by the desorbometer between t_1 and t_2, and the rate r_0 is extrapolated from t_1 using this curve constant. No change in testing procedures are required for application of this desorption theory, only in data reduction.

Time Constants

The time constant of a coal is a characteristic that reflects the rate of gas desorption from the particular coal. Upon compilation of the field data, an average time constant was calculated for each of three coal seams (Sub 3 seam, Castle Gate Mine; Rock Canyon seam, Soldier Creek Mine; B seam, Dutch Creek Mine) and input into the computer program to calculate estimated total gas contents based on initial desorption test data from the mines. The definition and testing of this constant is treated in detail in Plaizier (1990).

DISCUSSION OF RESULTS

Table 2 compares the results of total gas-content values for the three seams, as derived from laboratory measurements and from in situ estimates using the estimated time constant and desorbometer readings.

Some in situ estimates for the Sub 3 seam deviate from laboratory-measured gas content values by as much as 50% among all three methods. However, when comparing their respective sums, the percentage difference decreases significantly. This might be expected when using an estimated time constant derived by averaging all samples; however, for practical use in situ, the estimated constant should be verified by testing many more samples; however,

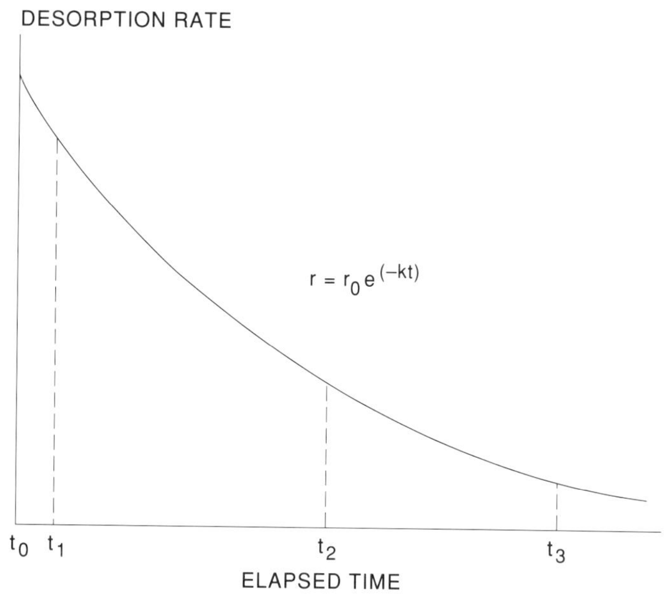

Figure 6. Desorption rate vs time, illustrating application of an exponential decay function solution. t_0 is the time at which drill bit cuts sample; t_1 is the time to insertion of sample into desorbometer; t_2 is the time to removal of sample from desorbometer and enclosure in a sealed flask; t_3 is the time to concentration measurement in a sealed flask and subsequent milling.

Table 1. Example record of field desorption test results.

Name of mine (test site):	Colorado No. 1 (Dutch Creek)
Location within mine:	Panel 102 Head Gate-B seam #5
Date of tests:	7/15/86
Adsorption constant:	6.7
Mine temperature (°C):	24.0
Mine atmospheric pressure (Pa):	70,897

Borehole No. 1

Sample No. 3

Ash content (%):	5.64	Borehole depth (m):	5.03
Sample weight (g):	3.40	Drill time (seconds):	32
Desorbometer constant:	1.233		

Data Reduction Function

Parameter	Power	Quadratic	Exponential
Total gas content (cc/g or m^3/t)	1.8	1.5	1.0
Total estimated gas content (cc/g or m^3/t)	2.8	1.4	1.4
Difference (%)	–54.2	11.8	–45.1
Gas lost while drilling (% of total)	52.4	42.8	8.1
Desorbed volume in situ, Q_1 (% of total)	54.9	45.8	12.9
Desorbed volume in container, Q_2 (% of total)	31.3	37.6	60.4
Desorbed volume after milling, Q_3 (% of total)	13.8	16.6	26.7
Coefficient of regression (r^2)	.947	.988	.939
Slope or K_t constant	1.36	0.13	0.28
Estimated time constant	10.83	37.89	7.17
Desorption constant (noncomparable)	170.43	–0.19	37.74

Time of Reading (minutes)	Desorbed Time (minutes)	Gas Volume (cc)	Desorption Rate (cc/min g)
2.783	0	0	—
3.267	0.483	0.025	15.903
3.733	0.467	0.025	16.471
4.467	0.733	0.025	10.482
5.400	0.933	0.025	8.235
6.400	1.000	0.025	7.686
7.917	1.517	0.025	5.068

for practical use in situ, the estimated constant should be verified by testing many more samples from other locations through the mine. Another explanation for the deviation perhaps could relate to difficulties in collecting samples. Because of the hardness of coal, drilling time and sample collection time during the first test far exceeded the drilling time on subsequent tests. As a result, an inadequate number of samples were collected for more complete analysis.

Estimated gas contents for the Rock Canyon seam consistently fall within 25% of laboratory-measured values using the quadratic and exponential functions. For cumulative totals among all three seams, the exponential decay function reflects the most consistency (3% to 6% difference) in estimated in situ values.

The results in Table 2 are insufficient for one to determine a preference in accuracy or reliability of any one desorption theory over the other two. However, more insight might be gained by plotting desorption-rate data and analyzing the desorption constant. In Figure 7, identical field data for the Rock Canyon seam have been plotted according to the respective requirements of the three desorption theories—that is, a semilog plot of desorption time and specific desorption rate for the exponential decay function; a log-log plot of desorption time and specific desorption rate for the power decay function; and square root of time and cumulative desorbed gas volume for the quadratic function.

When the data are subjected to regression analysis, as an indication of "best fit" to the expected theoretical curves, all theories perform equally well. Table 3 summarizes the results of the regression analyses by comparing coefficients of regression (r^2) and desorption constants (K_t) or "slope" for the various tests; 94% of the regression coefficients are calculated to be above .90.

Analysis by the quadratic theory generally yielded the highest coefficients of regression, but only in tests of the B seam were they significantly higher than by the other two methods. Based on regression coefficients, any of the three methods discussed could be used to model the desorption process.

The K_t value is a measure of the rate at which the total gas volume of the sample desorbs. The higher the K_t value, the faster the total gas volume will desorb from the sample. The K_t value obtained from the power-function analysis commonly is used in European coal mines to determine the risk of gas outbursts in locally within a working seam or for an entire seam (Janas, 1985). From the values listed in Table 3, it is apparent that each seam can and does have its own desorption characteristics.

The area of K_t determination is the area in which the three proposed methods differ most. Using the two log-arithmic functions (power and exponential), K_t values indicate a high gas-outburst potential for the Sub 3 seam, with less outburst potential for the B seam, and slight potential for the Rock Canyon seam. K_t values derived from the quadratic theory indicate that the Rock Canyon seam has the greatest gas outburst risk.

As a standard means of comparison, the standard deviations of the K_t values listed in Table 3 have been normalized by dividing the numerical value of the standard deviation by the average value.

SUMMARY

The total gas content values measured and calculated using the three data-reduction methods generally were within 20% of each other, but the power decay-function theory yielded consistently higher gas content values than the other two methods. The quadratic function fit the field desorption data slightly but not significantly better than the other two functions. The exponential theory resulted in overall better estimating accuracy using measurements taken at the face.

Table 2. Summary of laboratory-measured and estimated in situ (field desorbometer) gas-content values for Sub 3 coal seam, Castle Gate Mine, Utah; Rock Canyon coal seam, Soldier Creek Mine, Utah; and B coal seam, Dutch Creek Mine, Colorado.

Seam/Mine and Sample No.	Gas Content (cc/g)								
	Power Function			Quadratic Function			Exponential Function		
	Lab	In Situ	% Diff.	Lab	In Situ	% Diff.	Lab	In Situ	% Diff.
Sub 3 seam, Castle Gate Mine									
1	8.1	7.9	2	6.6	2.9	56	7.0	5.3	25
2	5.5	4.1	25	4.6	4.6	0	4.7	4.7	0
3	4.8	7.4	–54	4.0	5.4	–35	4.4	6.8	–53
Total	18.4	19.4	–5	15.2	12.9	15	16.1	16.8	–4
Rock Canyon seam, Soldier Creek Mine									
1	5.0	6.4	–28	4.5	5.0	–12	4.7	5.5	–19
2	4.2	3.6	14	3.8	2.6	33	3.8	2.9	23
3	5.4	5.3	9	5.0	4.2	16	5.1	4.6	10
4	4.6	6.0	–31	4.0	4.9	–22	4.4	5.3	–22
5	4.7	6.9	–46	4.0	5.0	–24	4.4	4.7	–7
6	4.9	6.2	–27	4.3	5.2	–21	4.6	4.7	–1
Total	28.8	34.4	–19	25.6	26.9	–5	27	27.7	–3
B seam, Dutch Creek Mine									
1	1.0	1.3	–31	1.0	0.6	44	0.5	0.6	–6
2	3.0	3.1	–4	2.4	2.4	–1	2.3	2.2	3
3	1.8	2.8	–54	1.5	1.4	12	1.0	1.4	–45
4	2.4	2.3	7	2.3	2.0	13	1.9	1.7	11
5	2.0	2.1	–4	1.9	1.6	14	1.6	1.6	0
6	3.4	2.5	25	3.0	2.5	16	2.8	1.8	35
7	3.5	2.9	19	3.3	5.2	–60	3.2	4.7	–47
8	2.7	2.6	5	2.4	3.1	–32	2.2	2.5	–13
Total	19.8	19.6	1	17.8	18.8	–6	15.5	16.5	–6

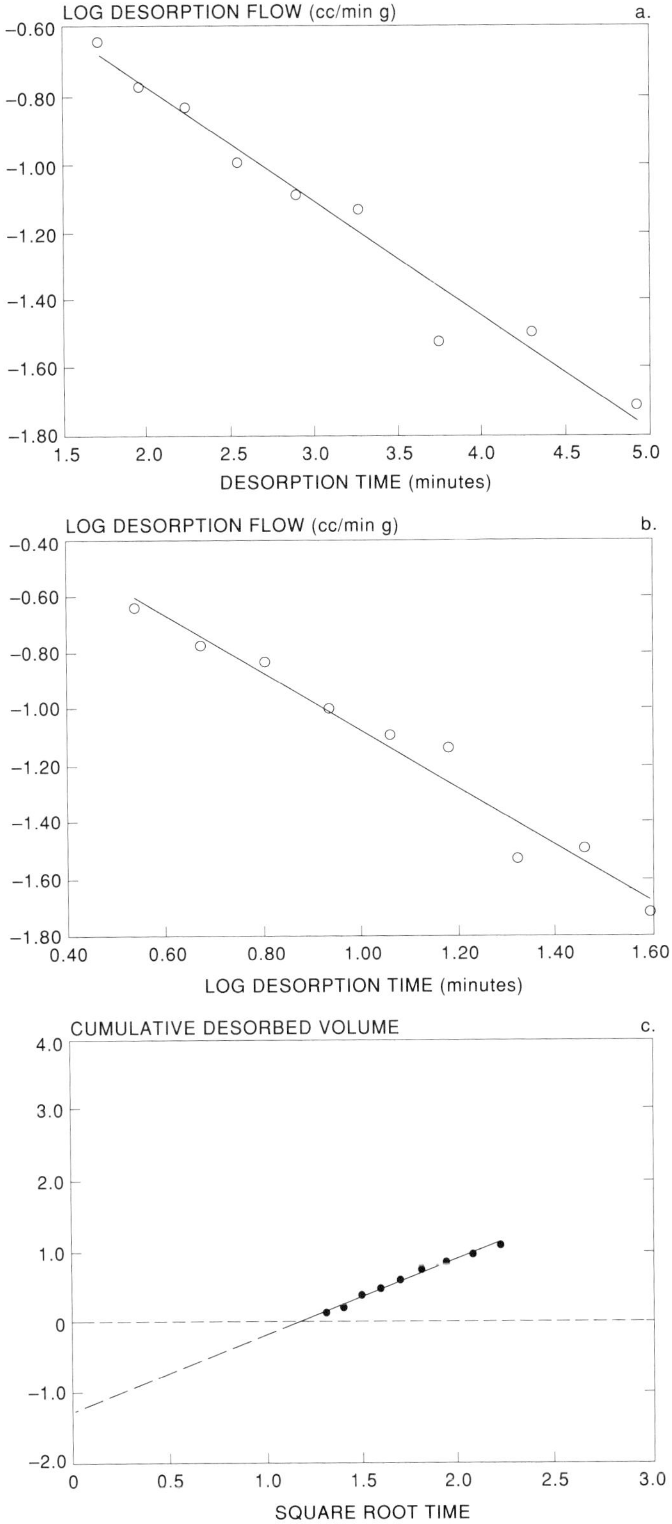

Figure 7. Desorption flow rate vs desorption time for the Rock Canyon seam (hole 1, sample 5), showing "best fit" linear regressions. *a.* Semilog plot according to the exponential decay function. *b.* Log-log plot according to the power decay function. *c.* Cumulative desorbed gas volume vs time according to the quadratic decay function.

ACKNOWLEDGMENTS

Thanks are due to the University of Utah Research Grant Committee for sponsoring this research work. Thanks also go to the U.S. Department of Energy for financial support necessary for preparation and writing this paper.

REFERENCES

Boxho, José, 1980, Firedamp drainage—Handbook for the coalmining industry in the European community: Coal Directorate of the Commission of the European Community, 415 p.

Hucka, B., Sommer, S.N., and Keith, A.C., 1989, Cleat and joint system evaluation and coal characterization of the B-seam, Dutch Creek Mine, Pitkin County, Colorado: Utah Geological and Mineral Survey, unpub. rept., 36 p.

Janas, H., 1979, Improved method for assessing the risk of gas/coal outbursts, *in* Mousset-Jones, P., ed., Proceedings, Second International Mine Ventilation Congress—Reno, Nevada, November 4–8, 1979: Society of Mining Engineers of AIME, p. 22–27.

McCulloch, C.M., 1975, Measuring methane content of bituminous coalbeds: U.S. Bureau of Mines Rept. of Investigations 8043, 22 p.

Plaizier, R.R., 1990, Methane desorption modelling and analysis for selected western coal fields: University of Utah, M.Sc. thesis, 207 p.

Table 3. Regression analysis summary of desorption data for the Sub 3 coal seam, Castle Gate Mine, Utah; Rock Canyon coal seam, Soldier Creek Mine, Utah; and B coal seam, Dutch Creek Mine, Colorado.

Seam/Mine and Sample No.	Desorption Function: Power		Quadratic		Exponential	
	r^2	K_t	r^2	K_t	r^2	K_t
Sub 3 seam, Castle Gate Mine						
1	0.945	2.11	0.948	0.36	0.984	0.71
2	0.996	1.42	0.987	0.56	0.992	0.63
3	0.881	1.71	0.970	0.86	0.947	0.62
Average	0.941	1.75	0.968	0.59	0.974	0.65
N Standard Deviation	0.062	0.20	0.021	0.42	0.025	0.08
Rock Canyon seam, Soldier Creek Mine						
1	0.998	0.59	0.971	0.99	0.981	0.24
2	0.983	0.69	0.999	0.82	0.989	0.17
3	0.993	0.82	0.996	1.11	0.964	0.21
4	0.994	0.94	0.991	1.37	0.981	0.22
5	0.963	1.02	0.993	1.09	0.964	0.34
6	0.992	0.87	0.978	1.21	0.993	0.33
Average	0.987	0.82	0.988	1.10	0.979	0.25
N Standard Deviation	0.013	0.20	0.011	0.17	0.012	0.28
B seam, Dutch Creek Mine						
1	1.000	1.07	0.998	0.08	0.993	0.33
2	0.976	1.42	0.981	0.27	0.988	0.46
3	0.947	1.36	0.988	0.13	0.939	0.28
4	0.971	0.94	0.999	0.22	0.952	0.31
5	0.956	1.09	0.993	0.22	0.943	0.27
6	0.949	1.29	0.987	0.30	0.977	0.47
7	0.912	0.74	0.992	0.65	0.937	0.23
8	0.776	0.97	0.989	0.38	0.723	0.32
Average	0.936	1.11	0.991	0.28	0.932	0.33
N Standard Deviation	0.075	0.21	0.006	0.64	0.093	0.27

Western United States Coal Gas-Content Correlations

Peet M. Sööt[1]

1. Northwest Fuel Development Inc., Portland, Oregon

ABSTRACT

Coal gas-content data from a variety of sources have been compiled to develop correlations between coal properties and gas contents for western United States coals. More than 500 data sets have been gathered for coals in Colorado, Montana, New Mexico, Oklahoma, Utah, Washington, and Wyoming. Preliminary evaluations show a correlation among depth, coal rank (as measured by fuel ratio and/or volatile matter), and gas content, with correlation coefficients as high as 0.75. Gas contents of low-moisture coals are more closely correlated with volatile matter, whereas gas content of other western coals have a higher correlation with depth.

The correlations of western United States coal gas contents are compared with results from coals in other parts of the United States and Australia. Significant differences are apparent with these coals.

INTRODUCTION

Gas content of coal can be measured from samples taken from cores. This can be an expensive and time-consuming task. By developing correlations between coal gas content and other more readily available parameters, one possibly could predict gas content without actually having to make a direct measurement. This also might allow one to review coal analyses from areas where no gas desorptions were performed and approximate possible gas content of the coal.

The development of a predictive correlation requires knowledge of the variables that may affect coal gas content. Realizing the adsorptive nature of methane in coal, one can forecast that pressure and coal characteristics would be primary variables. Coalbed methane literature often presents depth of coal burial as a representation of pressure. This may be suitable, but it should be used with the understanding that significant limitations may apply to this generality. Present burial depth may have little impact on coal gas content. Creedy (1988) has shown in Great Britain, for example, that the stratigraphic depth of coal is a better variable for indicating gas content. Stratigraphic depth provides an historical picture of the coal. Another caveat for the depth/pressure equality is the consideration of geothermal gradient. This would somewhat offset the expected higher gas contents at greater depths. Higher temperatures encountered at greater depths would lead to lower gas contents (Jolly and others, 1968). It is noteworthy that German coal mines today are encountering temperatures up to 140°F (60°C) at depths of 4,600 ft (1,400 m) (Engineering and Mining Journal, 1990). A final qualification is the fact that vertical pressure gradients near the earth's surface are not everywhere equal. One typically encounters gradients somewhat less than that created by simple hydrostatic head. However they are not consistent over wide geographic areas. Additional research may be warranted to detail this relationship as a basis for developing gas-content correlations.

Coal characteristics can be physical or chemical. Physical characteristics, such as pore size and distribution, control the areal dimensions over which methane is adsorbed onto the surfaces of coal particles. The chemical nature of the coal, as represented by proximate analyses for fixed carbon and volatile matter[2], may provide the molecular sites for methane adsorption. These physical and chemical characteristics are not necessarily independent variables. A number of researchers have shown that certain ranks of coal have similar physical characteristics of pore size and total surface area. For example, Gan and others (1972) and Nandi and Walker (1971) have noted that coals with carbon contents between 76% and 83% have total surface areas an order of magnitude greater than coals with carbon contents outside this range. The only exceptions to this conclusion are anthracites, whose carbon contents are greater than 92%.

A review of past research provides further indication of the important independent variables that needed to be considered in this work. Numerous writers have indicated a correlation between gas content and coal rank. Low-rank coals, such as subbituminous, hold only a relatively small amount of gas, typically less than 50 CF/T (1.6 cc/g). The gassiest coals are low-volatile bituminous coals, which can contain as much as 600 CF/T (19 cc/g). The higher rank

2. All references made in this paper to fixed carbon, volatile matter, moisture, ash, and heating value make use of terms described by ASTM methods for coal analysis.

coals, anthracite and meta-anthracite, hold somewhat less than the maximum of low-volatile bituminous coals. The inference of a maximum gas content for bituminous coal is based on *desorption* tests of field samples. Laboratory tests of gas *adsorption* in coal have noted a somewhat opposite effect. Several researchers have noted a minimum gas-adsorption capacity for coals with carbon contents of about 85% (Anderson and others, 1966; Moffat and Weale, 1955; Patching, 1970; Schwarzer, 1983). Gas-adsorption capacity is higher than this minimum in both higher and lower rank coals, but why this distinction occurs is not immediately apparent. It is sufficient for the present study, however, to identify coal rank as an important independent variable without determining the reason for the noted discrepancy.

Moisture content of coal may be a significant variable in gas content. Éttinger and others (1972) first observed that moist coal has a lower gas-adsorption capacity than dry coal. Kissell and Bielicki (1972) showed that adsorption capacity of the Pittsburgh coal seam drops by more than 30% with only 1.5% moisture as compared to coal samples with no moisture. The Pocahontas No. 3 coal seam exhibited a similar tendency. Joubert and others (1973, 1974) showed, however, that this dependence is not linear. Consequently, little, if any, effect between gas content and moisture content may exist above a certain "critical" moisture content.

DATA ACQUISITION

A significant database has been compiled in an attempt to develop the desired gas-content correlations. The U.S. Bureau of Mines has established one of the more extensive databases of coal gas content (Diamond and others, 1986). Because this published database gives ash content as the only coal characteristic, it was necessary to manually search the Bureau's records for more complete proximate analyses. State government reports in the western United States, on the other hand, include numerous proximate analyses. Noted references, such as Doelling and others (1979), Smith (1981), and Tremain (1980, 1984), have yielded a total database of more than 1,000 gas-content values within 540 data sets for coals in the western United States, as follows:

State	Data Sets
Colorado	250
Montana	2
New Mexico	8
Oklahoma	1
Utah	274
Washington	2
Wyoming	9

This study considers only the amount of gas contained in the coal. No effort has been made to identify or account for the type(s) of gas retained, although the database contains some information about gas composition. It is generally accepted that coal adsorbs carbon dioxide preferentially over methane, which has important economic implications to coalbed methane development. Furthermore, some observers have determined that the carbon dioxide in some areas is geologically younger than the methane, which may indicate a displacement of previously adsorbed methane. However, any correlations regarding relative content and age of adsorbed carbon dioxide in coal is beyond the scope of this investigation.

MULTIVARIATE REGRESSION ANALYSIS

A series of multivariate regression analyses were performed on the coal analytical data in an attempt to discover any statistically significant correlation among coal properties and gas content. The six primary independent variables considered in the regressions included coal depth, moisture content, ash content, volatile matter content, fuel ratio (ratio of fixed carbon to volatile matter), and heating value. Depth was considered to be somewhat reflective of the pressure exerted on the coal. The limitations to this assumption are described in some detail above. This variable was transformed for one case. Because laboratory coal adsorption isotherms are best described mathematically by an *exponential* relationship between gas content and pressure, one regression was performed with the natural logarithm of depth as the independent variable. Moisture content was included in light of the moisture effect observed by researchers as noted above. As a diluent in coal, ash adsorbs no gas; therefore, gas contents generally are corrected to an ash-free basis. The inclusion of ash as an independent variable was to determine any additional effect from this component. The remaining variables are indications of coal rank. ASTM results show that most of these variables have some correlation with coal rank. For example, coal rank has a direct correlation with fuel ratio and heating value and an inverse correlation with volatile matter. All rank-related variables are represented in the database on a moisture- and ash-free (maf) basis.

Two dependent variables were considered—total gas, and lost gas plus desorbed gas. Both were calculated on an ash-free basis.

Results

Regression A

Dependent variable:	Total gas, T (cc/g, ash free)
Independent variables:	Coal depth, D (ft); moisture content, M (%); ash content, A (%); volatile matter content, VM (%, maf); fuel ratio, R; heating value, HV (Btu/lb, maf)
Data set:	All western U.S. data sets (540)

The equation that best fits the data is as follows:

$$T = 0.00171D - 0.109M + 0.086A + 0.104VM + 3.86R + 0.000277HV - 12.97$$

The correlation coefficient for this equation is 0.723. The following table, prepared from the correlation matrix, shows the correlation coefficients for the individual variables:

Variable	Correlation Coefficient
Depth	0.513
Fuel ratio	0.467
Moisture	−0.396
Ash	0.360
Volatile matter	−0.322
Heating value	0.107

This shows that western coal gas content is most closely correlated to depth, with fuel ratio almost as strong an independent variable. The gas-content data plotted against depth (Figure 1) show a large amount of scatter and that the other variables are necessary to improve the correlation.

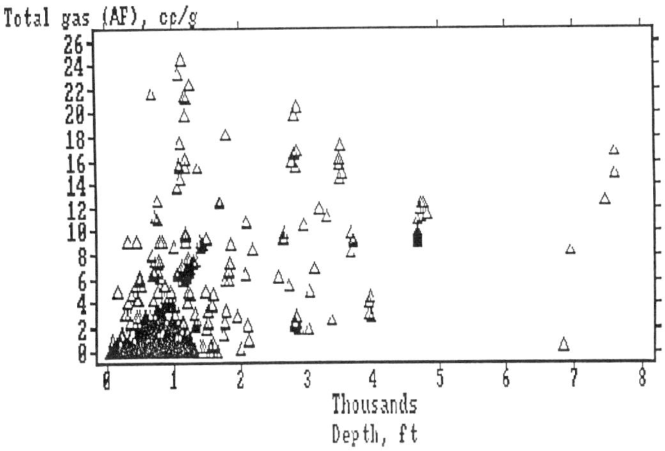

Figure 1. Total gas content vs depth for western United States coals.

The surprising result from the correlation matrix is that heating value has such a low correlation coefficient. The plots in Figures 2 and 3 show that heating value may not reflect coal rank in this database. In Figure 2, for example, heating value in general increases with percentage fixed carbon (which is equal to 100 – *VM*), but much more scatter in the data is apparent than was expected. In Figure 3 the fuel ratio actually declines as heating value increases from 13,200 Btu/lb to 15,000 Btu/lb (30,700 to 34,900 kJ/kg). The extremely low correlation between gas content and heating value is also apparent in the subsequent regressions.

Regression B

Dependent variable:	Total gas, *T* (cc/g, ash free)
Independent variables:	Natural log of depth, ln *D*; moisture content, *M* (%); ash content, *A* (%); volatile matter content, *VM* (%, maf); fuel ratio, *R*; heating value, *HV* (Btu/lb, maf)
Data set:	All western U.S. data sets (540)

The equation that best fits the data is as follows:

$$T = 2.24 \ln D - 0.0468M + 0.086A + 0.129VM + 4.44R + 0.000363HV - 29.42$$

The correlation coefficient for this equation is 0.716. The following table, prepared from the correlation matrix, shows the correlation coefficients for the individual variables:

Variable	Correlation Coefficient
ln depth	0.503
Fuel ratio	0.467
Moisture	−0.396
Ash	0.360
Volatile matter	−0.322
Heating value	0.107

This regression shows that the closest correlation with gas content is with the natural logarithm of depth, but it does

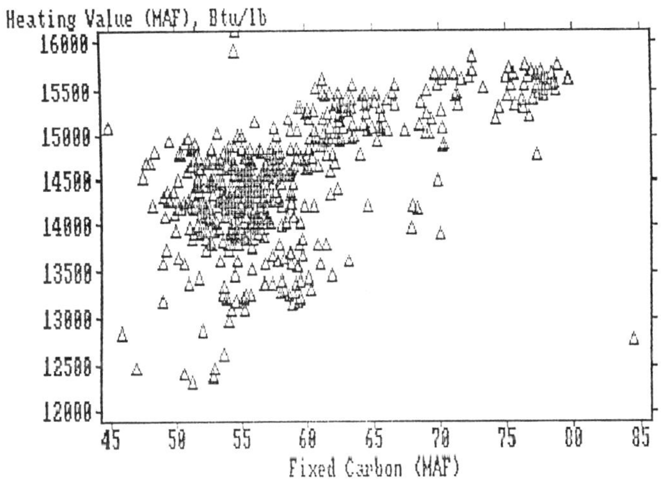

Figure 2. Heating value vs fixed carbon content for western United States coals.

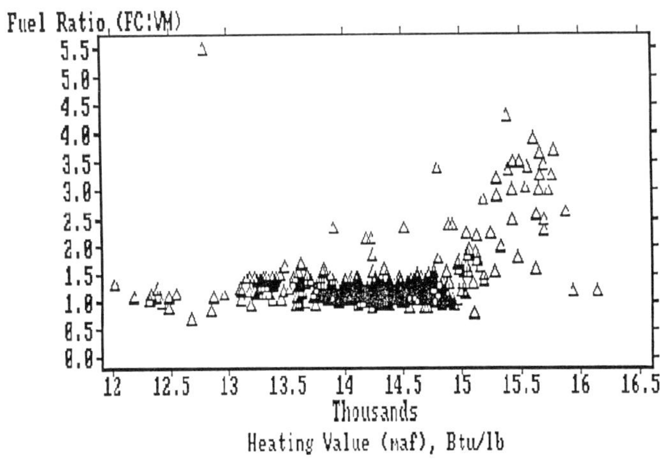

Figure 3. Fuel ratio vs heating value for western United States coals.

not correlate as well with gas content as does the simpler linear variable of depth.

Regression C

Dependent variable:	Total gas, T (cc/g, ash free)
Independent variables:	Coal depth, D (ft); moisture content, M (%); ash content, A (%); volatile matter content, VM (%, maf); fuel ratio, R; heating value, HV (Btu/lb, maf)
Data set:	Western U.S. data sets with moisture content ≤1.5% (88)

The equation that best fits the data is as follows:

$$T = 0.000878D - 0.524M + 0.219A - 0.0589VM + 1.94R + 0.00300HV - 45.79$$

The correlation coefficient for this equation is 0.655. The following table, prepared from the correlation matrix, shows the correlation coefficients for the individual variables:

Variable	Correlation Coefficient
Volatile matter	−0.510
Fuel ratio	0.486
Ash	0.396
Moisture	−0.260
Depth	0.231
Heating value	0.086

This regression shows that the most closely correlated variable for gas content in very low-moisture western coals is volatile matter. This is not surprising in light of the results for eastern coals, as described below. Lower moisture coals in the east show a strong correlation between gas content and volatile matter.

Regression D

Dependent variable:	Total gas, T (cc/g, ash free)
Independent variables:	Coal depth, D (ft); moisture content, M (%); ash content, A (%); volatile matter content, VM (%, maf); fuel ratio, R; heating value, HV (Btu/lb, maf)
Data set:	Western U.S. data sets with moisture content >1.5% (452)

The equation that best fits the data is as follows:

$$T = 0.00212D - 0.0962M + 0.0742A + 0.173VM + 4.65R + 0.000326HV - 18.06$$

The correlation coefficient for this equation is 0.755. The following table, prepared from the correlation matrix, shows the correlation coefficients for the individual variables:

Variable	Correlation Coefficient
Depth	0.644
Moisture	−0.382
Ash	0.281
Fuel ratio	0.195
Heating value	−0.025
Volatile matter	−0.024

In this regression depth most closely correlates with gas content in western coals of typical moisture content. It is considerably more important than the other variables.

Regression E

Dependent variable:	Lost + desorbed gas, LD (cc/g, ash free)
Independent variables:	Coal depth, D (ft); moisture content, M (%); ash content, A (%); volatile matter content, VM (%, maf); fuel ratio, R; heating value, HV (Btu/lb, maf)
Data set:	All western U.S. data sets (540)

The equation that best fits the data is as follows:

$$LD = 0.000513D - 0.140M + 0.0152A + 0.00858VM + 0.627R + 0.0000784HV + 1.38$$

The correlation coefficient for this equation is 0.369. The following table, prepared from the correlation matrix, shows the correlation coefficients for the individual variables:

Variable	Correlation Coefficient
Depth	0.255
Moisture	−0.286
Ash	0.151
Fuel ratio	0.176
Volatile matter	−0.126
Heating value	0.061

With such low values for both the individual variables and the entire equation, one must question whether any correlation exists here at all.

CORRELATION ANALYSIS

The existing database includes data sets for coals in the eastern United States, for which cursory analyses also were run. The results showed that eastern coals apparently are more dependent on volatile matter as the primary variable for describing total gas content. Moisture content in the eastern data sets had an extremely low correlation coefficient with gas content. One might generalize that gas content of low-moisture coals, whether they occur in the east or west, are more reliant on volatile matter than on depth.

The correlations developed from the western United States data sets then were applied to available Australian data (78 coal desorption tests) to see if the equations would be useful in predicting gas contents in another area of the world. The results were not encouraging. Figure 4 compares actual gas contents with gas-content values that would have been predicted by use of the correlation equation from regression A. Predicted values vary over a much more limited range than the actual values, although they do not exceed the actual values at either extreme. The reason for the lack of better correlation may, in part, be the fact that the Australian data did not include actual measured residual gas contents—those values were extrapolated from the desorption plots. Because the correlation for lost gas plus desorbed gas (regression E) for United States values was so poor, no attempt was made to correlate the respective Australian values with the equation from regression E.

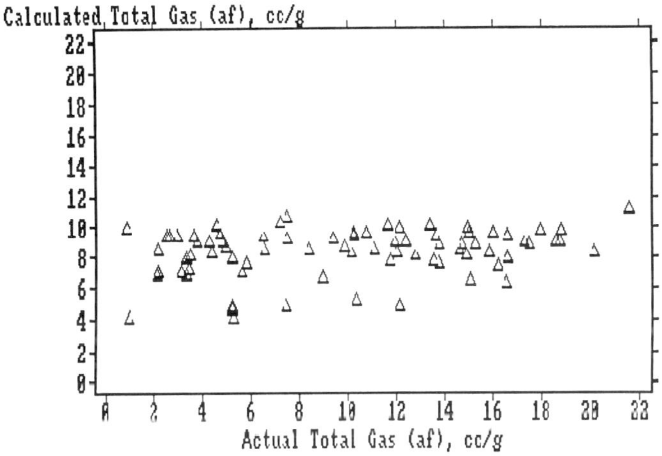

Figure 4. Actual total gas content (ash free) vs calculated gas content (ash free) for Australian coals.

CONCLUSIONS

Western United States coal gas contents are primarily dependent on depth, rank (fuel ratio and/or volatile matter), and moisture content. Correlation equations developed from multiple linear regressions during the present study are not precise enough to predict coal-gas contents in other areas. Additional investigation may help to refine these equations and methodology so that such a technique can be applied on a predictive basis, especially if future correlations are limited to smaller geographic areas.

Commercial coalbed methane most likely is associated with the production of lost gas plus desorbed gas. Wells drilled into coalbeds may never tap the residual gas held in the seam. In that regard, it is disconcerting that the correlation between lost gas plus desorbed gas and the available independent variables was not stronger.

The present study is an initial attempt at correlating gas content with various independent variables. The results indicate that the current assessment can provide some general observations, but it is apparent that the correlations developed here are as yet not accurate enough to be used as a method for predicting coal gas contents.

A significant limitation in this analysis is the lack of accuracy and comparability of the historic data. Most of the data were obtained by using the U.S. Bureau of Mines "direct" method of gas-content determination. Subsequent research has shown these results to be in error because of several factors, including (a) lost-gas extrapolation method, (b) adsorption of air during the methane desorption step, and (c) measurements taken without correcting to standard temperature and pressure (STP) conditions.

RECOMMENDATIONS

A similar statistical analysis should be performed with newer data sets that are known to be consistent in their methodology. This procedure would avoid the variability of data derived by the U.S. Bureau of Mines "direct" method. Additional data should also be acquired for further analysis. Subsequent work that can be performed should include the following tasks:

- Develop correlations for individual coal basins.
- Account for gas quality variations, such as carbon dioxide content.
- Develop stratigraphic depth data for as many of the available data sets as possible.
- Make additional segregations of the data into moisture-content groups to identify any possible occurrence of "critical" moisture contents.
- Improve the current regression analysis by using only a single independent variable for the description of rank and further analyze the database for possible exclusion of outliers.
- Analyze the reasons for poor correlation of lost gas plus desorbed gas in the present study.
- Review the geologic literature to determine if vertical pressure gradients can be defined for data in the existing database.
- Study the residuals from the present regression analyses to see if transformations of some independent variables are warranted.
- Develop an independent variable to reflect the geographic location of data in the existing database.
- Apply the developed correlations to other international data.
- Initiate work on a possible correlation between coal gas quality and coal characteristics and/or geographic location.

REFERENCES

Anderson, R.B., Bayer, J., and Hofer, L.J.E., 1966, Equilibrium sorption studies of methane on Pittsburgh seam and Pocahontas

No. 3 seam coal, *in* Given, P.H., ed., Coal science: American Chemical Society Publication 55, p. 386–399.

Creedy, D.P., 1988, Geological controls on the formation and distribution of gas in British coal measure strata: International Journal of Coal Geology, v. 10, no. 1, p. 1–31.

Diamond, W.P., La Scola, J.C., and Hyman, D.M., 1986, Results of direct-method determination of the gas content of U.S. coalbeds: U.S. Bureau of Mines Information Circular 9067, 95 p.

Doelling, H.H., Smith, A.D., and Davis, F.D., 1979, Methane content of Utah coals, *in* Coal studies: Utah Geological and Mineral Survey Special Studies 49, p. 1–43.

Engineering and Mining Journal, 1990, Cooling hot, deep German coal mines: Engineering and Mining Journal, v. 191, no. 2, p. 27–29.

Éttinger, I.L., Lidin, G.D., Dmitriev, A.M., and Zhupakhina, E.S., 1972, Systematic handbook for the determination of the methane content of coal seams from the seam pressure of the gas and the methane capacity of the coal: USSR Institute of Mining, Laboratory for the Prognosis and Control of Gas Emission [U.S. Bureau of Mines Translation 1505, as cited by Kissell, p. 25]

Gan, H., Nandi, S.P., and Walker, P.L., Jr., 1972, Nature of porosity in American coals: Fuel, v. 51, p. 272–277.

Jolley, D.C., Morris, L.H., and Hinsley, F.B., 1968, An investigation into the relationship between the methane sorption capacity of coal and gas pressure: Transactions, Institution of Mining Engineers, v. 94, p. 539–548.

Joubert, J.I., Grein, C.T., and Bienstock, Daniel, 1973, Sorption of methane in moist coal: Fuel, v. 52, no. 4, p.181–185.

Joubert, J.I., Grein, C.T., and Bienstock, Daniel, 1974, Effect of moisture on the methane capacity of American coals: Fuel, v. 53, no. 3, p.186–191.

Kissell, F.N., and Bielicki, R.J., 1972, An in situ diffusion parameter for the Pittsburgh and Pocahontas No. 3 coalbeds: U.S. Bureau of Mines Rept. of Investigations 7668, 13 p.

Moffat, D.H., and Weale, K.E., 1955, Sorption by coal of methane at high pressures: Fuel, v. 34, no. 4, p. 449–462.

Nandi, S.P., and Walker, P.L., 1971, Adsorption of dyes from aqueous solution by coals, chars, and activated carbons: Fuel, v. 50, p. 345–366.

Patching, T.H., 1970, The retention and release of gas in coal—A review: Canadian Institute of Mining and Metallurgy [CIM] Bulletin, v. 63, no. 703, p. 1302–1308.

Schwarzer, R.R., 1983, Variation in the quantity of methane adsorbed by selected coals as a function of coal petrology and coal chemistry: Draft Final Report under U.S. Department of Energy Contract DE–AC21–80MC 14219, 84 p.

Smith, A.D., 1981, Methane content of Utah coals progress report 1979–1980: Utah Geological and Mineral Survey Open-File Rept. 28, 9 p.

Tremain, C.M., 1980, The coal bed methane potential of the Raton Mesa coal region, Raton basin, Colorado: Colorado Geological Survey Open-File Rept. 80–4, 48 p.

Tremain, C.M., 1984, Colorado desorption samples—Descriptive statistics and gas prediction equations: Colorado Geological Survey Open-File Rept. 84–2, 84 p.

Composition, Distribution, and Origin of Fruitland Formation and Pictured Cliffs Sandstone Gases, San Juan Basin, Colorado and New Mexico*

Andrew R. Scott[1]
W.R. Kaiser[1]
Walter B. Ayers, Jr.[2]

1. Bureau of Economic Geology, The University of Texas at Austin, Austin, Texas
2. Formerly Bureau of Economic Geology, The University of Texas at Austin, Austin, Texas. Now with Taurus Exploration Inc., Birmingham, Alabama

ABSTRACT

One of the most active gas plays in the United States, the Fruitland Formation in the San Juan basin, contains an estimated 43 to 49 TCF (1,217 × 10^9 to 1,387 × 10^9 m^3) of coal gas in place. The Pictured Cliffs Sandstone, which underlies the Fruitland Formation, is a mature gas play that has produced more than 3 TCF (85 × 10^9 m^3) of natural gas. We have studied the composition of gases from these units to evaluate gas origin, migration, and possible communication between formations. Gas analyses from more than 470 Fruitland and 5,500 Pictured Cliffs wells were used. Histograms, ternary diagrams, and maps were made to evaluate variations among gases by producing lithology (coal and/or sandstone), formation, pressure regime, and regional occurrence.

The distribution of Fruitland coal gases is controlled by basin hydrology and coal rank. Gases produced from coalbeds in the overpressured area of the basin are chemically drier (C_1/C_{1-5} values greater than 0.97) and contain more carbon dioxide (commonly greater than 6%) than coal gases from the underpressured area (C_1/C_{1-5} values generally less than 0.94 and carbon dioxide content less than 3%). The wettest coal gases coincide with the liquid hydrocarbon-generating stage (vitrinite reflectance values from 0.50% to 0.78%) of the hydrogen-rich Fruitland coal. Gases from underpressured Fruitland coalbeds, Fruitland sandstone, and Pictured Cliffs Sandstone have similar compositions, indicating that gas composition alone cannot be used to determine gas origin in all parts of the basin. Analysis of gas composition suggests that Pictured Cliffs Sandstone gases may be derived from both the underlying Lewis Shale and locally from overlying Fruitland coalbeds.

The abrupt transition of gas composition across the overpressured/underpressured boundary, methane and carbon dioxide $\delta^{13}C$ values, and isotopic values of bicarbonate from formation water suggest that some Fruitland coalbed methane is biogenic in origin. Desorption of carbon dioxide from coal surfaces and exsolution from formation water as pressure decreases during production may account for the high carbon dioxide content (greater than 10%) of some coal gases in the northern part of the basin.

INTRODUCTION

Fruitland coalbeds and sandstones and Pictured Cliffs sandstones in the San Juan basin contain significant amounts of nonassociated natural gas. Fruitland coalbeds alone contain an estimated 43 to 49 TCF (1,217 × 10^9 to 1,387 × 10^9 m^3) of gas in place (Ambrose and Ayers, 1991), and annual production in 1990 is predicted to exceed 200 BCF (5.660 × 10^9 m^3). Cumulative gas production from Pictured Cliffs Sandstone reservoirs through 1988 was 3.1 TCF (87.7 × 10^9 m^3). Ultimate recovery is estimated to exceed 4.5 TCF (127.4 × 10^9 m^3) (R. Hugman, 1990, pers. comm.).

Pictured Cliffs shoreline sandstones overlie and intertongue with the marine Lewis Shale, and the base of the Pictured Cliffs Sandstone conventionally is placed at the base of an upward-coarsening sequence (Figure 1). The Pictured Cliffs Sandstone is overlain by the coal-bearing Fruitland Formation, which represents the continental facies deposited landward of the shoreline deposits. Temporary landward shifts of the paleoshoreline during an overall

* Published by permission of the Director, Bureau of Economic Geology, The University of Texas at Austin.

marine regression during the Late Cretaceous resulted in intertonguing of upper Pictured Cliffs sandstones (UP1 in Figure 1) with the Fruitland Formation in the northern part of the basin (Ambrose and Ayers, 1991). The Fruitland Formation is overlain conformably by the Kirtland Shale throughout most of the basin, but the Ojo Alamo Sandstone unconformably overlies or truncates the Fruitland Formation in the east (Ambrose and Ayers, 1991).

Fruitland and Pictured Cliffs gases are predominantly methane with variable amounts of heavier hydrocarbons (ethane, propane, butane, and pentane), carbon dioxide, and nitrogen. Natural gases are generated from organic matter during thermal maturation through biogenic, early thermogenic, and late thermogenic processes. Catagenesis is the process by which organic material is altered, primarily as the result of increasing temperature, to produce thermogenic gases and liquid hydrocarbons (Hunt, 1979). Biogenic methane is produced at relatively low temperature through the metabolic activity of methanogenic bacteria (methanogenesis). Nonassociated gases in Fruitland coalbeds, Fruitland sandstones, and Pictured Cliffs sandstones have been interpreted to be mainly thermogenic in origin (Rice, 1983). Early thermogenic gases are formed before and during the main stage of oil generation. Heavier hydrocarbons, including chemically wet gas and oil, are derived from the organic material when thermal maturation reaches the oil-generating stage. Dry thermogenic gases are released from terrestrial organic matter or are generated by the cracking of heavier hydrocarbons formed during the oil-generating stage in sapropelic or hydrogen-rich organic matter as temperature increases with burial depth.

Carbon dioxide can be generated by bacterial activity and during catagenesis. Methanogenic bacteria are capable of metabolizing organic acids through the process of acetate fermentation to produce carbon dioxide and methane (Carothers and Kharaka, 1980; Whiticar and others, 1986). Carbon dioxide also is derived from oxygen-bearing functional groups of the organic matter and is released during decarboxylation at relatively low levels of thermal maturity (Stach and others, 1975) before significant quantities of liquid hydrocarbons and thermogenic gases are generated. At higher temperature, decarboxylation of organic acids can generate both carbon dioxide and methane (Carothers and Kharaka, 1980). Molecular nitrogen in Fruitland and Pictured Cliffs gases probably is derived from the oxidation of ammonia (Rohrback and others, 1983) evolved from organic matter during catagenesis.

Variation in composition among gases produced from Pictured Cliffs sandstones, Fruitland sandstone and coal gases, or regionally within one unit, may prove useful in establishing gas origin and migration pathways and in performing cost analyses. Understanding the origin of gases is important in developing exploration strategies by allowing more favorable drilling sites to be selected based on gas-migration pathways. Furthermore, nonhydrocarbon components may be corrosive (carbon dioxide), and inert

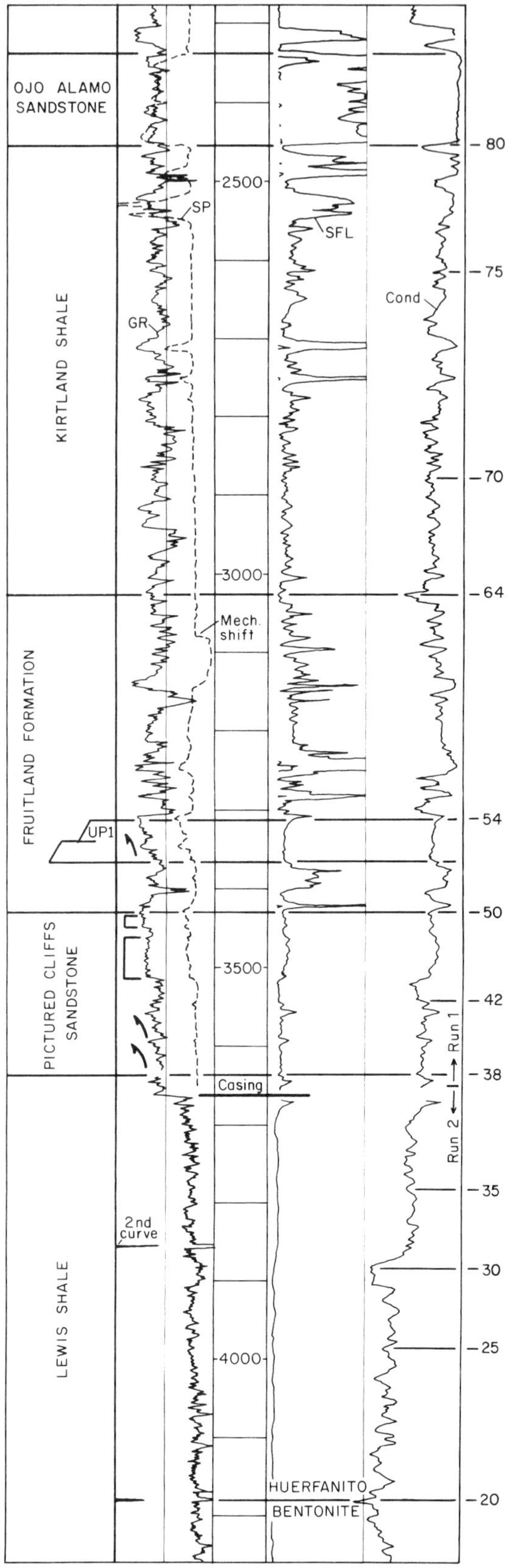

Figure 1. Type log of Upper Cretaceous stratigraphy in the San Juan Basin. An upper Pictured Cliffs sandstone (UP1) intertongues with the Fruitland Formation. From Ayers and others (1991).

gases (nitrogen) may significantly lower the heat content of the gas; both must be removed before the gas can enter the pipeline system.

The objectives of this study were to (1) determine any significant differences in gas composition among Fruitland coalbed, Fruitland sandstone, and Pictured Cliffs sandstone reservoirs; (2) determine any regional variations in gas composition among the same reservoir types; (3) compare Fruitland coal-gas composition with coal rank and basin hydrodynamics to evaluate factors controlling gas composition; and (4) determine if minor variations in gas composition can be used to determine the origin of these gases and possibly their migration pathways.

METHODOLOGY

Gas analyses from more than 5,500 Pictured Cliffs Sandstone wells and about 470 Fruitland Formation wells were used in this study. These data were obtained from pipeline companies, published reports, and Gas Research Institute cooperative wells. Gas samples generally were collected from the meter run, downstream from production equipment, or from the wellhead in nonproducing wells; collecting samples directly from the meter reduces the possibility of air contamination. Samples were analyzed with a gas chromatograph using a silica column.

Completion lithologies (sandstone and/or coal) for Fruitland wells were confirmed by the operator or were determined from completion reports and geophysical logs. If logs were unavailable for a well, then equivalent perforation intervals on well logs from adjacent wells were used to determine lithology. Wells whose producing zone and producing lithology could not be determined were designated as Basin Fruitland coal completions based on information provided by Dwight's Coal Bed Methane Report (1990a, 1990b). Wells were omitted if gases were commingled, if the producing lithology was uncertain, or if production could have been from both coalbeds and sandstone. For Pictured Cliffs Sandstone completions, a limited number of wells were examined to verify that the perforation interval actually was in the Pictured Cliffs Sandstone.

Gas contents discussed in this report are given in mole percent. The ratio of methane (C_1) to total hydrocarbons (C_{1-5}) was used to determine a gas-dryness index. We define gas dryness according to the following ranges in C_1/C_{1-5} values:

Dryness	C_1/C_{1-5}
Very dry	>0.99
Dry	0.94–0.99
Wet	0.86–0.94
Very wet	<0.86

These gas-dryness ranges are similar to those reported by Hanson (1990).

Isotopic data for methane, carbon dioxide, and water were collected from published reports. Bicarbonate is the dominant species in Fruitland Formation water, and $\delta^{13}C$ values for total dissolved carbonate species will herein be referred to as bicarbonate $\delta^{13}C$ values. All isotopic values are relative to the PDB (Peedee belemnite) standard.

COMPOSITION AND DISTRIBUTION OF FRUITLAND AND PICTURED CLIFFS GASES

Previous studies have suggested that Fruitland Formation and Pictured Cliffs sandstone gas composition and distribution are controlled by coal rank and source rock type (Rice and others, 1988, 1989). Furthermore, overpressuring in the Fruitland Formation was inferred to represent fossil geopressure associated with basin subsidence and thermal maturation, rather than basin hydrodynamics (Decker and Horner, 1987). However, our study suggests that compositional variations in Fruitland coal gases and regional distribution of the gases result from both thermal maturity and basin hydrology. The Fruitland Formation is a regional aquifer in which recharge occurs in the north, where thick northwest-trending coalbeds crop out along the Hogback monocline (Kaiser and others, 1991a). Because of vertical confinement and basinward pinchout of coalbeds and/or localized faulting (Ambrose and Ayers, 1991), the Fruitland is regionally overpressured in the northern part of the basin (Figure 2) (Kaiser and others, 1991a).

Fruitland coal rank ranges from subbituminous to low-volatile bituminous (Figure 3). Low-volatile bituminous coal occurs locally in the northern part of the basin. From this area of high maturity, coal rank decreases abruptly toward the northern basin margin. Along the Hogback monocline high-volatile C to high-volatile A bituminous isoreflectance lines parallel the basin margin. Southward, isoreflectance lines generally parallel the basin's structural contours. Ambrose and Ayers (1991) hypothesize the existence of a northwest-trending structural hingeline consisting of a normal fault zone downthrown on the northeast. Significantly, offset of coalbeds across this fault zone, and the southwestward pinchout of coalbeds, may have inhibited the southwestward migration of fluids derived from the thermally more mature area in the north. This hingeline would explain the closely spaced isoreflectance contours in the north (Figure 3, vitrinite reflectance values greater than 0.65%), as well as the southwestern boundary of overpressuring.

Previous studies have concluded that Fruitland coal gases are chemically distinct from Fruitland sandstone and Pictured Cliffs sandstone gases (Rice and others, 1988, 1989). Differences in gas composition between sandstone and coal reservoirs were inferred to result from different organic matter sources—Type III terrestrial kerogen in coalbeds produced chemically dry gas, whereas dispersed Type III kerogen in adjacent shale generated wetter gas. However, these studies were based on relatively few

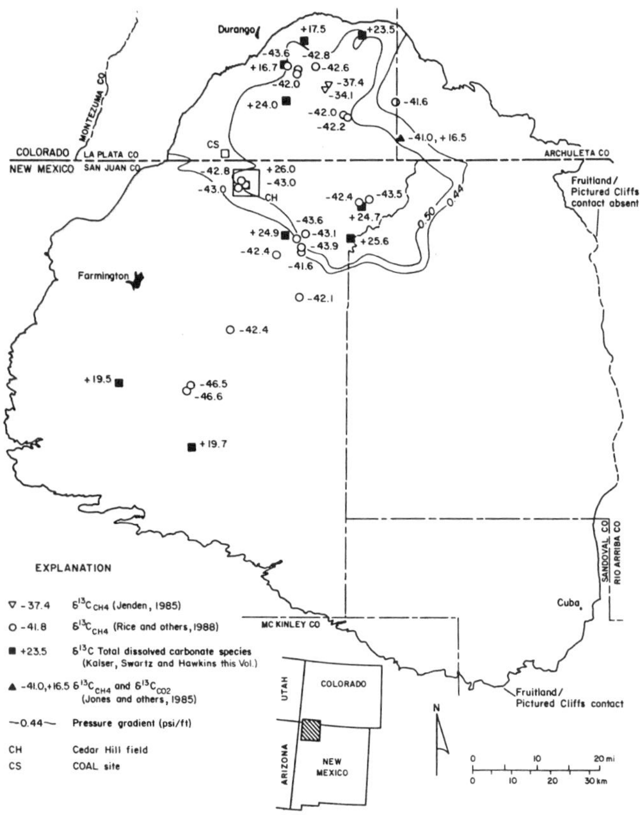

Figure 2. Regional pressure gradient and distribution of $\delta^{13}C$ values of methane and carbon dioxide produced from Fruitland coalbeds and $\delta^{13}C$ values of total dissolved carbonate species in Fruitland Formation waters, San Juan basin. The Fruitland is overpressured relative to hydrostatic pressure in the north (pressure gradient of 0.50 psi/ft); underpressured areas have pressure gradients less than 0.44 psi/ft.

samples, and extrapolation of their conclusions throughout the basin is not justified or supported by a regional database. In this study we show that gas composition varies more with coal rank and pressure regime than with producing lithology.

Carbon dioxide, ethane, and nitrogen were evaluated to characterize variations in gas composition among Fruitland coal, Fruitland sandstone, and Pictured Cliffs sandstone. These three components, rather than methane, were selected so that minor variations in gas composition could be recognized. If methane had been used as an end member, gas compositional data points would have clustered in one corner of the ternary diagram. Furthermore, a good correlation is apparent between methane content and these three components (Figure 4) and between ethane and heavier gas components (C_{3-5} hydrocarbons).

Coal gases from the overpressured part of the basin follow gas compositional trend A, which has a correlation coefficient of .99 (Figure 4a). These gases are composed predominantly of methane and carbon dioxide and only minor amounts of other hydrocarbon gases. Sandstone and coal gases from the northern part of the basin follow different trends. However, pressure and hydrochemical data indicate that Fruitland sandstones and coalbeds in the overpressured area are in regional hydraulic communication (Kaiser and others, 1991a, 1991b). Therefore, differences between sandstone and coal gas compositional trends in the north probably reflect the way gases are stored in, and released from, these reservoirs rather than noncommunication between reservoirs.

Fruitland coal and sandstones gases from the underpressured area generally follow trend B (correlation coefficient of .91; Figure 4b), which suggests possible communication between sandstone and coal in the southern part of the basin.

The relatively minor variation in coal-gas composition with time (Hale and Firth, 1988) suggests that the length of production should not be a major factor affecting gas compositional data and that gas compositional trends recognized in this study probably are valid.

Fruitland Formation Gases

Previous studies attributed the presence of wet gases produced from Fruitland coals to unspecified completion practices rather than to production directly from the coal (Rice and others, 1988, 1989). However, our study indicates that Fruitland coal is the source of wet coal gas. Further-

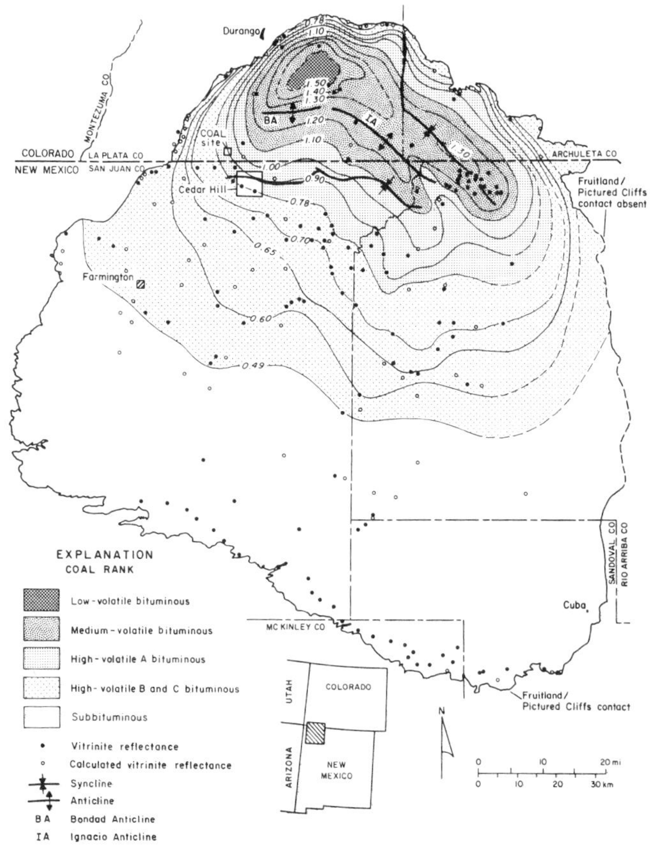

Figure 3. Fruitland Formation coal rank map. From Scott and others (1991). Structural axes are from the structure map of the Huerfanito Bentonite Bed (Ayers and others, 1991).

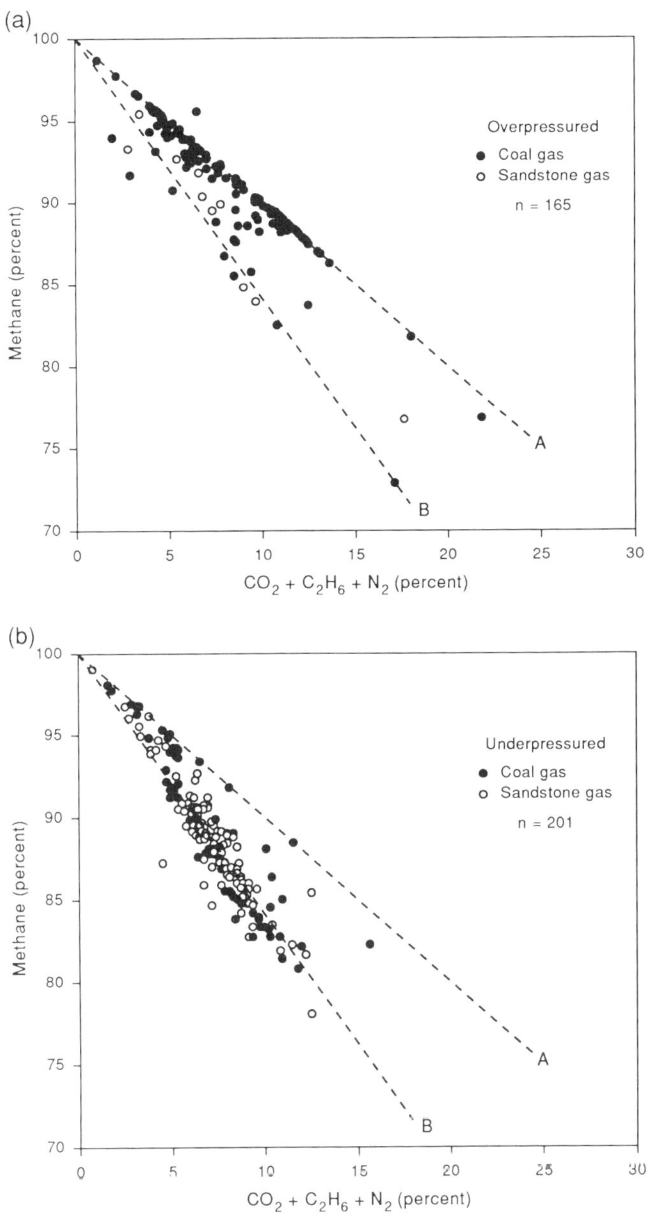

Figure 4. Methane content vs sum of carbon dioxide, ethane, and nitrogen content in Fruitland coal gas and Fruitland sandstone gas in (*a*) overpressured and (*b*) underpressured areas of the San Juan basin.

more, Fruitland coal and sandstone gas compositions are similar in the southern part of the basin, indicating that gas composition alone cannot be used to determine gas origin in this area.

Fruitland Sandstone Gases

Gases produced from Fruitland sandstone have values of C_1/C_{1-5} ranging from 0.80 to 1.00, with a mean of 0.90 (Table 1). Sandstone gases from the underpressured area are wetter than those from the overpressured area (mean C_1/C_{1-5} values of 0.90 and 0.94, respectively). Wet gases (C_1/C_{1-5} values commonly less than 0.91) occur in a northwest-trending band in the west-central part of the basin (Figure 5). The ethane content of sandstone gas ranges from less than 1% to nearly 11%, with gases from the underpressured region containing significantly more ethane than those from the overpressured region (Table 1, Figure 5).

Gases from overpressured and underpressured sandstone reservoirs have similar ranges of carbon dioxide content (Table 1). However, sandstone gases in the north generally contain more than 1% carbon dioxide (Figure 6) and have a mean carbon dioxide content of 2.2%.

Ternary diagrams of Fruitland sandstone gas composition reflect the relatively high ethane content (Figure 7). Gases from overpressured sandstones and UP1 have similar gas compositions and generally contain more carbon dioxide and less ethane than gases from underpressured Fruitland sandstones. Gases from some wells completed in sandstone but exhibiting coal-decline production behavior (Kaiser and others, 1991a) have proportionately higher carbon dioxide and/or nitrogen contents (Figure 7), indicating that these sandstones may have derived at least some gas from adjacent coalbeds. However, other wells with coal-decline production behavior do not follow this trend.

Fruitland Coal Gases

Analysis of Fruitland coal gases showed that although many coal gases are chemically dry, a significant number

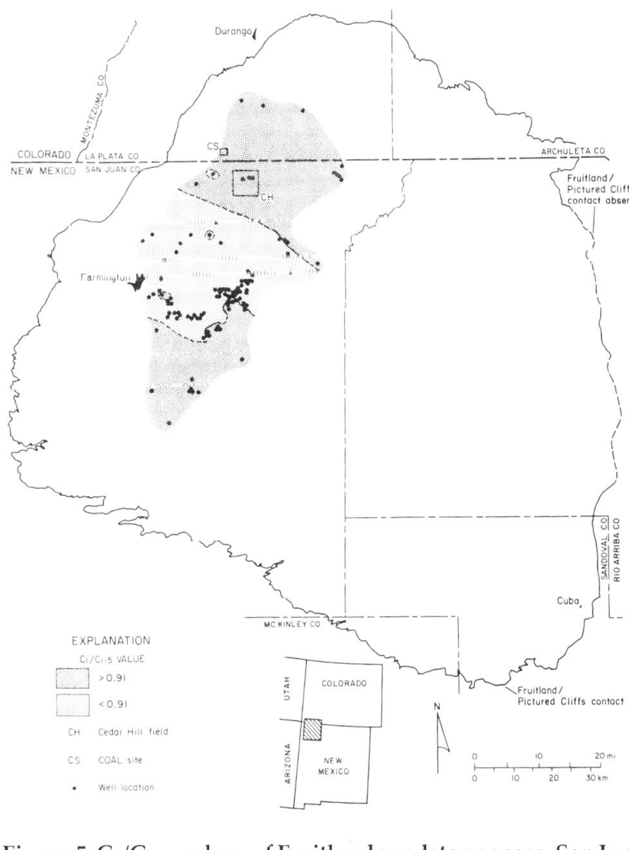

Figure 5. C_1/C_{1-5} values of Fruitland sandstone gases, San Juan basin.

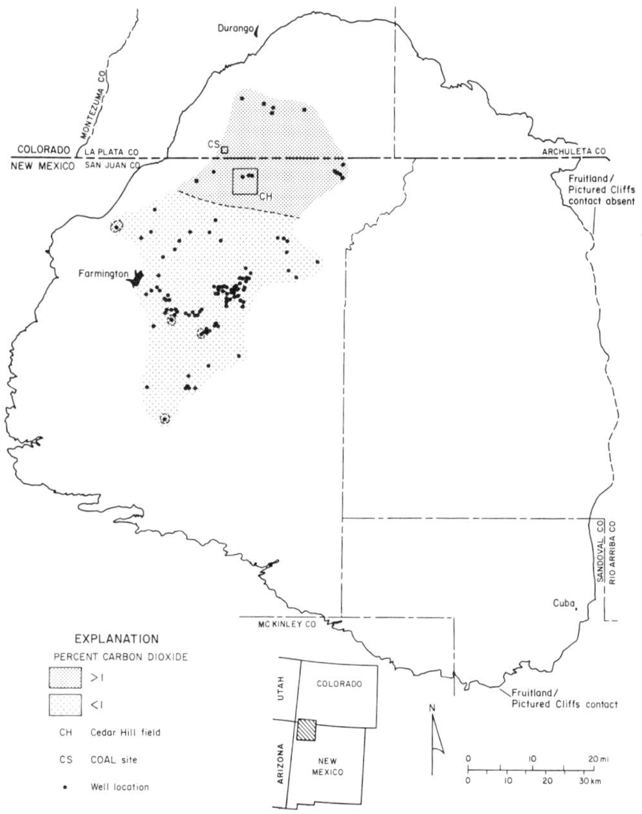

Figure 6. Carbon dioxide content of Fruitland sandstone gases, San Juan basin.

are chemically wet. The range of C_1/C_{1-5} values (0.77 to 1.00) is similar to that of Fruitland sandstone gases (Table 1). A west-northwest-trending band of relatively wet coal gases (Figure 8) occurs in the northern part of the underpressured southern basin. Coal gases from the overpressured area are chemically drier (mean C_1/C_{1-5} value of 0.98) than those from underpressured areas, which have a mean C_1/C_{1-5} value of 0.92. However, gases from underpressured coal and sandstone have similar values of mean C_1/C_{1-5} (0.92 and 0.90, respectively). Ternary diagrams also indicate that gases from overpressured coalbeds are generally drier and have significantly more carbon dioxide than gases from underpressured coalbeds, which are characterized by a higher proportion of ethane (Figure 9). Coal gases from the transition zone, located between overpressured and underpressured areas, have compositions between those of the two pressure regimes.

Carbon dioxide content in Fruitland coal gases ranges from less than 1% to more than 13% (Table 1). Coal gases with the highest carbon dioxide content (greater than 10%) are from the north-central part of the basin (Figure 10), which is further characterized by very dry gases (C_1/C_{1-5} values of 1.00) and highly productive coalbed methane wells. From here, carbon dioxide content decreases gradually northwestward and very abruptly southward. Coal gases in the south generally contain less than 1% carbon dioxide. Coal gases typically contain more carbon dioxide than either underpressured or overpressured sandstone gases (Table 1).

Histograms of coal-gas composition indicate major differences between overpressured and underpressured gases (Figure 11). Underpressured coal gases generally have C_1/C_{1-5} values less than 0.93 and carbon dioxide content less than 2%. Furthermore, gases from overpressured, thermally more mature coalbeds are significantly drier (C_1/C_{1-5} values generally greater than 0.97) and have more carbon dioxide (greater than 3%) than gases produced from underpressured coalbeds. The compositions of gases produced from underpressured coalbed and sandstone reservoirs are similar—both have relatively high ethane and generally minor carbon dioxide and nitrogen contents (Figures 7b and 9c). Histograms of coal and sandstone gases from the underpressured part of the basin are similar; these gases are chemically wet to very wet and contain only minor carbon dioxide (Figure 12).

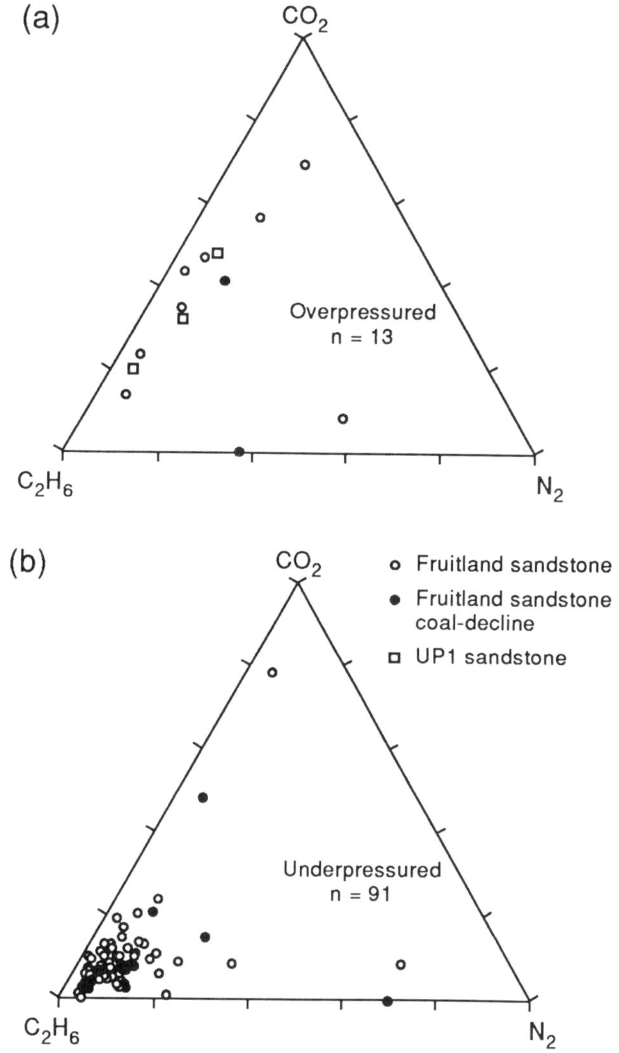

Figure 7. Ternary diagrams of ethane, carbon dioxide, and nitrogen composition of gases produced from Fruitland sandstone, Fruitland sandstone showing coal-decline behavior, and Pictured Cliffs sandstones. *a*. Overpressured area. *b*. Underpressured area.

Table 1. Composition of natural gases in Fruitland sandstone, Fruitland coalbeds, and Pictured Cliffs sandstones, San Juan basin.

	Fruitland Sandstone			Fruitland Coal			Pictured Cliffs Sandstone
Component	All (n = 101)	UP* (n = 91)	OP+ (n = 10)	All++ (n = 288)	UP* (n = 111)	OP+ (n = 157)	All (n = 1,553)
C_1/C_{1-5}							
x	0.90	0.90	0.94	0.96	0.92	0.98	0.88
S	0.04	0.03	0.04	0.05	0.04	0.03	0.04
C	0.04	0.04	0.05	0.05	0.05	0.03	0.04
min	0.80	0.80	0.86	0.77	0.83	0.85	0.75
max	1.00	1.00	1.00	1.00	1.00	1.00	1.00
Ethane (%)							
x	5.7	6.2	3.8	2.5	4.8	1.1	6.7
S	2.0	1.86	2.00	2.69	2.35	1.70	1.83
C	0.35	0.30	0.53	1.08	0.49	1.55	0.27
min	0.3	0.3	0.6	0	<0.1	0	<0.5
max	10.7	10.7	6.4	11.9	9.7	8.5	14.8
Carbon dioxide (%)							
x	0.8	0.5	2.2	4.5	1.4	6.4	0.7
S	0.73	1.86	0.87	3.90	1.54	3.71	0.64
C	0.93	0.31	0.41	0.87	1.10	0.58	0.91
min	0	0	<0.1	<0.1	<0.1	0.2	<0.1
max	3.4	3.4	3.2	13.5	11.4	13.5	10.1
Nitrogen (%)							
x	0.8	0.6	1.8	0.7	0.8	0.7	0.7
S	1.28	0.90	2.95	1.56	1.32	1.79	0.52
C	1.60	1.50	1.64	2.23	1.65	2.56	0.72
min	0.1	0.1	0.3	0	0	0	0
max	9.7	8.5	9.7	13.3	11.2	13.3	5.3

x = Arithmetic mean. * UP = Underpressured area.
S = Standard deviation. + OP = Overpressured area.
C = Coefficient of variation. ++ Includes 20 samples from transition zone.

Pictured Cliffs Sandstone Gases

Fruitland sandstone, Fruitland coal, and Pictured Cliffs Sandstone gases can have similar C_1/C_{1-5} values (Figures 13 and 14), indicating that gas composition alone cannot be used to determine gas origin in all areas of the basin. Pictured Cliffs gases show a normal distribution of C_1/C_{1-5} values (Figure 14) from 0.75 to 1.00 (mean 0.88); they may contain nearly 15% ethane (Table 1). The wettest Pictured Cliffs gases occur in the southeast (Figure 15). Elongate trends of relatively dry gas (C_1/C_{1-5} values greater than 0.85) parallel depositional strike in the southeastern corner of the basin, suggesting that localized variations in gas composition may be the result of (1) migration effects across shale/sandstone contacts, which would enrich the gases in methane relative to other components (Leythaeuser and others, 1979); or (2) migration of chemically dry gases southeastward through more permeable sandstone.

The ranges of carbon dioxide and nitrogen contents of Pictured Cliffs Sandstone gases also are similar to those in Fruitland sandstone gases. However, Fruitland coal gases generally are chemically drier and contain more carbon dioxide than Pictured Cliffs Sandstone gases (Table 1). Carbon dioxide content of Pictured Cliffs Sandstone gases generally is less than 1% (Figure 16). Carbon dioxide content is highest in the Cedar Hill field area (Figures 10 and 16); gases from the western part of the basin are chemically drier and contain more carbon dioxide (Figures 15 and 16). These areas of high carbon dioxide content may represent reservoir communication between Pictured Cliffs sandstones and Fruitland coalbeds and/or commingled production from the respective strata.

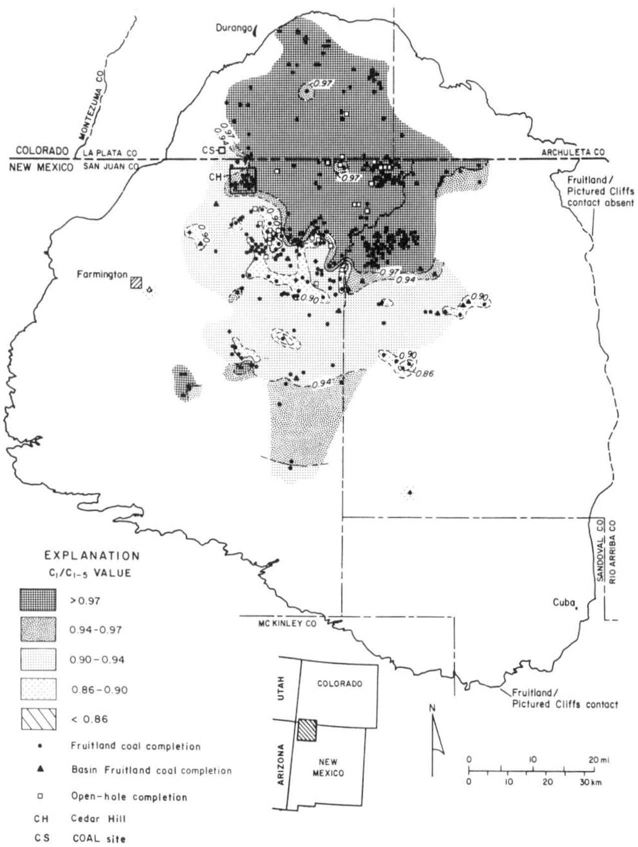

Figure 8. C_1/C_{1-5} values of Fruitland coal gases, San Juan basin.

RELATIONSHIP OF COAL-GAS COMPOSITION TO THERMAL MATURITY AND PRESSURE REGIME

Previous studies did not consider the influence of basin hydrodynamics on coal-gas composition (Rice and others, 1988, 1989). Interpretations of coal-gas composition and origin were based on the assumption that coal rank alone controlled gas composition. Coal typically is composed of type III kerogen, which traditionally is considered to be a major source of chemically dry gas; therefore, Fruitland coalbeds were assumed to generate only very dry gases (C_1/C_{1-5} values greater than 0.99; Rice and others, 1988; Figure 7 in Rice and others, 1989). However, Hanson (1990) reported that although many desorbed coal gases are dry to very dry, some are wet to very wet, with gas wetness depending on coal rank and maceral composition. Furthermore, Fruitland coal is hydrogen rich and may be the source of condensate produced from the coals (Rice and others, 1989).

Thermogenic gases, formed during catagenesis, become chemically drier with increasing temperature. Predictably, gases from the thermally more mature coals in the northern part of the basin are chemically dry to very dry. However, regional distribution of these gases (Figure 8) suggests that coal rank is not the only factor controlling the chemical composition of the gas. The eastern boundary of chemically dry coal gases crosses coal rank trends (Figure 3).

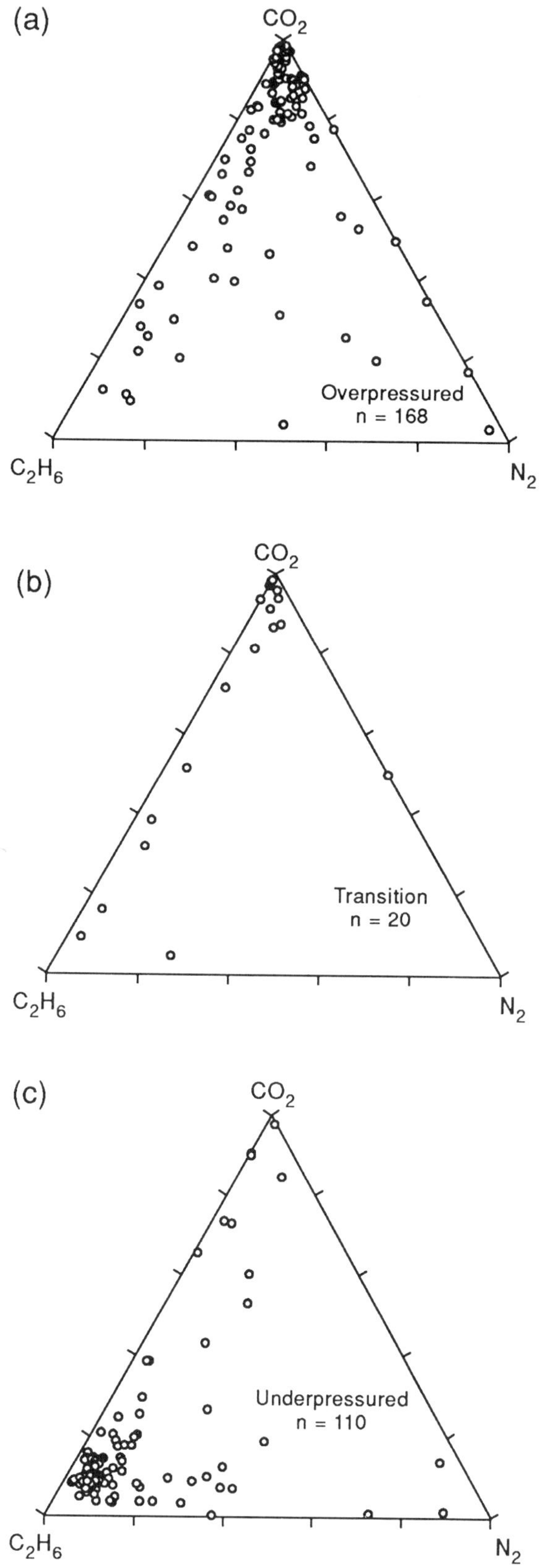

Figure 9. Ternary diagram of ethane, carbon dioxide, and nitrogen composition of Fruitland coal gases. *a.* Overpressured area. *b.* Transition zone between overpressuring and underpressuring. *c.* Underpressured area.

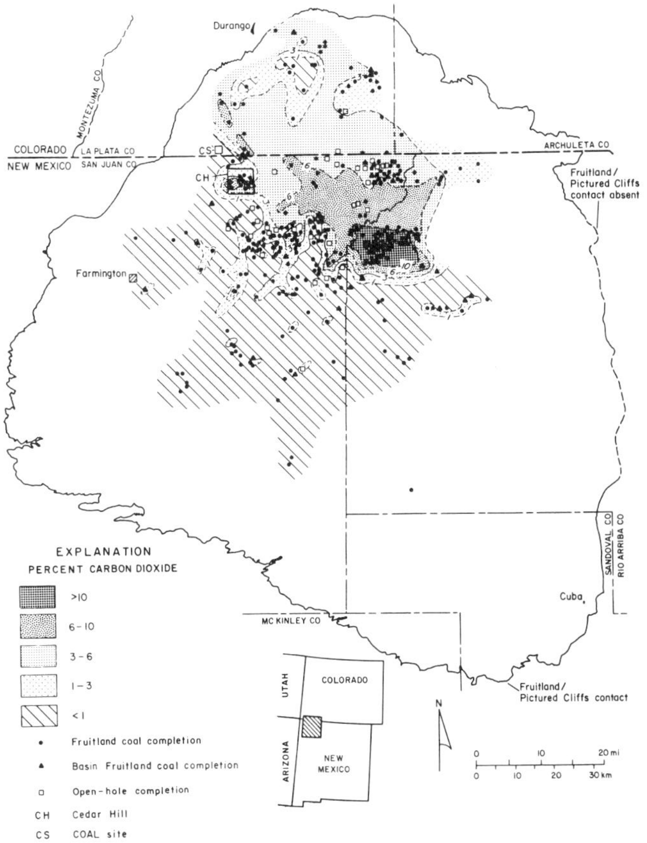

Figure 10. Carbon dioxide content of Fruitland coal gases, San Juan basin. Highest carbon dioxide content (greater than 6%) coincides with highest bottom-hole pressures (greater than 1,600 psi; Kaiser and others, 1991a) and regional overpressure (see Figure 16).

Furthermore, a gradational contact between very dry and wet coal gases paralleling vitrinite reflectance trends would be expected if coal rank alone controlled gas composition. Instead, the transition from very dry (C_1/C_{1-5} value of 1.00) to wet (C_1/C_{1-5} value of 0.87) coal gases is abrupt, occurring over a distance of less than 1.5 miles (2 km) along some parts of the overpressure/underpressure boundary (Figure 8). In fact, gas-composition trends coincide with the pressure regime in the Fruitland Formation (Kaiser and others, 1991a) rather than with coal rank (Figures 2 and 3). Therefore, basin hydrology is believed to be a major factor controlling coal-gas composition. Basinward migration of sufficient quantities of methane in Fruitland Formation waters could explain the presence of chemically dry gases along the overpressure/underpressure boundary, and some migration probably has occurred. However, methane is not very soluble in water, and large volumes of water moving through the coal would be required to transport significant amounts of methane. Furthermore, simple migration of methane from higher rank coals southward in formation water does not explain gas and water isotopic data. These isotopic data and the origin of the gases will be discussed below.

Previous studies have suggested that carbon dioxide content in coal gas decreases with increasing coal rank (Rice and others, 1988, 1989, 1990). However, Fruitland coal gases with the highest carbon dioxide content occur in the northern part of the basin where coal rank is high-volatile A bituminous. Gases from low- and medium-volatile bituminous coals in the north generally have lower carbon dioxide content (less than 3%) than some lower rank coals. However, carbon dioxide content does not follow vitrinite isoreflectance contours in that area (Figures 3 and 10). Carbon dioxide content coincides more with formation pressure (Kaiser and others, 1991a) and regional overpressure in the Fruitland than with coal rank (Figures 2,

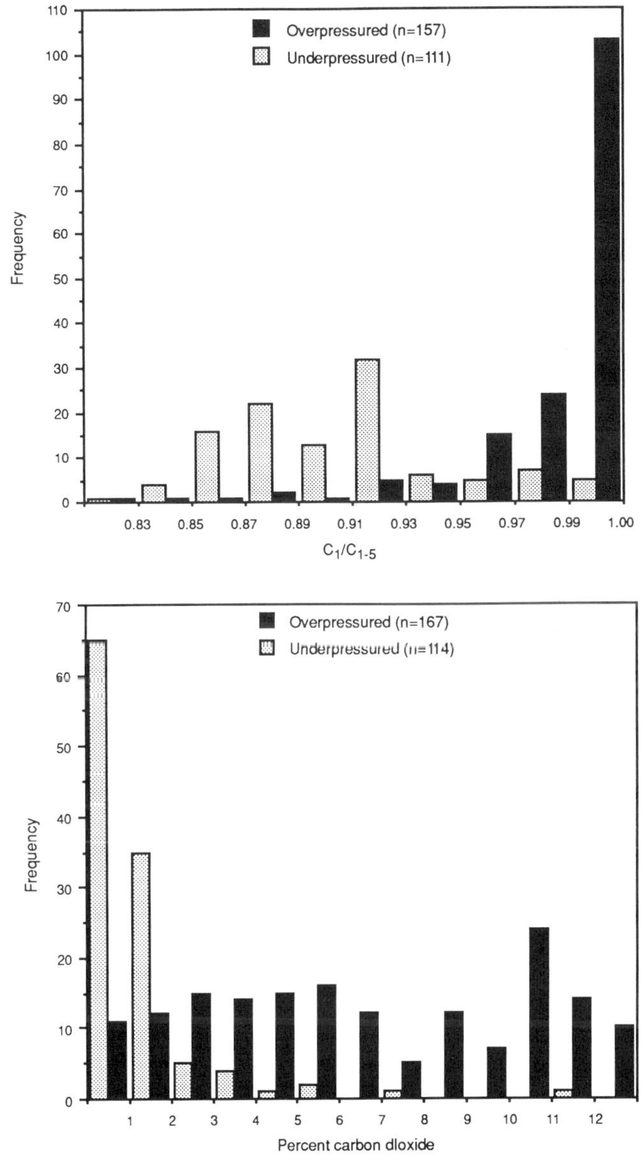

Figure 11. Histograms of C_1/C_{1-5} values and carbon dioxide content of gases from underpressured and overpressured Fruitland coalbeds, San Juan basin. Bimodal distribution of carbon dioxide in overpressured gases primarily reflects high carbon dioxide content of samples from the north-central part of the basin.

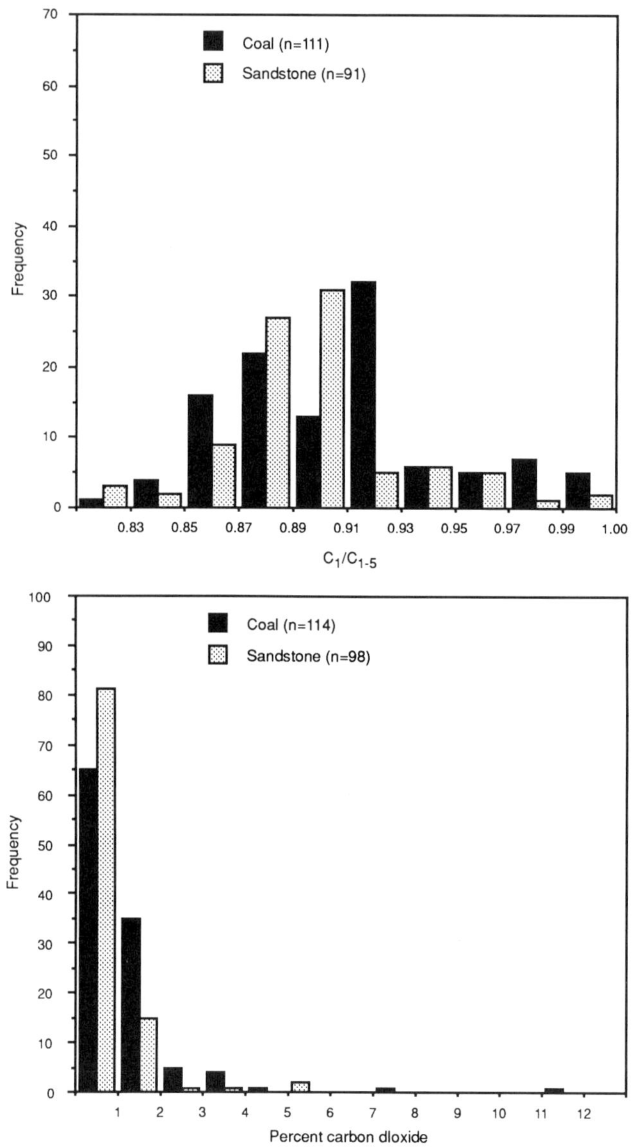

Figure 12. Histograms of C_1/C_{1-5} values and carbon dioxide content of gases from underpressured Fruitland coal and sandstone, San Juan basin.

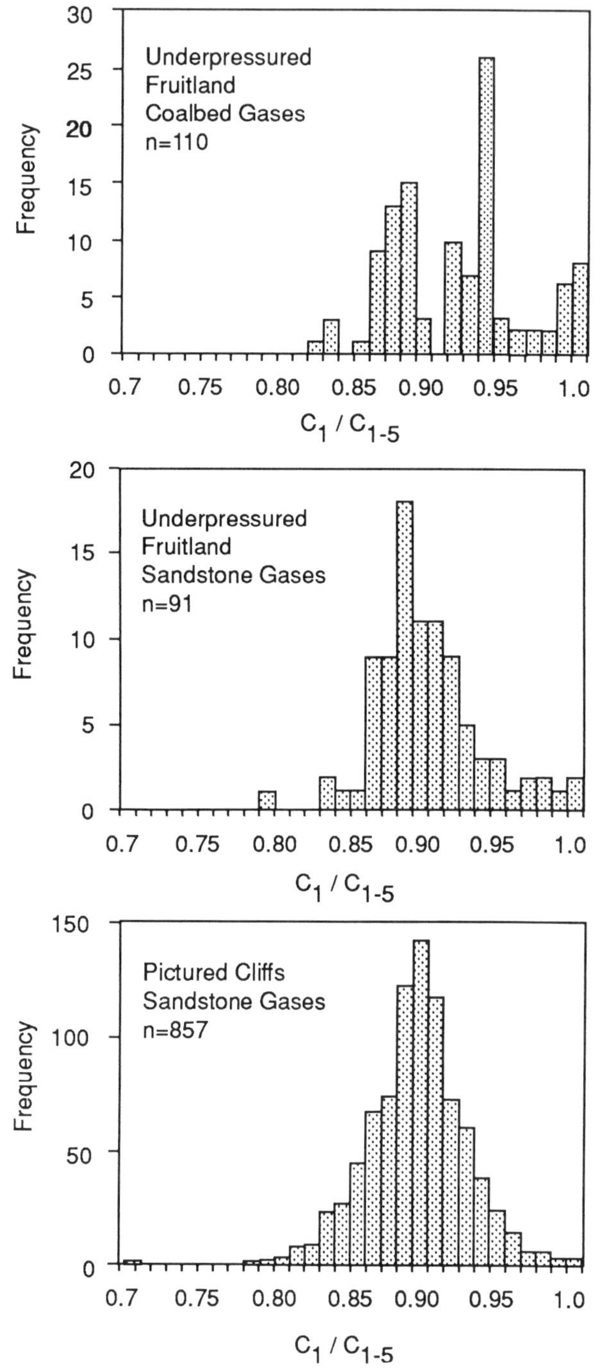

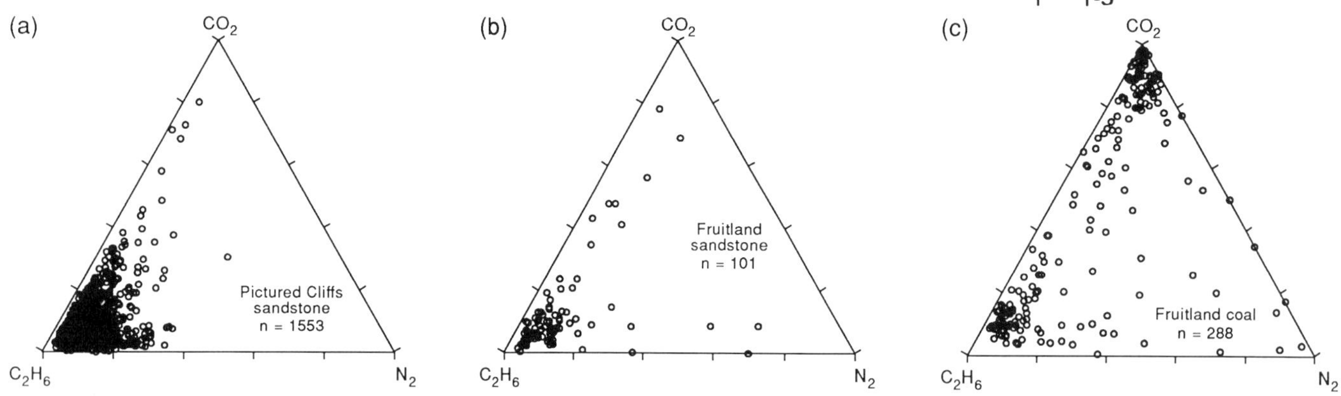

Figure 13. Ternary diagrams of ethane, carbon dioxide, and nitrogen composition of Pictured Cliffs and Fruitland gases.

3, and 10), indicating that carbon dioxide content in coal gases is controlled by both basin hydrology and coal rank.

ORIGIN OF FRUITLAND COALBED AND PICTURED CLIFFS SANDSTONE GASES

Fruitland Coal Gases

The quantity and composition of gas in a coalbed are determined by coal rank, formation pressure, ash content, and maceral composition. Rank and maceral composition, particularly the relative abundance of hydrogen-rich organic material, also control the types of gases produced from the coal. Coal generates progressively more thermogenic methane and less carbon dioxide and nitrogen during catagenesis, and the methane storage capacity of coal increases with rank; the density of methane molecules adsorbed onto coal surfaces can resemble that of liquid methane (Creedy, 1988).

Previous studies of gas wetness and carbon isotopic values of methane have suggested that most of the gas produced from the Fruitland Formation is thermogenic in origin; only a few gas samples in the southern basin were believed to be of possible biogenic origin (Rice and others, 1988, 1989). However, the distributions of carbon dioxide content and C_1/C_{1-5} values for Fruitland coal gases do not support these conclusions. Geochemical studies (Rice and others, 1989; Law and others, 1990) and the distribution of wet gases in the Fruitland Formation suggest that wet gases are indigenous to the coal. Although most of the gas produced from Fruitland coalbeds probably has a thermogenic origin, mixing of thermogenic and biogenic gases can better explain gas compositional distributions and $\delta^{13}C$ isotopic signatures of methane, total dissolved carbonate, and carbon dioxide found in the Fruitland Formation.

Coal traditionally has been thought of as a major source of dry gas, having little potential for generation of liquid hydrocarbons. However, hydrogen-rich coal is capable of generating liquid hydrocarbons (Khorasani, 1987). Furthermore, coals composed predominantly of vitrinite may contain submicroscopic particles of hydrogen-rich organic matter within the vitrinite. These particles would not be identified during microscopic study of the kerogen, leading to the interpretation that coal composed predominantly of vitrinite has little potential to generate wet gas (Bertrand, and others, 1986, and references therein). Results of Rock-Eval pyrolysis of Fruitland coals (Rice and others, 1989; Law and others, 1990) indicate that they are

Figure 14 (left). Histograms of gases from underpressured Fruitland coalbeds, Fruitland sandstones, and Pictured Cliffs sandstones. Pictured Cliffs gas data used in this histogram are from the same part of the basin as Fruitland gas data.

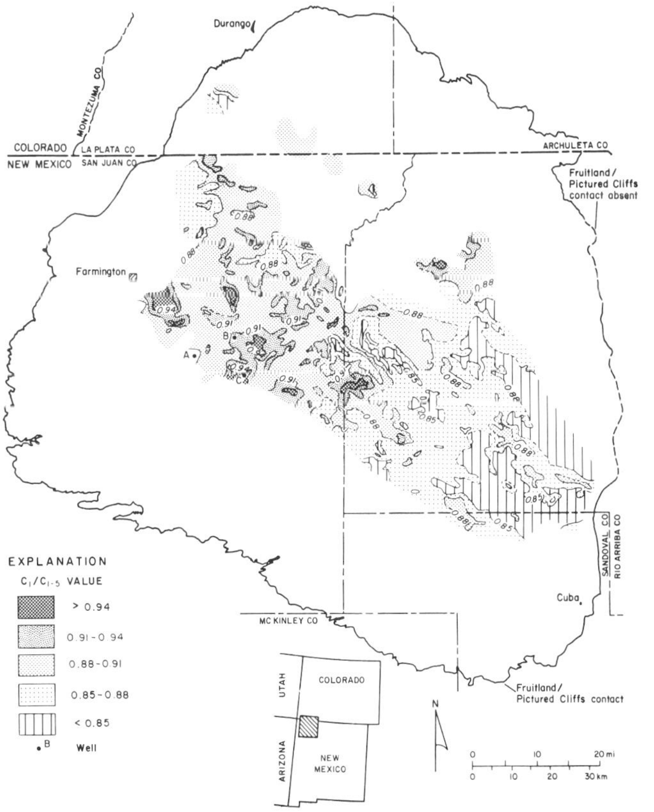

Figure 15. Distribution of C_1/C_{1-5} values of Pictured Cliffs Sandstone gases, San Juan basin, based on gas analyses from more than 5,500 wells.

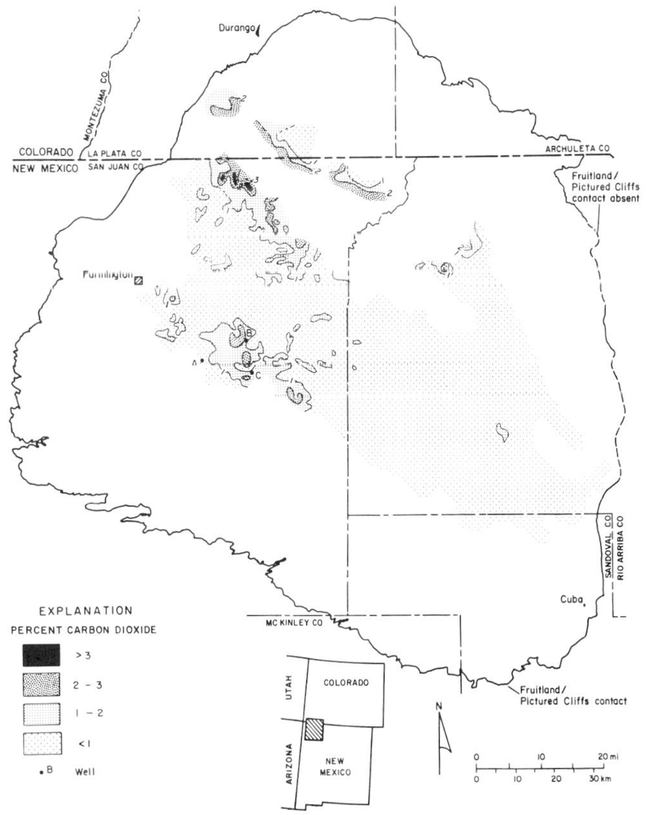

Figure 16. Carbon dioxide distribution in Pictured Cliffs Sandstone gases, San Juan basin, based on gas analyses from more than 5,500 wells.

hydrogen rich and have moderate potential for generating liquid hydrocarbons. Saturated-hydrocarbon distributions on gas chromatograms of Fruitland coal and carbonaceous shale extracts are similar to condensates produced from Fruitland coalbeds, indicating that the condensates originated from the coals (Rice and others, 1989). From these observations, Fruitland coalbeds are capable of producing wet gas, and areas of wet coal gas in the southern part of the basin are directly related to coal rank.

The generation of dry thermogenic gas from coal typically begins at vitrinite reflectance values of 0.7% (Meissner, 1987), which coincides with the southwestern limit of dry coal gases (Figure 8), suggesting that dry gases in the northern San Juan basin are thermogenic in origin, as suggested previously (Rice and others, 1988, 1989). However, if Fruitland coalbed methane is entirely thermogenic, $\delta^{13}C$ values of methane should become less negative (–55‰ to –15‰) with increasing thermal maturation (Schoell, 1983; Jenden, 1985). Instead, the isotopic signature of methane from Fruitland coal varies little throughout the basin and does not change significantly with coal rank (Figure 2). Either the process of thermogenic gas generation from coal is significantly different than gas generation from other types of organic matter, resulting in a relatively narrow range of $\delta^{13}C$ methane values over a wide range of coal ranks, or Fruitland coal gases may be mixtures of thermogenic and biogenic gases.

By comparison, biogenic methane $\delta^{13}C$ values range from less than –90‰ to –40‰ (Jenden, 1985), indicating that biogenic and thermogenic gases can have overlapping methane $\delta^{13}C$ isotopic values in the range of –55‰ to –40‰. Biogenic gases generated by carbonate reduction and/or acetate dissimilation can produce both chemically very dry gas (C_1/C_{1-5} values of 1.00) and isotopically heavy bicarbonate (Carothers and Kharaka, 1980). Furthermore, bacterial alteration of chemically wet gas can remove nearly all heavier gas components, leaving a chemically dry gas that resembles thermogenic gas (James and Burns, 1984). Therefore, the composition and distribution of coal gases and the isotopic composition of bicarbonate from formation water were considered in addition to methane $\delta^{13}C$ values to determine the origin of Fruitland coalbed methane.

Regional variations in carbon dioxide content and chemical composition (C_1/C_{1-5} values) of coal gases coincide with regional overpressure in the Fruitland, indicating a relationship between basin hydrology and gas composition (Figures 2, 8, and 10). Waters from the northern, overpressured area of the basin are dominantly sodium bicarbonate, characterized by high alkalinity and low chloride content (Kaiser and others, 1991a). $\delta^{13}C$ values for the bicarbonate are isotopically heavy, ranging from +16.7‰ to +26.0‰ (Kaiser and others, 1991a). These values are consistent with Jones and others' (1985) reported $\delta^{13}C$ values of +16.2‰ and +16.8‰ for carbon dioxide produced from the Glover No. 1 well in southwestern Archuleta County, Colorado. Isotopically heavy values such as these can be derived from dissolution of isotopically heavy carbonate cement and/or bacterial activity.

Dissolution of isotopically heavy carbonate cement could produce the range of bicarbonate isotopic values found in Fruitland Formation waters. Carbonate cement occurs within Fruitland coal cleats (Tremain and others, 1991, and this volume), but the coal and/or adjacent sandstones contain insufficient carbonate cement to account for the large amount of carbon dioxide in some coal gases. Dissolution of large quantities of isotopically very heavy carbonate cement ($\delta^{13}C$ greater than or equal to +26‰) would be required to explain the $\delta^{13}C$ values of total dissolved carbonate found in Fruitland Formation waters (maximum +26.0‰). Furthermore, carbonate dissolution would have to be demonstrated; precipitation of isotopically heavy carbonate in the cleat system may have occurred rather than dissolution of carbonate cement. Presently, these waters, under in situ conditions, are supersaturated with respect to all carbonate minerals. Therefore, bacterial degradation of organic acids and/or reduction of carbon dioxide by methanogenic bacteria may have produced the isotopically heavy carbon dioxide and bicarbonate in the Fruitland Formation.

A combination of thermogenic and biogenic gases can explain both methane $\delta^{13}C$ values and the heavy isotopic signature of total dissolved carbonate species in Fruitland Formation waters. Methanogenic degradation of short-chained organic acids through acetate dissimilation would produce both positive $\delta^{13}C$ values for total dissolved carbonate and methane depleted in ^{13}C by 74% to 65% relative to the bicarbonate at present in situ formation temperatures of 30°C to 54°C (Friedman and O'Neil, 1977). Furthermore, acetate dissimilation probably is the major source of biogenic gas in fresh-water environments (Whiticar and others, 1986, and references therein). Biogenic carbonate reduction of carbon dioxide also produces isotopically heavy bicarbonate and isotopically light methane. A mixture of 85% thermogenic methane ($\delta^{13}C$ of –42‰) and 15% biogenic methane ($\delta^{13}C$ of –48‰) would produce methane with an average isotopic value of –43‰. The $\delta^{13}C$ value of the corresponding bicarbonate would be +23‰, assuming a fractionation factor of 71% between the biogenic methane and bicarbonate. These isotopic values are similar to isotopic ranges seen in Fruitland coal gases and formation waters (Figure 2). The less positive $\delta^{13}C$ values of total dissolved carbonate (+16.7‰ and +17.5‰) in the northwestern part of the basin (Figure 2) may indicate dilution of isotopically heavy bicarbonate by fresh ground water containing isotopically light bicarbonate derived from atmospheric and/or soil sources.

Although the carbon in biogenic methane is derived from carbon dioxide or organic material, the hydrogen originates from formation waters. Methanogenic bacteria using carbon dioxide as a carbon source derive all the hydrogen for methane formation from the water, whereas acetate-fermenting bacteria derive only about 25% of the hydrogen from water (Whiticar and others, 1986). Therefore, carbon- and hydrogen-isotope fractionation during methanogenesis results in the generation of isotopically light biogenic methane that is depleted in ^{13}C and in deuterium (D). Because of the enrichment of ^{12}C and ^{1}H in the methane, formation waters become progressively enriched in ^{13}C and D and become isotopically heavier. The close correlation of δD with $\delta^{13}C$ in Fruitland Formation waters (Figure 17a) also suggests that some Fruitland coalbed methane is biogenic in origin (Kaiser and others, 1991a), particularly in that no correlation was found between δD and salinity to suggest mixing of fresh and connate waters. Furthermore, Fruitland Formation waters generally plot above the meteoric water line (Figure 17b), suggesting isotopic exchange between hydrogen gas and formation water (Kaiser and others, 1991a).

northern part of the basin where coal rank is higher; thermogenic gas from lower rank coals may have $\delta^{13}C$ values similar to biogenic methane.

The presence of biogenic methane in coal gases is supported in part by studies of gas desorption from Fruitland coal, which reported that desorbed methane from coal samples was isotopically heavier than methane produced from nearby wells (Rice and others, 1988). This suggests that $\delta^{13}C$ values of produced gases could represent a mixture of biogenic methane, occurring as free or dissolved gas in the cleat system, and thermogenic methane desorbed from the coal surfaces. The presence of free gas in some wells may be indicated by a geometric mean value of 195 MCFD (5,518 m^3/day) for initial production that is significantly larger than a mean value of 93 MCFD (2,632 m^3/day) for maximum annual production (Kaiser and others, 1991b).

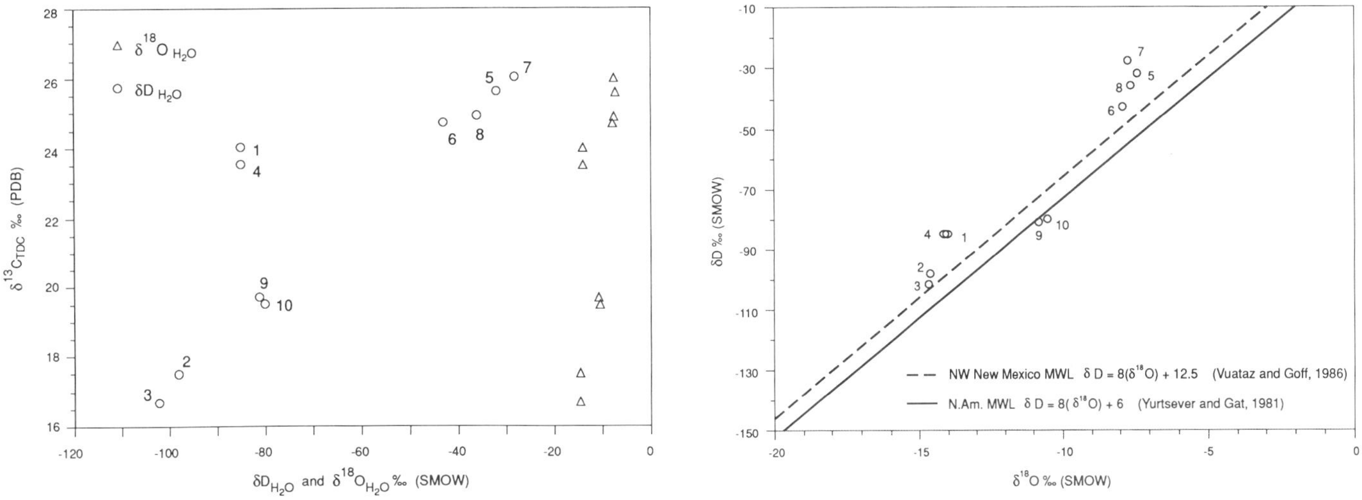

Figure 17. Stable-carbon isotope values for Fruitland Formation water. *a.* Formation water δD and $\delta^{18}O$ values increase with increasing $\delta^{13}C$ values. *b.* Northern and southern waters plot above the meteoric water line (MWL), suggesting isotopic exchange with hydrogen gases. From Kaiser and others (1991a).

Adsorption of methane onto coal particle surfaces is controlled by the amount of methane in the gas phase (Puri and Yee, 1990). Therefore, as the concentration of biogenic methane increases in the cleat system, methane will be adsorbed, assuming the maximum methane storage capacity has not been reached. If the concentration of biogenic methane is insufficient for adsorption to occur, methane remains as free gas in the cleat system or is dissolved in the formation water. As pressure decreases during production, adsorbed biogenic methane would be released together with thermogenic methane. During initial production, the gas would have a relatively large biogenic component (isotopically light methane), whereas subsequently desorbed gas would become progressively enriched in thermogenic methane (isotopically heavier). If this hypothesis is correct, $\delta^{13}C$ values of the produced methane should become less negative with time, as the relative proportions of the two gases change. The change in methane $\delta^{13}C$ values with time would be most pronounced in the

Rice and others (1988, 1989) suggested that dry coal gases in the southern San Juan basin may be biogenic in origin; the light isotopic signature of the methane ($\delta^{13}C$ values of –46.6‰ to –46.5‰) results from acetate dissimilation instead of carbonate reduction during methanogenesis. $\delta^{13}C$ values of +19.5‰ and +19.7‰ for bicarbonate in Fruitland water produced from two wells in this area support the conclusion that the gases are biogenic. However, some relatively dry gases in the southwestern part of the basin may be of early thermogenic origin, as suggested by Rice and others (1989).

As shown in Figure 10, coal gases in the overpressured area of the basin may contain significant amounts of carbon dioxide (greater than 3%). This area of high carbon dioxide content coincides with highest bottom-hole pressures, regional overpressure in the Fruitland, and medium- to high-volatile A bituminous coal. High production rates and large quantities of carbon dioxide (locally greater than 10%) in the gases produced from coalbeds in this area may be a function of several factors, including (1) presence of thick coalbeds (which increase the resource) and folds (which increase fracture porosity and permeability in the coal; Ayers and Zellers, 1991), (2) coal rank, which may control the amount of carbon dioxide and organic acids in the system, (3) a general southeastward movement of aqueous transported organic compounds, bicarbonate, and methane, (4) permeability controls restricting lateral flow, and (5) bacterial degradation of organic compounds to produce significant amounts of isotopically heavy carbon dioxide and isotopically light biogenic methane.

Carbon dioxide is very soluble in water, and a significant proportion of the isotopically heavy carbon dioxide produced by acetate dissimilation probably was dissolved in formation waters; bicarbonate may have been subsequently transported southeastward (basinward). Because formation pressure increases basinward, progressively more carbon dioxide could be dissolved, and total alkalinity would increase. As reservoir pressure is reduced during production, carbon dioxide exsolves from the formation water and is produced with the gas. Therefore, the large proportion of carbon dioxide in coal gases from the overpressured area is a result of carbon dioxide desorption from coal surfaces and exsolution from formation waters as reservoirs pressures decrease during production.

Pictured Cliffs Sandstone Gases

Previous studies suggest that Pictured Cliffs Sandstone gases were derived from Fruitland coalbeds (Choate and others, 1984; Rice and Threlkeld, 1986) or from dispersed type III kerogen in the underlying Lewis Shale (Rice and others, 1988, 1989, 1990). Significant quantities of early thermogenic gas can be generated from type III organic material at low levels of thermal maturity (Galimov, 1988). Methane $\delta^{13}C$ values of Pictured Cliffs Sandstone gases (–44.3‰ to –39.1‰; Rice, 1983; Rice and others, 1988) suggest that Pictured Cliffs Sandstone gases are thermogenic in origin. Although significant volumes of Pictured Cliffs gas probably were derived from the underlying Lewis Shale, some could have migrated from overlying Fruitland coalbeds, and mixing of gases from these two sources is possible.

The wettest Pictured Cliffs Sandstone gases (C_1/C_{1-5} values less than 0.88) occur in the southeast corner of the basin (Figure 15). These wet gases may reflect lower levels of thermal maturity and/or differences in kerogen type. The fact that the highest nitrogen contents in Pictured Cliffs Sandstone gases also occur in this area may reflect gases derived predominantly from sapropelic organic material (containing more nitrogen-rich proteins relative to terrestrial organic matter) in the Lewis Shale. Alternatively, the wet gases in the southeast may represent gases derived only from the Lewis Shale, whereas Pictured Cliffs gases from other parts of the basin may represent a mixture of gases that migrated into the sandstone reservoirs from adjacent coalbeds and from the Lewis Shale.

Choate and others (1984) suggested that Pictured Cliffs reservoirs are sourced by Fruitland coalbeds, where pressure differences between the two formations result in downward migration of chemically drier coal gases. Relatively dry (C_1/C_{1-5} values greater than 0.88) sandstone gases occur in the northwestern part of the basin where thick basal coalbeds rest directly on top of Pictured Cliffs sandstones. Dip-elongate bands of coal are not as thick or abundant in the southeast (Ambrose and Ayers, 1991). The driest Pictured Cliffs gases (C_1/C_{1-5} values greater than 0.91), which occur in a dip-elongate band in eastern San Juan County, may result from completion practices (dual completion in Fruitland coalbeds and Pictured Cliffs Sandstone) or from downward migration from Fruitland coalbeds. This trend of dry gas coincides with dip-elongate bands of basal Fruitland coalbeds that may exceed 20 ft (6 m) in thickness (Ambrose and Ayers, 1991). The fact that C_1/C_{1-5} values of coal gases in western Rio Arriba County are similar to those from underlying Pictured Cliffs Sandstone gases also suggests possible local communication between the two reservoirs.

Hydraulic communication between basal Fruitland coalbeds and Pictured Cliffs Sandstone in the west-central part of the basin also is indicated by similar water chemistries and hydraulic heads (Kaiser and others, 1991a). However, the similarity of underpressured Fruitland coalbed and Pictured Cliffs Sandstone C_1/C_{1-5} values (Figures 9c, 13, and 14) and gas-composition distribution (Figures 8, 10, 15, and 16) indicate that gas composition alone does not necessarily distinguish coalbed from sandstone gases in parts of the southern basin. More detailed analytical methods, including methane, hydrogen, and carbon isotopic analyses and isotopic analyses of the wet gas components of both Fruitland and Pictured Cliffs gases on a basinwide scale, probably are required to establish the origin and migration pathways of Pictured Cliffs Sandstone gases.

CONCLUSIONS

1. Wet coal gases (C_1/C_{1-5} values less than 0.94) in the southern part of the San Juan basin probably are indigenous to the hydrogen-rich Fruitland coal and, therefore, do not result from completion practices and/or migration of gas from adjacent shale. Chemically dry to very dry coal gases in some parts of the southern basin most likely are biogenic in origin, although some early thermogenic methane may be present.

2. The similar composition of Pictured Cliffs Sandstone, Fruitland sandstone, and underpressured Fruitland coal gases in the southern part of the basin make determination of gas origin, based on gas composition alone, difficult or impossible. However, minor differences in carbon dioxide content and gas wetness in some areas may allow coal gases to be distinguished from sandstone gases locally.

3. Basin hydrology and coal rank control Fruitland coal-gas composition. The effect of basin hydrology on coal gas

composition is evidenced by the abrupt transition from carbon dioxide-rich, chemically dry to very dry gases in the northern, overpressured part of the basin to carbon dioxide-poor, chemically wet gases in the southern, underpressured area.

4. Bacterial degradation of organic acids by acetate dissimilation and/or the microbial reduction of carbon dioxide are required to produce the isotopically heavy bicarbonate ($\delta^{13}C$ values of +16.7‰ to +26‰) found in Fruitland Formation waters. The presence of free and dissolved biogenic methane in the coal cleat system and/or biogenic methane adsorbed onto coal particle surfaces may explain the relatively constant $\delta^{13}C$ methane values in the northern part of the basin, reported in earlier studies.

5. The large amounts of carbon dioxide in gases produced from coalbeds in the overpressured part of the basin may result from a combination of factors. Isotopic data suggest that much of the carbon dioxide in Fruitland coal gas is biogenic in origin but that some may originate from the formation water. We suggest that bacterially generated carbon dioxide was dissolved in formation waters and transported basinward. As reservoir pressure is reduced during production, carbon dioxide exsolves from the formation water and mixes with carbon dioxide, methane, and other gases desorbed from the coal.

6. Pictured Cliffs Sandstone gases are thermogenic in origin and probably were derived from the underlying Lewis Shale and, possibly, from overlying thick, basal Fruitland coalbeds in the western part of the basin. The wettest Pictured Cliffs gases, which occur in the southeast, may reflect a change in source-rock type (greater marine influence) and/or indicate gases derived entirely from the Lewis Shale. Chemically drier Pictured Cliffs gases elsewhere may represent a mixture of gases derived from both overlying coal and underlying shale.

ACKNOWLEDGMENTS

We thank Jeff Peace and George Lippman of El Paso Natural Gas Company, Brent W. Hale of Northwest Pipeline Corporation, Tom Hemler of Jerome P. McHugh and Associates, and Ernie Busch of the New Mexico Oil Conservation Commission for providing the gas data used in this study. Carol M. Tremain of the Colorado Geological Survey and Ernie Busch provided cumulative Fruitland coalbed methane production estimates. We also thank William B. Hanson of Amoco Production Company and Peter D. Jenden of Chevron Production Research for their valuable comments concerning coal-gas chemistry. Drafting was provided by the cartographic staff of the Bureau of Economic Geology under the direction of R.L. Dillon. We thank Tucker F. Heintz for reviewing the manuscript. This publication was authorized by the Director, Bureau of Economic Geology, The University of Texas at Austin. Research was funded by the Gas Research Institute under contract no. 5087-214-1544.

REFERENCES

Ambrose, W.A., and Ayers, W.B., Jr., 1991, Geologic controls on coalbed occurrence, thickness, and continuity, Cedar Hill field and the COAL site, *in* Ayers, W.B., Jr., and others, Geologic and hydrologic controls on the occurrence and producibility of coalbed methane, Fruitland Formation, San Juan basin: Gas Research Institute, Topical Rept. GRI–91/0072 (August 1987–July 1990), p. 47–68.

Ayers, W.B., Jr., and Zellers, S.D., 1991, Geologic controls on Fruitland coal occurrence, thickness, and continuity, Navajo Lake area, San Juan Basin, *in* Ayers, W.B., Jr., and others, Geologic and hydrologic controls on the occurrence and producibility of coalbed methane, Fruitland Formation, San Juan basin: Gas Research Institute, Topical Rept. GRI–91/0072 (August 1987–July 1990), p. 69–94.

Bertrand, Philippe, Behar, Françoise, and Durand, Bernard, 1986, Composition of potential oil from humic coals in relation to their petrographic nature: Organic Geochemistry, v. 10, nos. 1–3, p. 601–608.

Carothers, W.W., and Kharaka, Y.K., 1980, Stable carbon isotopes of HCO_3^- in oil-field waters—Implications for the origin of CO2: Geochimica et Cosmochimica Acta, v. 44, no. 2, p. 323–332.

Choate, R., Lent, J., and Rightmire, C.T., 1984, Upper Cretaceous geology, coal, and the potential for methane recovery from coalbeds in San Juan Basin—Colorado and New Mexico, *in* Rightmire, C.T., Eddy, G.E., and Kirr, J.N., eds., Coalbed methane resources of the United States: AAPG Studies in Geology, Series 17, p. 185–222.

Creedy, D.P., 1988, Geologic controls on the formation and distribution of gas in British coal measure strata: International Journal of Coal Geology, v. 10, no. 1, p. 1–31.

Decker, A.D., and Horner, D.M., 1987, Origin and production implications of abnormal coal reservoir pressure, *in* The 1987 Coalbed Methane Symposium Proceedings—University of Alabama, Tuscaloosa, November 16–19, 1987: School of Mines and Energy Development, Gas Research Institute, and U.S. Mine Safety and Health Administration, p. 51–62.

Dwight's Oil and Gas Reports, 1990a, Natural gas well production histories: Colorado Statewide Coal Bed Methane Report, v. 1, 373 p.

Dwight's Oil and Gas Reports 1990b, Natural gas well production histories: New Mexico Statewide Coal Bed Methane Report, v. 1, 853 p.

Friedman, Irving, O'Neil, J.R., and Fleischer, Michael, 1977, Compilation of stable carbon isotope fractionation factors of geochemical interest: USGS Professional Paper 440–KK, 12 p.

Galimov, E.M., 1988, Sources and mechanisms of formation of gaseous hydrocarbons in sedimentary rocks, *in* Schoell, M., ed., Origin of methane in the earth: Chemical Geology, v. 71, nos. 1–3, p. 77–95.

Hale, B.W., and Firth, C.H., 1988, Production history of the San Juan Unit No. 6 well, northern San Juan basin, New Mexico, *in* Fassett J.E., ed., Geology and coal-bed methane resources of the northern San Juan basin, Colorado and New Mexico: Rocky Mountain Association of Geologists Guidebook, p. 199–204.

Hanson, W.B., 1990, Chemistry of western interior USA coal-bed gases based upon desorption of subsurface coal samples (abs.): AAPG Bulletin, v. 74, no. 8, p. 1326.

Hunt, J.M., 1979, Petroleum geochemistry and geology: San Francisco, W.H. Freeman and Co., 617 p.

James, T.A., and Burns, B.J., 1984, Microbial alteration of subsurface natural gas accumulations: AAPG Bulletin, v. 68, no. 8, p. 957–960.

Jenden, P.D., 1985, Analysis of gases in the earth's crust: Gas Research Institute Rept. GRI–85/0106, 110 p.

Jones, A.H., Kelkar, S., Bush, D., Hanson, J., Rakop, K., Ahmed, U., Holland, M., Tibbitts, G., Owen, L.B., and Bowman, K.C., 1985, Methane production characteristics of deeply buried coalbed reservoirs: Gas Research Institute, Final Rept. GRI–85/0033 (February 1982–December 1984), 176 p.

Kaiser, W.R., Ayers, W.B., Jr., Ambrose, W.A., Laubach, S.E., Scott, A.R., and Tremain, C.M., 1991b, Geologic and hydrologic characterization of coalbed methane production, Fruitland Formation, *in* Ayers W.B., Jr., and others, Geologic and hydrologic controls on the occurrence and producibility of coalbed methane, Fruitland Formation, San Juan basin: Gas Research Institute, Topical Rept. GRI–91/0072 (August 1987–July 1990), p. 273–301.

Kaiser, W.R., Swartz, T.E., and Hawkins, G.J., 1991a, Hydrology of the Fruitland Formation, San Juan Basin, *in* Ayers, W.B., Jr., and others, Geologic and hydrologic controls on the occurrence and producibility of coalbed methane, Fruitland Formation, San Juan basin: Gas Research Institute, Topical Rept. GRI–91/0072 (August 1987–July 1990), p. 195–241.

Khorasani, G.K., 1987, Oil-prone coals of the Walloon Coal Measures, Surat Basin, Australia, *in* Scott, A.C., ed., 1987, Coal and coal-bearing strata—Recent advances: Geological Society Special Publication 32, p. 303–310.

Law, B.E., Anders, D.E., and Michael, G.E., 1990, Use of Rock-Eval pyrolysis and vitrinite reflectance data in characterizing type and maturity of organic matter in coal, upper Cretaceous Fruitland Formation, San Juan Basin, New Mexico and Colorado (abs.): AAPG Bulletin, v. 74, no. 8, p. 1333.

Leythaeuser, D., Schaefer, R.G., Cornford, C., and Weiner, B., 1979, Generation and migration of light hydrocarbons (C_2–C_7) in sedimentary basins: Organic Geochemistry, v. 1, p. 191–204.

Meissner, F.F., 1987, Mechanisms and patterns of gas generation and expulsion—Migration and accumulation associated with coal measures Green River and San Juan Basins Rocky Mountain region, USA, *in* Doligez, Brigitte, ed., Migration of hydrocarbons in sedimentary basins: Paris, Éditions Technip, p. 79–112.

Puri, R., and Yee, D., 1990, Enhanced coalbed methane recovery: Society of Petroleum Engineer Paper 20732 [presented at SPE Annual Technical Conference and Exhibition, September 23–26, New Orleans, Louisiana].

Rice, D.D., 1983, Relation of natural gas composition to thermal maturity and source-rock type in San Juan Basin, northwestern New Mexico and southwestern Colorado: AAPG Bulletin, v. 67, no. 8, p. 1199–1218.

Rice, D.D., Clayton, J.L., and Pawlewicz, M.J., 1989, Characterization of coal-derived hydrocarbons and source-rock potential of coal beds, San Juan Basin, New Mexico and Colorado, U.S.A.: International Journal of Coal Geology, v. 13, p. 597–626.

Rice, D.D., and Threlkeld, C.N., 1986, Comparison of natural gases produced from Upper Cretaceous Fruitland Formation coal beds and adjacent sandstone reservoirs, San Juan basin, New Mexico and Colorado (abs.): AAPG Bulletin, v. 70, no. 5, p. 638.

Rice, D.D., Threlkeld, C.N., Vuletich, A.K., and Pawlewicz, M.J., 1988, Identification and significance of coal-bed gas, San Juan Basin, northwestern New Mexico and southwestern Colorado, in Fassett, J.E., ed., Geology and coal-bed methane resources of the northern San Juan basin, Colorado and New Mexico: Rocky Mountain Association of Geologists Guidebook, p. 51–59.

Rice, D.D., Threlkeld, C.N., Vuletich, A.K., and Pawlewicz, M.J., 1990, Nonassociated gas potential of San Juan Basin considerable: Oil & Gas Journal, v. 88, no. 33 (August 13), p. 60–61.

Rohrback, B.G., Peters, K.E., Sweeney, R.E., and Kaplan, I.R., 1983, Ammonia formation in laboratory simulated thermal maturation—Implications related to the origin of nitrogen in natural gas, *in* Bjorøy, M., ed., Advances in organic geochemistry, 1981: New York, John Wiley and Sons, p. 819–823.

Schoell, Martin, 1983, Genetic characterization of natural gases: AAPG Bulletin, v. 67, no. 12, p. 2225–2238.

Scott, A.R., Kaiser, W.R., and Ayers, W.B., Jr., 1991, Thermal maturity of Fruitland coal and composition and distribution of Fruitland Formation and Pictured Cliffs sandstone gases, *in* Ayers, W.B., Jr., and others, Geologic and hydrologic controls on the occurrence and producibility of coalbed methane, Fruitland Formation, San Juan basin: Gas Research Institute, Topical Rept. GRI–91/0072 (August 1987–July 1990), p. 243–270.

Stach, E., Mackowsky, M.T., Teichmüller, M., Taylor, G.H., Chandra, D., and Teichmüller, R., 1975, Stach's text book of coal petrology, 2d ed.: Stuttgart, Gebrüder Borntraeger, 428 p.

Tissot B., and Welte, D., 1978, Petroleum formation and occurrence: Berlin, Springer-Verlag, 521 p.

Tremain, C.M., Laubach, S.E., and Whitehead, N.H., III, 1991, Coal fracture (cleat) patterns in Upper Cretaceous Fruitland Formation, San Juan Basin, Colorado and New Mexico, *in* Ayers, W.B., Jr., and others, Geologic and hydrologic controls on the occurrence and producibility of coalbed methane, Fruitland Formation, San Juan basin: Gas Research Institute, Topical Rept. GRI–91/0072 (August 1987–July 1990), p. 97–117.

Whiticar, M.J., Faber, E., and Schoell, M., 1986, Biogenic methane formation in marine and freshwater environments, CO_2 reduction *vs.* acetate fermentation—Isotope evidence: Geochimica et Cosmochimica Acta, v. 50, no. 5, p. 693–709.

Yurtsever, Y., and Gat, J.R., 1981, Atmospheric waters, *in* Gat, J.R., and Gonfiantini, R., eds., Stable isotope hydrology—Deuterium and oxygen-18 in the water cycle: International Atomic Energy Agency Technical Report Series 210, p. 103–142.

Vuataz, F.D., and Goff, F., 1986, Isotope geochemistry of thermal and nonthermal waters in the Valles Caldera, Jemez Mountains: Journal of Geophysical Research, v. 91, p. 1835–1854.

Influence of Coal Composition and Rank on Fracture Development in Fruitland Coal Gas Reservoirs of San Juan Basin

Jay C. Close[1]
Matthew J. Mavor[1]

1. Resource Enterprises Inc., Salt Lake City, Utah

ABSTRACT

In the San Juan basin of Colorado and New Mexico, Fruitland Formation coal cleat frequency and aperture width generally increase as a function of several interdependent factors—increasing coal rank; decreasing coal bulk density; increasing vitrain layer abundance; decreasing vitrain layer thickness; and increasing frequency of vitrain and nonvitrain interbedding.

We hypothesize that the genesis of Fruitland cleats depends on the presence of high-volatile A bituminous and greater rank coals that contain a significant frequency of discrete vitrain/nonvitrain interbeds. We propose that the presence of layer macroheterogeneity and permeability in Fruitland coals can be qualitatively recognized through interpretation of open-hole high-resolution bulk density and microlog suites. This paper summarizes core fracture description, geophysical log, coal characterization, and drillstem test data for eight coal-gas wells in the San Juan basin to support this hypothesis. This hypothesis may in turn be applied to Fruitland coal-gas development programs.

INTRODUCTION

The purpose of this paper is to summarize nonstressed core fracture description and formation evaluation data collected by the Gas Research Institute (GRI) from coal-gas wells in the San Juan basin, and to interpret these data to support an hypothesis about fracture (cleat) genesis. Cleat permeability is the reservoir parameter that most affects commercial viability of a coal-gas well, given sufficient gas in-place resources, reservoir pressure, and favorable (rapid) gas diffusion and desorption characteristics. Knowledge of the occurrence and genesis of cleats, which represent permeability avenues for Darcy flow of gas and formation water, is essential for predicting fluid productivity in Fruitland coal reservoirs.

The Fruitland coal-gas play has matured over the last 20 years from largely exploration and delineation into a development phase. Consequently, the boundaries of the region of high production potential are reasonably well known. Commercial Fruitland coal-gas production at current gas prices is largely confined to the northern San Juan basin in areas of thick, gassy coals of high-volatile A bituminous rank or greater that are slightly underpressured to overpressured and that are essentially free of significant folding and faulting.

Coal cleats are extensional fractures envisioned to have formed primarily in vitrain layers as a result of active coalification processes and fluid pressure exerted during tectonic stressing events. Vitrain/nonvitrain interbeds, which have different mechanical properties, were subjected to different strain magnitudes during these tectonic episodes, which also favored fracture genesis. For differences in incremental and total strain in these layers to have been present, Fruitland coals must have undergone significant lithification before fracture genesis.

For the Fruitland wells that were studied, a qualitative correspondence was found between (a) greater cleat frequency, greater cleat aperture width, and increasing frequency of vitrain/nonvitrain interbedding; and (b) greater well test absolute permeability estimates, changes in high-resolution bulk-density log responses, and positive separation on high-resolution vertically focused microlaterologs (such as Schlumberger's Microlog™, as used in this paper).

On the basis of these observations, we hypothesize that the genesis of many open and interconnected Fruitland coal cleats depends on the presence of high-volatile A bituminous or greater rank coals that contain a high frequency of discrete vitrain/nonvitrain interbeds.

Based on available log, fracture description, and formation evaluation data, we also propose a technique by which significant layer macroheterogeneity and permeability in Fruitland coal zones can be recognized, respectively, by sharp changes in bulk density on open-hole high-resolution bulk-density logs and by prominent positive separation on microlog traces. This log-interpretation technique has not, however, been completely quantified for permeability estimation in part because of lack of differentiation among Fruitland coal lithotypes on the bulk-density logs and difficulties in normalizing microlog data.

COMMENT

The proposed technique for inferring the presence of Fruitland coal permeability is not quantitative and is based on interpretation of a limited set of observations. The technique must be evaluated further over areas of the basin beyond those in which the research wells are located. Furthermore, the technique cannot and should not be used in place of thorough geologic and engineering analysis of all available data for the acreage under consideration and in no way precludes the need for complete and accurate formation evaluation of a well (Mavor and others, 1990) drilled in a newer area of the Fruitland (or any other) coal-gas play. A complete formation evaluation data set always is prudent for reservoir simulation of fluid productivity, reserve estimation, and economic projections.

The reader also will note that the suggested permeability identification technique really is a simple application of available technology for conventional reservoirs. The method for analyzing microlog data to infer mud-filtrate invasion, and thus the presence of permeability, in fresh-water-based mud-drilled wells was publicly introduced by Schlumberger about 40 years ago (Doll, 1953; see also Doll, 1950, 1951; Asquith, 1982). Many operators who target Fruitland coals routinely run microlog and/or other resistivity logs to infer mud-filtrate invasion as part of formation evaluation procedures. Other methods for inferring areas with greater coal-gas reservoir permeability have been described by Close and others (1990a) and Close and Dutcher (1990). The influence of layer heterogeneity on fracturing in rocks also has been known for more than a century (Pollard and Aydin, 1988). However, we believe that synergistic integration of microlog and bulk-density logs (which can be used for qualitative recognition of permeability and layer macroheterogeneity in Fruitland coal zones), presentation of specific Fruitland cleat description data with respect to other formation evaluation data, and processes involved in Fruitland cleat development, is a new contribution to the field.

RESEARCH DATABASE

The database for this study is derived from eight GRI cooperative research wells in the San Juan basin (Figure 1; Table 1) (Mavor and Close, 1989a–e; Close and others, 1990b). Of the seven data sets for Fruitland wells, six wells are located in the northern San Juan basin north of a line that separates T30N and T29N in New Mexico (about the latitude of southern Navajo Lake and the southern part of the prolific Meridian "400" Fruitland field). The seventh Fruitland well is located approximately 20 miles (32 km) due south of the Meridian field.

The eighth well is a Menefee coal-gas test along the northwestern basin margin west-northwest of the Meridian 400 field. The Menefee Formation is part of the upper Mesaverde Group (Upper Cretaceous), which, in ascending order, consists of the Point Lookout Sandstone, Menefee Formation, and Cliffhouse Sandstone. The Cliffhouse is, in turn, overlain by the Lewis Shale, Pictured Cliffs Sandstone, and Fruitland Formation.

The methods used for collecting the fracture description and formation evaluation data are presented in Mavor and others (1990). All data are available from the authors.

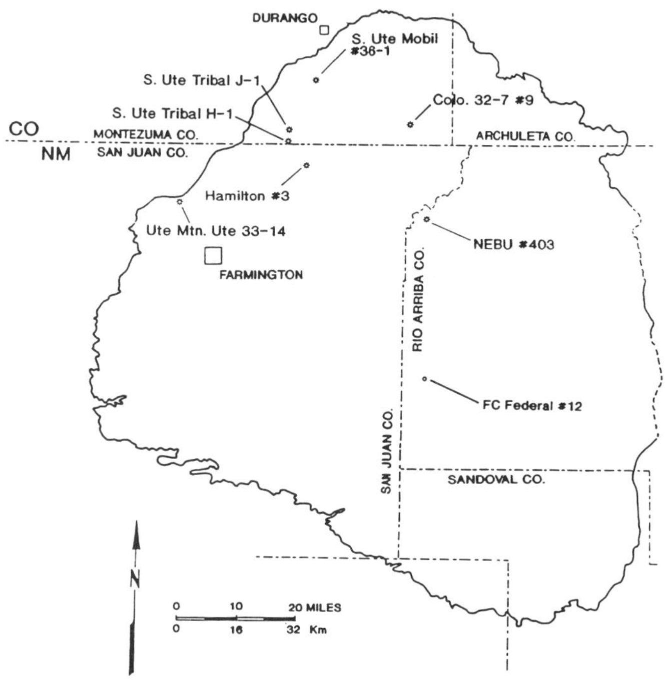

Figure 1. Location of GRI cooperative research wells in the San Juan basin.

CLEAT TERMINOLOGY

A "cleat" is defined in this paper as "a natural fracture surface in coal along which there has been basically imperceptible movement associated with mode I opening." Mode I opening is the extensional separation of fracture surfaces along a fracture propagation front (cf., Pollard and Aydin, 1988), as shown schematically in Figure 2. Fracture surfaces during mode I opening are displaced in the direction of least compressional (greatest tensile) stress. A "fracture" is defined as any natural break or discontinuity in a rock caused by stresses exceeding the strength of that rock (modified from Kulander and others, 1990).

Face (primary) and butt (secondary) cleats are the two main fracture systems in San Juan basin coal. Butt cleats terminate against more laterally continuous face cleats.

Table 1. GRI cooperative research wells in San Juan basin.

Well	Operator	Location Sec.	Twp	Rge	County	State
Colorado 32–7 No. 9	Mobil Oil	4	32N	7W	La Plata	CO
FC Federal 12	Mesa Limited Partnership	33	26N	7W	Rio Arriba	CO
Hamilton 3	Mesa Limited Partnership	30	32N	10W	San Juan	NM
Northeast Blanco Unit 403	Blackwood and Nichols	9	30N	7W	Rio Arriba	NM
Southern Ute Mobil 36–1	McKenzie Methane	36	34N	10W	La Plata	CO
Southern Ute Tribal H–1	Amoco Production	18	32N	10W	La Plata	CO
Southern Ute Tribal J–1	Amoco Production	7	32N	10W	La Plata	CO
Ute Mountain Ute 33–14	Wintershall Corp.	33	31N	14W	San Juan	NM

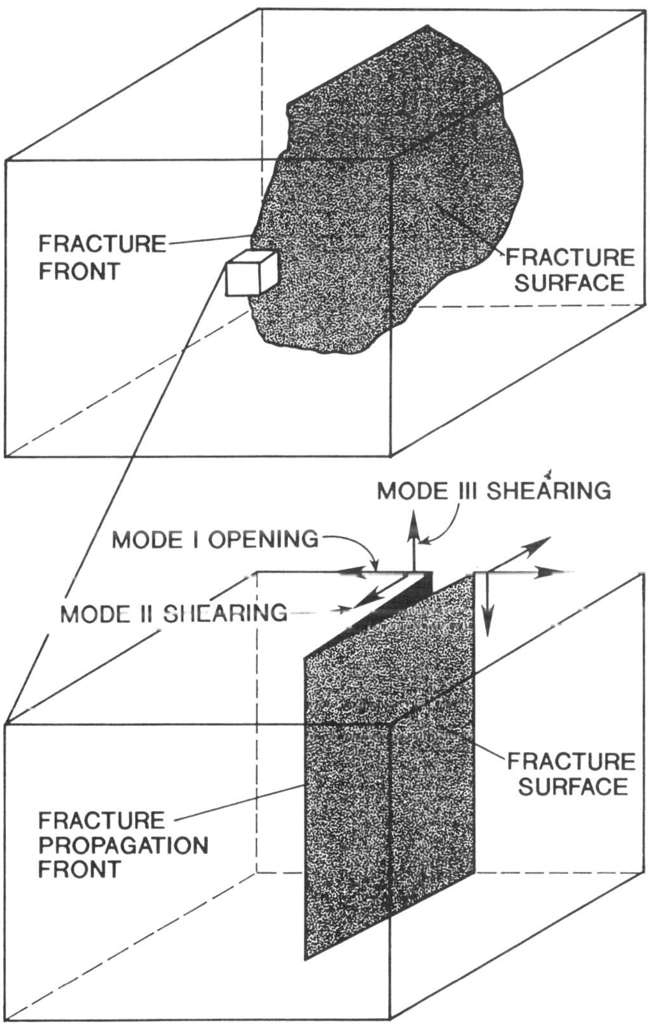

Figure 2. Schematic illustration showing modes of fracture genesis. After Pollard and Aydin (1988).

Third- and fourth-order cleats, in order of decreasing lateral continuity, are common in Fruitland cores. These fractures terminate at face and butt cleats. Fruitland coals commonly contain more than two principal cleat systems.

THE PARADOX OF FRACTURED COALS

Interestingly, fractures in noncoal strata (sandstone, siltstone, and shale) are present in the eight core sets studied but are much less abundant than cleats; the findings are supported by descriptions of Fruitland and Menefee rocks in roadcuts and outcrops in the northern San Juan basin. These observations are curious for several reasons. First, quantitative laboratory rock mechanics data for coal and noncoal Fruitland cores indicate that the coal is much less competent (Jones and others, 1988). Second, based on these data and numerous outcrop investigations, more competent rocks typically have greater propensity for failure via brittle fracturing, and fracture frequency is greater in the more competent rocks (Billings, 1972; Hobbs and others, 1976; Davis, 1984; Nelson, 1985; Pollard and Aydin, 1988). Third, given that the San Juan basin has undergone regional tectonic deformation after Fruitland sediment deposition and lithification, one would expect Fruitland coals to exhibit less fracture development than the surrounding competent clastic rocks. Therefore, the presence of Fruitland coal cleats, which are more numerous than fractures in adjacent competent noncoal rocks, is a paradox.

We will address this paradox by attempting to answer four important questions about cleat-genesis. First, exactly where in Fruitland coals are the cleats? Second, what are the specific geometries and spatial relationships among these cleats? Third, what processes acting over what time frame may have been responsible for Fruitland cleat genesis? Fourth, what log responses can an operator expect in a permeable or impermeable Fruitland coal zone and why?

Many authors simply refer to coal cleats as "extensional fractures" that formed as a result of "tectonic stresses," "compactional processes shortly after deposition," or "coalification," and devote no further discussion to cleat occurrence, geometry, genesis, or permeability recognition. This is surprising, given that cleat permeability exerts such strong control on the success or failure of coal-gas production programs.

OBSERVATIONS

Our hypothesis on cleat genesis in Fruitland coal reservoirs is developed from fracture observations (Terra Tek Inc., 1988a, 1988b, 1989a, 1989b, 1990a, 1990b) and formation evaluation data from seven Fruitland coalbed methane wells and one Menefee well. Table 2 summarizes the fracture description data, including the number of cleats, cleat dip angle, height, aperture width, and frequency, for six of the Fruitland cores (two cores were nearly devoid of cleats).

Coal Cleat

Cleat description data for nonstressed whole core and slabbed polished blocks, together with coal chemical rank, vitrinite reflectance, maceral composition, and lithotype information, indicate that most Fruitland cleats occur in discrete vitrain layers in coals of high-volatile A to medium-volatile bituminous rank (Figure 3). Fruitland coal core samples of high-volatile C bituminous rank from the FC Federal 12 well, located 20 miles (32 km) south of the Meridian 400 field (Figure 1), were virtually devoid of cleats.

Figure 3. Polished core block from the Southern Ute Mobil 36–1 Fruitland coal well at depth of approximately 2,414 ft, showing face cleats in a vitrain band. Core diameter is 3.5 in. (8.9 cm).

The Menefee coal samples, on the other hand, are high-volatile B and A bituminous, vitrinite rich, mostly clarain (mixed vitrinite and mineral matter) lithotype, virtually devoid of cleats, and exhibit conchoidal fractures.

The occurrence of Fruitland cleats can be recognized at various scales—(a) entire coal seam vs surrounding noncoal strata; (b) parts of a coal seam that are more than 1 ft (0.3 m) thick; (c) hand samples 1 ft (0.3 m) or less in length; (d) vitrain bands 1 to 2 in. (2.5 to 5 cm) thick; (e) megascopically visible vitrain bands less than 1 in. (2.5 cm) thick; (f) vitrain bands distinguishable through a binocular microscope under magnifications of one to 25 diameters; and (g) vitrinite group macerals that are resolvable through a coal petrographic microscope with oil immersion lenses under standard magnifications up to 500 diameters.

Table 2. Natural fracture description data for six Fruitland coal research wells in San Juan basin. Data include fracture descriptions from whole cores and polished blocks; consequently, the total number of descriptions is greater than the number of cleats described for the orientation rose diagrams in Figure 5.

Well	Dip Angle (°)	Cleat Aperture Height (cm)	Cleat Aperture Width (cm)	Cleat Frequency (cm^{-1})
Hamilton 3				
Minimum	20	0	0.01	1
Maximum	90	14	0.21	49
Average	86	4	0.06	3
No. of descriptions	111	112	112	110
Northeast Blanco Unit 403				
Minimum	70	0	0.01	1
Maximum	90	20	0.10	20
Average	89	3	0.02	6
No. of descriptions	191	191	190	183
Southern Ute Mobil 36–1				
Minimum	80	0	0.01	0
Maximum	90	4	1.38	40
Average	86	1	0.06	10
No. of descriptions	151	151	151	151
Colorado 32–7 No. 9				
Minimum	35	0	0.01	1
Maximum	90	11	0.21	28
Average	76	4	0.05	7
No. of descriptions	49	49	48	49
Southern Ute Tribal H				
Minimum	40	0	0.01	0
Maximum	90	60	0.30	20
Average	89	3	0.02	6
No. of descriptions	168	163	163	168
Southern Ute Tribal J				
Minimum	80	0	0.01	0
Maximum	90	20	0.07	20
Average	89	2	0.02	5
No. of descriptions	318	296	312	318

Mineral Matter

A key difference between the Fruitland and Menefee coal cores is that mineral matter (approximately equivalent to "ash" in this case) in Fruitland samples occurs in discrete layers in sharp contact with coal layers irrespective of rank, whereas mineral matter in Menefee coal is dispersed and rarely present in discrete layers. "Ash" is what remains of the coal after combustion during routine proximate analysis. Fruitland cleat development is largely confined to those zones within the coal that have low ash content and that lay adjacent to higher ash layers, as shown in Figure 4.

The Menefee samples come from a well (Ute Mountain Ute 33–14) on the Hogback monocline (Figure 1) along the northern basin margin. The core is similar in rank, maceral composition, coal chemistry, and depth to Fruitland cores from other areas such as Cedar Hill field. However, the Cedar Hill Basal Fruitland pool is characterized by less than 200 ft (60 m) of true structural relief, which is far less than that along the monocline. One then would expect Menefee coal from the folded terrane to be more fractured and permeable. However, the opposite was found—fluid could not even be injected into the Menefee coal, whereas Fruitland absolute permeability in the Hamilton 3 well at Cedar Hill is more than 5 md (Mavor and Close, 1989a).

Cleat Geometry and Spatial Relationships

Fruitland face and butt cleats in whole cores are vertical to within 10° of vertical (Table 2), and most cleats are perpendicular to vitrain/nonvitrain surfaces. Face and butt cleat trends also are orthogonal, as shown in Figure 5. Butt cleats, which usually measure less than 1 in. (2.5 cm) long, terminate against face cleats, which range in length from approximately 1 in. (2.5 cm) to 3.5 in. (8.9 cm, core diameter).

Third-order and fourth-order Fruitland cleats in whole-core samples occur in vitrain bands, have rather variable lengths (but less than face cleat lengths), dip directions, and dip angles. Those that are vertical or nearly vertical (and that may be open under burial stress) usually are orthogonal. The complexity of Fruitland cleats is illustrated in Figure 6.

Fruitland cleat height at the megascopic (hand sample) scale is controlled by vitrain layer thicknesses that range from approximately 2 in. (5 cm) to slightly less than 0.1 in. (0.3 cm). The cleats commonly terminate at boundaries with nonvitrain coal lithotypes (Figure 3). However, several cleats can transect vitrain/nonvitrain layers and extend as high as 24 in. (61 cm).

Fruitland cleat frequency, which increases generally as a function of decreasing vitrain layer thickness, varies from 0.25/in. to 100/in. (0.1/cm to 40/cm) and averages 15/in. (6/cm). The frequency of cleats that less commonly transect vitrain/nonvitrain layers varies from an average of 0.3/in. to 1/in. (0.1/cm to 0.4/cm). Fruitland cleat frequency also increases as a function of increasing frequency of vitrain/nonvitrain interbedding.

Fruitland face and butt cleat aperture widths vary from 0.0004 to 0.05 in. (0.001 to 0.138 cm), and average 0.015 in. (0.038 cm). These values are biased toward the high end because the information was obtained through examination of nonstressed cores. By comparison, the smallest size (100 mesh) sand proppant used in hydraulic fracture treatments in Fruitland coals is 0.0059 in. (0.015 cm) in diameter.

Menefee cleats in core, when present, occur in vitrain layers less than 1 in. (2.5 cm) thick and are orthogonal to vitrain/nonvitrain surfaces and to one another. These cleat heights also are controlled by vitrain band thickness. Cleat lengths usually are less than 1 in. (2.5 cm), and cleat frequency averages 1/in. (0.4/cm). Aperture widths are similar to those for Fruitland coal. Observations of similar rank Menefee coal in roadcuts and abandoned mine entries near Durango, Colorado, are in general agreement with core descriptions.

Minerals and Macerals in Cleats

Calcite is the most common mineral present in Fruitland coal cleats. Pyrite and siderite occasionally are present. Also present in cleat apertures is exsudatinite, an orangish-yellow secondary (migrated) coal maceral that originates from resinite macerals common in Cretaceous Western Interior coals. It is somewhat common in Fruitland coal

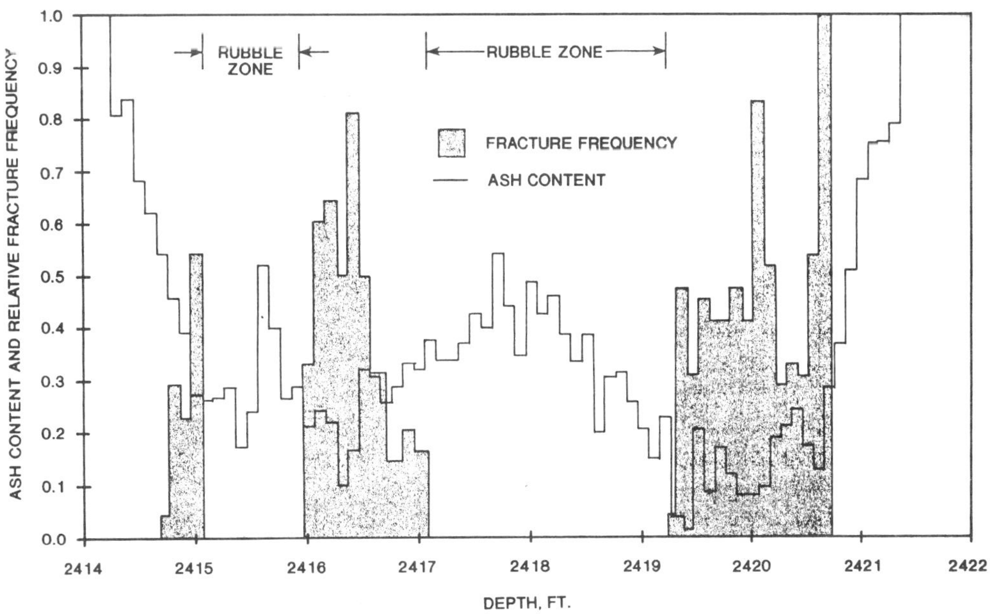

Figure 4. Comparison of ash content and relative fracture frequency for Fruitland coal cores from the Southern Ute Mobil 36–1 well.

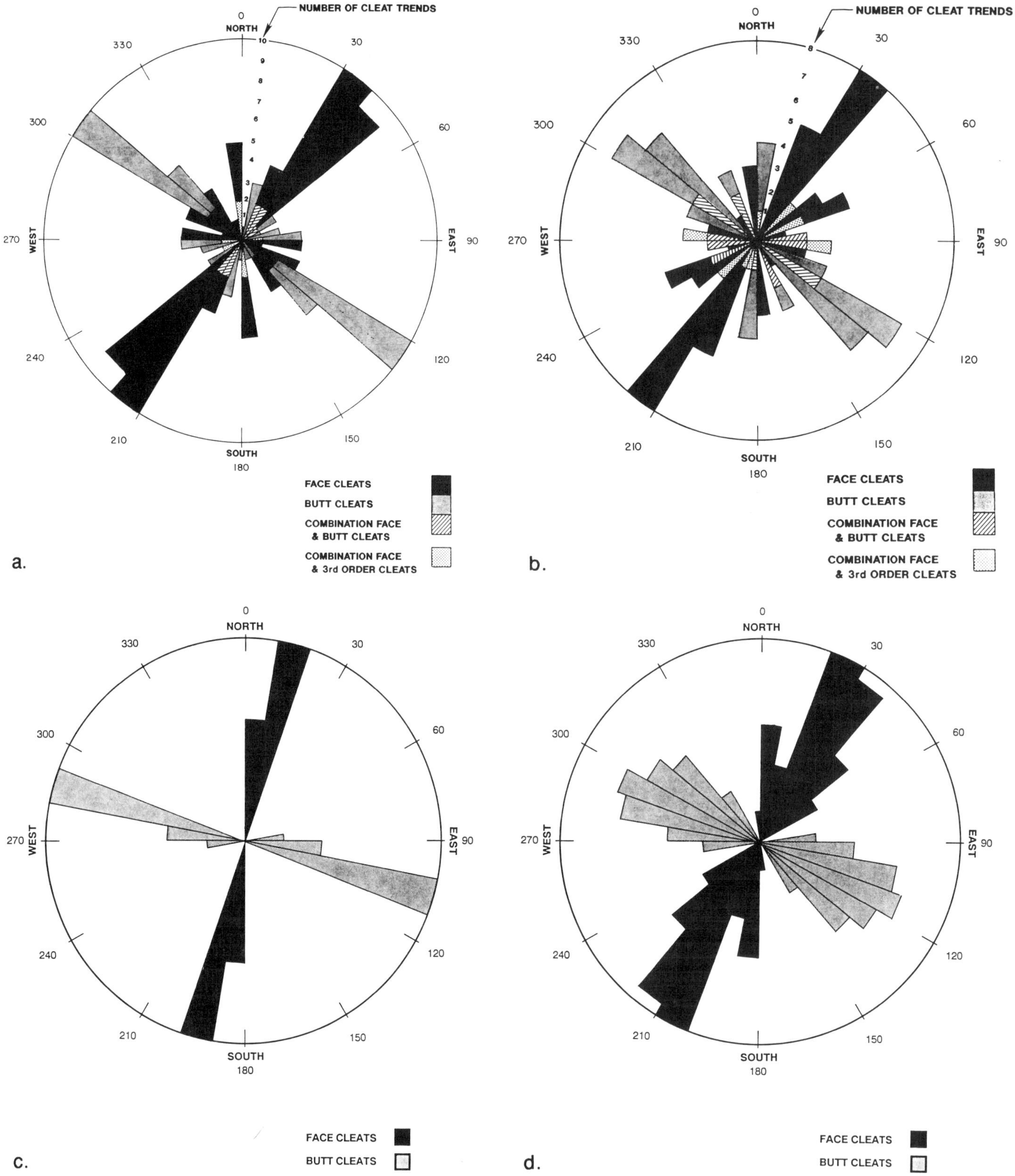

Figure 5. Rose diagrams of Fruitland coal cleat orientation. ***a.*** **Hamilton 3 well.** ***b.*** **Northeast Blanco Unit 403 well.** ***c.*** **Colorado 32–7 No. 9 well.** ***d.*** **Southern Ute Tribal H–1 well.**

samples that have a vitrinite reflectance of less than approximately 1.2% (lower end of medium-volatile rank). Geochemical typing for one well (Kruge, 1989) suggests that the resinite and exsudatinite macerals, which typically comprise about 5% of high-volatile bituminous Fruitland coal, are probably significant sources of paraffin that is

Figure 6. Cleats in Fruitland coal core from the Southern Ute Tribal H–1 well at approximately 3,209.5 ft (975.5 m). Core diameter is 3.5 in. (8.9 cm).

produced from Fruitland wells in the west-central San Juan basin.

Cleat Occurrence and Bulk Density

Comparison of cleat description and computerized axial tomography (CAT) scan bulk-density data (Terra Tek Inc., 1988c, 1988d, 1989c, 1989d, 1990c, 1990d) for six of the Fruitland cores indicate that cleats occur in discrete vitrain bands that have approximate densities between 1.2 g/cc and 1.75 g/cc. An example of the relationship between face cleat frequency and CAT scan bulk density is shown in Figure 7. These data in turn suggest that zones in Fruitland coals of high-volatile A bituminous and greater rank that may have significant cleat development can be initially inferred through inspection of calibrated open-hole high-resolution bulk-density logs.

Cleat Trends and Major Structures

Fruitland face and butt cleats typically are oriented perpendicular and parallel, respectively, to major fold axes that were created by inferred regional paleotectonic compression following Fruitland sediment deposition. Interpretation of a structure map on the top of the Pictured Cliffs Sandstone (Ayers, 1990) indicates that the primary synclinal axis of the San Juan basin trends northwest. The cleat-orientation diagrams in Figure 5 indicate that face cleats trend predominantly northeast, which is parallel to the inferred Laramide-style paleotectonic compressive stress direction. Vertical or near-vertical third- and fourth-order cleats that are orthogonal to one another often are oriented 45° to the trends of face and butt cleats. These features formed after the face and butt cleats, perhaps the result of west-directed regional compression related to middle Tertiary to Quaternary extension of the Rio Grande Rift east of the San Juan basin.

Little variability was found in the trend of a specific cleat type, which further suggests paleotectonic stress control on cleat genesis. Cleats that formed as a result of other processes, such as compaction, probably would not show such regularity in trends (or dip angles; see data in Table 2).

Cleat orientations in cores differ from those determined along outcrops in the northern San Juan basin. Care must be exercised in any comparisons because the data come from different structural domains (Hogback monocline vs basin interior) and from coals of distinctly different rank.

Volcanic Ash Beds

Megascopic core description, x-ray diffraction, and scanning electron microscope/energy-dispersive x-ray analyses indicate that a selected set of tonsteins (volcanic ash beds in coal), bentonites, and shale layers in the Fruitland consist mostly of quartz and plagioclase feldspar, with subordinate kaolinite and mixed-layer clays. Tonsteins and bentonites can most consistently be recognized on modern, calibrated open-hole high-resolution bulk-density logs, since the layers typically have high gamma ray readings (often more than 200 API units), bulk densities of 2.2 to 2.4 g/cc, and thicknesses less than 1 ft (0.3 m). Natural gamma ray spectroscopy log traces also indicate that tonsteins and bentonites contain thorium and uranium concentrations well above background levels.

Relationship to Stratigraphy and Structure

Interpretation of large-scale (detailed) structure maps constructed on tonsteins and bentonites (which are highly reliable time-correlative stratigraphic and structural markers) and analysis of fracture-description data indicate that significant cleat frequency is not dependent on the presence of folds and faults. Commercial Fruitland wells often are located in areas with essentially no folding and faulting.

Conversely, interpretation of the same maps, together with well test permeability and fluid-production data, indicate that some Fruitland wells that are located on major folds such as the Ignacio anticline and that have sufficient gas in-place resources, reservoir pressure, and desorption characteristics for potentially commercial production, are noneconomic gas producers (Dhir and others, this volume). Therefore these areas possibly could be subject to greater horizontal or near-horizontal stresses that are constricting cleat apertures.

CLEAT GENESIS

Processes

Based on the above observations, we hypothesize that significant Fruitland cleat development is dependent on

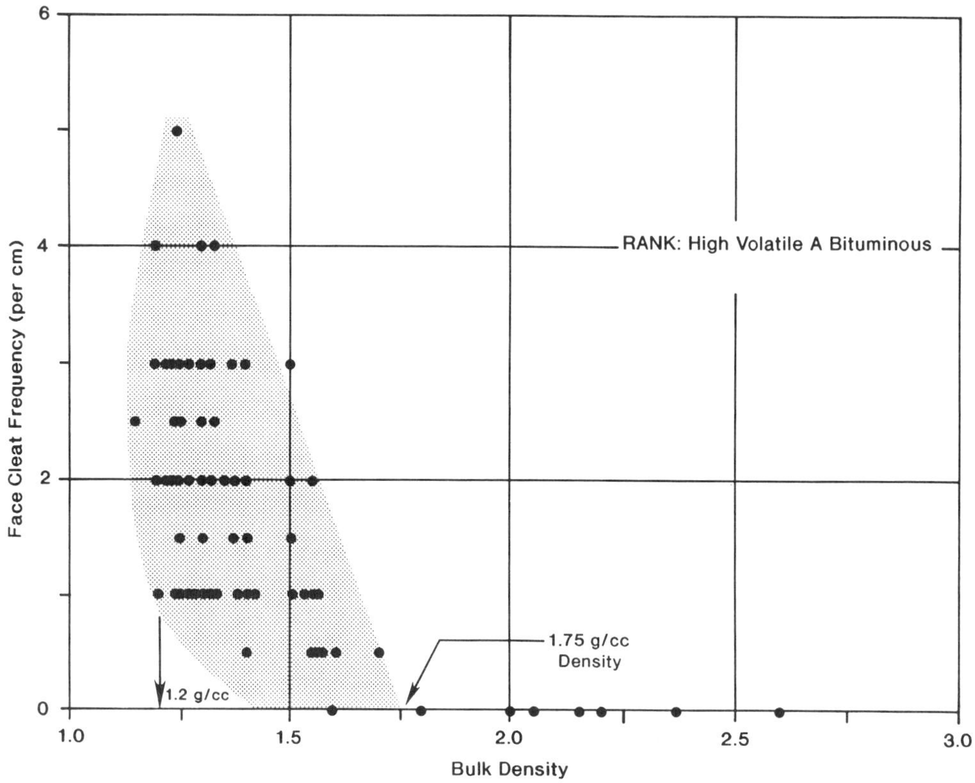

Figure 7. Face cleat frequency vs CAT scan bulk density for cores from the Hamilton 3 well.

the presence of high-volatile A bituminous and greater rank coals that contain a high frequency of vitrain and nonvitrain interbeds. The rank of high-volatile A bituminous is commonly known in the oil and gas industry as the onset of the petroleum generation window. Cleats in vitrain bands are extensional features envisioned to have formed after lithification and during active coalification as a result of progressive moisture loss, devolatilization, and gas generation that occurred simultaneously with episodes of tectonic compression. We suggest that in situ pore-fluid stresses, which exceeded the tensile strength of vitrain layers during coalification, also were required for cleat development. During active coalification, the pore fluids could have been expelled into adjacent vitrain bands as a result of clay dewatering from interbedded shale, which was occurring simultaneously with expulsion of water from the vitrain.

Timing

A significant degree of lithification and coalification (through burial metamorphism), tectonic deformation, and Fruitland cleat genesis could have occurred in as few as 3.4 million years (about 5% of the age of the Fruitland Formation). In other words, Fruitland cleat genesis could have been extremely rapid in geologic terms.

This hypothesis concerning pre-Laramide Fruitland coalification directly challenges the commonly held belief that high temperatures associated with magmas that formed the San Juan Mountains north of the San Juan basin were responsible for the observed Fruitland coal rank pattern (cf., Eddy, 1984). The reasoning for this inference relates to (1) distribution of coal rank and (2) age and stratigraphic relationships among Upper Cretaceous and Tertiary strata.

First, on Ayers' (1990) revised Fruitland coal-rank map, high-volatile C, B, and A bituminous isorank lines are "folded" along the deformed northern basin margin. Therefore a significant degree of Fruitland coalification must have preceded Fruitland deformation. Determination of relative age of coalification with respect to deformational events is explained in Hacquebard and Donaldson (1974) and Pearson and Grieve (1985).

Second, the age of the top of the Kirtland Shale, which overlies the Fruitland Formation in the northeastern San Juan basin, is estimated to be 73.2±0.7 Ma (Fassett, 1987). This then is the oldest age for the McDermott Member (base) of the Animas Formation, which overlies the Kirtland. (Cretaceous rocks become younger toward the northeast, following regression of the Late Cretaceous seaway.) The McDermott Member, the youngest Cretaceous unit in the basin, is believed to be early Maastrichtian (Newman, 1987), and the latest early Maastrichtian is thought to be 70.5 Ma (Palmer, 1984). Therefore the age of the top of the McDermott, and thus the top of the Cretaceous in the northeastern San Juan basin, is estimated at 70.5 Ma.

An erosional disconformity separates Cretaceous and Tertiary rocks in the San Juan basin (Fassett and others, 1987; Sikkink, 1987; Ayers and others, 1988, 1990; Ayers, 1990). Folded and eroded McDermott and older rocks lay beneath Tertiary rocks that have not undergone significant Laramide deformation. One can then bracket early deformation to a post-Fruitland sedimentation and pre-Tertiary time frame. The time interval between (a) Fruitland deposition, lithification, and coalification, and (b) early Laramide deformation that influenced Fruitland cleat genesis and the Cretaceous/Tertiary disconformity, is therefore estimated to be 2.7±0.7 million years (73.2±0.7 Ma to 70.5 Ma), or no more than 3.4 million years and possibly as little as 2 million years.

EXAMPLES OF LOG INTERPRETATION

Three case studies are presented to illustrate the use of the suggested log-interpretation techniques in Fruitland coal. The first technique is the application of high-resolution bulk-density logs, on which layer macroheterogeneity in the coal is manifested as sharp changes (>0.25 g/cc) in density. The second technique is the identification of prominent positive separation on calibrated microlog responses as an indication of the presence (or absence) of permeability.

Microlog positive separation is detected when the *micronormal* trace, which measures formation resistivity at a penetration depth of 2 to 4 in. (5 to 10 cm), reads higher than the *microinverse* trace, which measures mud cake resistivity. Mud cake sometimes can be detected through inspection of the caliper trace for wellbores that are in gauge—borehole diameter will be slightly less through zones with mud cake. Positive separation occurs only if the following resistivity contrast is present at reservoir temperature:

mud cake resistivity > drilling mud resistivity
> mud filtrate resistivity

Note that the microlog is an open-hole log run in wells drilled with fresh-water-base mud and is most effective when the borehole is in gauge, and the well of interest was drilled in a balanced to overbalanced condition.

A zone of low to negligible permeability can qualitatively be recognized on the microlog trace when the magnitude of positive separation is minimal and when the resistivity of the zone of interest is at least ten times that of the drilling mud (Asquith, 1982) at reservoir temperature.

The geometry of the microlog pad device (which contacts only a small part of the mud cake), when considered with the overall microlog response during repeat passes in a permeable Fruitland coal zone, indicates that a large number of open cleats must control near-wellbore mud-filtrate invasion and fluid deliverability.

Southern Ute Mobil 36–1

The Southern Ute Mobil 36–1 well is located about 10 miles (16 km) south-southwest of Durango, Colorado, in the northwestern San Juan basin. The coal zones in this well consist of medium-volatile bituminous coal and mostly shale; a tonstein occurs at a depth of 2,470 ft (751 m) (Mavor and Close, 1989c). On the bulk density log/microlog suite shown in Figure 8, note the numerous sharp fluctuations in bulk density in the coal zones and the significant positive separation on the microlog trace. The fact that the positive microlog separation is restricted to the coal seams indicates mud-filtrate invasion only into the coals and not into adjacent shale and sandstone. One therefore can infer that the shale and sandstone are impermeable and that development of fractures and permeability is confined to the coals.

Based on coal rank and log signatures for this well, one could infer a high degree of interbedding and speculate that the coal reservoirs of interest have significant permeability. An open-hole drillstem test (DST) of the interval between 2,392 and 2,454 ft (727 and 746 m) was completed to obtain data for estimation of reservoir pressure, permeability, and damage to near-wellbore permeability. Analysis of these data indicated that the coal zone between 2,408 and 2,452 ft (732 and 745 m) has an absolute permeability of 36 md, which validates the log interpretation.

Fracture description data indicate an average coal cleat dip of 86°, ranging from 80° to 90°, (based on 151 observations); average (nonstressed) cleat aperture widths of 0.002 in. (0.006 cm), ranging from 0.0004 to 0.05 in. (0.001 to 0.138 cm); and average cleat frequency of 25/in. (10/cm), ranging from 0.8/in. to 102/in. (0.3/cm to 40/cm). The coal therefore has excellent cleat development, and one would expect general concordance among the fracture description data, high rank, micrulog response, and high well-test absolute permeability estimate.

FC Federal 12

The FC Federal 12 well is located about 40 miles (64 km) southeast of Farmington, New Mexico, and about 20 miles (36 km) south of the Meridian 400 field. Coal from this well is high-volatile C bituminous. On the log suite (Close and others, 1990a) in Figure 9, note the numerous sharp fluctuations in bulk density in the coal zone between 2,226 and 2,246 ft (677 and 683 m), which suggest significant layer macroheterogeneity. Responses to repeat logging passes with the microlog showed very limited to no positive separation in the coals and formation resistivities more than ten times than that of the drilling mud at reservoir temperature. Both suggest very low permeability. Analysis of drillstem test data from the interval 2,200 to 2,252 ft (669 to 685 m) yielded an absolute permeability estimate of 0.004 md, which confirms the log interpretation. Coal core from this well is virtually devoid of cleats, and thus the low well-test permeability estimate is not surprising.

Although it is only one data point, this example underscores the hypothesis that Fruitland cleat genesis is dependent on the presence of layer macroheterogeneity and high-volatile A bituminous or greater coal rank. Although the coal from this well exhibits well developed layer macroheterogeneity, it has not reached high-volatile A bituminous rank and has low permeability.

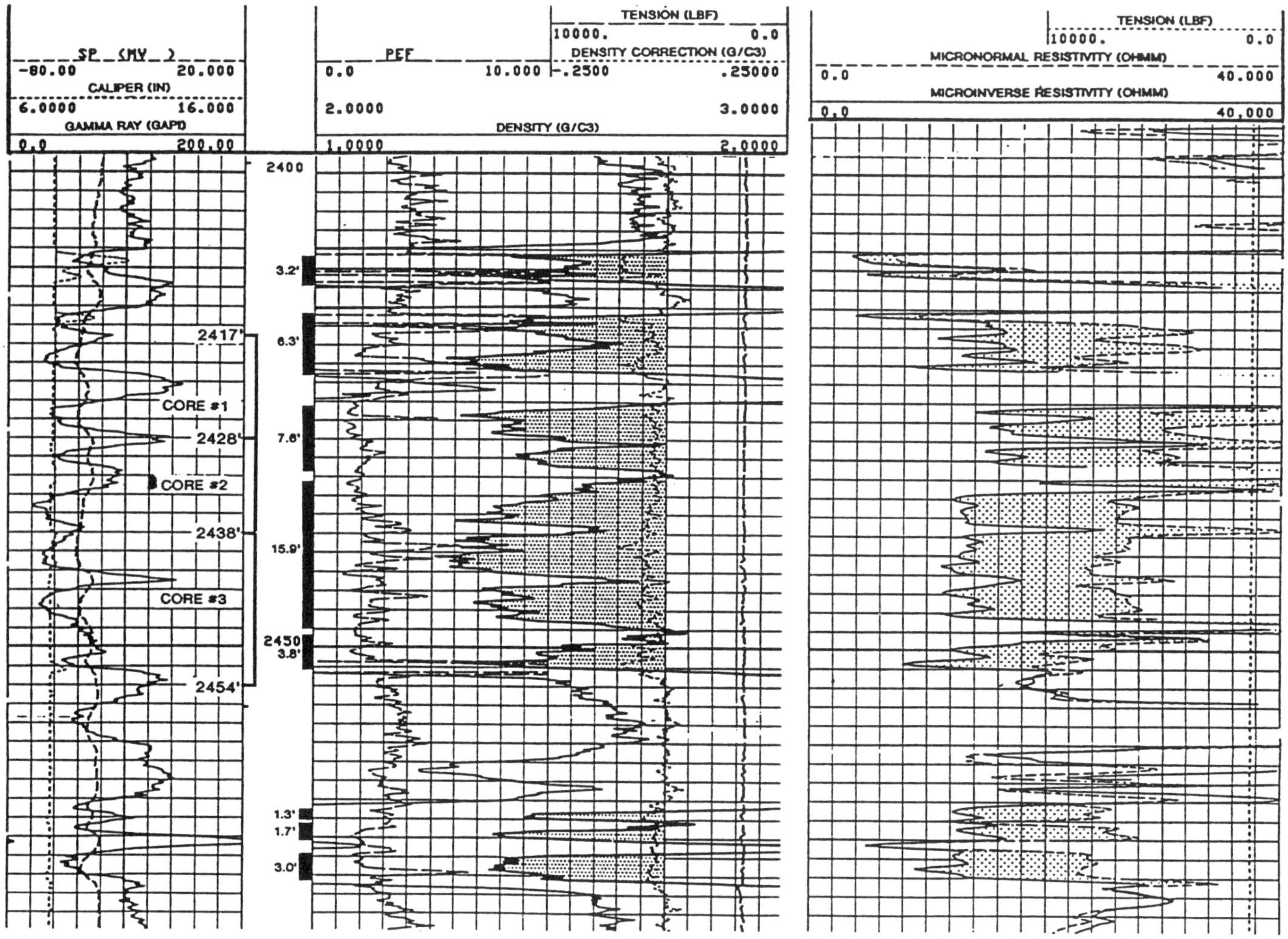

Figure 8. Gamma ray, bulk density, and microlog log suite for the Southern Ute Mobil 36–1 well.

Colorado 32–7 No. 9

The Colorado 32–7 No. 9 well is located about 20 miles (36 km) southeast of Durango, Colorado (Mavor and Close, 1989d). Inspection of the high-resolution bulk-density log (Figure 10) indicates that significant layer macroheterogeneity is present. One would expect good cleat development given this density log response and rank of the coal (medium-volatile bituminous). However, the microlog shows only minor positive separation in the coal zones, which suggests low permeability. Analysis of drillstem test data for the interval 2,876 to 2,946 ft (874 to 896 m) yielded an absolute permeability estimate of 1.75 md, which confirmed the log interpretation.

This example illustrates well that not all northern San Juan basin Fruitland acreage positions that have thick gassy coals, sufficient reservoir pressure, and favorable diffusion and desorption characteristics are economically producible at current gas prices (Dhir and others, this volume). The case study also highlights the importance of running the microlog and drillstem tests as part of the formation evaluation effort even in an area of the basin that has high potential.

Note also that the Colorado 32–7 No. 9 well is located on the southwestern flank of the Ignacio anticline, a significant intrabasinal structure with which, according to many, significant cleat development can be associated. Folding is envisioned to have created significant fracturing in zones of maximum flexural magnitude and rate of change, as identified through curvature analysis. Yet, the permeability of Fruitland coal at this well site is quite low; thus, curvature analysis must be used with care. Possibly this well is situated in an area that is characterized in part by greater horizontal stress, which has decreased cleat aperture widths and reduced permeability.

SUMMARY

The presence or absence of favorable fracture (cleat) permeability is a key factor that will strongly influence the success of future Fruitland development projects in areas that have favorable reservoir characteristics.

The most abundant low-cost type of publicly available information for a widespread area of the Fruitland play is the geophysical log. To make full use of these available

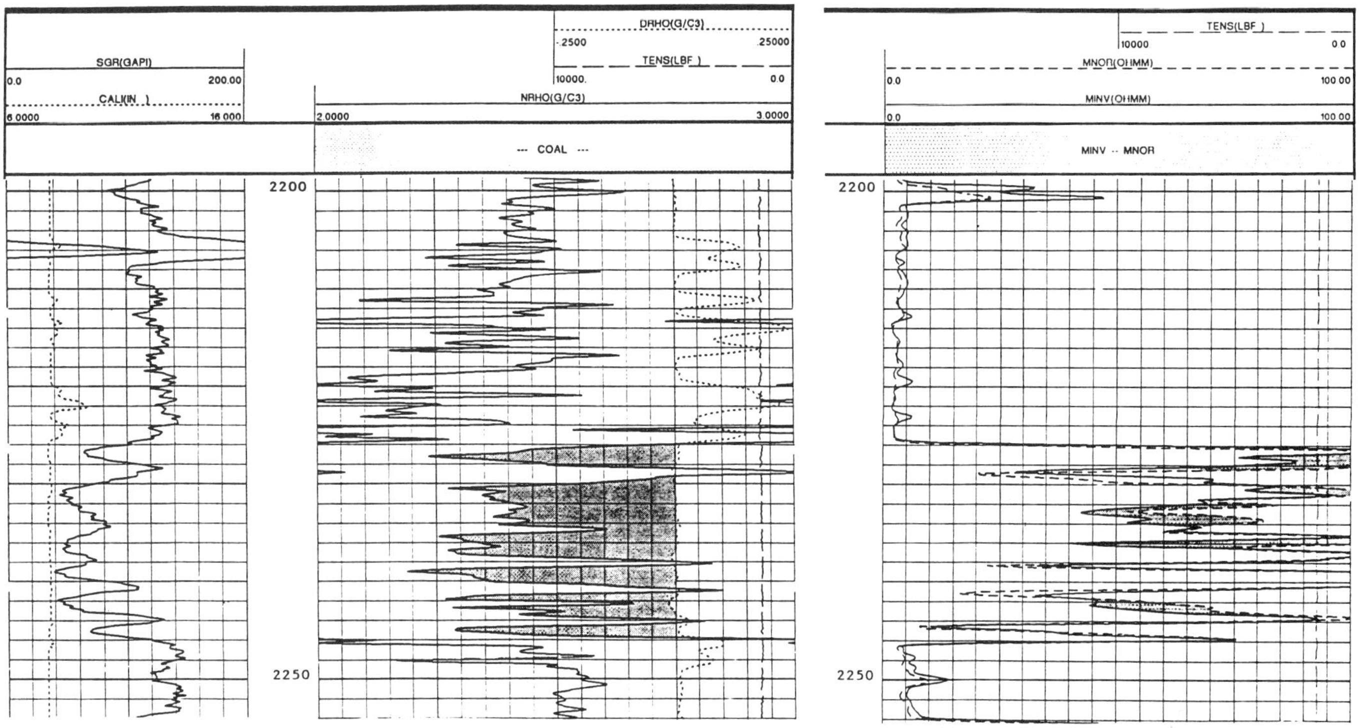

Figure 9. Gamma ray, bulk density, and microlog log suite for the FC Federal 12 well.

logs, we suggest that the microlog be used to infer the presence of fresh-water-based drilling-mud invasion into permeable Fruitland coal zones. Microlog technology has been available for about 40 years, and many operators run this log in Fruitland and other deeper wells as part of formation evaluation efforts.

In this paper we have analyzed a detailed set of fracture description and formation evaluation data to understand Fruitland cleat occurrence and genesis, and have discussed calibration of microlog responses to cores from coal zones of variable permeability. Comparison of microlog traces, fracture description data, and drillstem test data indicate that microlog responses in wells with commercial and noncommercial Fruitland permeability are different and recognizable. We caution, however, that the proposed log-interpretation technique never be used in place of rigorous formation evaluation efforts for Fruitland (or any) coal wells, since complete data sets are absolutely essential for accurate reservoir simulation of fluid productivity, reserve estimation, and economic projections.

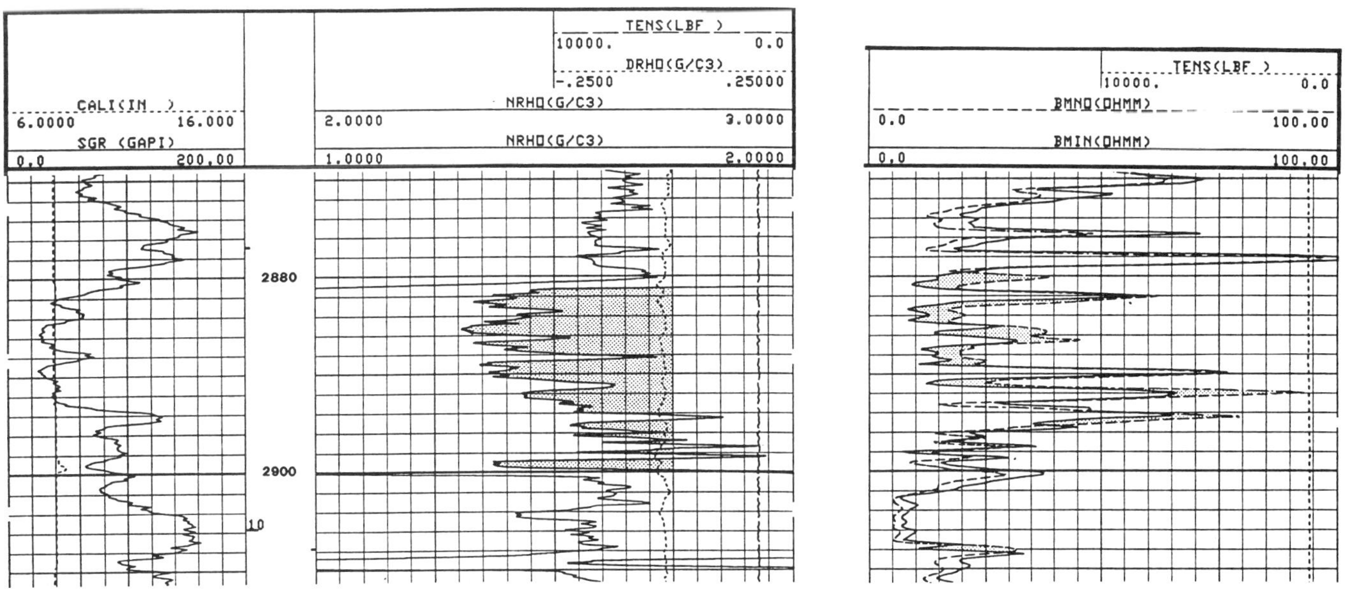

Figure 10. Gamma ray, bulk density, and microlog log suite for the Colorado 32–7 No. 9 well.

ACKNOWLEDGMENTS

Funding for this research was provided by the Gas Research Institute through its Western Cretaceous Coal Seam Project (contract no. 5088-214-1657), which is sincerely acknowledged. We also thank the various operators in the San Juan basin for allowing cooperative research to be performed in conjunction with their efforts.

REFERENCES

Asquith, G.B., 1982, Basic well log analysis for geologists: AAPG Methods in Exploration Series, 216 p.

Ayers, W.B., Jr., 1990, Geologic evaluation of critical production parameters for coalbed methane resources, San Juan Basin: Gas Research Institute, Natural Gas Supply Project Advisors Meeting guidebook, Methane from Coal Deposits Project Area, September 17–19, Durango, Colorado.

Ayers, W.B., Jr., Kaiser, W.R., Ambrose, W.A., Swartz, T.E., Laubach, S.E., Fisher, W.L., Tremain, C.M., and Whitehead, N.H., III, 1990, Geologic evaluation of critical production parameters for coalbed methane resources, part 1—San Juan basin: Gas Research Institute, Annual Rept. GRI–90/0014.1 (August 1988–July 1989), 175 p.

Ayers, W.B., Jr., Kaiser, W.R., Swartz, T.E., Zellers, S.D., and Scanlon, A.H., 1988, Geologic evaluation of critical production parameters for coalbed methane resources, part 1—San Juan Basin: Gas Research Institute, Annual Rept. GRI–88/0332.1 (August 1987–July 1988), 100 p.

Billings, M.P., 1972, Structural geology: Englewood Cliffs, N. J., Prentice Hall, 606 p.

Close, J.C., and Dutcher, R.R., 1990, Prediction of permeability trends and origins in coal-bed methane reservoirs of the Raton Basin, New Mexico and Colorado, *in* Bauer, P.W., Lucas, S.G., Mawer, C.K., and McIntosh, W.C., eds., Tectonic development of the southern Sangre de Cristo Mountains, New Mexico: New Mexico Geological Society 41st Annual Field Conference Guidebook, p. 387–395.

Close, J.C., Mavor, M.J., Dhir, R., Pratt, T.J., and Hughes, B., 1990b, Western Cretaceous Coal Seam Project preliminary evaluation of the co-operative research well FC Federal #12 operated by Mesa Operating Limited Partnership: Preliminary report submitted to Gas Research Institute under Contract 5088–214–1657.

Close, J.C., Mavor, M.J., and McBane, R.A., 1990a, Importance, genesis and recognition of fracture permeability in Fruitland coalbed methane reservoirs of the northern San Juan Basin, Colorado and New Mexico: Petroleum Society of CIM/Society of Petroleum Engineers Paper CIM/SPE 90–106 [presented at the International Technical Meeting, Calgary, Alberta, Canada, June 10–13, 1990]

Davis, G.H., 1984, Structural geology of rocks and regions: New York, John Wiley, 492 p.

Doll, H.G., 1950, The microlog—A new electrical logging method for detailed determination of permeable beds: American Institute of Mining and Metallurgical Engineers Transactions, v. 189, p. 155–164 [reprinted in Foster, N.H., and Beaumont, E.A., compilers, 1990, Formation evaluation I—Log evaluation: AAPG Treatise of Petroleum Geology, Reprint Series 16, p. 77–86].

Doll, H.G., 1951, The laterolog—A new resistivity logging method with electrodes using an automatic focusing system: American Institute of Mining and Metallurgical Engineers Transactions, v. 192, p. 305–316 [reprinted in Foster, N.H., and Beaumont, E.A., compilers, 1990, Formation evaluation I—Log evaluation: AAPG Treatise of Petroleum Geology, Reprint Series 16, p. 65–76].

Doll, H.G., 1953, The microlaterolog: American Institute of Mining and Metallurgical Engineers Transactions, v. 198, p. 17–32 [reprinted in Foster, N.H., and Beaumont, E.A., compilers, 1990, Formation evaluation I—Log evaluation: AAPG Treatise of Petroleum Geology, Reprint Series 16, p. 87–102].

Eddy, G.E., 1984, Conclusions on the development of coalbed methane, *in* Rightmire, C.T., Eddy, G.E., and Kirr, J.N., eds., Coalbed methane resources of the United States: AAPG Studies in Geology, Series 17, p. 373–375.

Fassett, J.E., 1987, The ages of the continental, Upper Cretaceous, Fruitland Formation and Kirtland Shale based upon a projection of ammonite zones from the Lewis Shale, San Juan Basin, New Mexico and Colorado, *in* Fassett, J.E., and Rigby, J.K., Jr., eds., The Cretaceous-Tertiary boundary in the San Juan and Raton Basins, New Mexico and Colorado: GSA Special Paper 209, p. 5–16.

Fassett, J.E., Lucas, S.G., and O'Neill, F.M., 1987, Dinosaurs, pollen and spores, and the age of the Ojo Alamo Sandstone, San Juan Basin, New Mexico, *in* Fassett, J.E., and Rigby, J.K., Jr., eds., The Cretaceous-Tertiary boundary in the San Juan and Raton Basins, New Mexico and Colorado: GSA Special Paper 209, p. 17–34.

Hacquebard, P.A., and Donaldson, J.R., 1974, Rank studies of coals in the Rocky Mountains and Inner Foothills Belt, Canada, *in* Dutcher, R.R., Hacquebard, P.A., Schopf, J.M., and Simon, J.A., eds., Carbonaceous materials as indicators of metamorphism: GSA Special Paper 153, p. 75–94.

Hobbs, B.E., Means, W.D., and Williams, P.F., 1976, An outline of structural geology: New York, John Wiley, 571 p.

Jones, A.H., Bell, G.J., and Schraufnagel, R.A., 1988, A review of the physical and mechanical properties of coal with implications for coal-bed methane well completion and production, *in* Fassett, J.E., ed., Geology and coal-bed methane resources of the northern San Juan basin, Colorado and New Mexico: Rocky Mountain Association of Geologists Guidebook, p. 169–181.

Kruge, M.A., 1989, Molecular marker analysis of wax precipitates from the Mesa Hamilton No. 3 Well, Cedar Hill Field, San Juan Basin, New Mexico: Report to Resource Enterprises Inc., Salt Lake City, Utah, for the GRI Western Cretaceous Coal Seam Project.

Kulander, B.R., Dean, S.L., and Ward, B.J., Jr., 1990, Fractured core analysis—Interpretation, logging, and use of natural and induced fractures in core: AAPG Methods in Exploration Series 8, 88 p.

Mavor, M.J., and Close, J.C., 1989a, Western Cretaceous Coal Seam Project evaluation of the cooperative research well Hamilton #3 operated by Mesa Operating Limited Partnership: Gas Research Institute, Topical Rept. GRI–90/0400, 40 p.

Mavor, M.J., and Close, J.C., 1989b, Western Cretaceous Coal Seam Project evaluation of the cooperative research well Northeast Blanco Unit operated by Blackwood & Nichols Co. Ltd.: Gas Research Institute, Topical Rept. GRI–90/0041, 46 p.

Mavor, M.J., and Close, J.C., 1989c, Western Cretaceous Coal Seam Project evaluation of the cooperative research well Southern Ute Mobil 36–1 operated by McKenzie Methane Corporation: Gas Research Institute, Topical Rept. GRI–90/0042, 18 p.

Mavor, M.J., and Close, J.C., 1989d, Western Cretaceous Coal Seam Project evaluation of the cooperative research well Colorado 32–7 #9 operated by Mobil Oil Corporation: Gas Research Institute, Topical Rept. GRI–90/0043, 21 p.

Mavor, M.J., and Close, J.C., 1989e, Western Cretaceous Coal Seam Project final project summary, January 1, 1988, to August 31, 1989, research: Gas Research Institute, Final Rept. GRI–90/0044, 28 p.

Mavor, M.J., Close, J.C., and McBane, R.A., 1990, Formation evaluation of exploration coalbed methane wells: Petroleum Society of CIM/Society of Petroleum Engineers Paper CIM/SPE 90–101 [presented at International Technical Meeting, Calgary, Alberta, Canada, June 10–13, 1990].

Nelson, R.A., 1985, Geologic analysis of naturally fractured reservoirs: Houston, Gulf Publishing Co., 320 p.

Newman, K.R., 1987, Biostratigraphic correlation of Cretaceous-Tertiary boundary rocks, Colorado to San Juan Basin, New Mexico, *in* Fassett, J.E., and Rigby, J.K., Jr., eds., The Cretaceous-Tertiary boundary in the San Juan and Raton Basins, New Mexico and Colorado: GSA Special Paper 209, p. 151–164.

Palmer, A.R., compiler, 1984, The Decade of North American Geology, geologic time scale: GSA Map and Chart Series MC–50.

Pearson, D.E., and Grieve, D.A., 1985, Rank variation, coalification pattern and coal quality in the Crowsnest coalfield, British Columbia: Canadian Institute of Mining and Metallurgy [CIM] Bulletin, v. 78, no. 881, p. 39–46.

Pollard, D.D., and Aydin, A., 1988, Progress in understanding jointing over the past century : GSA Bulletin, v. 100, no. 8, p. 1181–1204.

Sikkink, P.G.L., 1987, Lithofacies relationships and depositional environments of the Tertiary Ojo Alamo Sandstone and related strata, San Juan Basin, New Mexico and Colorado, *in* Fassett, J.E., and Rigby, J.K., eds., The Cretaceous-Tertiary boundary in the San Juan and Raton Basins, New Mexico and Colorado: GSA Special Paper 209, p. 81–104.

Terra Tek Inc., 1988a, Fracture analysis of cores from Mesa Petroleum Ltd.'s Hamilton #3 well: Salt Lake City, Utah, Terra Tek Core Services, File 88730.

Terra Tek Inc., 1988b, Fracture analysis of cores from Blackwood and Nichols Ltd., Northeast Blanco Unit #403 well, Rio Arriba County, New Mexico: Salt Lake City, Utah, Terra Tek Core Services, File 89802C.

Terra Tek Inc., 1988c, CAT scan analysis performed on the Hamilton #3 well, San Juan Basin: Salt Lake City, Utah, Terra Tek Geoscience Services, Project 89–118x.

Terra Tek Inc., 1988d, CAT scan analysis performed on Blackwood & Nichols Co., Ltd., Northeast Blanco Unit #403: Salt Lake City, Utah, Terra Tek Geoscience Services, Project 89–170x.

Terra Tek Inc., 1989a, Fracture analysis of core from McKenzie Methane Southern Ute Mobil 36–1 well, La Plata County, Colorado: Salt Lake City, Utah, Terra Tek Core Services, File 5572.

Terra Tek Inc., 1989b, Fracture analysis of core No. 4 from Mobil Exploration and Producing U.S. Inc.'s Colorado 32–7 No. 9 well, La Plata County, Colorado: Salt Lake City, Utah, Terra Tek Core Services, File 5353.

Terra Tek Inc., 1989c, CT analysis performed on core from the McKenzie Methane Southern Ute Mobil 36–1 Research well: Salt Lake City, Utah, Terra Tek Geosciences, Project 3562.

Terra Tek Inc., 1989d, CT analysis performed on core from Mobil Colorado 32–7 #9 well: Salt Lake City, Utah, Terra Tek Geoscience Services, Project 3562.

Terra Tek Inc., 1990a, Fracture analysis of a Fruitland coal core from Amoco Production Company's Southern Ute Tribal H #1 well, La Plata County, Colorado: Salt Lake City, Utah, Terra Tek Geosciences Group, File 5639.

Terra Tek Inc., 1990b, Fracture analysis of Fruitland coal cores from Amoco Production Company's Southern Ute Tribal J #1 well, La Plata County, Colorado: Salt Lake City, Utah, Terra Tek Geoscience Services, File 5638.

Terra Tek Inc., 1990c, CT analysis performed on core from the Southern Ute Tribal H–1 well: Salt Lake City, Utah, Terra Tek Geoscience Services, Project 5639.

Terra Tek Inc., 1990d, CT analysis performed on core from the Southern Ute Tribal J–1 well: Salt Lake City, Utah, Terra Tek Geoscience Services, Project 5638.

West Spanish Peak
Giant Staircase

Predicting Cleats in Coal Seams from Mineral and Maceral Composition with Wireline Logs

David J. Johnston[1]
Patrick L. Scholes[1]

1. Schlumberger Well Services, Aurora, Colorado

ABSTRACT

Wireline logs have been used to identify coals and have been calibrated in certain basins to provide grade and rank. With the addition of a geochemical logging suite and computer software that will simultaneously solve log-response equations for various minerals, it is now possible to determine the mineralogy of the inorganic matter and the macerals of the organic material. It will be shown in this paper that by knowing the mineral and maceral compositions of a coal, a relationship can be derived as to whether or not a coal is cleated.

In this paper such a relationship is presented for a test well in the San Juan basin. The research at this site was sponsored by the Gas Research Institute.

INTRODUCTION

Traditional wireline logs such as resistivity, density, and neutron have been used in certain coal basins to identify and grade different coal seams. Table 1 shows typical coal responses for these logs. The problem with using only these logs is that wells with identical log responses can differ widely in their rates of gas production. More log information therefore is needed to evaluate coal seams for gas producibility. To enable this determination, a method of evaluating coal seams for permeability and gas content is necessary.

Permeability of coal is directly related to the presence and quality of its cleat system. As the number of cleats per unit length of coal increases, permeability also increases, assuming that the cleats are open and of uniform width. Various wireline logs respond to formation permeability and are useful in predicting permeability in coal seams (Ludwick and Bratton, 1989; Mullen, 1988).

Table 1. Coal log responses for resistivity, density, and neutron wireline logs.

Log	Response
Resistivity	High (depends on thickness, usually >10 ohm-m)
Density	Low (<1.75 g/cc)
Neutron	High porosity (>60 PU)
Sonic	High Δt (120 μs/ft)
Gamma ray	Low (<75 API)
Photoelectric factor	Low (0.17-0.25 barns/electron)

Measurement of spontaneous potential (SP) can be one such indicator of coal permeability. Limitations of the SP log are that mud cake must be formed to provide a distinct potential difference, and salinities of the mud filtrate and formation water must differ (Schlumberger, 1989). In wells drilled with either fresh water, formation water, or air, no mud cake forms, and the SP cannot reliably be used. In wells where mud cake is formed by the drilling fluid, errors can arise if the coal seams are not logged immediately after drilling through them. In such cases, by the time the hole is logged, little potential difference between the mud cake and the coal may remain because of deep invasion of the drilling fluid into the cleat system. This then may cause a highly permeable zone to be missed or a zone of low permeability to appear favorable.

The microlog is a second device that can be used as a coal permeability indicator. In principle, the micrlog also requires mud cake to indicate permeability, but it is sensitive to borehole rugosity (Schlumberger, 1989). When no mud cake exists, microlog measurements may yield useful information about lithology or borehole condition, but no quantitative interpretation currently is possible. Furthermore, because of potential borehole damage due to drilling, the microlog should be used with caution.

The Stoneley wave of a sonic log is a third wireline measurement that can be used as a coal permeability indicator. The Stoneley wave is attenuated when logging permeable zones (Ludwick and Bratton, 1989). In coals, however, the wave is always attenuated. Coal, because of its low density, absorbs the sound wave. Unless the sound energy is amplified to a level that can overcome this natural attenuation, the difference in response of high-permeability and low-permeability coals is small.

The method proposed in this paper is to identify the mineral and maceral composition of a coal and to relate this to a cleat rating and cleat porosity. Cleat rating is an empirical relationship based on core and well performance data. The cleat porosity measurement, which is based on mineral-matter content and bed resistivity, is used with the cleat rating to verify whether cleats are open or closed. Coal macerals will be used only to help determine if a coal has a tendency to cleat; in future work maceral composition will be used in determining gas content.

This method of using minerals and macerals to determine the degree of cleating has been applied to more than 50 wells and has been calibrated with core data for mineral and maceral content and cleat intensity.

MINERAL MATTER IN COAL

Mineral matter that is found in a coal is the result of several independent processes (Carmichael, 1989). The first is what is inherent in the original plant material. The second originates from detrital and air-fall minerals that were deposited in the peat swamps. Most such minerals found in coals consist of clays, together with some quartz and feldspar. The third way that minerals can be deposited in a coal is through percolation of water, from which secondary minerals may precipitate. Secondary minerals generally include calcite, kaolinite, and pyrite.

Knowing the processes by which mineral matter accumulated in a coal is very important. If the mineralogy can be determined for a coal seam, the minerals can be categorized as detrital or secondary. For secondary minerals to be present, the coal must exhibit cleats or tectonically induced fractures. This does not mean, however, that if no secondary minerals are present, then no cleats are present. Similarly, in some cases percolating waters no not precipitate minerals in the cleats. In these cases, the cleat porosity that can be determined from the mineral matter will be used. This will be dealt with in a later section.

MACERAL COMPOSITION

Coal consists of three principal maceral groups—vitrinite, inertinite, and liptinite (exinite) (Dyrkacz and Horwitz, 1982; Dyrkacz and others, 1984; Levine, 1987). Each maceral group refers to a particular type of plant material that was deposited in the swamp. In the laboratory macerals are identified by a point-count method on a polished surface. Because each maceral has a particular chemical composition, a geochemical logging tool can be used to identify macerals, as will be explained in a later section. Table 2 lists chemical composition of maceral types for various ranks of coal.

From laboratory data, it has been determined that the maceral with the greatest tendency to cause cleating is vitrinite, and the least is inertinite. From core data, it has been determined that a high percentage of inertinite in a coal will reduce the tendency of a coal to cleat.

GEOCHEMICAL LOGGING TOOL

The advantage of adding a geochemical logging tool to the logging suite is that the tool can supply key measurements in determining mineral-matter content, rank, and maceral composition of a coal. As an added benefit, a detailed analysis of the mineralogy of surrounding beds can be made at the same time.

The geochemical logging tool provides two modes of measuring data (Schlumberger, 1988; Gilchrist and others, 1982, 1983; Hertzog, 1978; Westaway and others, 1980; Roscoe and Grau, 1985). In the *capture* mode, the elements that are resolved are aluminium, silicon, calcium, iron, sulfur, hydrogen, potassium, thorium, uranium, gadolinium, and chlorine. Carbon and oxygen are obtained in the *inelastic* mode.

Because carbon, hydrogen, oxygen, and sulfur are the primary elements in coal, the geochemical tool is a reliable coal identifier. Table 3 shows that the amounts of carbon, hydrogen, and oxygen vary with rank of the coal. As rank increases, the percentage of carbon increases, and that of hydrogen and oxygen decrease. The logging tool thus can be used to determine coal rank.

Table 2. Maceral chemical composition for various ranks of coal.

		Composition (%)				
Coal Rank	Maceral	Sulfur	Hydrogen	Carbon	Nitrogen	Oxygen
Brown coal	Vitrinite	1.1	5.5	57.9	0.2	35.3
Lignite	Vitrinite	1.4	4.6	73.0	0.6	20.4
Subbituminous	Vitrinite	0.2	5.0	75.5	0.4	18.9
High-volatile bituminous	Inertinite	0.5	2.8	92.6	0.4	3.7
High-volatile bituminous	Vitrinite	1.1	5.5	84.5	1.6	7.3
High-volatile bituminous	Liptinite	0.4	10.4	85.6	0.4	3.2
Medium-volatile bituminous	Inertinite	3.0	2.9	93.0	0.5	0.6
Medium-volatile bituminous	Vitrinite	1.2	5.2	88.9	1.4	3.3
Low-volatile bituminous	Inertinite	0.3	3.3	93.5	0.5	2.4
Low-volatile bituminous	Vitrinite	0.7	4.9	90.3	1.4	2.7
Subbituminous	Inertinite	1.0	4.3	82.6	0.5	11.6

Table 3. Carbon, hydrogen, and oxygen content for different ranks of coal.

	Composition (%)						
Substance/Rank	Volatile Matter	Fixed Carbon	Sulfur	Hydrogen	Carbon	Nitrogen	Oxygen
Cedar wood	90	10	Trace	6	52	Trace	42
Peat wood	75	25	0.5	6	57	0.5	36
Brown coal	67	33	1.0	5.5	58	0.5	35
Lignite	49	51	1.0	5.5	71	0.5	22
Subbituminous	40	60	Trace	5	75.5	0.5	19
High-volatile bituminous	36.5	63.5	1	5	82	2	10
Medium-volatile bituminous	24	76	1	5	90	1	3
Low-volatile bituminous	17	83	1	5	91	1	2
Anthracite	3	97	1	2	95	1	1

If the rank of a coal is known, it is now possible to determine coal maceral content with the geochemical log measurements and density data. As Figure 1 illustrates, however, maceral determination is not possible for all ranks of coal. When the coalification process approaches the rank of anthracite, the chemical compositions of maceral types become nearly identical. Medium-volatile bituminous and lower rank coals provide enough resolution in chemical composition to allow determination of maceral groups with wireline logs.

MINERAL MATTER AND RANK DETERMINATION

Before mineral-matter content can be determined, log data first must be corrected for environmental effects (Schweitzer and others, 1982). The next step is to determine log responses for each mineral (Ellis and others, 1988). In addition, the type of coal encountered in the wellbore must be known before log responses for that coal can be determined.

When all log responses have been determined, the data are input to a computer program that simultaneously solves for mineral-matter content and coal volume in one model and for mineralogy and fluids in a second model (Hertzog, 1978). The log measurements that are used come from the capture mode of the geochemical logging tool and include density, neutron, resistivity, and photoelectric adsorption cross sections. The results of the first run through the computer then are combined with the inelastic-mode measurements (carbon and oxygen) to solve for the rank of the coal.

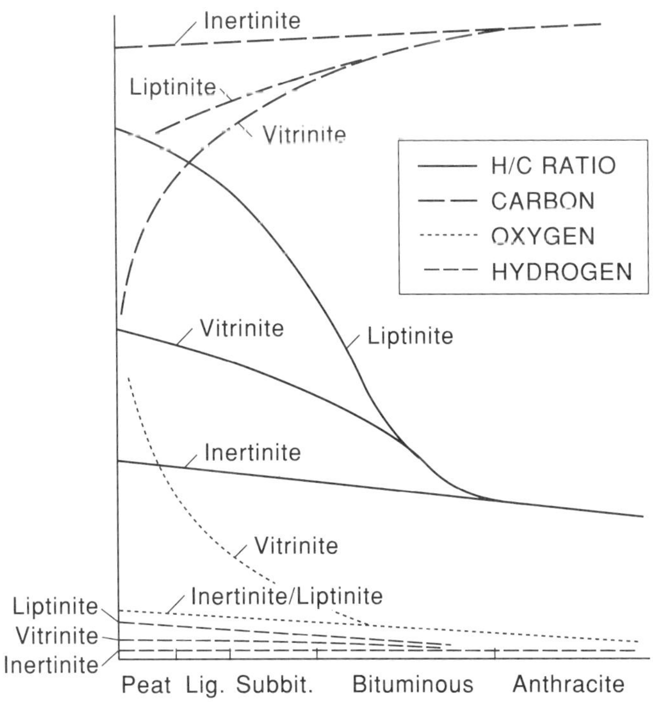

Figure 1. Maceral chemical composition vs coal rank.

MACERAL DETERMINATION

As shown in Figure 1, the chemical composition of each maceral group changes with rank until compositions become nearly identical at about the rank of anthracite. Assuming that coal seams of the same geological age in a given area of a basin are of about the same rank, then identical log responses for maceral type can be used for those coals. If the coals are of different rank, new end points are necessary. Ideally, the model would incorporate variable maceral end points. Using rank of the coal seam as input, the model then would generate the maceral parameters to be used. This capability has yet to be implemented in the current coal-analysis model. In the example described later, the coals are at or near a rank of high-volatile A bituminous, for which the density and chemical ranges or end points of the three maceral groups are given in Table 4.

Macerals are determined in the same way that rank is determined, except that the input data are corrected for mineral-matter contribution. For example, in the case of the density measurement, the product of mineral volume and mineral density is subtracted from the density measurement. This was done for all data inputs. With the corrected data, maceral composition then was determined.

Table 4. Maceral responses for a high-volatile A bituminous coal.

Maceral	Density (g/cc)	Carbon (wt %)	Oxygen (wt %)	Hydrogen (wt %)
Liptinite	1.1–1.2	82–86	3–5	7–10
Vitrinite	1.2–1.35	80–85	7–10	5–6
Inertinite	1.35–1.55	84–90	3–4	2–3

CLEAT DETERMINATION

Before cleats can be determined using mineral and maceral contents, mineral-maceral relationships first must be established for coals that are not cleated. From core data it has been found that the volume of clastic minerals in coal can affect the tendency of coal to cleat. For example, mixtures of abundant clay minerals and quartz tend to bind the coal, inhibiting cleating. Shown below are minimum volumes of inertinite and inorganic components that are used in determining whether or not a coal will be cleated.

Ash volume > 45%
Clay volume > 25%
Quartz volume > 10%
Inertinite volume > 40%

In the San Juan basin, where secondary minerals occur within the coal cleat system, the presence of calcite or pyrite is a good indication that a coal is cleated.

With these relationships, a method of rating the intensity of cleating was developed. Although this rating is purely empirical, it was, whenever possible, tied to core data and well performance. The method determines the volumes of minerals that could cause a coal not to cleat and the volumes of secondary minerals. With this mixture of minerals, one can determine whether a coal is weakly cleated or strongly cleated.

CLEAT POROSITY

While working with the available core data and with wells in other basins, it became necessary to develop a method of calculating coal cleat porosity. Because not all coals have secondary minerals in their cleats, another way of determining cleat intensity was needed. Furthermore, one should be able to determine to what degree secondary minerals fill the cleat system and therefore affect permeability.

To resolve the cleat porosity, it was assumed that the cleats were filled (saturated) with water. In cases where the cleats are not saturated, this method will not work. This is not that disadvantageous because in wells drilled with water or water-based mud, invasion of drilling fluid into the cleats, which originally were gas saturated, allows the method to work. In this technique a shallow resistivity device is used to provide good vertical resolution. A deep resistivity measurement cannot be used because of thin-bed effects. The best measurement comes from a high-resolution dipmeter device calibrated to shallow resistivity. Because the resistivity measurement used in the calculation is for a shallow depth of investigation, the resistivity of the water in the cleat system is assumed to be the mud-filtrate resistivity. Cleat porosity then is calculated by the following equation:

$$C_{SFL}^{0.5} = \phi_c C_{mf}^{0.5} + \phi_{bw(i)} C_{SFLclayi}^{0.5}$$

where C_{SFL} is conductivity of the shallow resistivity device; ϕ_c is cleat porosity; C_{mf} is conductivity of the mud filtrate; $\phi_{bw(i)}$ is porosity of bound water associated with clay$_{(i)}$; and $C_{SFLclay(i)}$ is conductivity of the bound water of clay$_{(i)}$.

LOG EXAMPLE

As an example of cleat and mineral-matter analysis, log data are presented for a GRI cooperative research well in the San Juan basin. Figure 2 shows the responses of typical wireline logs for four coal seams, identified as A, B, C, and D. Tables 5 and 6 (at the end of the paper) are examples of the various log measurements for coal seams A and B.

Simple analysis of the four coals can be made with the traditional logs. First, as the gamma ray reading decreases in a coal, it can be interpreted that clay content of the coal also is decreasing. However, this type of interpretation can be misleading because of the presence of nonclay radioactive material, such as in potassium feldspar, and because of changing clay mineral types. Otherwise, if one assumes that composition of the mineral mixture does not change from coal to coal, but only the volume proportion within the coal, then application of the gamma ray as a coal grade response can be used with some degree of confidence.

The same is true for the density measurement. If mineral composition is constant, and only the volume proportions change in each coal, then density also can be used as a coal grade response. In Figure 2, coals B and D appear to be the seams with higher grade.

Proximate analysis results from the wireline density, neutron, and resistivity measurements of the four coals are plotted in Figure 3a. This determination uses the assumptions stated above in its calculations and finds seam B to be a higher grade coal than seam A. A comparison of log-derived results and core-test results is shown in Figure 3b.

Geochemical wireline data are plotted in Figure 4. From this plot, one can easily distinguish the coal seams on the basis of both hydrogen and carbon yields. Silicon, calcium, iron, titanium, gadolinium, potassium, and thorium are responding to mineral matter in the coal.

Direct permeability measurements are presented in Figure 5. The shaded areas under the Stoneley wave trace are zones that indicate permeability. As stated previously,

Figure 2 (right). Typical logging suite in a San Juan basin coal-gas well.

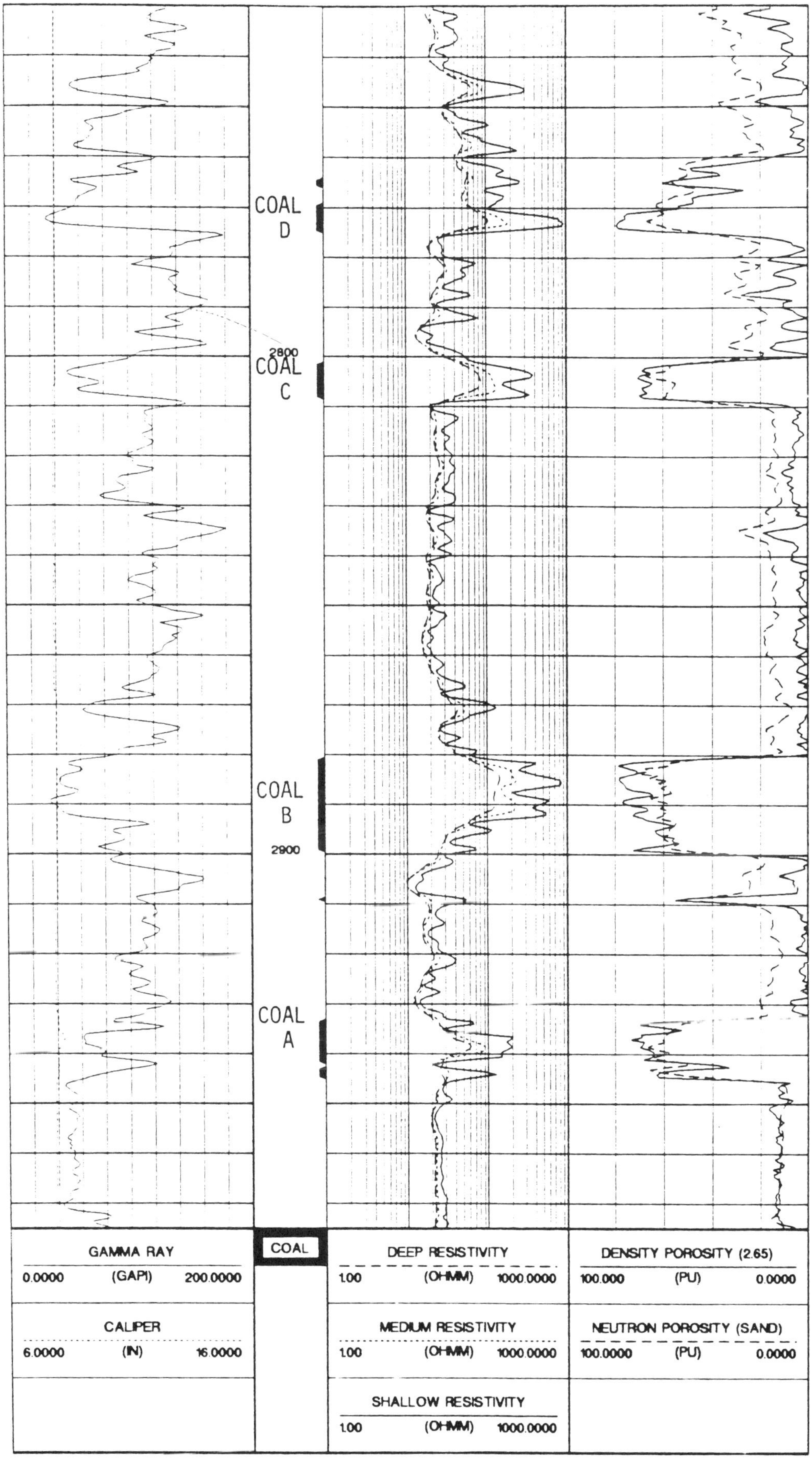
COAL
D
2800
COAL
C
COAL
B
2900
COAL
A
GAMMA RAY
0.0000 (GAPI) 200.0000
CALIPER
6.0000 (IN) 16.0000
COAL
DEEP RESISTIVITY
1.00 (OHMM) 1000.0000
MEDIUM RESISTIVITY
1.00 (OHMM) 1000.0000
SHALLOW RESISTIVITY
1.00 (OHMM) 1000.0000
DENSITY POROSITY (2.65)
100.000 (PU) 0.0000
NEUTRON POROSITY (SAND)
100.0000 (PU) 0.0000

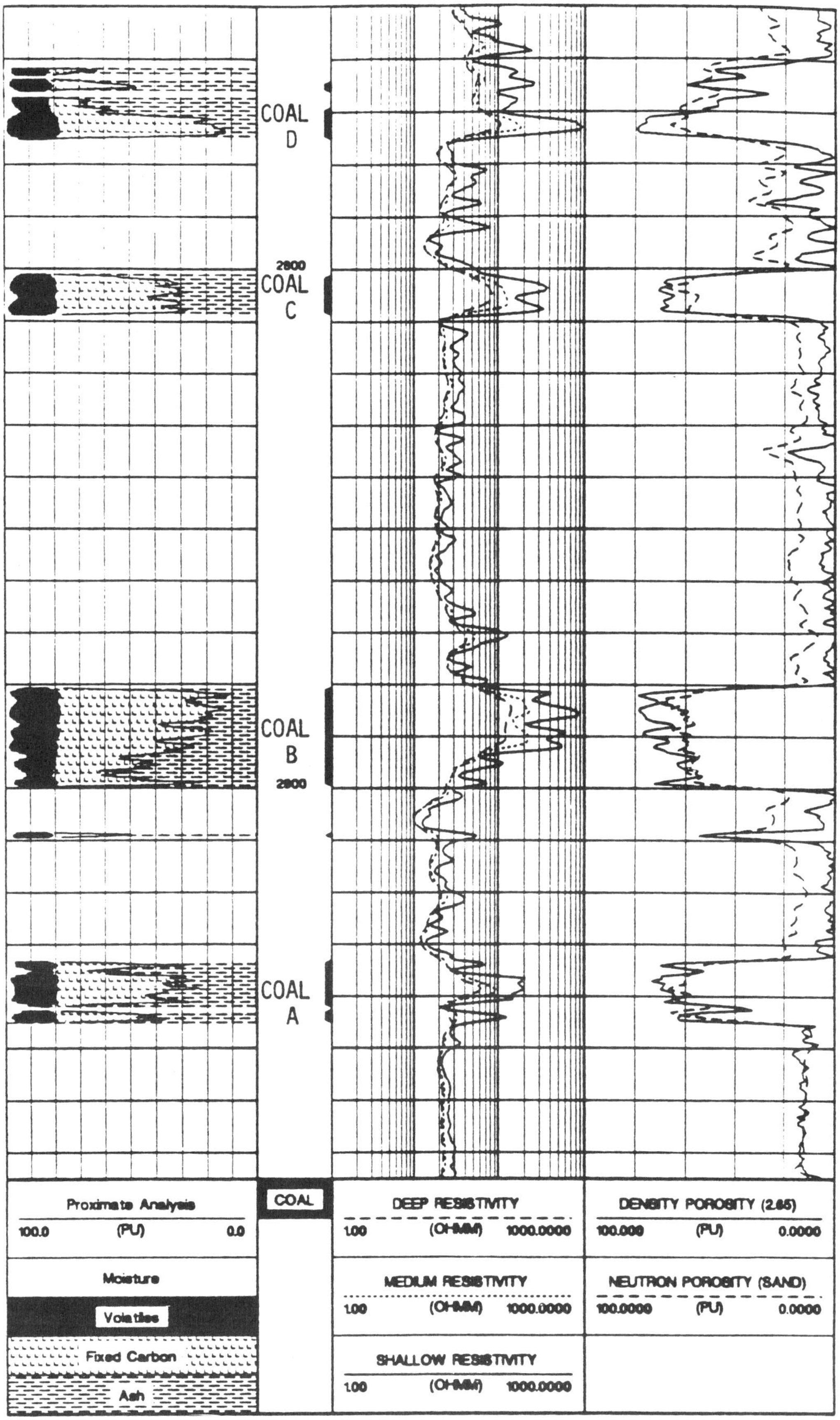

Figure 3. Proximate analysis results of four coal seams derived from (*a*, above) wireline density, neutron, and resistivity measurements. *b* (next page). Comparison of log-derived and core-test results. Scale for ash, fixed carbon, and volatiles content is

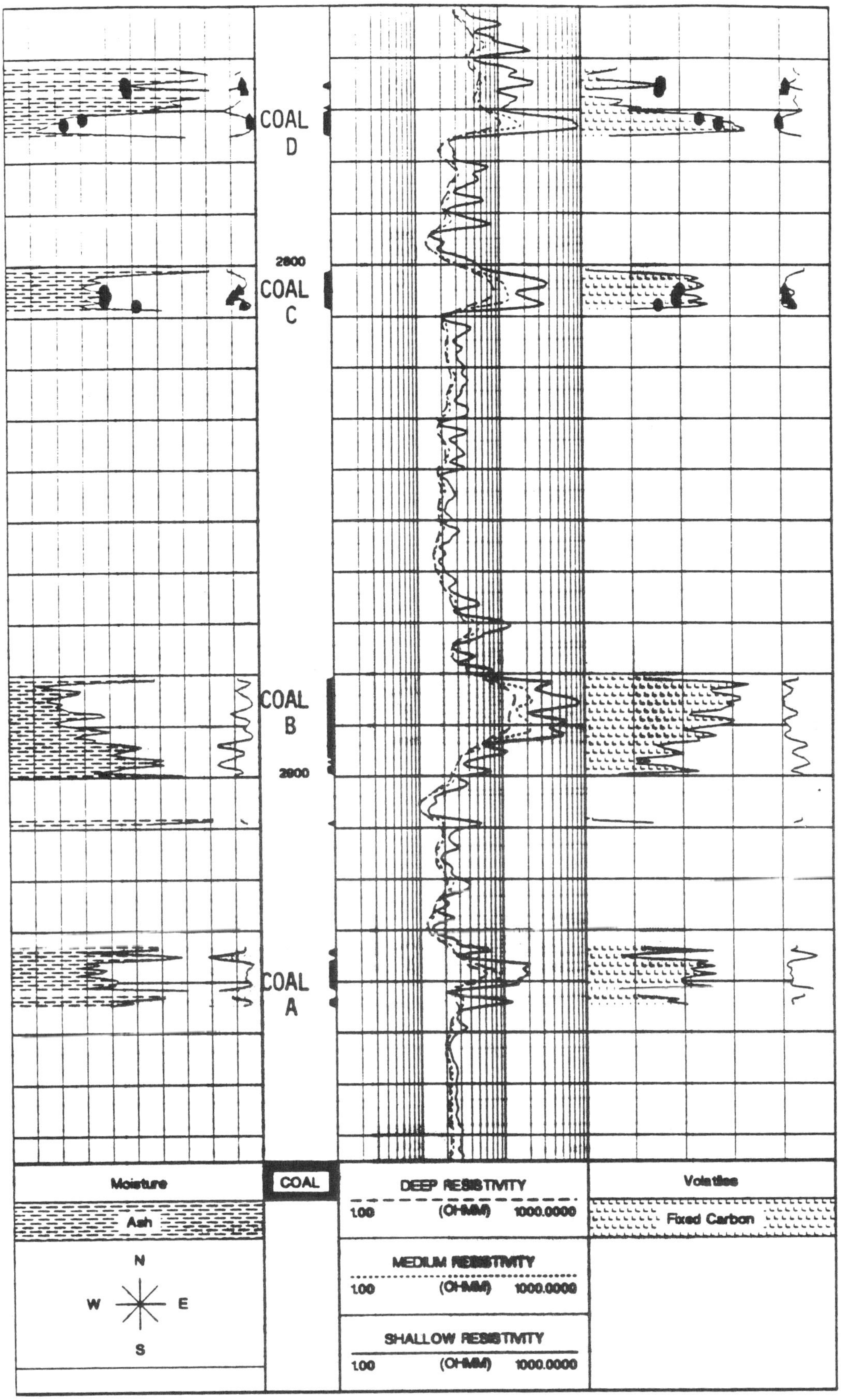

0 to 100%; moisture content is 0 to 25%.

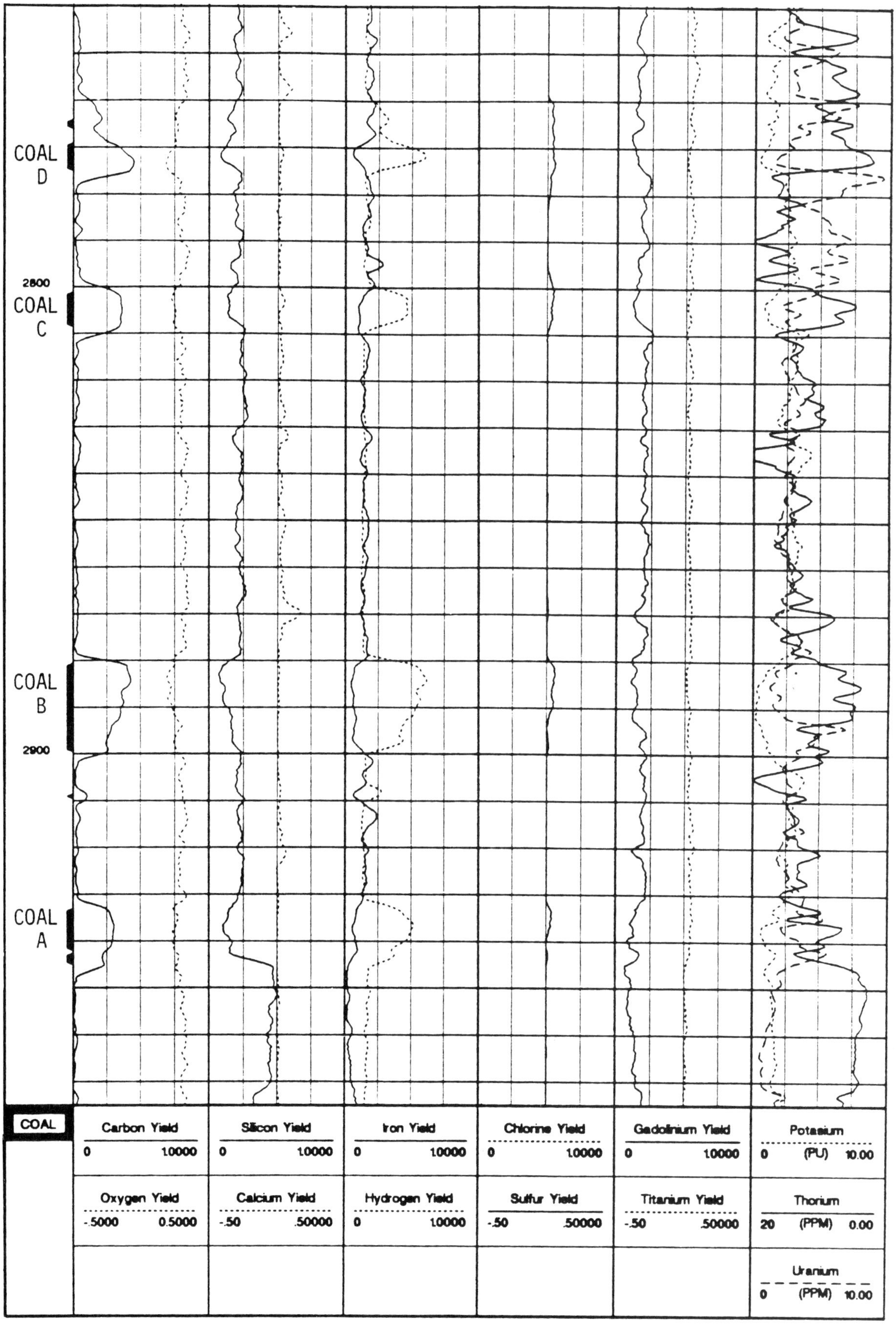

Figure 4. Geochemical relative yields and natural gamma ray measurements.

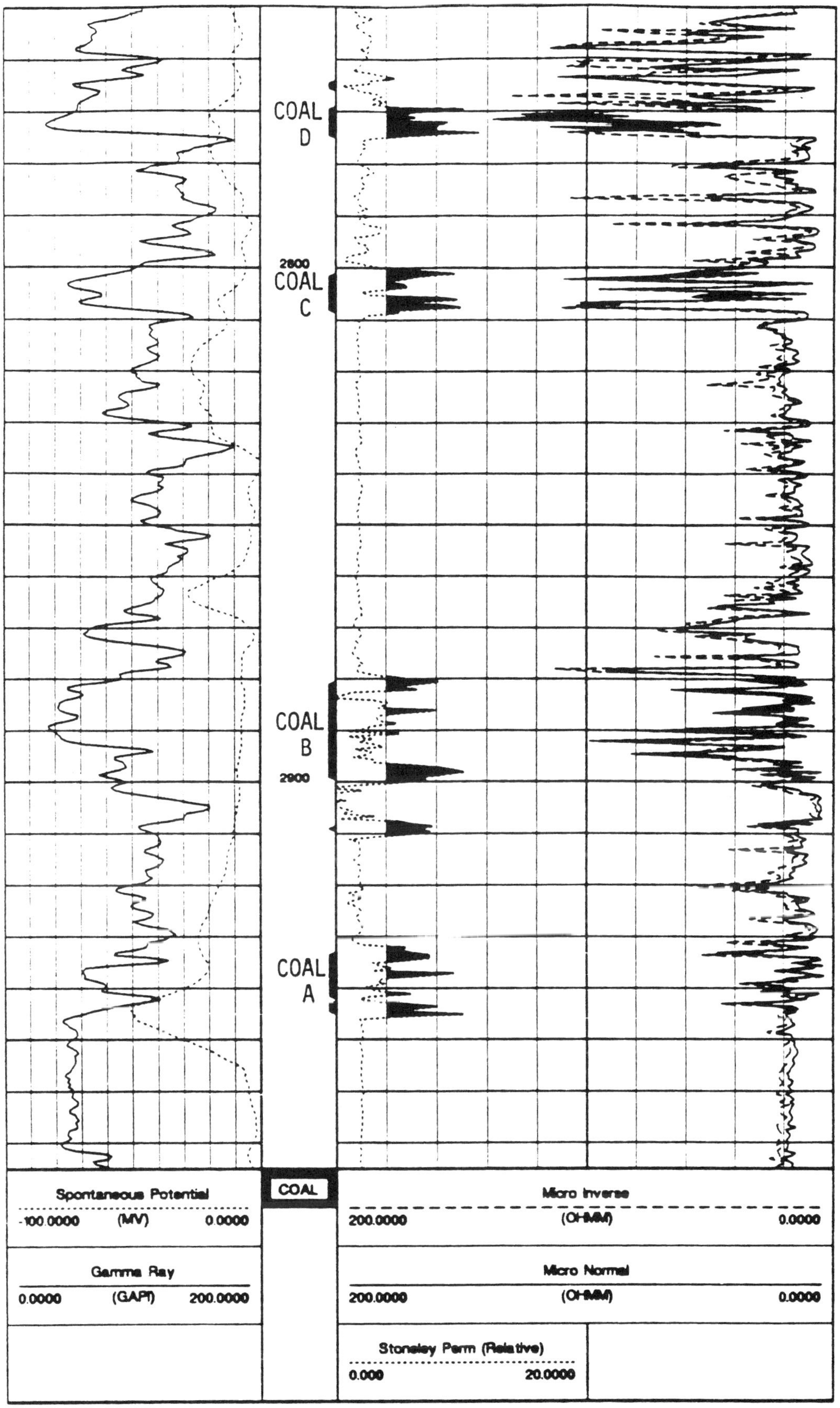

Figure 5. SP, microlog, and sonic (Stoneley wave) wireline measurements used directly to predict coal permeability.

it is difficult to quantify how each zone will perform based on these measurements. Each zone looks permeable from the micrologand Stoneley wave. The spontaneous potential indicates permeability only in coal A.

Figure 6a shows the results using the geochemical logging data with the other wireline measurements for mineral-matter and rank determination for each coal. The core data for coals C and D are plotted with these results in Figure 6b. The match with the computed mineral volume is excellent. Because of thin-bed effects using the geochemical and resistivity device, some of the fixed-carbon and moisture data points do not match in coal D. This is not a major problem because the determination of macerals, cleat rating, and cleat porosity depend mainly on the volumes of mineral matter.

The coal maceral computation results, which are presented in Figure 7, show that the primary maceral group is vitrinite. Figure 8 shows the cleat rating based on mineral-matter and maceral volumes as inputs on the left, and the cleat-porosity computation based on clay volumes and the shallow resistivity device on the right. Excellent agreement can be seen between high cleat porosity and higher cleat rating in the coal horizons. This confirms that the cleats determined for each coal are still open and not filled with secondary minerals.

To summarize the permeability of the coalbeds in this well, both the cleat rating and cleat porosity are taken into consideration. Coal A is the least permeable of all four seams portrayed in these illustrations. Although this coal exhibits one thin zone of permeability, it has, for the most part, no cleat porosity and either a high clay volume or a high total ash volume. Coal B is the most permeable of the seams as a result of its total thickness of cleated area. Coals C and D are about the same in terms of permeability. In terms of potential gas production, the in situ gas content for each coal must be taken into consideration.

CONCLUSIONS

With the addition of geochemical logging tools, coal evaluation can go beyond the proximate analysis. Geochemical logging data can be used to determine mineralogy in the ash fraction of the proximate analysis. With this information, coals that have identical proximate analysis, but produce gas at different rates, can be differentiated by mineral composition. Mineral composition can be important in terms of the secondary minerals that may be found in the coal cleat system.

Furthermore, with the geochemical logging tool, it is now possible to derive maceral-group composition of coal from wireline logs. The amount of gas that is available to enter the cleat system by means of diffusion is a function of

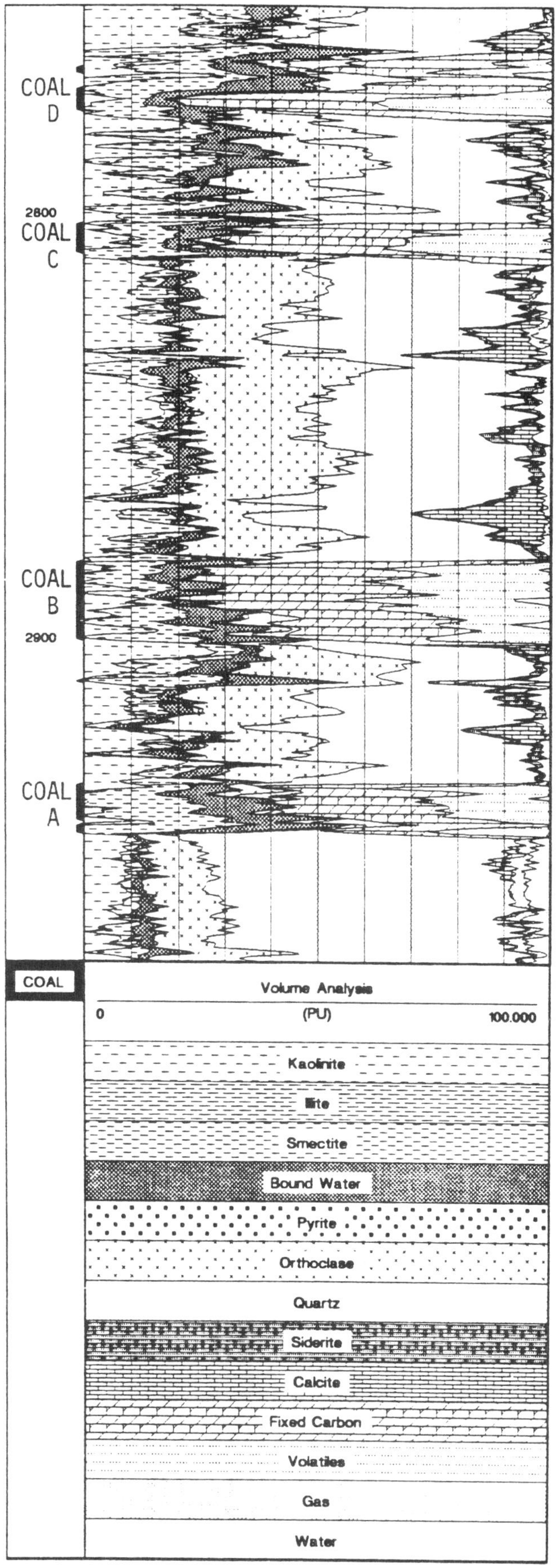

Figure 6. *a* (right). Computer-derived mineral-matter content (volume %) and coal rank using conventional and geochemical wireline log data. *b* (next page). Comparison of log-calculated and core-derived mineralogy, fixed carbon, and moisture content (weight %).

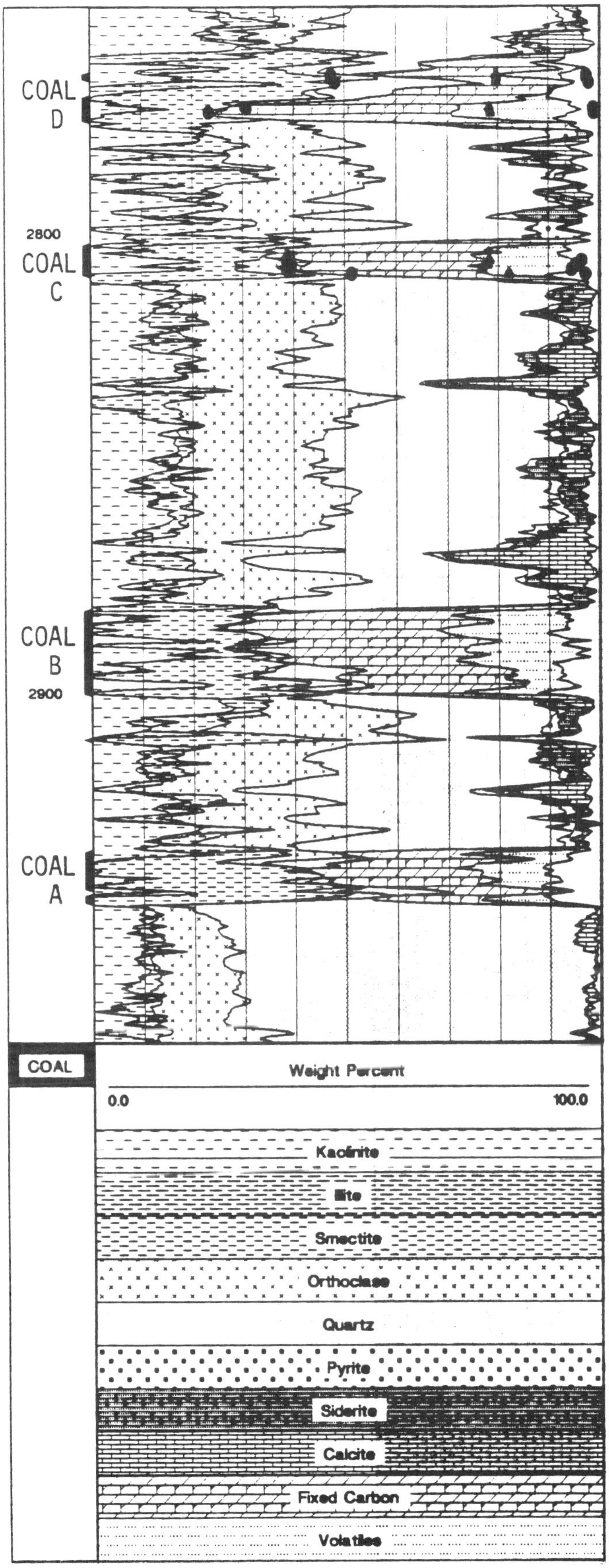

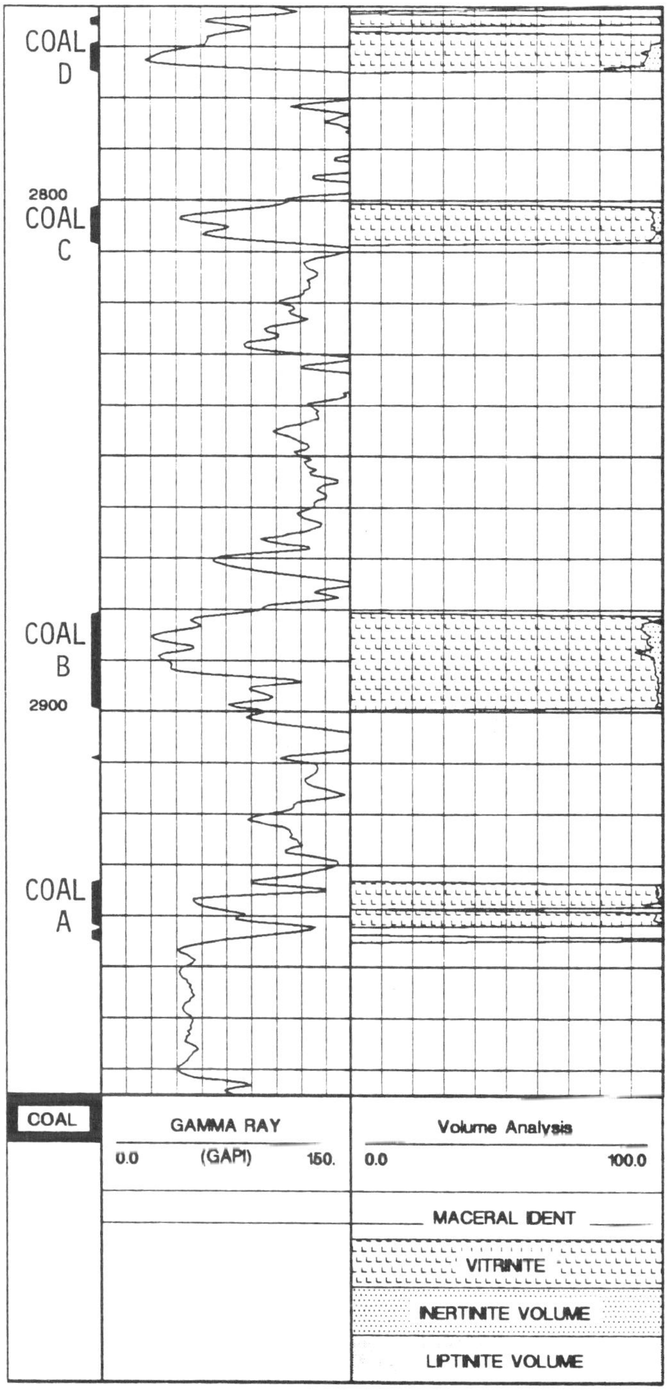

Figure 7 (above). Maceral volumes computed from ash-corrected density, carbon, oxygen, and neutron using computer-derived mineral-matter content.

the amount of adsorbed gas in the microporosity and molecular structure of the coal macerals. The volume of this gas is limited by inorganic materials in the coal and by maceral composition. In the future, equations to calculate in situ gas content will be established that incorporate the volumes of minerals and macerals in the calculation.

With the identification of minerals and macerals, an empirical relationship has been determined to provide a rating of the degree of cleating in coal. With clay analyses

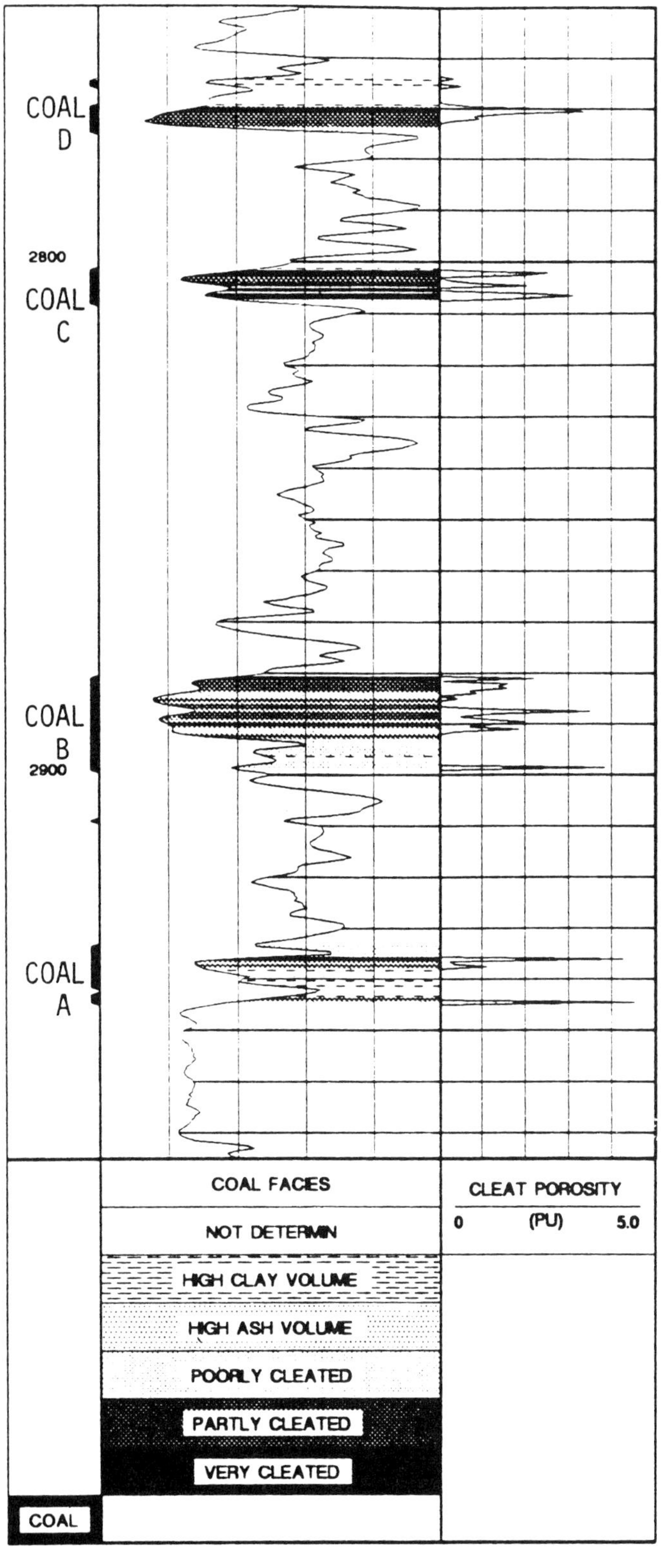

Figure 8. Computer-generated cleat rating (left) and calculated cleat porosity using shallow conductivity (right).

a cleat porosity can be calculated using the highest resolution resistivity device available. Cleat rating and cleat porosity are important guides in predicting the potential for coalbed methane production. Since coal permeability is a function of total cleat porosity, cleat rating and cleat porosity will indicate whether a coal seam is capable of producing gas by identifying potential permeable intervals.

REFERENCES

Carmichael, R.S., ed., 1989, Practical handbook of physical properties of rocks and minerals: Boca Raton, Fla., CRC Press, p. 102–109.

Dyrkacz, G.R., Bloomquist, C.A.A., and Solomon, P.R., 1984, Fourier transform infrared study of high-purity maceral types: Fuel, v. 63, no. 4, p. 536–542.

Dyrkacz, G.R., and Horwitz, E.P., 1982, Separation of coal macerals: Fuel, v. 61, no. 1, p. 3–12.

Ellis, D., and others, 1988, Mineral logging parameters—Nuclear and acoustic: The Technical Review, v. 36, no. 1, p. 38–52.

Gilchrist, W.A., Jr., Quirein, J.A., Boutemy, Y.L., and Tabanou, J.R., 1982, Application of gamma ray spectroscopy to formation evaluation, *in* Twenty-Third Annual Logging Symposium Transactions—July 6–9, 1982, Corpus Christi, Texas: Society of Professional Well Log Analysts, v. 1, Paper B.

Gilchrist, W.A., Jr., Rogers, L.T., and Watson, J.T., 1983, Carbon/oxygen interpretation—A theoretical model, *in* Twenty-Fourth Annual Logging Symposium Transactions—July 27–30, 1983, Calgary, Alberta, Canada: Society of Professional Well Log Analysts, v. 2, Paper FF.

Hertzog, R., 1978, Laboratory and field evaluation of an inelastic-neutron-scattering and capture gamma ray spectroscopy tool: SPE Paper 7430.

Levine, J.R., 1987, Influence of coal composition on the generation and retention of coalbed natural gas, *in* The 1987 Coalbed Methane Symposium Proceedings—University of Alabama, Tuscaloosa, November 16–19, 1987: School of Mines and Energy Development, Gas Research Institute, and U.S. Mine Safety and Health Administration, p. 15–18.

Ludwick, L., and Bratton, T., 1989, Applications of Stoneley permeability in the Four Corners area: North Region Interpretation Symposium.

Mullen, M.J., 1988, Log evaluation in wells drilled for coal-bed methane, *in* Fassett, J.E., ed., Geology and coal-bed methane resources of the northern San Juan basin, Colorado and New Mexico: Rocky Mountain Association of Geologists Guidebook: p. 113–124.

Quirein, J., Kimmanau, S., LaVigne, J., Singer, J., and Wendel, F., 1986, A coherent framework for developing and applying multiple formation evaluation models, *in* Twenty-Seventh Annual Logging Symposium Transactions—June 9–13, 1986, Houston, Texas: Society of Professional Well Log Analysts, v. 1, Paper DD.

Roscoe, B.A., and Grau, J., 1985, Response of the carbon/oxygen measurement for an inelastic gamma ray spectroscopy tool: SPE Paper 14460 [presented at 60th Annual Fall Technical Conference and Exhibition, Las Vegas, Nevada].

Schlumberger, 1988, Geochemical logging: Schlumberger Publication SMP–9120.

Schlumberger, 1989, Log interpretation principles: Schlumberger Publication SMP–7017.

Schweitzer, J.S., Manente, R.A.,and Hertzog, R.C., 1982, Gamma ray spectroscopy tool environmental effects: SPE Paper 11144 [presented at 57th Annual Fall Technical Conference and Exhibition, New Orleans, Louisiana].

Westaway, P., Hertzog, R.C., and Plasek, R.E., 1980, The gamma spectrometer tool inelastic and capture gamma-ray spectroscopy for reservoir analysis: SPE Paper 9461.

Table 5. Example wireline log measurements for coal seam A.

Depth (ft)	Gamma Ray (API)	Resistivity Shallow (ohm-m)	Resistivity Medium (ohm-m)	Resistivity Deep (ohm-m)	Density (g/cc)	Neutron Porosity (sand) %	Photoelectric Factor (barns/electron)	Caliper (in.)
2,944.9	62.625	56.953	30.865	27.903	1.919	29.120	2.131	7.964
2,944.6	66.168	75.438	31.503	27.303	1.623	39.185	1.937	7.995
2,944.3	71.457	101.156	31.836	26.414	1.617	46.655	1.669	8.000
2,943.9	77.098	120.266	32.033	25.564	1.654	53.287	1.618	8.000
2,943.6	82.738	99.250	31.855	25.198	1.625	58.347	2.727	8.000
2,943.4	89.375	74.875	31.176	25.147	1.660	62.399	2.910	8.000
2,943.1	97.023	47.125	29.994	25.409	1.759	65.448	2.169	8.000
2,942.8	105.609	33.391	28.047	25.844	1.958	62.946	2.237	8.000
2,941.9	118.977	23.414	23.469	26.811	2.035	51.094	2.170	8.000
2,941.6	114.223	30.188	23.000	26.933	1.880	52.314	1.634	8.000
2,941.3	102.875	49.141	24.188	27.344	1.659	56.970	1.510	8.000
2,940.9	90.617	74.969	26.449	28.010	1.476	61.777	1.449	8.000
2,940.6	80.375	122.844	32.422	29.660	1.552	63.901	1.296	8.000
2,940.4	75.701	146.028	41.695	32.418	1.592	63.741	1.300	8.000
2,940.1	76.590	144.581	54.273	36.293	1.597	61.301	1.462	8.000
2,939.8	77.864	144.021	68.125	42.203	1.600	59.448	1.522	8.000
2,939.4	78.829	146.688	80.906	48.449	1.596	58.459	1.614	8.000
2,939.1	77.863	164.563	87.098	54.316	1.564	61.230	1.962	8.000
2,938.9	75.246	180.281	89.925	58.393	1.519	63.887	2.909	8.052
2,938.6	70.973	193.813	89.400	60.682	1.463	66.432	4.456	8.155
2,938.3	66.094	189.133	84.938	59.531	1.496	65.949	3.168	8.137
2,937.9	62.024	180.680	79.102	57.012	1.541	64.542	1.726	8.102
2,937.6	62.626	171.563	70.367	51.086	1.553	61.533	2.345	8.102
2,937.4	62.374	173.109	62.492	45.344	1.529	60.202	3.443	8.155
2,937.1	61.268	185.328	55.484	39.773	1.467	60.545	5.023	8.263
2,936.8	61.147	194.906	49.891	36.406	1.495	59.848	4.786	8.227
2,936.4	61.396	191.000	44.773	33.797	1.529	59.127	4.114	8.174
2,936.1	62.506	122.438	40.609	32.766	1.501	59.325	3.083	8.188
2,935.9	69.344	73.000	36.602	31.898	1.568	58.923	2.211	8.141
2,935.6	81.914	42.594	32.738	31.195	1.728	57.922	1.498	8.034
2,935.3	100.938	32.262	29.672	30.000	1.797	57.205	2.154	8.008
2,934.9	118.855	27.910	27.067	28.648	1.828	56.460	2.831	8.000
2,934.6	124.727	33.234	25.970	27.007	1.754	55.287	2.239	8.000
2,934.4	123.212	44.891	25.211	25.410	1.641	54.365	1.758	8.000
2,934.1	114.391	62.922	24.789	23.854	1.490	53.690	1.387	8.000
2,933.8	99.078	64.648	24.266	22.648	1.607	52.129	1.136	8.000
2,933.4	85.396	60.172	23.742	21.564	1.793	49.548	1.094	8.000
2,933.1	86.362	41.063	23.320	20.729	2.049	42.767	1.967	8.000

Table 6. Example wireline log measurements for coal seam B.

Depth (ft)	Gamma Ray (API)	Resistivity Shallow (ohm-m)	Resistivity Medium (ohm-m)	Resistivity Deep (ohm-m)	Density (g/cc)	Neutron Porosity (sand) %	Photoelectric Factor (barns/electron)	Caliper (in.)
2,899.3	86.906	58.516	26.273	24.438	1.684	43.030	3.020	8.059
2,898.8	77.641	68.070	26.922	24.742	1.557	53.897	2.166	8.102
2,898.3	77.320	50.906	27.641	25.813	1.615	56.583	2.219	8.102
2,897.8	85.613	35.193	28.242	27.664	1.707	54.653	1.726	8.059
2,897.3	92.895	35.115	28.648	29.367	1.751	56.229	1.504	8.008
2,896.8	94.795	42.398	29.195	30.320	1.751	59.502	1.673	8.000
2,896.3	93.250	55.612	30.852	31.078	1.713	59.685	1.785	8.000
2,895.8	89.387	82.422	34.234	32.648	1.615	58.012	1.600	8.000
2,895.3	86.330	107.234	39.969	35.625	1.599	57.172	1.357	8.000
2,894.8	93.102	98.047	47.875	40.250	1.639	57.462	1.340	8.000
2,894.3	107.859	71.031	54.234	46.516	1.680	56.358	1.582	8.000
2,893.8	112.707	63.328	57.031	53.406	1.664	54.959	1.646	8.000
2,893.3	96.406	94.594	60.422	59.641	1.574	55.920	1.467	8 000
2,892.8	72.094	229.625	71.813	65.688	1.519	60.229	1.405	8.000
2,892.3	55.230	436.875	100.938	73.469	1.485	61.859	1.448	8.000
2,891.8	47.313	536.838	149.219	85.688	1.464	60.093	1.430	8.000
2,891.3	44.230	482.625	196.375	102.719	1.531	58.890	1.268	8.000
2,890.8	42.284	394.469	220.938	116.688	1.552	59.821	1.322	8.020
2,890.3	39.677	396.500	225.500	121.344	1.481	60.326	1.522	8.031
2,889.8	36.625	481.875	221.125	120.750	1.413	58.595	1.551	8.012
2,889.3	34.699	561.938	208.750	119.344	1.381	60.577	1.590	8.000
2,888.8	37.426	488.750	189.188	118.906	1.401	64.307	1.935	8.000
2,888.3	45.293	311.313	169.313	120.156	1.494	63.338	2.402	8.000
2,887.8	53.730	220.766	156.875	121.813	1.552	58.850	2.226	8.000
2,887.3	57.453	248.375	150.750	123.688	1.487	57.133	1.544	8.000
2,886.8	53.082	386.500	147.375	127.563	1.419	59.404	1.769	8.000
2,886.3	46.078	647.625	159.375	132.500	1.392	59.964	2.573	8.000
2,885.8	42.928	826.516	196.688	135.938	1.389	60.579	2.500	8.070
2,885.3	42.624	818.250	227.750	138.688	1.375	60.278	1.632	8.121
2,884.8	43.125	710.688	222.375	140.750	1.348	62.054	1.332	8.129
2,884.3	45.424	522.250	206.188	139.938	1.376	64.215	1.482	8.078
2,883.8	52.191	368.688	201.438	134.313	1.433	67.010	1.287	8.000
2,883.3	59.244	279.844	195.500	123.219	1.467	68.814	1.269	8.000
2,882.8	60.563	255.328	176.813	108.375	1.422	67.101	1.886	8.141
2,882.3	54.434	308.969	142.90	91.875	1.343	63.129	3.984	8.289
2,881.8	49.478	375.125	104.844	76.781	1.369	58.346	3.922	8.273
2,881.3	54.332	269.875	78.406	65.500	1.585	55.750	1.953	8.215
2,880.8	68.539	106.813	63.328	56.703	1.920	48.315	2.246	8.141

Cleat Detection in Coalbeds Using the Microlog

Michael J. Mullen[1]

1. Halliburton Logging Services, Farmington, New Mexico

ABSTRACT

One of the most important pieces of information in developing a coalbed methane prospect is the ability to determine the degree of cleat development in the coal. Cleat development is one of the critical factors for an economic coalbed methane well. Coal may contain a large quantity of adsorbed gas, but unless permeable pathways exist to the wellbore, that gas may not be producible. Cleat development also influences the design of completion type and stimulation procedure.

The microlog is a wireline logging tool that historically has been used to identify permeable formations. This paper proposes a technique, using the microlog, to quantify the degree of cleating in coalbeds as well cleated, moderately cleated, or poorly cleated. The microlog curves were normalized between the resistivity of the mud and 30 times mud resistivity to define a range of cleat types by the midpoint of positive separation between the micronormal and microinverse curves, creating a microlog ratio. For wells drilled without a low water-loss mud system, microlog "porosity" was calculated from the micronormal curve and used as a determinant for cleat type identification. The interpretation method was derived by correlating microlog response to production records and test data of wells producing from coalbeds in the San Juan basin. The proposed limits for cleat identification from the microlog are as follows:

Cleat Type	Microlog Ratio	Microlog "Porosity"
Well cleated	<30%	>8%
Moderately cleated	30%–60%	4%–8%
Poorly cleated	>60%	<4%

INTRODUCTION

Recent drilling activity in the San Juan basin of New Mexico and Colorado has been dominated by the development of wells to produce gas from coalbeds. Coalbeds of the Fruitland Formation once were considered a nuisance in drilling to deeper Cretaceous sandstones, but they are now the target of much attention both by major oil companies and by independents. It is estimated that Fruitland coalbeds contain nearly 50 TCF ($1,415 \times 10^9$ m^3) of gas in place within 219 billion short tons (198.7 billion t) of coal (Kelso and Wicks, 1988). The gas potential and size of this reservoir make the understanding of coalbeds as gas reservoirs a worthwhile pursuit.

Many factors control the production of gas from coalbeds. Coal rank, reservoir pressure, permeability, and completion techniques are some of the major controlling factors. The method of obtaining the most complete coalbed information is to core every coal in every well. This obviously is impractical and uneconomical. Thus, studies that relate wireline log data to core and production data are of immense value in predicting production from new wells drilled into coal-bearing formations.

Traditional log analysis (which focuses on deriving porosity and water saturation) and conventional reservoir modeling are not applicable to coalbeds because of the unusual mechanical and reservoir properties of coal. Thus, new analysis techniques are being developed to deal with this type of reservoir (Mullen, 1988, 1989). Algorithms have been proposed to derive coal rank and gas-content information from log data, and work is continuing to develop other methods whereby wireline logs can be used to identify and evaluate other coalbed characteristics, such as permeability and cleat development, that affect gas production.

This paper discusses a technique for quantifying one such coal characteristic—degree of cleating. By comparing logging data with production information on several wells, a relationship was developed between coalbed permeability and degree of cleating.

The wireline tool historically used to identify permeable zones is the microlog. The microlog is a good indicator of mud cake buildup on permeable formations. This study will demonstrate recent analysis techniques to quantify the microlog response in coalbeds for determination of the degree of cleating. The greater the degree of cleating, the greater the possibility of bringing in a productive well from a coalbed.

MICROLOG THEORY

The microlog is a pad-type resistivity tool (Figure 1) that usually is run in combination with the Spectral Density[2] tool. Three measurement electrodes on the pad are forced against the borehole wall when the powered caliper is opened. Two resistivity measurements are made with the pad electrodes—(1) 2-in. (5-cm) normal resistivity is measured between the bottom electrode and the top electrode; and (2) a 1-in. (2.5-cm) inverse or lateral measurement is made using the potential difference between the top and middle electrodes. The "micronormal" has a depth of investigation of 4 in. (10 cm), and the "microlateral" (or microinverse) has a depth of investigation of about 1.5 in. (4 cm).

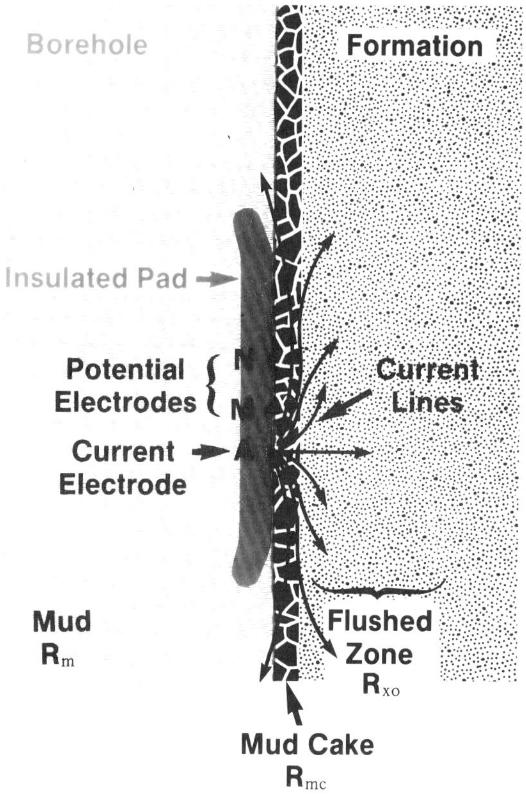

Figure 1. Microlog pad configuration in a borehole. After Gearhart Industries (1983).

In conventional reservoirs permeable zones exhibit positive separation between the micronormal and microlateral curves of the microlog trace. Curve separation is said to be positive when micronormal resistivity is higher than microlateral resistivity. This difference in resistivity is caused by drilling-mud filtration loss into a permeable zone and formation of a mud cake on the borehole wall. The micronormal, because of its greater depth of investigation, responds somewhat to mud cake resistivity, Rm_c, but more to flushed-zone resistivity, R_{xo} (Figure 1). On the other hand, the microlateral, because of its shallower depth of investigation, responds more to mud cake resistivity and only slightly to formation resistivity. In tight formations, both microlog curves read high resistivity values because of the absence of mud filtration loss. In shale both resistivities read low values, with the micronormal curve often reading lower than the microlateral (Hilchie, 1979; Bateman, 1985).

When microlog resistivities are less than 20 to 30 times mud resistivity (measured at formation temperature), positive separation may be interpreted as an indication of permeability (Hilchie, 1979; Bateman, 1985). For this study a usable cutoff value of microlog resistivity was selected to be 30 times mud resistivity.

COALBED MODEL FOR MICROLOG INTERPRETATION

The model currently in use to describe coalbeds is that of a highly fractured formation with orthogonal sets of fractures separating small blocks of matrix material. Fracture or cleat width is very narrow in most coalbeds—a fracture width of 1/64 in. (0.4 mm) is considered large (Strickland, W.R., 1989, pers. comm.). Classically, the cleat system consists of two or more sets of subparallel fractures, both of which are oriented approximately normal to bedding. One set of fractures, the face cleat, usually is dominant, with individual cleat planes being relatively straight and persistent. Butt cleats or end cleats normally are subperpendicular to the face cleats but are relatively discontinuous, curved, and often terminate against face cleat surfaces (Levine and Deul, 1989).

As mud filtrate enters the coalbeds, mud cake builds up on the faces of the fractures. When the logging tool traverses a cleated coalbed, most of the mud cake on the borehole wall is wiped away, but some remains in the near-wellbore region of the cleat system. This causes the positive separation observed between the micronormal and microlateral curves.

The typical microlog response observed in coalbeds can be divided into four categories (Figure 2). First, in noncleated or very poorly cleated coalbeds, both microlog curves measure high resistivity values, and minimal positive separation is apparent between the micronormal and microlateral. Second, in poorly to moderately cleated coalbeds, some positive separation is observed between the micronormal and microlateral curves, with a trend toward higher resistivity values. Third, in well cleated coalbeds, positive separation is observed between the micronormal and microlateral curves, with a trend toward lower resistivity values. Fourth, in thin-bedded or laminated coal sequences, the microlog curves often exhibit poor to moderate cleating indication. These coalbeds usually require fracture stimulation to be productive.

2. Mark of Halliburton Logging Services.

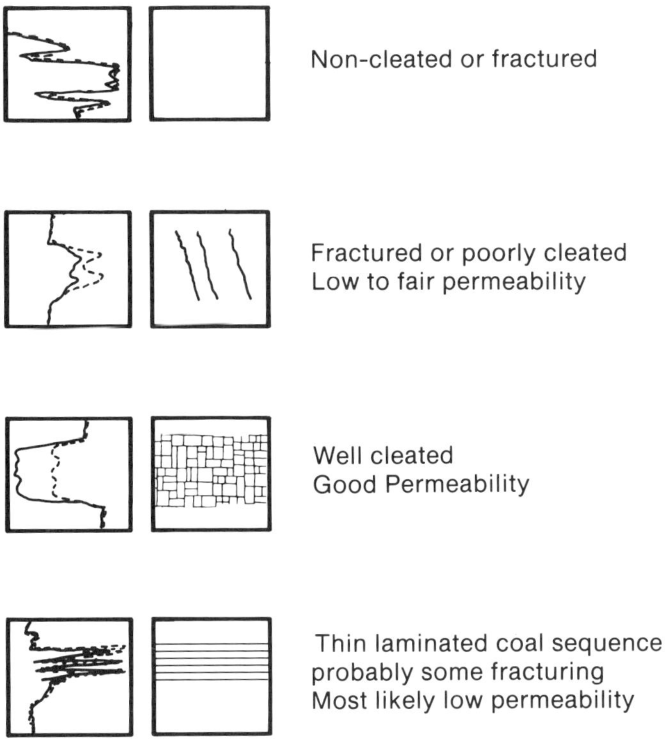

Figure 2. Typical microlog tool responses in coalbeds.

CLEAT DETERMINATION

Traditionally, cleat determination in a coalbed has been made by examining core. Because it is not possible to core every well, the microlog is used to help identify the presence of cleating in coalbeds and to quantify the relative degree of cleating. The model discussed above basically states that noncleated coalbeds will have high microresistivity values, and well cleated coalbeds will have lower microresistivity values due to mud filtration loss into the coalbeds. Positive separation usually is observed between the micronormal and microlateral curves to highlight cleated coalbeds. The degree of positive separation observed is more a function of mud properties than of cleat development. The mechanism that controls mud filtration loss into a coal is the development of the cleat system. Therefore, the closer the positively separated microlog resistivities read to the mud resistivity, the better developed is the cleat system in the coalbed. Consequently, the higher the microlog resistivity value, the poorer the cleat system development. The maximum limit of microlog resistivity values that can be used in this interpretive model is 30 times mud resistivity. This cutoff value then defines the "tight" end of a scale to quantify microlog resistivity values; the low end of the scale corresponds to the mud resistivity.

By this procedure the log analyst can rescale the microlog into a ratio (I) of mud resistivity (zero end of the scale) and the cutoff resistivity value (100% end of the scale), according to the following equation:

$$I = \frac{(R - R_m)}{(R_c - R_m)}$$

where R_m is mud resistivity corrected to bottom-hole temperature; R_c is cutoff resistivity ($30R_m$); and R is the average or midpoint value of microlateral resistivity (R_{ml}) and micronormal resistivity (R_{mn}), that is, $R = (R_{ml} + R_{mn})/2$.

The microlog ratio then is independent of mud resistivity, and consistent cutoff values of the ratio can be established to quantify the microlog for cleat-type determination. The microlog resistivity value used for this determination is the midpoint (R) of the positive separation between the micronormal and microlateral curves. Based on field experience in the San Juan basin, the cutoff values proposed for cleat-type identification are as follows:

Cleat type	Microlog Ratio
Well cleated	<30%
Moderately cleated	30%–60%
Poorly cleated	>60%

Some wells in the San Juan basin are drilled with fresh water or an air-mist mud system. In these wells, casing is set to the top of the Fruitland Formation, and a completion unit is used to drill the coal section. The well then is allowed to "blow out" if possible. This is advantageous in identifying well cleated coalbeds but is not practical for every area being developed for coal-gas potential. The wells are completed open hole through an uncemented, preperforated liner and produced without stimulation. If such a well does not produce naturally, it usually is logged with wireline instruments before a liner is cemented. Then the coals are perforated and fracture stimulated. In wells drilled using this method, the mud system is such that mud cake usually is not built. However, permeable low-pressure coalbeds will be invaded by water. Micronormal resistivity then can be used as an approximation of R_{xo} to calculate a "porosity" from the chart in Figure 3. Microlog "porosity" determined from this chart is not considered an accurate measure of porosity in the coalbed but is used only as a relative indicator for cleat-type identification. Microlog porosities calculated for the wells described in this paper were crossplotted with the microlog ratio to determine porosity cutoff values for cleat-type identification (Figure 4). These proposed cutoffs are as follows:

Cleat Type	Microlog Porosity
Well cleated	>8%
Moderately cleated	4%–8%
Poorly cleated	<4%

EXAMPLES OF CLEAT DETERMINATION BY THE MICROLOG

The following six examples are offered in support of the observations outlined above. Calculated microlog ratios

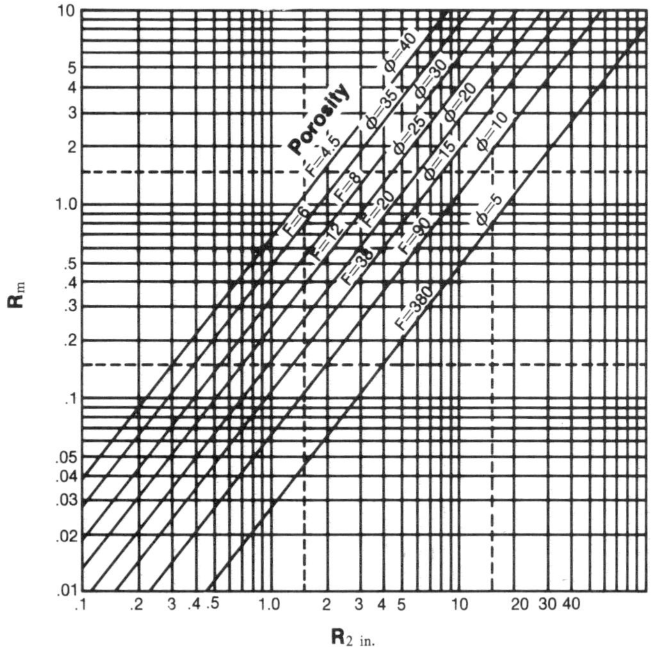

Figure 3. Empirical microlog porosity (Schlumberger Well Services, 1952).

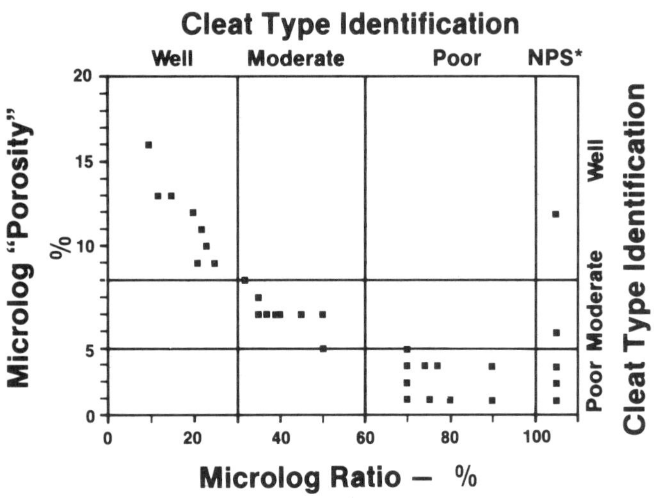

Figure 4. Crossplot of microlog ratio vs microlog "porosity" showing cleat type interpretation.

and the corresponding cleat development categories have been added to each of the microlog traces shown in the accompanying illustrations. Intervals in which positive separation occurs between the micronormal and microlateral curves have been highlighted. Cleat-type footage and production or test results for the example wells are compiled in Table 1.

Well 1

As indicated by the microlog in Figure 5, well 1 has 47.5 ft (14.5 m) of well cleated coal, 15 ft (4.6 m) of moderately cleated coal, and 6 ft (1.8 m) of poorly cleated coal in 11 beds, the thickest of which is 10 ft (3 m). This well was perforated and fracture stimulated. During a two-week production test, this well recorded production of 33 MSCFD (9,339 m^3/day) and 660 BWPD (105 m^3/day) on pump. Production from this well supports the microlog interpretation that classified the coalbeds as well cleated.

Well 2

As indicated by the microlog in Figure 6, well 2 contains 24 ft (7.3 m) of well cleated coal and 2 ft (0.6 m) of moderately cleated coal in six coalbeds, the thickest of which is 12 ft (3.7 m). The lower three coalbeds, containing 12 ft (3.7 m) of well cleated coal, were perforated and fracture stimulated. During a production test the well had daily production of 160 MSCFD (4,528 m^3/day) and 440 BWPD (70 m^3/day) on pump. This production supports the microlog interpretation, which suggests that the coal is well cleated.

The caliper log indicates that the upper 12 ft (3.7 m) of coal has washed out. Because the microlog, which here indicates well cleated coal, is a pad device, it is sensitive to borehole rugosity, and its measurements may be questionable in washed-out zones.

Well 3

Well 3 has 33 ft (10 m) of poorly cleated coal, as indicated by the microlog in Figure 7. Twelve coalbeds are present, the thickest of which is 8 ft (2.4 m). No positive separation is observed between the micronormal and microlateral curves. In every coalbed the microlateral curve reads

Table 1. Summary of cleat type and net cleated footage for six example wells.

Well Number	Net Cleated Thickness (ft) Well	Moderate	Poor	Production or Test Data
1	47.5	15	6	330 MSCFD, 660 BWPD.
2	24.0	2	–	Total well.
2-Completed	12.0	–	–	160 MSCFD, 440 BWPD.
3	–	–	33	30-50 MSCFD, no water.
4	–	4	21	Total well.
4-Completed	–	–	13	Nonproductive.
5	–	3	37	Would not produce naturally.
6	–	3	4	Would not produce naturally.

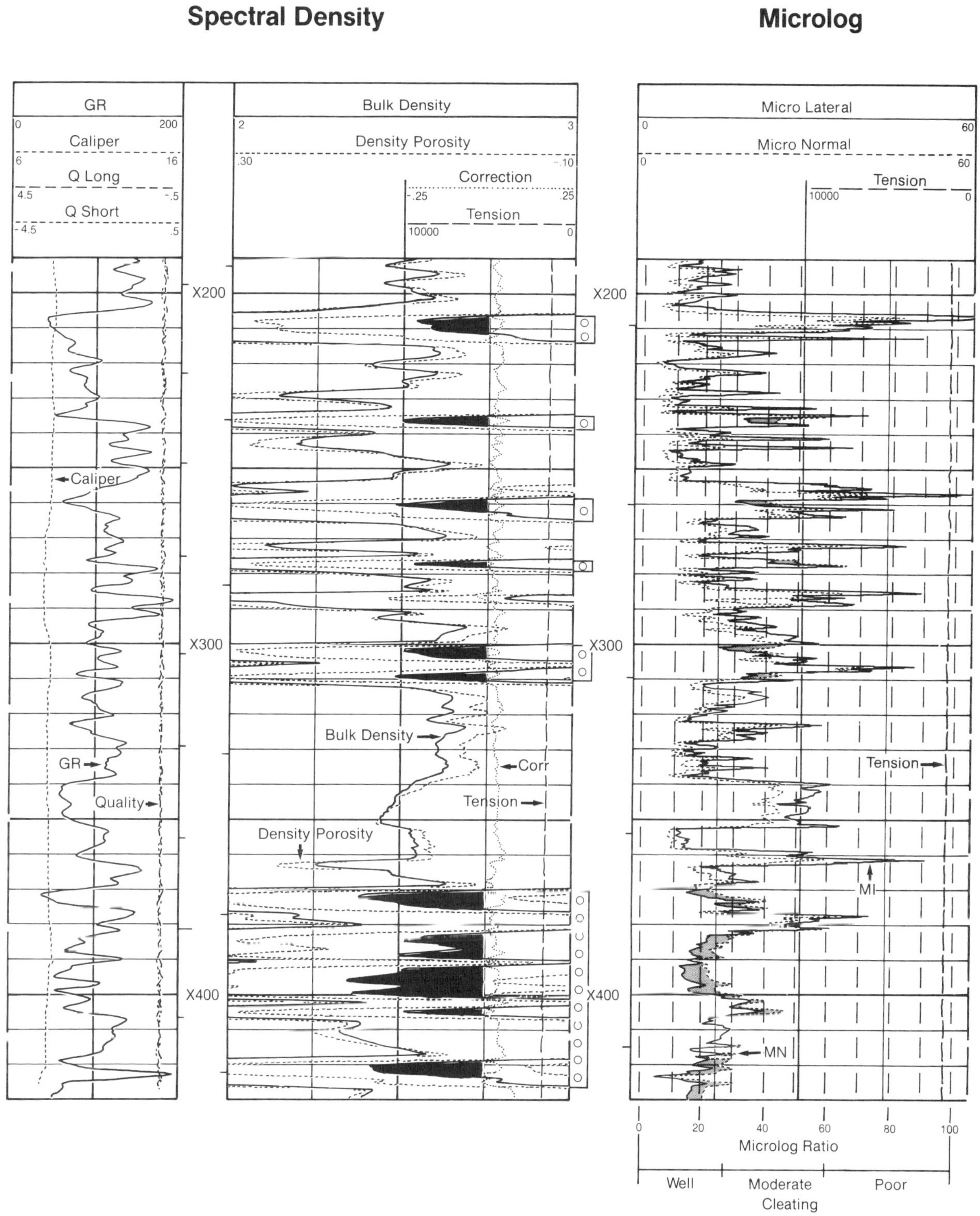

Figure 5. Spectral Density log and microlog for example well 1. R_m = 1.9 ohm-m at 135°F.

resistivity values higher than the cutoff of 30 times mud resistivity. This characteristic also indicates low permeability or poor cleating. The well was perforated and fracture stimulated in the coalbeds. Current daily production is reported as 30 to 50 MSCFD (849 to 1,415 m^3/day) with no water. This production agrees with the proposed microlog interpretation model. Microlog "porosity" calculated in this well varies from 1% to 3%, also indicating poor cleat development.

Well 4

From the microlog in Figure 8, well 4 contains 4 ft (1.2 m) of moderately cleated coal and 21 ft (6.4 m) of poorly

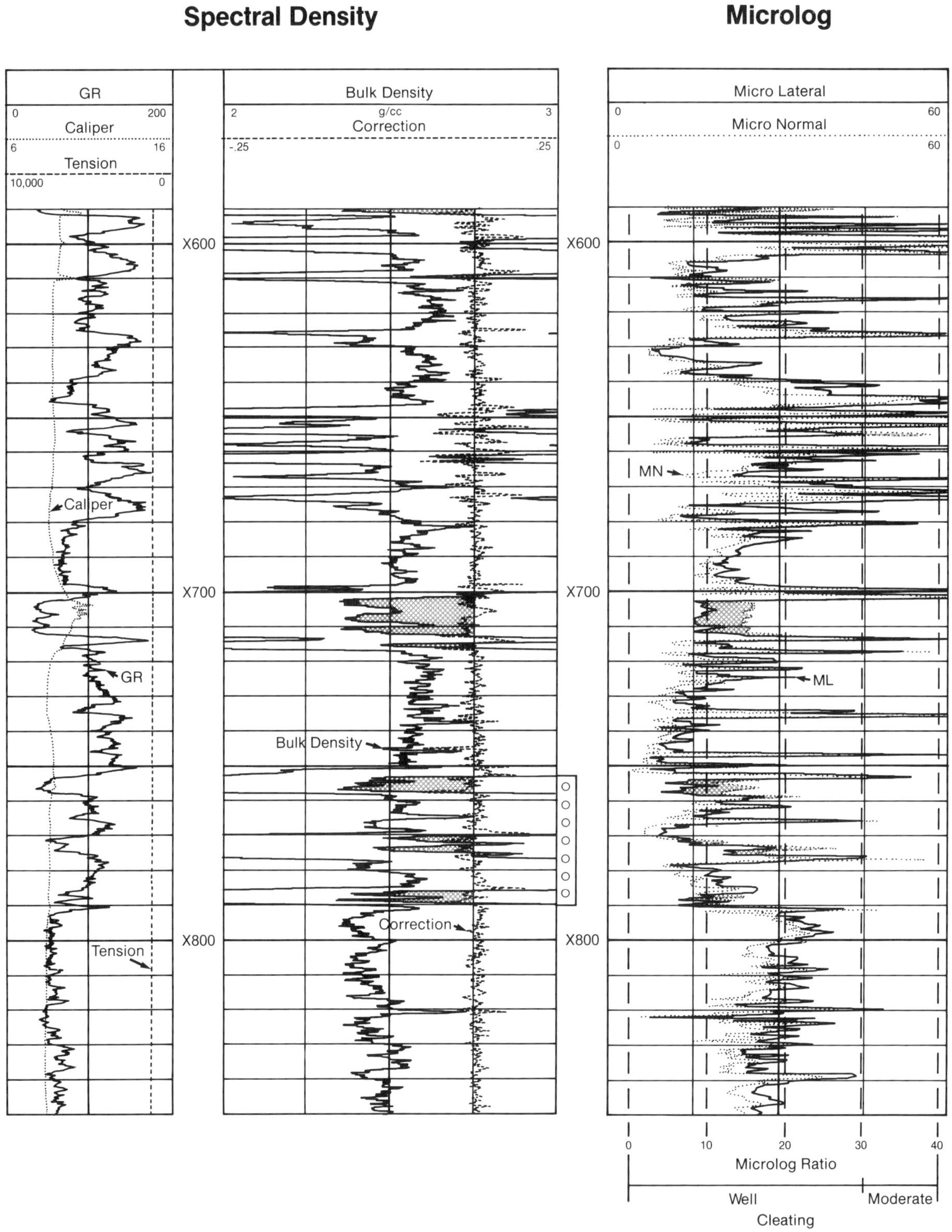

Figure 6. Spectral Density log and microlog for example well 2. R_m = 4.96 ohm-m at 120°F.

cleated coal in six coalbeds of maximum thickness 8 ft (2.4 m). This well was perforated and fracture stimulated in 13 ft (4 m) of poorly cleated basal coal. The frac job screened out four times in this zone, and no production was established. Production and log analysis agree with the interpretation that the coal is poorly cleated. In addition, desorption rates from a whole core in this zone were very slow, also indicating poorly cleated coal.

Well 5

The microlog of well 5 indicates 3 ft (0.9 m) of moderately cleated coal and 37 ft (11.3 m) of poorly cleated coal

(Figure 9); the interval contains 12 coalbeds, the thickest of which is 14 ft (4.3 m). Six of these coalbeds, comprising 22 ft (6.7 m) of net coal, show no positive separation on the microlog curves, and microlog "porosity" was calculated at 1% to 3%. These results indicate poor cleat development. Because an offset well produced less than 50 MSCFD (1,415 m^3/day) naturally, well 5 was drilled with mud, perforated, and fracture stimulated. Offset production seems to confirm the microlog interpretation that the coal has poor cleat development.

Well 6

Well 6 contains 47 ft (14.3 m) of poorly cleated coal in 11 coalbeds, the thickest being 10 ft (3 m). This well was drilled with an air-mist, and following drilling the hole was filled with water prior to logging. For four of the coalbeds, 13 ft (4 m) of net coal, no positive separation was observed on the microlog curves (Figure 10). Positive separation would not, however, be expected in this well due to the lack of a good mud drilling system. The positive separation that appears in the other coalbeds should be ignored because it occurs only in the higher resistivity range above the cutoff resistivity value. Microlog "porosity" calculated in this well indicates 3 ft (0.9 m) of moderately cleated coal and 44 ft (13.4 m) of poorly cleated coal. Open-hole drillstem testing of this interval yielded a permeability value of 1.5 md, but the well would not produce gas naturally. The test data in this well agree with the microlog interpretation that the coals have poor cleating.

FACTORS THAT AFFECT THE MICROLOG

Many factors govern the resistivity measured by the microlog and the degree of positive separation observed on the log trace. Although these factors are not the focus of this study, they are mentioned here as a precaution when quantifying microlog resistivity. The optimum range for mud resistivity is from 1 to 3 ohm-m at bottom-hole temperature. Low mud resistivities compress the resistivity of the microlog curves; high mud resistivities enhance microlog resistivities. This is why the microlog must be

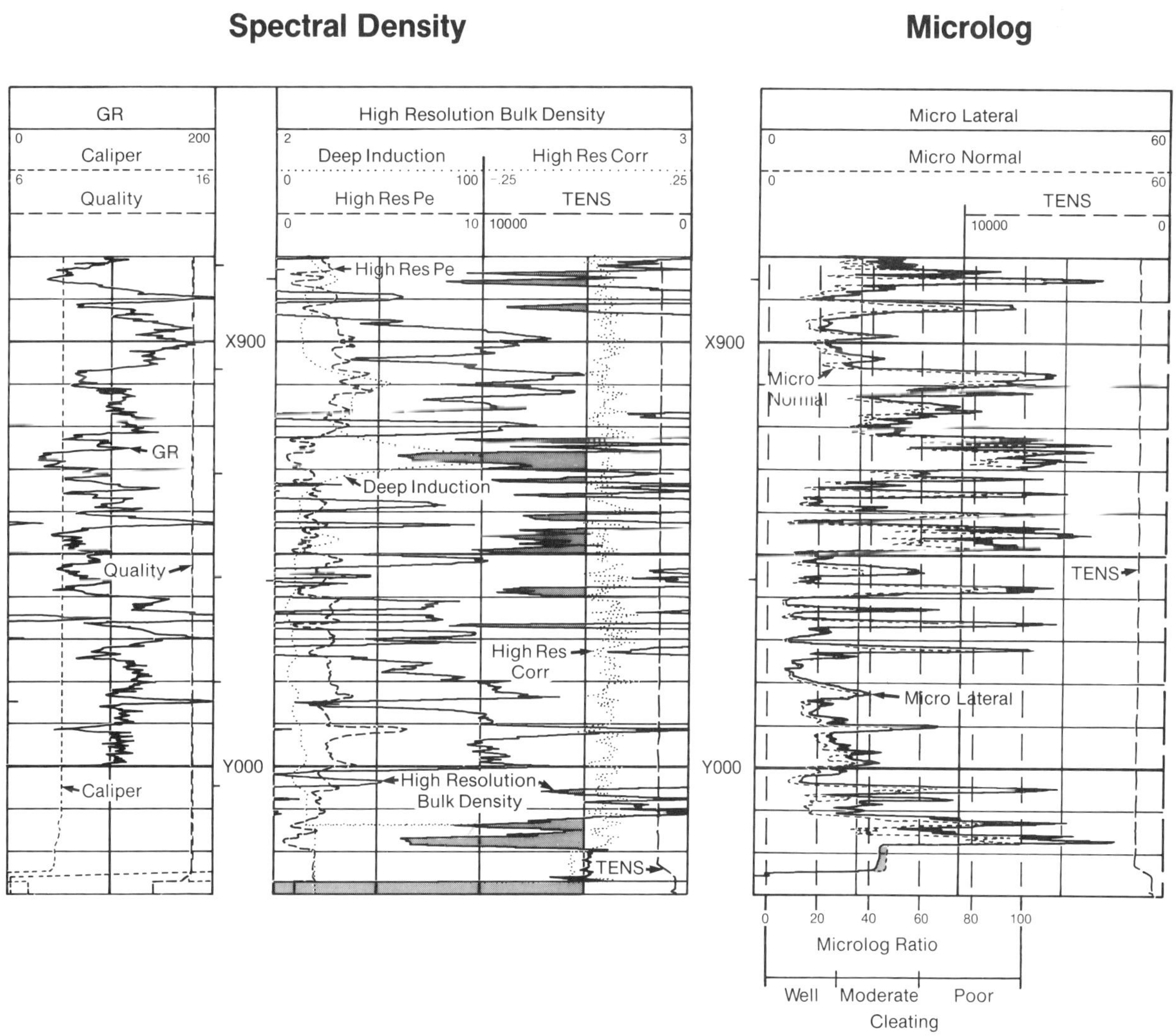

Figure 7. Spectral Density log and microlog for example well 3. R_m = 1.3 ohm-m at BHT.

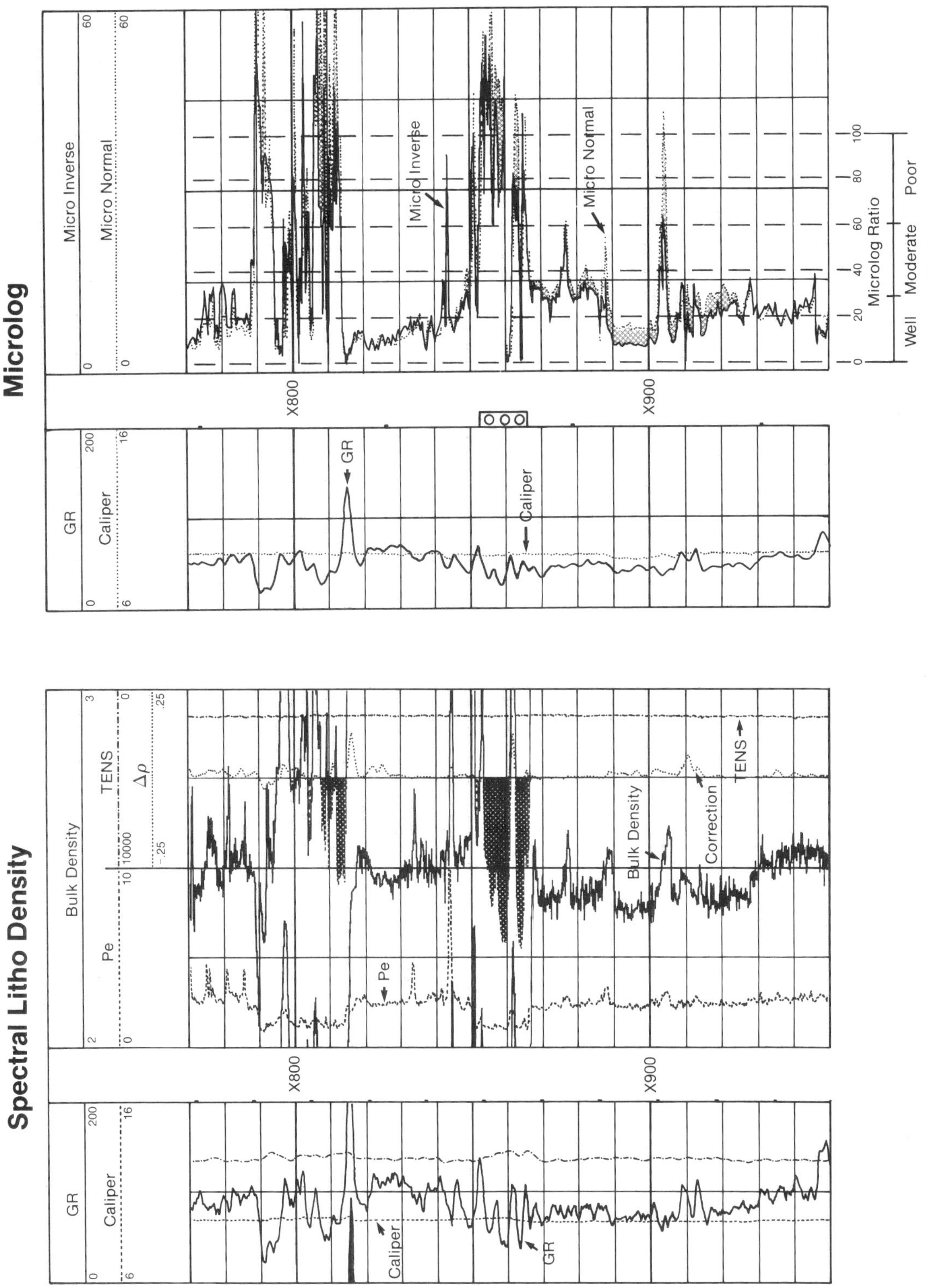

Figure 8. Spectral Litho Density log and micrologue for example well 4. R_m = 1.4 ohm-m at BHT.

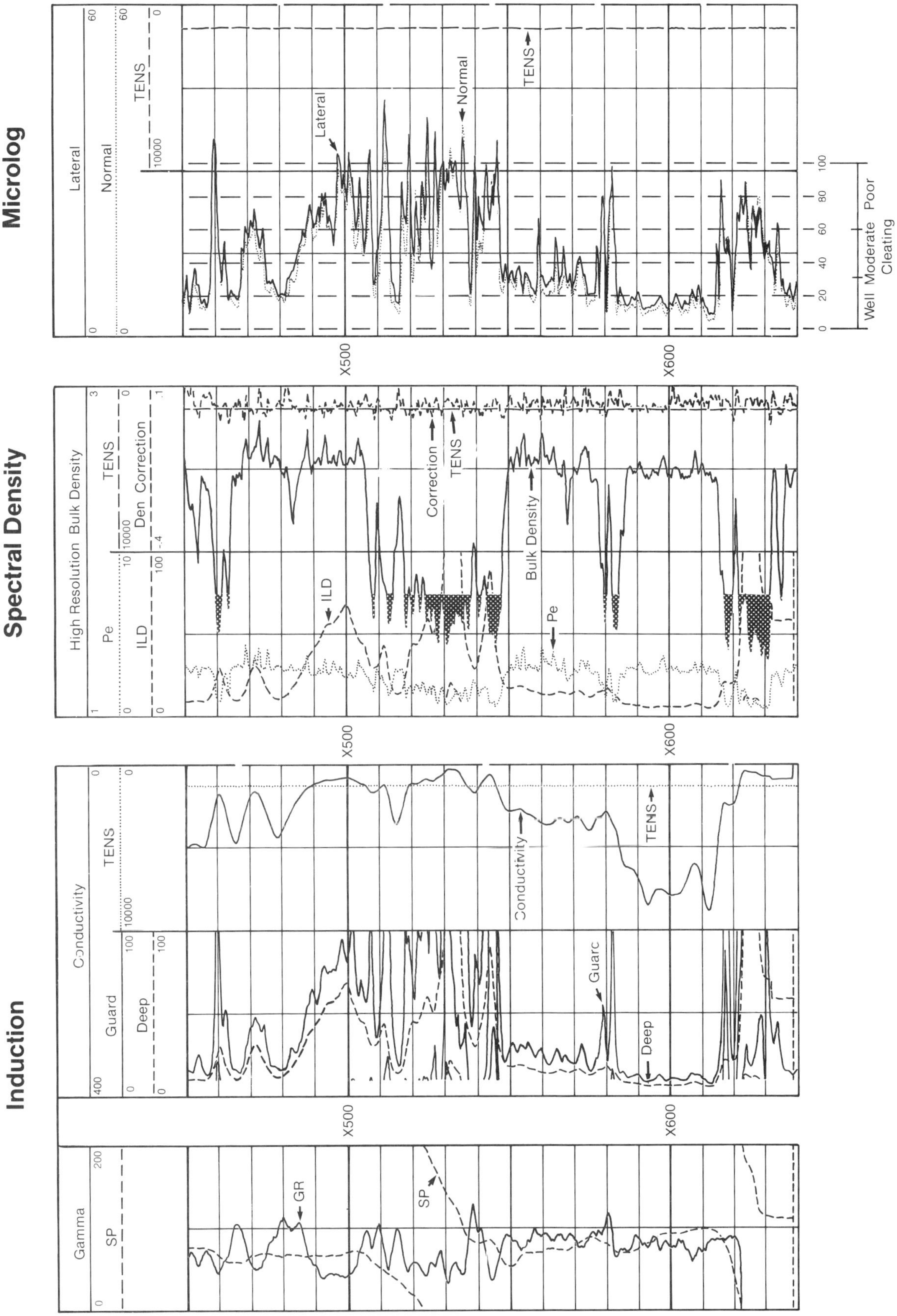

Figure 9. Induction log, Spectral Density log, and microlog for example well 5. R_m = 1.02 ohm-m at BHT.

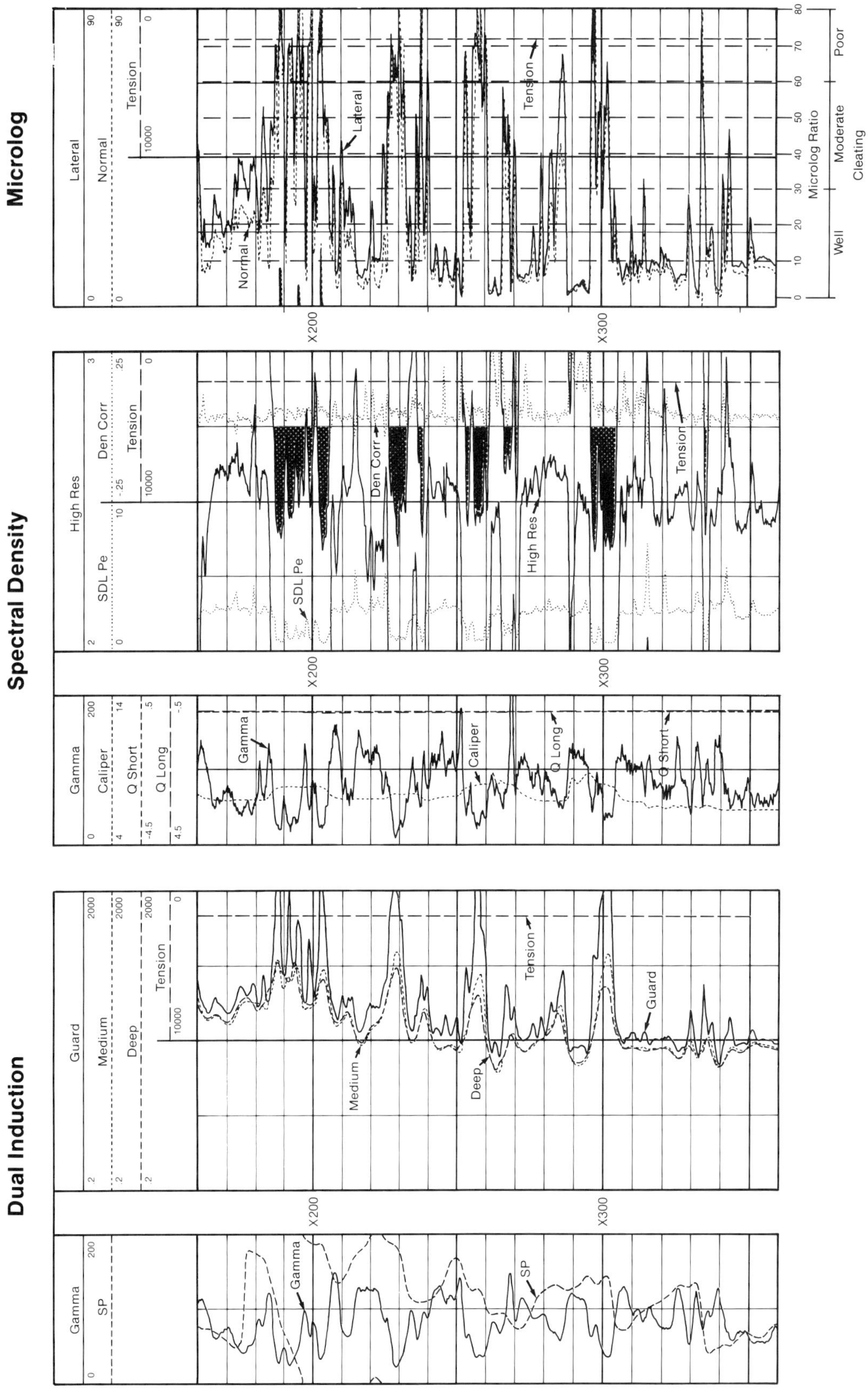

Figure 10. Dual Induction log, Spectral Density log, and microlog for example well 6. R_m = 2.5 ohm-m at 124°F.

normalized before being interpreted for cleat-type determination.

Mud cake thickness is controlled by mud properties and formation permeability. Microlog interpretation is limited to a mud cake thickness less than 0.5 in. (1.27 cm) in conventional applications. Mud cake thickness in coalbeds usually is very thin, but mud invasion into the coalbeds usually is sufficient for valid microlog interpretation.

In most wells, no adverse effects on the microlog have been noticed due to mud additives. However, some additives may cause a thin mud cake to form through the entire wellbore. This results in positive microlog curve separation over large intervals in the well.

Finally, because microlog measurements are made with a pad device, borehole rugosity is the condition that most often clouds microlog interpretation. Over rugose intervals, microlog measurements are questionable, and reliable analysis may not be possible.

CONCLUSIONS

The microlog tool has been used for many years as an indicator of permeable zones. This paper demonstrates that the microlog can be interpreted to yield relative cleat-type identification in coalbed methane exploration and development. Cleated coalbeds can be identified from the microlog by two methods. In wells drilled with a low water-loss mud system, cleated coalbeds exhibit positive separation between the microlog curves, which usually will read low resistivity values. In wells drilled with water, the porosity is calculated from the micronormal curve to indicate the degree of cleating. For a fraction of the cost of coring, this procedure gives the operator useful information about the coal for completion design or future drilling techniques. As a word of caution, because the microlog is a pad-type device, the quality of cleat interpretation from the microlog will be subject to the condition of the borehole wall.

REFERENCES

Bateman, R.M., 1985, Open-hole log analysis and formation evaluation: Boston, IHRDC, p. 270.

Gearhart Industries, 1983, Study guide—Micro Resistivity, WS-677: Ft. Worth, Texas, Gearhart Industries, p. MEL–14.

Hilchie, D.W., 1979, Old electrical log interpretation: Golden, Colo., Douglas W. Hilchie Inc., p. 103–117.

Kelso, B.S., and Wicks, D.E., 1988, A geologic analysis of the Fruitland Formation coal and coalbed methane resource of the San Juan Basin, southwest Colorado and northwest New Mexico, *in* Fassett, J.E., ed., Geology and coal-bed methane resource of the northern San Juan basin, Colorado and New Mexico: Rocky Mountain Association of Geologists Guidebook, p. 69–80.

Levine, J.R., and Deul, M., 1989, Introduction to coal petrology with application to coalbed methane research and development: University of Alabama, Short Course Notes, p 3.5.

Mullen, M.J., 1988, Log analysis in wells drilled for coalbed methane, *in* Fassett, J.E., ed., Geology and coal-bed methane resources of the northern San Juan basin, Colorado and New Mexico: Rocky Mountain Association of Geologists Guidebook, p. 113–124.

Mullen, M.J., 1989, Coalbed methane resource evaluation from wireline logs in the northeast San Juan Basin—A case study: SPE Paper 18946 [presented at the Low Permeability Reservoir Symposium and Exhibition, Denver, Colo., March 6–8].

Schlumberger Well Services, 1952, Interpretation charts for Microlog (Document 5): Houston, Schlumberger Well Services.

Cavity Stress-Relief Method for Recovering Methane from Coal Seams

Denes Masszi[1]

1. D. Masszi Consulting Services Ltd., Toronto, Ontario

ABSTRACT

Stimulating gas flow from coal seams is vital to the economical production of methane. The most commonly used stimulation method is hydrofracturing, the purpose of which is to increase permeability of the coal matrix, creating a long channel with high permeability. This method does not always deliver the expected results. Consequently, in 1979 the author developed a new coalbed stimulation method based on recognition of the fact that approximately 90% of the methane distributed in a coal seam is in adsorbed form and only 10% is in free stage. To increase gas production, not only permeability of the coal matrix but also the rate of desorption must be increased. The cavity stress-relief method that was developed takes advantage of the effects of depth-dependent triaxial compression, under which the coal is subjected, to increase coal seam permeability and significantly increase the rate of diffusion.

INTRODUCTION

Methods of extracting methane from coal can be classified into two categories, based on whether they are carried out within a mine or independent of a mine. In the first method, as a procedure to increase mine safety, methane is drained from coal that is being mined or that will be mined in the near future.

The second method, *virgin coal demethanation* (VCD), is most often used to extract gas from coal that will be mined within five or more years or from coal that may never be mined. The VCD method was developed in North America fifteen years ago, mostly in response to the energy shortages of the 1970s. At that time methane in coal was recognized as a significant resource that could supplement conventional oil and gas.

The VCD system combines three technologies—coal, oil and gas, and hydrogeology. In this method a borehole is drilled from the surface to intercept one or more coal seams. The contact between the coal and the borehole is kept as large as possible. The borehole usually is drilled about 30 ft (10 m) below the lowermost coal. A pump is installed there to remove the water. Pumping the water reduces the hydrostatic pressure on the coal, allowing methane to desorb and move toward and up the wellbore under a pressure gradient.

Because most coals have low permeability and usually occur in place in large blocks, wells generally have low productivity unless the coal is stimulated. In early work of the U.S. Bureau of Mines, which initiated the program in demethanation, the need arose to improve well production. Hydraulic fracturing was attempted because it was a convenient, available technique. This technique, introduced in the early 1950s, was designed to enhance the productivity of oil and gas wells by creating a large, conductive fracture in the reservoir. Since its introduction, development of the technique has been directed toward enhancing this effect and improving control on fracture size, length, and direction.

The production rate of oil and gas wells is limited only by permeability. That is, if infinite permeability could be induced in the reservoir containing the hydrocarbons, all the production could be obtained almost instantaneously. This is not true of production from coal demethanation wells.

Using the hydrofrac method in coal solved what was believed to be the cause of low gas production—that is, low permeability of the coal. The desired effect was obtained when the well was hydraulically fractured, thus "proving" the validity of the theory. It is my intent to show that the effect of hydraulic fracturing is not what is commonly thought to occur, and that once the actual mechanism responsible for improved gas production via an induced fracture is understood, a better stimulation technique will be developed.

EXPERIENCE WITH HYDRAULIC FRACTURES

Since 1977 I have carried out more than eighteen hydraulic fracturing operations of variable size, depth of application, injection fluid, and presence or absence of propping agents to optimize their application to virgin coal demethanation. When the results of these treatments were analyzed, it was found that the effects of the fractures were generally greater than could be ascribed solely to an increase in permeability, even assuming the long fractures predicted by the service companies. Through hydrogeologic techniques such as pump tests with observation wells, a significant amount of work was done to determine the permeability increase resulting from a fracture. I concluded that another mechanism also was operating, and it is now believed that, in most cases, this mechanism is more significant than increased permeability. In an attempt to characterize this phenomenon, a bibliographic study was undertaken of the results of theoretical studies, laboratory work, and mine methane drainage related to methane desorption from coal over the last 35 years. Laboratory tests also were undertaken at the University of Calgary and University of Manitoba to better understand the mechanisms of methane desorption and fracture propagation in coal.

THEORETICAL BACKGROUND

Unlike conventional gas production, production of coal gas is controlled by both permeability and diffusion (or desorption). Either may be the limiting factor at a specific time. As shown in Figure 1, the rate of gas production initially is limited by permeability (during the period of unloading). However, after a certain time the concentration gradient decreases to such an extent that the rate of desorption is less than the fracture flow capacity, and diffusion becomes the limiting factor. Thus, it becomes obvious that to maintain a prolonged large flow from a well, not only must permeability be improved (the supposed effect of a fracture), but diffusion also must be enhanced. To design a stimulation process specifically for demethanation, one should consider the factors that affect desorption of gas from coal.

Factors Affecting Desorption of Gas from Coal

The volume or rate of desorption of gas from coal is affected principally by the following physical parameters:

Factors affecting volume:

1. Pressure
2. Moisture content of coal

Factor affecting rate:

3. Particle size of coal

Factors affecting volume and rate:

4. Temperature
5. Permeability

Pressure

The volume of methane that can be retained by coal at two different equilibrium levels differing only in their pressure is the amount of gas desorbable between those two pressure levels for a specific coal. The amount of gas desorbable per unit pressure drop is not linear but is greater at lower pressures (Figure 2). The rate at which gas is released is not infinite but controlled.

Moisture Content

The quantity of methane adsorbed in coal is inversely proportional to the inherent moisture content (Figure 3). Above 5% moisture the gas content is virtually unaffected. Thus, if moisture content could be reduced to less than 5%, more methane could be desorbed.

Particle Size

Particle size affects desorption rate because of the distance through which the gas must diffuse. One measure of evaluating desorption-rate effects is to compare the theoretical time for 90% of the methane to desorb from coal particles of various sizes:

Particle Diameter	Time for 90% Desorption of Methane
1 μm	4.65 seconds
10 μm	10 minutes
100 μm	13 hours
1 mm	1 month
1 cm	15 years
1 m	150,000 years

The desorption effects among the smaller particle sizes (7 to 170 mesh) are apparent from the laboratory test results shown in Figure 4. Particle size therefore has an enormous effect on gross desorption rate. It follows that the volume of gas produced by a VCD borehole could be greatly enhanced if coal grain size could be reduced.

Temperature

Temperature affects both the volume of methane retained by the coal and the rate at which it desorbs. Figure 5 shows

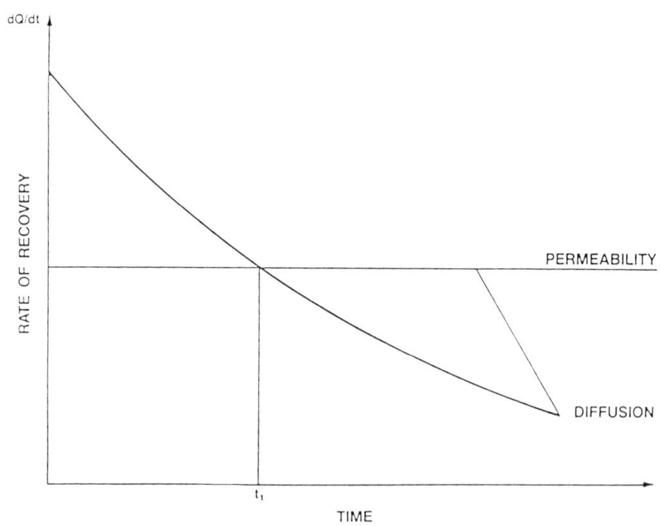

Figure 1. Interaction among gas recovery mechanisms in coal. After T.M. Dosher.

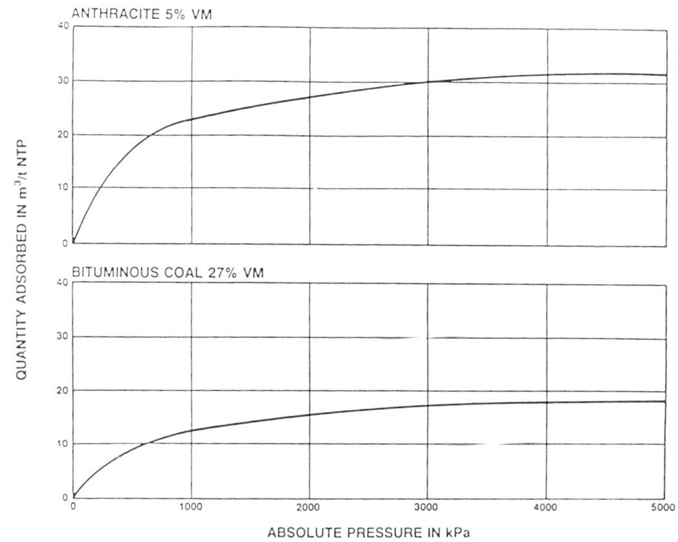

Figure 2. Relationship between pressure and adsorption for anthracite and bituminous coal.

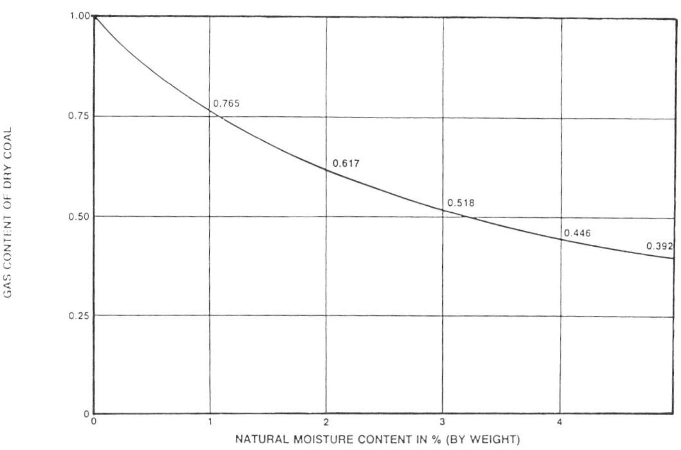

Figure 3. Effect of moisture content on methane content of coal.

the effects of temperature on the volume of methane retained in the coal under a particular set of conditions. The higher the temperature, the less the methane can be retained by the coal. Furthermore, the rate of desorption is directly proportional to temperature. Thus, if the temperature of the coal reservoir can be increased, then added benefits of increased volume and rate would ensue. The reverse also is true.

Permeability

The permeability of coal controls the rate at which desorbed methane can escape. Thus, as discussed above, it acts as a buffer between internal and external pressures on the coal.

One can conclude that the change in particle size exerts the most significant effect on desorption rate. In view of the above data, one then can evaluate the effect of hydraulic fracturing as applied to coal and attempt conceptually to design a better stimulation technique.

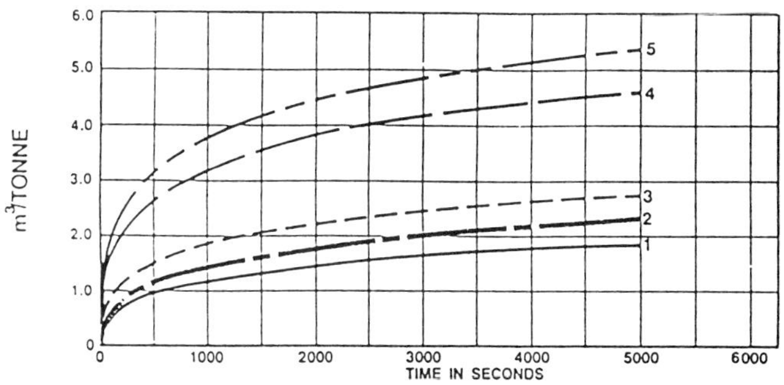

Curve	Particle Size (mesh)
1	–7 + 14
2	–14 + 25
3	–25 + 45
4	–45 + 80
5	–80 +170

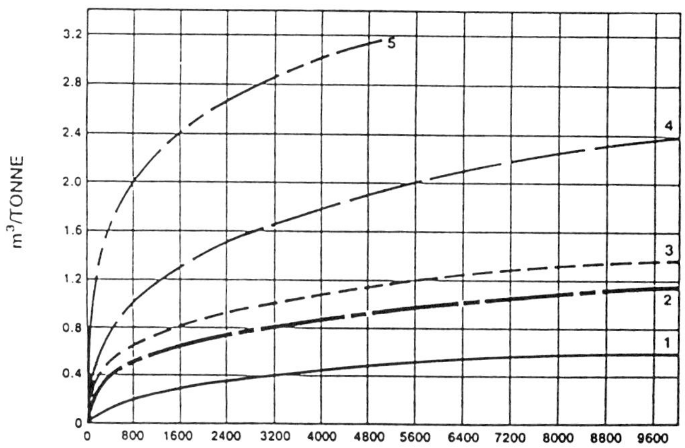

Curve	Particle Size (mesh)
1	–7 + 18
2	–18 + 25
3	–25 + 45
4	–45 + 80
5	–80 +170

Figure 4. Relationship between coal particle size and quantity of methane desorbed from Alberta (upper graph) and Nova Scotia (lower graph) coals. Desorption from dry samples at 21°C.

HYDRAULIC FRACTURING

In the subsurface, coal is subjected to triaxial compression principally due to overburden pressure. When fluid is pumped into the formation at a pressure sufficient to overcome both the compressive stress and the tensile strength of the coal, a fracture is created. When this happens, coal in the vicinity of the fracture is crushed, forming small tensional fractures perpendicular to the large main fracture. Fractures parallel to the induced fracture are closed. With the use of proppants, coal in the vicinity of the large fracture will remain compressed. It is probable that creation of the tensional fractures rather than the large fracture is more beneficial to a VCD hole in the long term. The large fracture enhances the unloading period, while the numerous smaller tensional fractures reduce the gas diffusion path and,

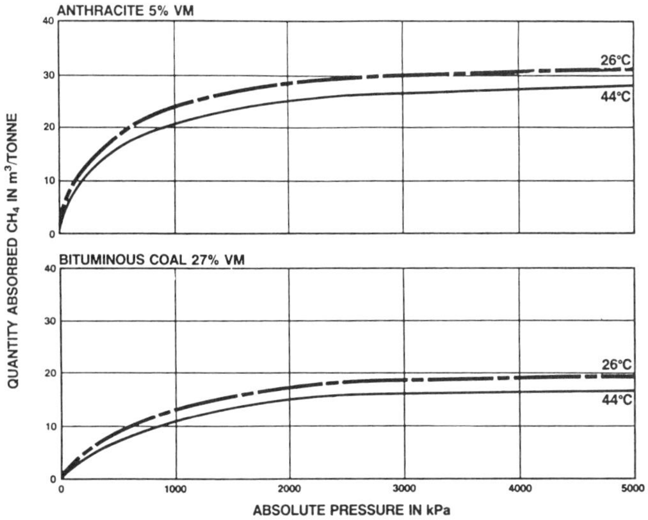

Figure 5. Relationship between temperature and methane adsorption in anthracite and bituminous coal.

therefore, reduce the time required to desorb a particular block of coal. A reduction in time implies increased well productivity.

A NEW STIMULATION PROCESS

With a better understanding of the factors affecting desorption of methane from coal, we undertook to develop an improved stimulation technique, which, from experience in methane drainage in coal mines, is based on the phenomenon of stress relief to cause an unsupported cavity. The cavity should be developed in, partly in, or proximal to the coal seam and should be unsupported and designed to be as collapsible as possible.

As illustrated sequentially in Figure 6, immediately after an opening is made, restraint is removed from one direction, and redistribution of the in situ triaxial stress system occurs. After excavation the vertical stress, which no longer acts above the excavation, is partially transferred to each side, resulting in increased unidirectional loading accompanied by reduction in pressure above the excavation. The area in which the stress is reduced near the open face is known as the destressed or relaxed zone, whose size varies according to the size and shape of the cavity. Because increased loading in this zone may exceed the load-carrying capacity of the coal, the coal fails, causing the abutment zone to propagate farther into the cavity. Consequently, the size of the relaxed zone increases, and redistribution of stresses thus leads to fracturing. The degree of fracturing depends on vertical loading (overburden) and strength of the coal within the relaxed zone. Permeability in the relaxed zone will be increased.

The unsupported cavity in the coal seam has an additional effect. Coal from the seam moves toward the cavity under the influence of the redistributed compression. As the coal moves, its physical characteristics change. Coal particle are broken into smaller sizes. Rate of methane desorption from smaller coal particles is significantly higher.

The effect of the developed cavity commonly is used to increase gas production in underground methane drainage systems, to protect against gas outbursts, or to overcome the problems of high-pressure zones encountered during tunnel development.

Thus, the net effect of cavity formation and its subsequent collapse is to produce a highly permeable zone filled with

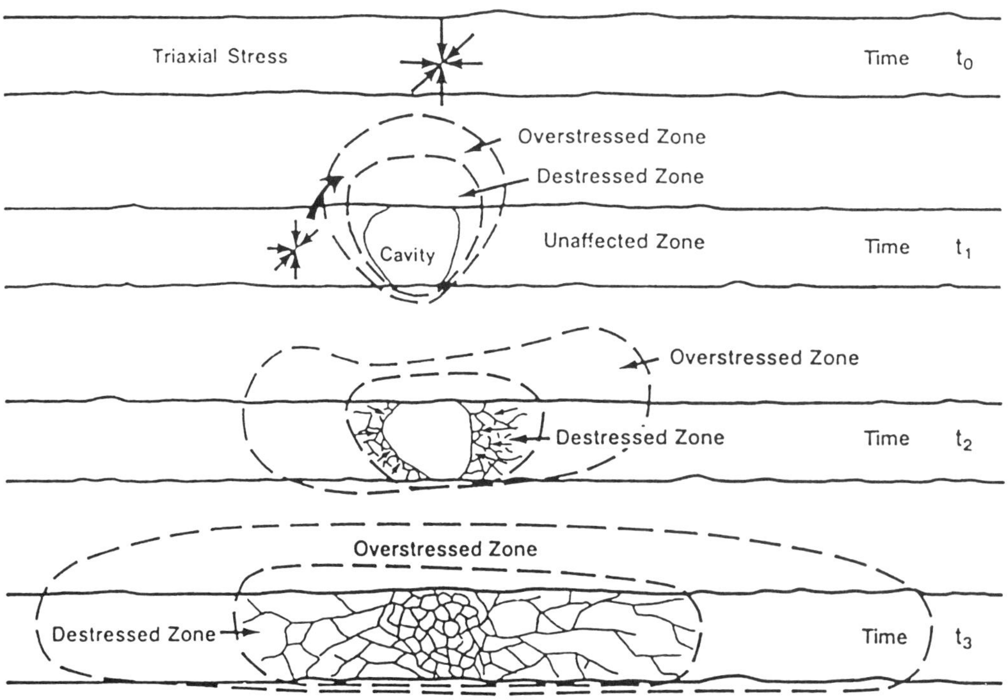

Figure 6. Simplified cross sections of triaxial stress redistribution in a coal seam through time caused by creation of a cavity.

Table 1. Comparison of hydraulic fracturing and cavity stress-relief.

Hydraulic Fracturing	Cavity Stress-Relief
1. Injects fluid into coal, which reduces rate of desorption.	1. No injected fluid.
2. Produces compression in one direction, thereby closing the set of natural fractures.	2. Releases stress so that all natural fractures remain open.
3. Increases permeability along one or two directions.	3. Increases permeability radially around wellbore. Effect can extend as far into the formation as an induced fracture.
4. Increases surface area by producing tension cracks perpendicular to induced fracture.	4. Breaks up a large volume of coal radially around wellbore, with fracturing around coal particles.
5. Affected by geologic inhomogeneities, such as faults and highly permeable horizons, producing high leakoff.	5. Is not as affected by geologic inhomogeneities, unless adjacent formations are weaker than coal.
6. Does not adversely affect mining.	6. Assists mining by destressing coal. Most important in cases susceptible to outburst.
7. Increased depth increases costs and injection pressure.	7. Increased depth assists the effect and may eliminate need for cavity-making equipment because of natural collapse of coal.

fine-grained coal, a condition ideal for gas production and transmission. An increase in temperature of the coal and reduction in its inherent moisture content would improve the productivity of a VCD well, but imposing those conditions is not considered practical at this time. A comparison is given in Table 1 between hydraulic fracturing and the new stimulation method, which has been called *cavity stress-relief*, or CSR. The technique has been patented worldwide by Nova Corporation of Alberta, and ownership has been transferred to the author.[2]

FIELD EXPERIENCE

To support the theoretical expectations of the CSR method, two boreholes in Alberta, Canada, were stimulated. Both were located close to conventionally stimulated wells so as to minimize any effects of varying geological parameters. Gas production from the wells was compared to evaluate the efficiency of the CSR-based stimulation.

The cavity can be created in various ways. In one method the coal washes out during drilling. Because of its somewhat higher friability at greater depth, coal can be seen to cave or spall into the borehole during drilling. Circulating drilling fluid washes out more coal. Methane liberated from the coal can cause spontaneous outbursts, which breaks up additional coal and expands the relaxed zone around the wellbore.

Another method is the use of high-pressure water jets to cut the coal and compressed air to remove the cuttings (Figures 7 and 8). With more friable coal, less equipment and time are needed to produce the desired cavity. The CSR technique was tested in two VCD holes, each located near a VCD hole that had been hydraulically fractured to compare the two stimulation techniques. One test well was very shallow to determine any possible minimum depth to the CSR method. There is no theoretical maximum depth. Depth is controlled mostly by economics.

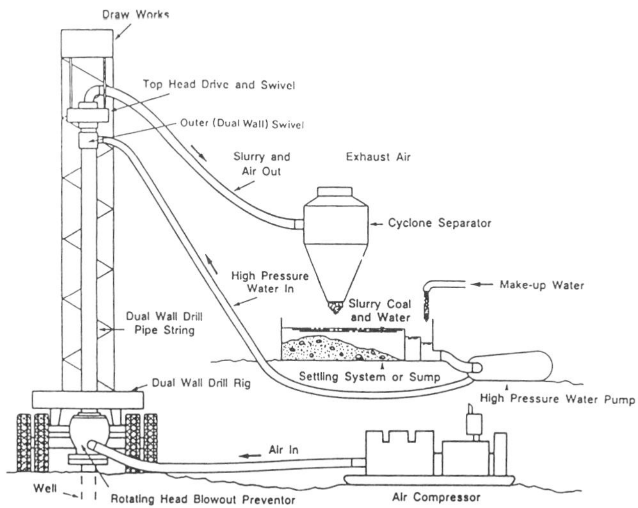

Figure 7. Surface equipment for creating a cavity in coal seams by means of high-pressure water jetting.

2 Canadian Patent No. 1,140,457. United States Patent No. 4305464.

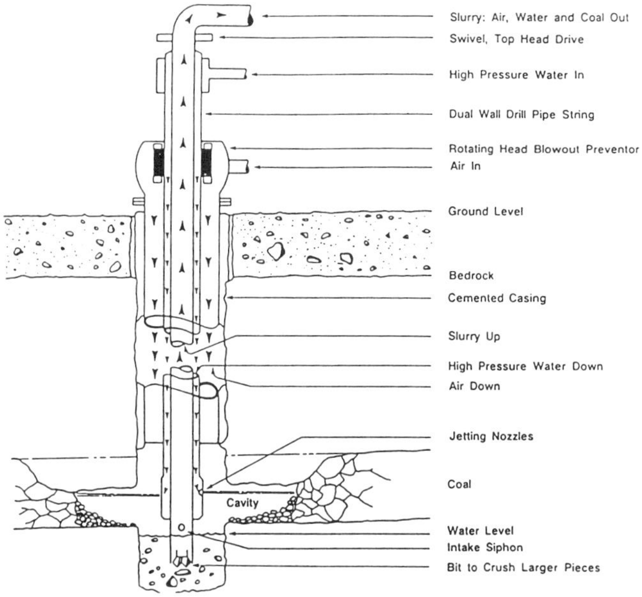

Figure 8. Borehole configuration for creating a cavity in coal seams by means of high-pressure water jetting.

The first hole stimulated by the CSR method was the Canmore 3 (C3) well, located about 420 ft (128 m) from the Canmore 1 (C1), which had been hydraulically fractured and was producing. Both holes were 1,200 ft (366 m) deep. C1 was stimulated with 1,176 bbl (187 m^3) of 75% nitrogen foam, and 40,000 lb (18,181 kg) of 100-mesh and 40,000 lb (18,181 kg) of 10/40-mesh sand. Breakdown pressure was 1,900 psig (13,100 kPa). C3 was "stimulated" by removing 88 short tons (80 t) of coal.

The second application of the CSR method was done on the Riverside 1A well (R1A), located approximately 340 ft (104 m) from the Riverside 4 (R4) well, which has been hydraulically fractured. R4 was stimulated twice. First, the well was fractured with 396 bbl (63 m^3) of water with no visible breakdown. Five months later, it was fractured with 881 bbl (140 m^3) of 90% foam, again with no visible breakdown pressure. R1A was stimulated by removing 55 short tons (50 t) of coal. Both holes were 150 ft (46 m) deep.

GAS PRODUCTION

All four wells were completed and brought into production in the same manner. Through the first four months of production and after the unloading period, C3 produced an average 90 MCFD (2,547 m^3/day), while C1 produced an average 35 MCFD (991 m^3/day). During this production period, severe winter weather caused the C3 wellhead to freeze, greatly restricting gas flow. This produced considerable back pressure in the coal reservoir. Therefore, the given rate is believed to be a minimum relative to the potential of the well. The average gas production for R1A over a period of eight months and after the unloading period was 4 MCFD (113 m^3/day), while that of R4 over the same period was 1 MCFD (28 m^3/day).

COST OF CAVITY MAKING

The total cost of stimulating the deep C3 borehole was \$128,000 CDN in 1982. Because of the new design, inexperience in its use, and modifications that were required in the field, creating the cavity took three times longer than anticipated and was achieved in two attempts. The shallower R1A stimulation cost \$56,000 CDN. In contrast, the costs of hydraulic fracturing were \$65,000 CDN for C1 and \$30,000 CDN for R4. The most expensive item in creating the cavity is the dual-wall drillpipe (Figure 8), which is needed for high-pressure water-jet cutting and which constituted 50% to 60% of the cost of the C3 stimulation. This high cost is caused by the one-off nature of research. Furthermore, because C3 and R1A were stimulated with a different rig than was used to drill them, they incurred higher mobilization and demobilization charges.

CONCLUSIONS

In underground coal mines, cavity stress-relief has been used during the past 25 years to increase gas production and eliminate gas outbursts. This experience and our field tests corroborate the theoretical projection that the cavity stress-relief technique of reservoir stimulation is superior to hydraulic fracturing, particularly in deep VCD wells where coal is friable and has low permeability (<0.1 μd).

Initial capital costs of the CSR method are high if the dual-wall drillpipe method is applied to only one well during the development stage, but the method will become less expensive than hydraulic fracturing once it is applied routinely.

REFERENCES

Boxho, José, 1980, Firedamp drainage—Handbook for the coalmining industry in the European community: Coal Directorate of the Commission of the European Community, 415 p.

Bachelor, F.W., 1982, The effects of particle size, moisture content and temperature on desorption of methane from coal: University of Calgary, Dept. of Chemistry Study, July 16.

Doscher, T.M., Kuuskraa, V.A., and Hammerschaib, Edgar, 1981, The controlling production mechanism of methane gas from coalbeds: Energy Sources, v. 5. no. 1, p. 71–84.

Kahil, A., and Masszi, D., 1984, Cavity stress relief methods to stimulate demethanation boreholes: Proceedings, SPE/DOE/GRI Unconventional Gas Recovery Symposium.

Thimons, E.D., and Kissell, F.N., 1973, Diffusion of methane through coal: Fuel, v. 52, no. 4, p. 274–280.

Coal Geology and Coalbed Methane Potential of Comox and Nanaimo Coal Fields, Vancouver Island, British Columbia

C. Gwyneth Cathyl-Bickford[1]

1. Westwater Mining Ltd., Calgary, Alberta

ABSTRACT

Coalbeds of eastern Vancouver Island have been worked by underground mines since 1849. Considerable effort and ingenuity have been demanded of the miners, as they face difficult structural conditions and high gas contents of coals, which locally attain thicknesses of 66 ft (20 m).

The two principal coal fields, Comox and Nanaimo, contain coals of Santonian and Campanian (Late Cretaceous) age. Associated with the coals are soft, organic-rich mudstone and porous, water- and gas-bearing sandstone and conglomerate. Much of the coal above a depth of 1,476 ft (450 m) has been mined out, but deep drilling indicates that coals persist to depths of 3,280 to 4,920 ft (1,000 to 1,500 m), well within the depth range of established coalbed methane production technologies. Gas could also be recovered from or stored in old mine workings.

Attractive features of the Vancouver Island coal-gas play include large potential market base, high gas content of the coals, and good seismic response of the coalbeds. Drawbacks include structural complexity of certain coal deposits, propensity of some coalbeds to split and pinch out, and well-organized environmental groups opposed to any new industrial development.

INTRODUCTION

The coal fields of eastern Vancouver Island have long been notorious for difficult mining conditions. High gas contents of the coals, together with the local propensity for spontaneous outbursts from mine faces and borehole walls, have been major impediments to large-scale development. With the advent of practical technology for coal demethanation in the 1980s, gas contained in coals on Vancouver Island has become a potential commercial source of energy and an exploration target in its own right.

Coal mining on Vancouver Island commenced in 1849 and reached its peak development in 1921. The last large deep mines closed in 1947 and 1952, largely because of ventilation problems related to high gas content of the coals. Mining since then has been concentrated along the edges of the coal fields, where the coals lie under shallower cover. Most of the mining has taken place in two principal fields—Comox in the north, and Nanaimo in the south (Figures 1 and 2). Most of the readily accessible shallow coal has been mined out, and remaining reserves lie at depths below 1,475 ft (450 m). Deep drilling and Vibroseis reflection profiling in these areas have shown that coals may extend to depths as great as 4,920 ft (1,500 m). Although such deep coals are beyond the foreseeable reach of conventional underground mines, they are well within the depth range of established coal-gas production technology.

GEOLOGIC SETTING

Major coal deposits of Vancouver Island, including Comox and Nanaimo coal fields, are of Late Cretaceous age. The coal measures are part of the Nanaimo Group, which crops out along the southeastern and northwestern edges of the Cretaceous- to Tertiary-age Georgia basin. As shown on Figure 1, Georgia basin is elongate to the northwest and largely underlies the Strait of Georgia between the British Columbia mainland and Vancouver Island. A substantial area of eastern Vancouver Island and a smaller part of the Puget Lowlands province of Washington State are underlain by as much as 13,124 ft (4,000 m) of Nanaimo Group sedimentary rocks (Bustin and England, 1991).

Sedimentary rocks of the Nanaimo Group are of Turonian to Maastrichtian age and have been deformed by folds and faults of late Eocene to early Oligocene age. During this early Tertiary deformation, laccoliths and stocks were emplaced along the southwestern margin of Georgia basin. The southern half of the basin underwent transpressional folding and thrusting; most of the structures display southwestward vergence, although northeastward back-thrusts occur locally in the upper Nanaimo Group. Northeastward back-thrusts are well developed in the vicinity of the Nanaimo River coal-gas prospect (Figure 3). Northern Georgia basin underwent transtensional extension, resulting in a basin-and-range structural configuration.

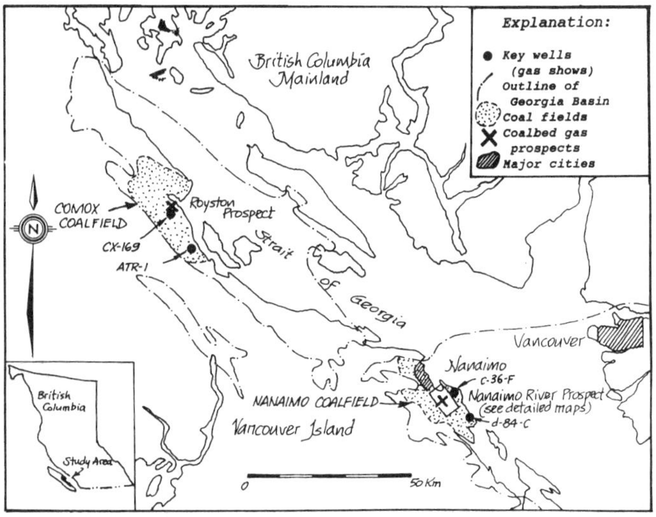

Figure 1. Location map of Comox and Nanaimo coal fields, Vancouver Island, British Columbia.

The stratigraphic nomenclature of Cretaceous rocks in Georgia basin recently has been reviewed by Bickford and Kenyon (1988), England (1989), and Cathyl-Bickford and Hoffman (1991). Tables 1 and 2 summarize the lithostratigraphy of component formations in the lower, coal-bearing part of the Nanaimo Group in the Nanaimo and Comox coal fields, respectively.

REGIONAL COAL DISTRIBUTION

Although coals occur at numerous stratigraphic levels within the Nanaimo Group, the major coalbeds are found in the Comox, Extension, and Pender Formations, in the lower half of the group. Coalbeds of the lower Nanaimo Group are closely associated with organic-rich mudstone and oil shale, some of which are potential hydrocarbon source beds (Bustin and England, 1991).

Coals of the Pender and Extension Formations are present only in Nanaimo coal field, while coals of the Comox Formation are present in both Comox and Nanaimo coal fields. Little is known about Comox Formation coals in Nanaimo coal field, as they were only recently discovered and appear to be confined to deeper parts of the field.

COALS OF NANAIMO COAL FIELD

Two major coalbeds occur in the Newcastle Member of the Pender Formation (Table 1)—the Douglas seam near the middle, and the Newcastle seam at the base, 49 to 82 ft (15 to 25 m) below the Douglas seam. These two coalbeds may coalesce in the southeastern part of the field.

The Douglas seam is the thickest of the coalbeds at Nanaimo. Its average thickness is 8.2 ft (2.5 m), although it characteristically pinches and swells rapidly and locally is as thick as 69 ft (21 m). The Douglas coal was deposited on the landward (southern) side of a storm beach and is, in

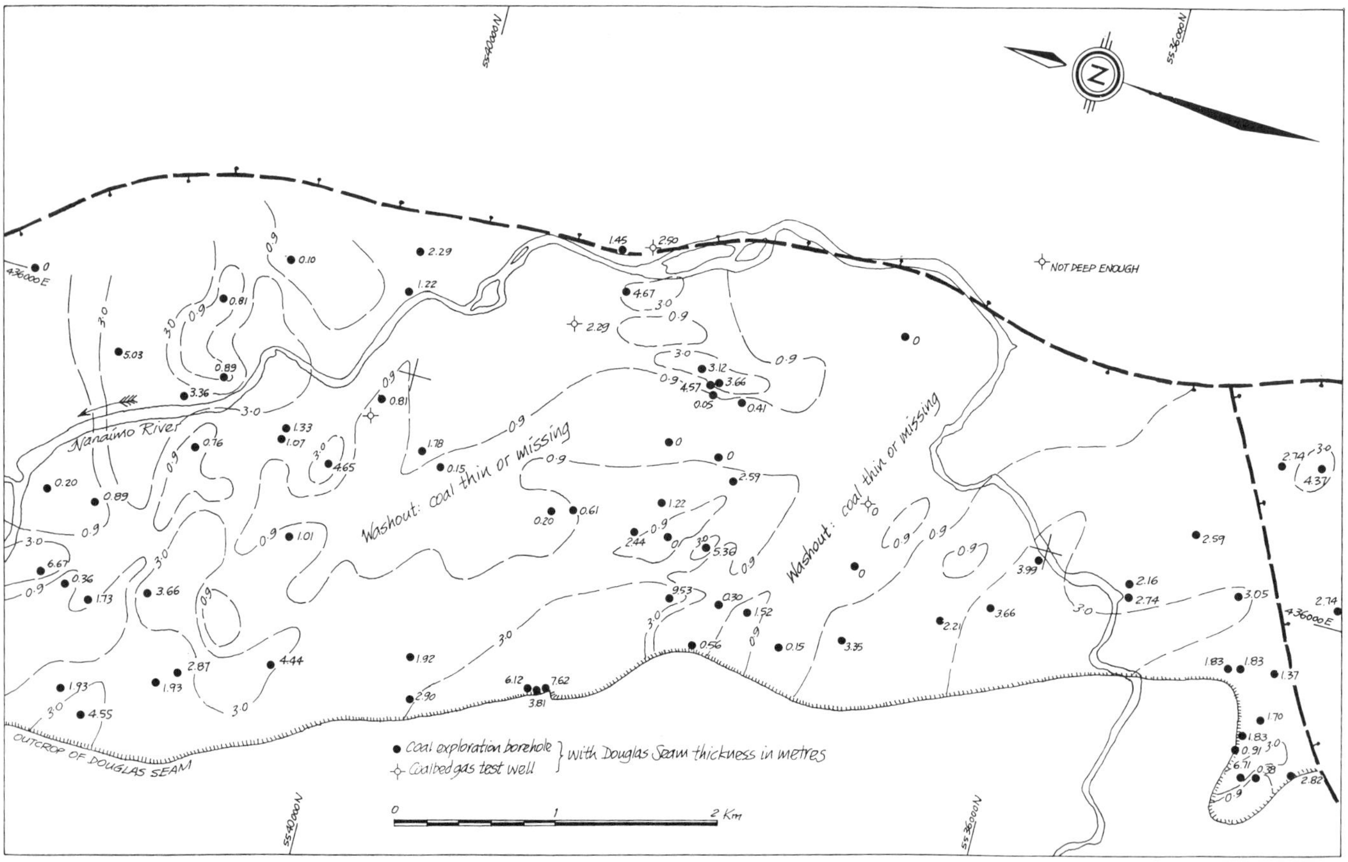

Figure 2. Isopach map of Douglas coal seam, Nanaimo coal field, Vancouver Island.

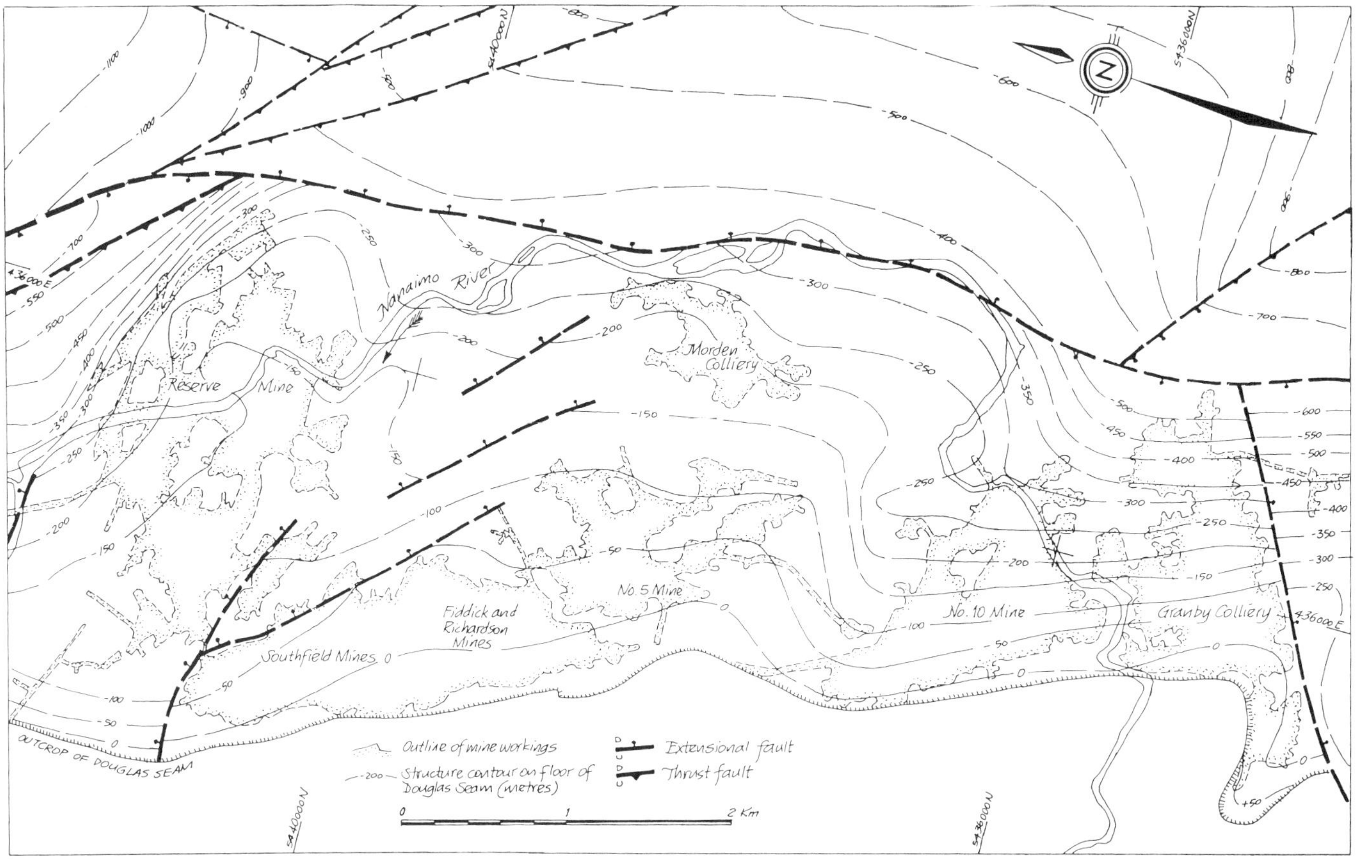

Figure 3. Structure contour map of floor of Douglas coal seam, Nanaimo coal field, Vancouver Island.

turn, bounded to the south by carbonaceous mudstone that was deposited in a protected delta-front lagoon. Isolated patches of coaly mudstone within the main body of the Douglas seam may have been deposited in ephemeral lakes and ponds, while more extensive washouts of the coal represent the courses of streams that drained the swamps (Figure 2). The Douglas seam generally is sheared in central and southern Nanaimo coal field and is therefore friable and often devoid of recognizable cleat systems.

The Newcastle seam is the thinnest coalbed in the field, averaging 3.9 ft (1.2 m) thick. It contains numerous mudstone and siltstone partings. The Newcastle seam was deposited in a salt-marsh environment overlying delta-front mud flats and, in turn, overlain by gravelly beach deposits. In contrast to the Douglas seam, the Newcastle seam usually is harder and blockier and exhibits a well developed cleat system.

One major coalbed, the Wellington seam, occurs in the Extension Formation at Nanaimo. Thickness of the Wellington seam, 3.9 to 8.9 ft (1.2 to 2.7 m), varies rapidly because of irregular paleotopography of its floor (Bickford, 1989). The bed pinches out to the east and generally does not extend eastward beyond the outcrop of the overlying coalbeds. The Wellington seam accumulated in an extensive coastal-plain swamp that was traversed by meandering and braided streams of varying widths and erosional capabilities. Stream channels generally were unstable, resulting in deposition of numerous clastic partings within the seam. The Wellington seam is hard and blocky, with well developed cleats often filled with calcite and pyrite.

All three major coalbeds at Nanaimo are high-volatile B to A bituminous in rank, with mean random vitrinite reflectance values of 0.64% to 0.72%. The maturation pattern of Nanaimo coals is indicative of postdeformational coalification (Kenyon and Bickford, 1989).

COALS OF COMOX COAL FIELD

Three major coalbeds occur in the Comox Formation (Table 2) of Comox coal field (in descending order)—No.1, No.2, and No.4 seams. No.1 seam lies about 82 ft (25 m) above the base of the Dunsmuir Member and ranges in thickness from 2.9 to 7.9 ft (0.9 to 2.4 m). It consists of numerous thin beds of bright coal interbanded with laterally persistent sandstone and siltstone partings. No.1 seam probably was deposited on the landward side of a sandy barrier island that occasionally was overtopped by storm surges.

No.2 seam lies at the top of the Cumberland Member and ranges in thickness from 2.6 to 7.2 ft (0.8 to 2.2 m). It is the most persistent coalbed within the field. It accumulated along the landward shores of coastal lagoons that were

Table 1. Lithostratigraphy of Nanaimo coal field, Vancouver Island.

Age	Formation/Member	Lithology	Thickness (ft)	Thickness (m)
Late Campanian	De Courcy Formation	Sandstone; minor conglomerate; shale	984–1,804	300–550
	Cedar District Formation	Shale; minor sandstone.	1,017–1,640	310–500
	Protection Formation			
	McMillan Member	Sandstone and siltstone.	197–295	60–90
Early to Late Campanian	Reserve Member	Siltstone, sandstone; minor coal.	131–197	40–60
Early Campanian	Cassidy Member	Sandstone and conglomerate.	262–344	80–105
	Pender Formation			
	Newcastle Member	Shale, conglomerate and coal. Douglas Seam near middle; Newcastle Seam at base.	164–197	50–60
	Cranberry Member	Sandstone and siltstone; shale.	492–640	150–195
	Extension Formation			
	Millstream Member	Conglomerate; minor coal.	394–492	120–150
	Northfield Member	Siltstone and sandstone; coal. Wellington Seam at base.	33–98	10–30
	East Wellington Formation	Sandstone.	26–148	8–45
Santonian to Early Campanian	Haslam Formation	Shale and siltstone.	213–984	65–300
Santonian	Comox Formation			
	Dunsmuir Member	Sandstone; minor coal.	0–394?	0–120?
	Cumberland Member	Sandstone, shale and coal.	0–197?	0–60?
Turonian to Santonian?	Benson Member	Conglomerate and red shale.	0–492	0–150

Table 2. Lithostratigraphy of Comox coal field, Vancouver Island.

Age	Formation/Member	Lithology	Thickness (ft)	Thickness (m)
Late Campanian	Trent River Formation			
	Oyster River Member	Sandstone; minor conglomerate and shale.	20?	50?
Early Campanian	Royston Member	Mudstone, siltstone.	492–722	150–220
	Tsable Member	Mud-matrix conglomerate; pebbly shale.	16–459	5–140
	Browns Member	Sandstone.	33–394	10–120
	Puntledge Member	Siltstone, shale.	328–492	100–150
	Comox Formation			
	Dunsmuir Member	Sandstone; minor shale and coal. No.1 coalbed near base.	394–820	120–250
Santonian	Cumberland Member	Sandstone, shale, coal. No.2 coalbed near top; No.4 coalbed near base.	98–492	30–150
Turonian?	Benson Member	Conglomerate; sandstone; red shale.	0–984	0–300

sheltered by sandy barrier islands. Partings in the No.2 seam are finer grained and less continuous than those in the No.1 seam, suggesting that its barrier islands provided more effective shelter from the open seaway.

No.4 seam lies near the base of the Cumberland Member and ranges in thickness from 3.9 to 14.8 ft (1.2 to 4.5 m). It formed in irregular alluvial-plain backswamps between stream levees and steep-sided basement hills. Stream channels were stable, and splits in the No.4 seam accordingly are rare; washouts along avulsed stream channels are narrow and deep. The No.4 seam grades laterally into stony coal and coaly mudstone near the basement hills.

All Comox coals are blocky and hard and have well developed cleat systems. Rank ranges from high-volatile A bituminous to anthracite, with higher ranks confined to narrow belts 0.6 to 1.9 miles (1 to 3 km) wide adjacent to Tertiary laccoliths. Mean random vitrinite reflectance values of Comox coals range from 0.59% to 0.99% throughout most of the field but are as high as 3.21% near the laccoliths (Kenyon and Bickford, 1989).

PETROGRAPHY OF VANCOUVER ISLAND COALS

Although proximate analyses are variable, petrographic compositions of Nanaimo Group coals are fairly consistent. All have high vitrinite content, which is consistent with their generally bright appearance in hand specimen. Tables 3 and 4 present analytical data for the principal coals of Nanaimo and Comox coal fields.

Table 3. Typical proximate analyses of Douglas coal seam (%, air-dried basis), Nanaimo coal field, Vancouver Island.

Location	Moisture	Volatile Matter	Fixed Carbon	Ash	Sulfur	Parr Mineral Matter
Reserve Mine	1.9	35.2	51.8	11.1	2.04	13.11
Morden Mine	1.6	36.0	54.0	8.4	0.70	9.46
No.10 Mine	1.1	29.98	56.3	12.6	0.47	13.87

Table 4. Typical petrographic analyses of Nanaimo and Comox coals (volume percent).

Coalbed	Vitrinite	Inertinite	Liptinite
Douglas	90–98	2–5	0–5
Newcastle	81–95	4–19	0–2
Wellington	75–96	3–25	0–3
Comox No.1	81–94	5–17	2–4
Comox No.2	76–86	14–24	0–1
Comox No.4	82–87	12–13	1–5

EXPLORATION AND PRODUCTION HISTORY

More than 1,200 coal exploration boreholes have been drilled in the coal fields of Vancouver Island. Approximately 900 of these were drilled in the vicinity of Comox coal field. Most of the remainder were drilled in Nanaimo coal field, although virtually every outlier of the Nanaimo Group on Vancouver Island and on the Gulf Islands has been prospected to some extent.

Several good shows of coal gas and associated coal-sourced "conventional" gas have been discovered, mostly in the Royston area of Comox coal field (Cathyl-Bickford, 1990). Locations of four key wells with gas shows are shown on Figure 1. Side notes on the driller's log of borehole CX–189 at Royston (Wall, 1945) provide insight into the hazards of drilling through gas-bearing coal measures:

> "Exceptionally heavy outburst of gas occurred at 6 P.M. January 31st [1945] while drilling at 1,372 feet. Gas ignited and destroyed machine and building."

The fire burned out of control for twelve hours and required the combined forces of three fire brigades to extinguish it. Borehole CX–189 subsequently was completed by the landowner as a farm gas well.

A short distance south of the Royston well, the Alvensleben Tsable River ATR–1 coal exploration borehole encountered gassy coal at a depth of 1,805 ft (550 m) and was abandoned in 1914 because of "excessive gas pressure." Like the Royston well, ATR–1 subsequently was put into service as an unlicensed gas well to serve a nearby forestry camp. ATR–1 continued to produce gas until 1984, when its casing was sheared by a landslide.

Several minor gas shows were encountered in the course of other coal exploration drilling in Nanaimo coal field, but none were as spectacular as those at Royston. Nevertheless, the long-documented gassy nature of the Douglas seam at Nanaimo (Touhey, 1923; Wilson and Henderson, 1927; Baird, 1944) led Novacorp International Consulting Ltd. to focus on this seam in its initial program of coalbed methane exploration on Vancouver Island. In 1984 Novacorp drilled 14 boreholes, of which seven were located near the Nanaimo River (Murray, 1986). Five of the Nanaimo River wells reached the Douglas seam, and coal cores from two wells were recovered for desorption tests. No commercial production was established because of lack of funding (Kahil, 1990).

In 1986 British Petroleum drilled two "conventional" gas test wells southeast of Nanaimo (Figure 1). The first well, the Harmac c–36–F, encountered the Douglas seam at a depth of 2,000 ft (610 m) but was plugged and abandoned after drillstem testing indicated the presence of salt water in a sandstone reservoir.

British Petroleum's second well, Yellow Point d–84–C, encountered more than 65 ft (20 m) of interbedded shale and coal, probably correlative with the Douglas seam, at a depth of 3,317 ft (1,011 m). This well continued into the Comox Formation, where it encountered a gassy coal at least 10 ft (3 m) thick at a depth of approximately 5,150 ft (1,570 m). The coal burst into the wellbore, causing its abandonment because of stuck drillpipe. The Yellow Point well is significant in that it discovered coal in the Comox Formation in Nanaimo coal field. Prior to drilling this well, Comox coals were not believed to occur that far south in Georgia basin.

Since 1986 both Novacorp and British Petroleum have relinquished their petroleum and natural gas exploration permits in the Nanaimo area because of failure to make a commercial discovery. Several other mining and gas companies since have expressed interest in the Nanaimo River and Royston coal-gas prospects, and new geological surveys have been conducted with the objective of selecting sites for further drilling at Nanaimo.

The complex isopach and structure patterns of the Douglas seam along the Nanaimo River (Figures 2 and 3) have been well established by mining, and such rapid variation in thickness must be kept in mind when planning coal-gas development. The Douglas seam remains the major exploratory target at Nanaimo, although deeper coals of the Comox Formation also have attracted attention. Drilling depths to the Douglas seam at Nanaimo range from 1,148 to 3,937 ft (350 to 1,200 m); Comox coals would lie at depths of 2,625 to 5,906 ft (800 to 1,800 m).

GAS CONTENT

Gas-related mining problems were particularly acute in southern Nanaimo coal field (Baird, 1944), where the Douglas seam was notoriously prone to spontaneous outbursts. Following several fatalities among the miners, the Chief Inspector of Mines prohibited further development work southward into the seam (Wilkinson, 1926; British Columbia Minister of Mines, 1927). Table 5 presents typical gas-emission rates for some of the deep coal mines on Vancouver Island, as reported by the British Columbia Minister of Mines.

In evaluating gas content of a coalbed on the basis of mine-air analyses, one should remember that gas-emission rates from coal mines are a function of coal's desorbable gas content, permeability of the coal and its roof and floor strata, and rate of coal production from the mine. Kissell and others (1973) have suggested that in extensively worked, deep coal mines, a fairly constant relationship exists between desorbable gas content of virgin coal and rate of gas emission from the mine. They consider that six to nine times as much gas is emitted from such mines as could be accounted for by direct desorption from the mined coal. The increased gas emission probably arises, in part, to migration of gas through the coalbed to the working faces and to emission of gas from roof and floor strata bounding the worked-out seam.

Desorption tests of coal cores from boreholes drilled near Morden Colliery show the uncorrected desorbable gas content of the Douglas seam to be 192 to 384 CF/T (6 to 12 cc/g) (Kenyon and Murray, 1990; Proudlock, 1990). In contrast, methane-emission rates from Morden Colliery average 763 CF/T (23.8 cc/g) of mined coal, which is approximately two to four times as high as the desorbable gas content of the cores. This disparity suggests that either the Douglas seam or its bounding strata may be relatively impermeable, compared with the coalbeds studied by Kissell and others.

Table 5. Gas-emission rates for Vancouver Island deep coal mines.

Mine	Field	Year	Coal Seam	Average Depth (ft)	Average Depth (m)	Emission Rate (from coal mined) (CF/T)	Emission Rate (from coal mined) (cc/g)
Reserve	Nanaimo	1918	Douglas	984	300	282	8.8
Morden	Nanaimo	1918	Douglas	820	250	763	23.8
No. 5	Comox	1918	No.1	656	200	551	17.2
No. 5	Comox	1935	No.2	984	300	4,104	128.1
No. 4	Comox	1918	No.4	820	250	2,067	64.5

Although Comox mines had higher gas-emission rates than Nanaimo mines, no spontaneous outbursts were reported from the Comox operations. Several factors may have contributed to this apparent anomaly. First, Comox coals are stronger and have well defined cleats, in contrast to Nanaimo coals, which tend to be sheared and friable. Second, Comox mines generally were worked by longwall; Nanaimo mines used the room-and-pillar method. Touhey (1923) reported that the incidence of outbursts appeared to be reduced following a change from room-and-pillar to modified longwall methods at Granby Colliery in Nanaimo. Finally, Comox coals are closely associated with thick, porous sandstone, which may have drained gas from the coalbeds. The Douglas seam at Nanaimo generally is underlain and overlain by relatively impermeable siltstone and mudstone. Such impermeable roof and floor strata may increase the likelihood of overpressuring in the coalbeds (Decker and Horner, 1987).

Gas from the Douglas seam contains 95% methane, 4.5% heavier hydrocarbons, and 0.5% carbon dioxide. This is acceptable quality for delivery to a gas-transmission system (Kenyon and Murray, 1990).

OUTLOOK FOR COAL-GAS DEVELOPMENT ON VANCOUVER ISLAND

Encouraging factors for coal-gas development include the availability of high-quality reflection seismic data, which were acquired when oil companies explored for natural gas on eastern Vancouver Island and offshore in Georgia Strait. The fact that major coalbeds can be imaged by these seismic profiles simplifies assessment of their distribution and depth. That some coals are gassy, as had been demonstrated by previous mining operations, has now to some extent substantiated by desorption tests. Porous conglomerate and sandstone are closely associated with the coalbeds, particularly in Comox coal field. The occurrence of such clastic reservoirs adjacent to a gassy coal may enhance the economics of coal-gas production by allowing diffusion of gas from the coal into these clastic reservoirs if a pressure differential is established (Wyman, 1984). A gas transmission and distribution network that is nearing completion will afford the potential of ready access into a large market area.

Potential difficulties include the activities of well financed and vocal antidevelopment lobbying groups on Vancouver Island. Second, because the Nanaimo River gas prospect, in particular, lies in cottage country, surface rights for drilling and production could be costly to acquire. Coals of the Nanaimo Group are stratigraphically and structurally complex. The splitting, pinching, and faulting of the coals that has frustrated miners for many years can be expected as well to frustrate drillers and gas production engineers.

The development potential of coalbed gas prospects on eastern Vancouver Island ultimately will depend on local residents' perceptions of natural gas as an environmentally acceptable and relatively low-priced heating fuel, compared to the cost of competing fuels such as propane, electricity, and firewood.

ACKNOWLEDGMENTS

Financial and logistical support for this study was provided by the British Columbia Geological Survey, under the auspices of the Canada-British Columbia Mineral Development Agreement. Brian Young of the British Columbia Archives and Records Service found the driller's log, and Barbara Lemke at the Cumberland Mining Museum retrieved the fire brigade's account of the Royston well fire. I thank my colleagues at Westwater Mining for their insight, arguments, and technical nitpicking. For bravely drilling two wildcat wells into the Nanaimo coal measures, I thank my former colleagues in the Frontier Group of British Petroleum. Lastly, I thank Georgia Hoffman and Kate Slater for their patience in translating my work into technical English.

REFERENCES

Baird, H., 1944, Blowouts at Canadian Collieries (D) Ltd. No.10 Mine, South Wellington, Vancouver Island: Paper presented to Canadian Institute of Mining and Metallurgy.

Bickford, C.G.C., 1989, Geology, mining conditions and resource potential of the Wellington coal bed, Georgia Basin, *in* Geological fieldwork, 1988: British Columbia Ministry of Energy, Mines and Petroleum Resources, Paper 1989–1, p. 553–558.

Bickford, C.G.C, Hoffman, G.L., and Kenyon, C., 1990, Geological investigations in the coal measures of the Oyster River, Mount Washington and Cumberland Areas, Vancouver Island, *in* Geological fieldwork, 1989: British Columbia Ministry of Energy, Mines and Petroleum Resources, Paper 1990–1, p. 431–437.

Bickford, C.G.C., and Kenyon, C., 1988, Coalfield geology of eastern Vancouver Island, *in* Geological fieldwork, 1987, British Columbia Ministry of Energy, Mines and Petroleum Resources, Paper 1988–1, p. 441–450.

British Columbia Minister of Mines, 1927, Annual report for the year 1926: British Columbia Minister of Mines.

Bustin, R.M., and England, T.D.J., 1991, Petroleum source rock potential of the Nanaimo Group, western margin of the Georgia Basin, southwestern British Columbia, *in* Current research, part A: Geological Survey of Canada Paper 91–1A, p. 143–145.

Cathyl-Bickford, C.G., 1989, Coal-mining geology of the Comox and Extension Formations (Upper Cretaceous) of Vancouver Island, *in* Advances in western Canadian coal geoscience—Forum proceedings: Alberta Research Council Information Series 103, p. 355.

Cathyl-Bickford, C.G., 1990, Habitat of natural gas in the Comox and Trent River Formations, Georgia Basin, Vancouver Island (abs.): Bulletin of Canadian Petroleum Geology, v. 38, no. 1, p. 159.

Cathyl-Bickford, C.G., and Hoffman, G., 1991, Geology and coal resources of the Nanaimo Group in the Alberni, Ash River, Cowie Creek and Parksville areas, Vancouver Island, *in* Geological Fieldwork, 1990: British Columbia Ministry of Energy, Mines and Petroleum Resources, Paper 1991–1, p. 381–385.

Decker, A.D., and Horner, D.M., 1987, Origin and production implications of abnormal coal reservoir pressure, *in* The 1987 Coalbed Methane Symposium Proceedings—University of Alabama, Tuscaloosa, November 16–19, 1987: School of Mines and Energy Development, Gas Research Institute, and U.S. Mine Safety and Health Administration, p. 51–62.

England, T.D.J., 1989, Lithostratigraphy of the Nanaimo Group, Georgia Basin, southwestern British Columbia, *in* Current research, part E: Geological Survey of Canada Paper 89–1E, p. 197–206.

Kahil, A.A., 1990, History and experience of Canadian companies in coal demethanation, *in* Coal bed methane in Alberta—What's it all about?: Alberta Research Council Information Series 108.

Kenyon, C., and Bickford, C.G.C., 1989, Vitrinite reflectance study of Nanaimo Group coals of Vancouver Island, *in* Geological fieldwork, 1988: British Columbia Ministry of Energy, Mines and Petroleum Resources, Paper 1989–1, p. 543–552.

Kenyon, C., and Murray, D.K., 1990, Coal-bed methane potential of Vancouver Island coalfields (abs.): AAPG Bulletin, v. 74, no. 5, p. 693.

Kissell, F.N., McCulloch, C.M., and Elder, C.H., 1973, The direct method of determining methane content of coalbeds for ventilation design: U.S. Bureau of Mines Report of Investigations 7767, 17 p.

Muller, J.E., and Atchison, M.E., 1971, Geology, history and potential of Vancouver Island coal deposits: Geological Survey of Canada Paper 70–53.

Murray, D.K., 1986, Evaluation of the exploration permits and freehold lease, and vicinity, of Novacorp International Consulting Ltd. on Vancouver Island, British Columbia: Unpublished report.

Proudlock, P.J., 1990, An overview of Albertan and Canadian demethanation experience, technology advances and development potential, *in* Coal bed methane in Alberta—What's it all about?: Alberta Research Council Information Series 108.

Rice, G.S., 1922, Report on blowouts at Granby Colliery No.1, Cassidy, Vancouver Island: British Columbia Ministry of

Energy, Mines and Petroleum Resources, Inspection and Engineering Branch Open-File Report.

Touhey, J.B., 1923, Outbursts of gas and coal at Cassidy, B.C.: Coal Age, v. 23, p. 519–522.

Wall, W.H., 1945, Driller's log of Canadian Collieries (Dunsmuir) Limited Royston borehole Cx–189: British Columbia Ministry of Energy, Mines and Petroleum Resources, Coal Section Open-File Report.

Wilkinson, G., 1926, Special report on conditions in No.7 South Section, Granby No.1 Mine, Cassidy, B.C.: British Columbia Ministry of Energy, Mines and Petroleum Resources, Inspection and Engineering Branch Open-File Report.

Wilson, R.R., and Henderson, Robert, 1927, Outbursts of gas and coal at Cassidy Colliery, Vancouver Island, British Columbia: American Institute of Mining Engineers Transactions, v. 75, p. 583–591.

Wyman, R.E., 1984, Gas resources in Elmworth coal seams, *in* Masters, J.A., ed., Elmworth—Case study of a deep basin gas field: AAPG Memoir 38, p. 173–187.

Delineation of Prospective Coalbed Methane Trends in Western and Central Washington State

Steven P. Pappajohn[1]
Terry E. Mitchell[1]

1. GeoTrends Inc., Kent, Washington

ABSTRACT

Middle to late Eocene nonmarine coal-bearing sediments of the Puget Group and correlative formations are preserved in the structural subbasins of western and central Washington State and represent some of the thickest (>18,000 ft [5,500 m]) fluvial-deltaic sequences in North America. The subbasins are areally distinct, tectonically generated, and characterized by complex stratigraphic and structural relationships. The distribution of coal rank, however, follows a predictable pattern for approximately 40 miles (64 km) on both the eastern and western flanks of the north-south-trending Cascade Range without regard for subbasin boundaries or structural fabric. Late Eocene through Miocene Cascade Range-related plutonism upgraded coal rank throughout the coal-bearing sequences. The overprinting of a thermal maturity aureole on existing burial-generated coal rank trends created wide fairways of coalbed methane-prospective bituminous coals.

In most subbasins the combination of surface geology, coal exploration coreholes, coal mining maps and records, coal analyses, and conventional petroleum exploration well data provide significant information on coal quality and rank, coalbed stratigraphy and distribution, and indications of methane. Recent and on-going published and proprietary studies on the geology, coal rank/vitrinite reflectance trends, and tectonism of the region indicate considerable coalbed methane potential in high-graded areas of each subbasin. An initial attempt to test coalbed methane production in an area of the prospective trend has yielded encouraging results, including multiple (three to five) coalbed horizons with approximately 40 to 50 ft (12 to 15 m) of total net coal, methane desorption measurements averaging 423 SCF/T (13.2 cc/g), and gas analyses averaging 950 Btu/CF (35,396 kJ/m^3). An in-place resource base potential of more than 18 BCF/mi^2 (197 × 10^6 m^3/km^2) of pipeline-quality methane is indicated.

INTRODUCTION

Regional Geologic and Tectonic Setting

The western Washington coal region extends along the western flank of the Cascade Range within the Puget downwarp structural province (Figure 1). The entire trend of the downwarp generally is believed to be the eastern part of a forearc setting that evolved during early middle Eocene time (Figure 2). The geology of western Washington evolved in response to dynamic interactions between converging oceanic and continental plate systems during the Tertiary, which resulted in complex structural and stratigraphic relationships throughout the region (Snavely and others, 1980; Wells and others, 1984; Armentrout and Suek, 1985; Engebretson and others, 1985; Gwilliam and others, 1990). The depositional and tectonic evolution of western Washington resulted in several modifications to physical coal characteristic trends, most notably the great areal extent of coal measures, repetitive nature of coal horizons in the stratigraphic column, thickness of individual coalbeds, and a wide range in thermal maturation of hydrocarbon source lithologies, including coal.

The present distribution and thermal maturation trends of the coal basins are a direct result of post-middle Eocene igneous activity and uplift of the Cascade Range in response to episodic changes in plate-motion dynamics (Figures 3 and 4). The western part of the north-south-trending belt of coal basins, which formed in progradational, nearshore to delta-plain environments, became separated geographically from its eastern nonmarine, fluvial facies as development of the Cascade Range commenced in late Eocene time (Mooney and Weaver, 1989; Choate and Johnson, 1980). A prolific outpouring of flood basalt (Columbia River Basalt) east of the Cascades during Miocene time buried most of the fluvial facies, but coal deposits in western Washington were buried only by sporadic volcanic and glacial deposits.

Geologic History

The broad coastal plain upon which coal-bearing sediments were deposited developed primarily during late Eocene time (Vine, 1968; Bukovic, 1974, 1979; Johnson, 1982, 1985, 1987). In response to nearly static conditions of sediment deposition and tectonic subsidence, an extensive

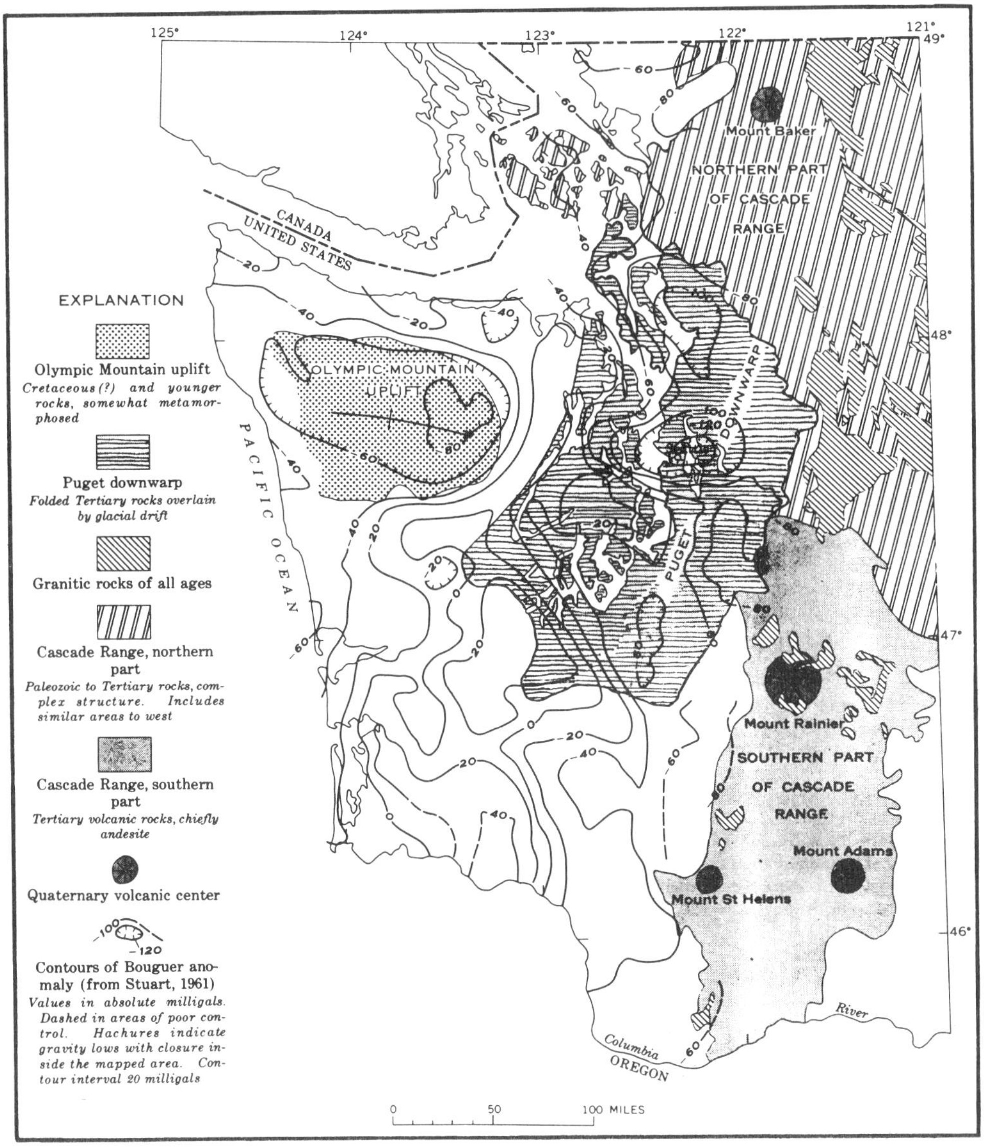

Figure 1. Generalized geology, tectonic provinces, and Bouguer gravity of western Washington. Modified from Vine (1969).

system of coalescing deltas, created by westward-flowing streams, developed in association with lush forests and poorly drained swamps. Sediment sources lay predominantly to the east in crystalline terranes of eastern Washington and Idaho (Bukovic, 1979; Johnson, 1988). Based on the occurrence of coal-bearing Eocene sediments indicated by surface and subsurface sampling, this coastal plain/deltaic environment extended from the Canadian border west of the Cascade Range to the central Oregon coast on the south, and from Puget Sound on the west to the Columbia River basin east of the Cascade Range. The climate during Eocene time was characterized as humid and subtropical.

The coastal plain was interrupted by constructional volcanic hills in the present-day Puget Lowlands area during the early stages of the ancestral Cascade Range (Bukovic, 1979). Several eruptive centers that developed within the deltaic trend during late Eocene time both overwhelmed and redirected arkosic sedimentation locally until continued regional subsidence resulted in the re-establishment of prevolcanic sediment-distribution patterns (Walsh and Phillips, 1988). Deep-marine embayments along the margin of, and within(?), the plain resulted in deposition of areally restricted turbidite and marine shale sequences periodically during Eocene time in response to local variations in sedimentation and subsidence rates.

In latest Eocene (Refugian) and early Oligocene time, approximately synchronous with a marine transgression, the Cascade magmatic arc became active. Extensive volcanism choked the sediment supply with tuffaceous debris and ended deltaic deposition. The Oligocene section grades from nonmarine, subalkalic to calc-alkalic volcanic rocks and impermeable volcaniclastic conglomerate on the eastern margin of the Puget trough to deep-marine tuffaceous

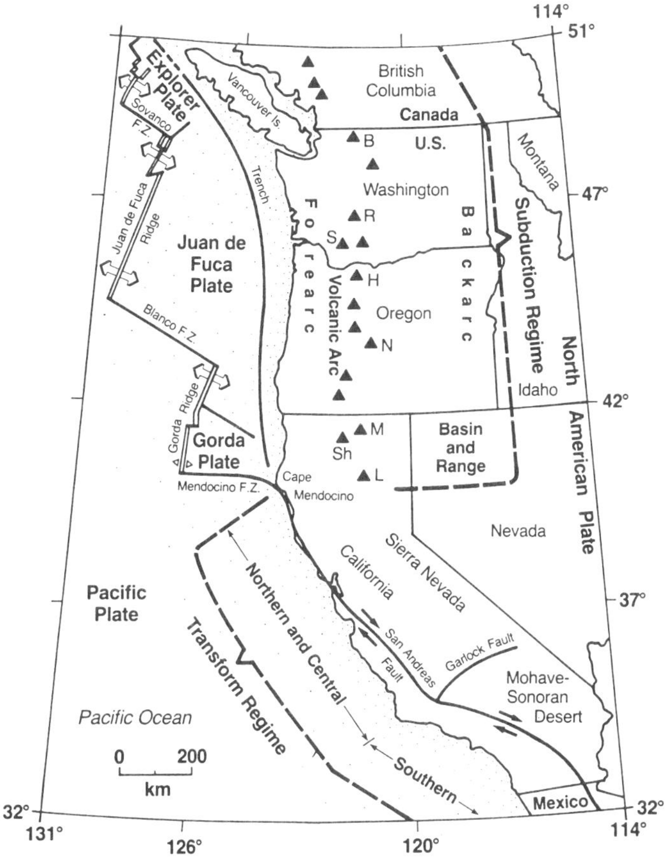

Figure 2. Regional tectonic features of the Pacific Northwest. After Mooney and Weaver(1989).

siltstone in the subbasins. Concurrently, related igneous bodies were emplaced along the trend of the present Cascade Range, and they appear to have raised the rank of coal-bearing sediments relatively rapidly on both a regional and local basis.

During Miocene time, vast sheets of basaltic lava flows were extruded east of the Cascade Range, burying a 62,000-mi^2 (160,580-km^2) area of coal-bearing fluvial sediments to a depth of as much as 10,000 ft (3,050 m) in places. West of the Cascades, Miocene sediments were less affected by volcanic activity. Eocene sediments are overlain sporadically by Miocene and Pliocene volcanogenic rocks that grade westward into terrigenous conglomerate and then marine clastic rocks. The renewed igneous activity that accompanied uplift of the Cascades began in Pliocene time and continues today, as manifested by one active (Mount St. Helens) and four dormant stratovolcanoes (Mount Baker, Mount Rainier, Mount Adams, and Glacier Peak). Continental glaciation covered the area of the Puget downwarp during Pleistocene time, followed by formation of alpine glaciers that are still abundant on the higher peaks.

Stratigraphy, Age, and Depositional Environments

A thick sequence of early and middle late Eocene (equivalent to the Ulatisian and Narizian benthic

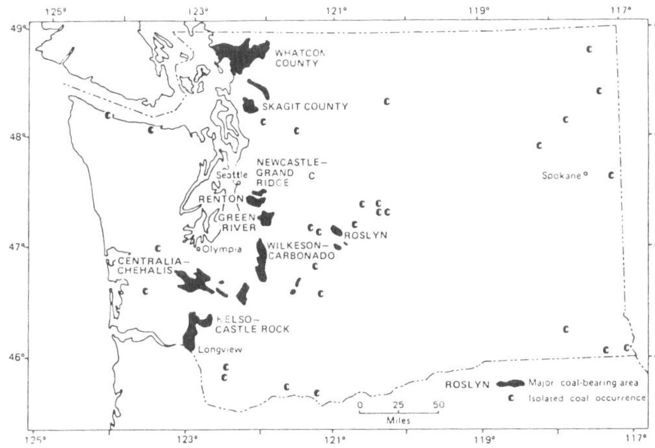

Figure 3. Major coal-bearing areas and coal occurrences of Washington. After Walsh and Phillips (1989).

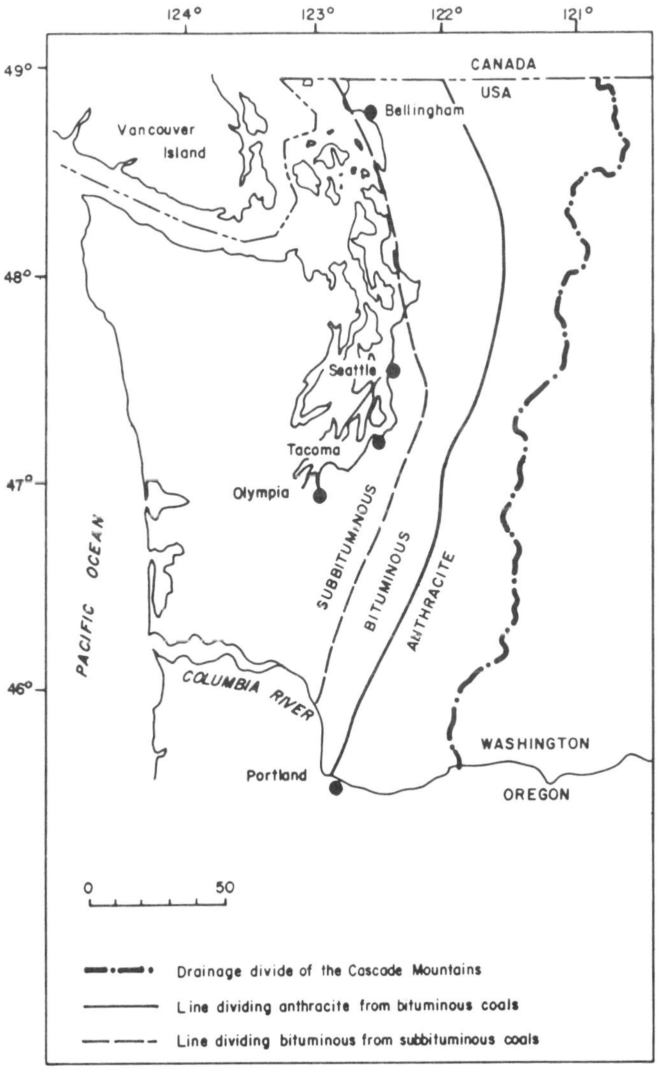

Figure 4. Coal rank distribution in western Washington. After Choate and Johnson (1980).

foraminiferal stages), nonmarine and marginal-marine, coal-bearing sediments underlies much of the Puget downwarp in western and central Washington from near the Canadian border in the north to the Columbia River in the south (Walsh and Lingley, 1991). Known throughout the region as the Puget Group (and local correlative formations such as the Chuckanut, Carbonado, and Skookumchuck), this sequence consists of repetitive interbeds of arkosic sandstone, siltstone, carbonaceous shale, and numerous coalbeds (Figure 5). The base of the Puget Group is not exposed and has never been encountered in drilling.

Prolific rates of fluvial-deltaic sedimentation in the rapidly subsiding depositional and structural subbasins during Eocene time resulted in the creation of a sequence of nonmarine, arkosic, coal-bearing strata up to 18,000 ft (5,500 m) thick in places (Johnson, 1988). Coalbeds ranging in thickness from several inches to more than 45 ft (14 m) occur throughout the Eocene stratigraphic section (Larson, 1985). A range of coal depositional environments from the coalescing Eocene deltas are seen repeatedly in stratigraphic sections exposed in outcrops and encountered in drilling throughout the subbasin trend. Such environments include lower delta, nearshore swamps (thicker coalbeds) and middle and upper delta/interdistributary regions (thinner coalbeds) within the deltaic plain.

Compositional analyses of Puget Group sandstones indicate that the provenance of the sediment was the granitic and metamorphic terranes of what is now eastern Washington and Idaho (Vine, 1969; Evans, 1989). The position of the Pacific coast shoreline during Puget Group deposition was approximately coincident with the position of the present-day Puget lowlands through the Everett-Seattle-Tacoma-Olympia trend. The general vicinity of the Chehalis-Kelso-Longview area probably represents more distal parts of the depositional system (Mackin and Cary, 1965). In northwestern Washington the paleoshoreline probably extended northwest to somewhere west of Bellingham.

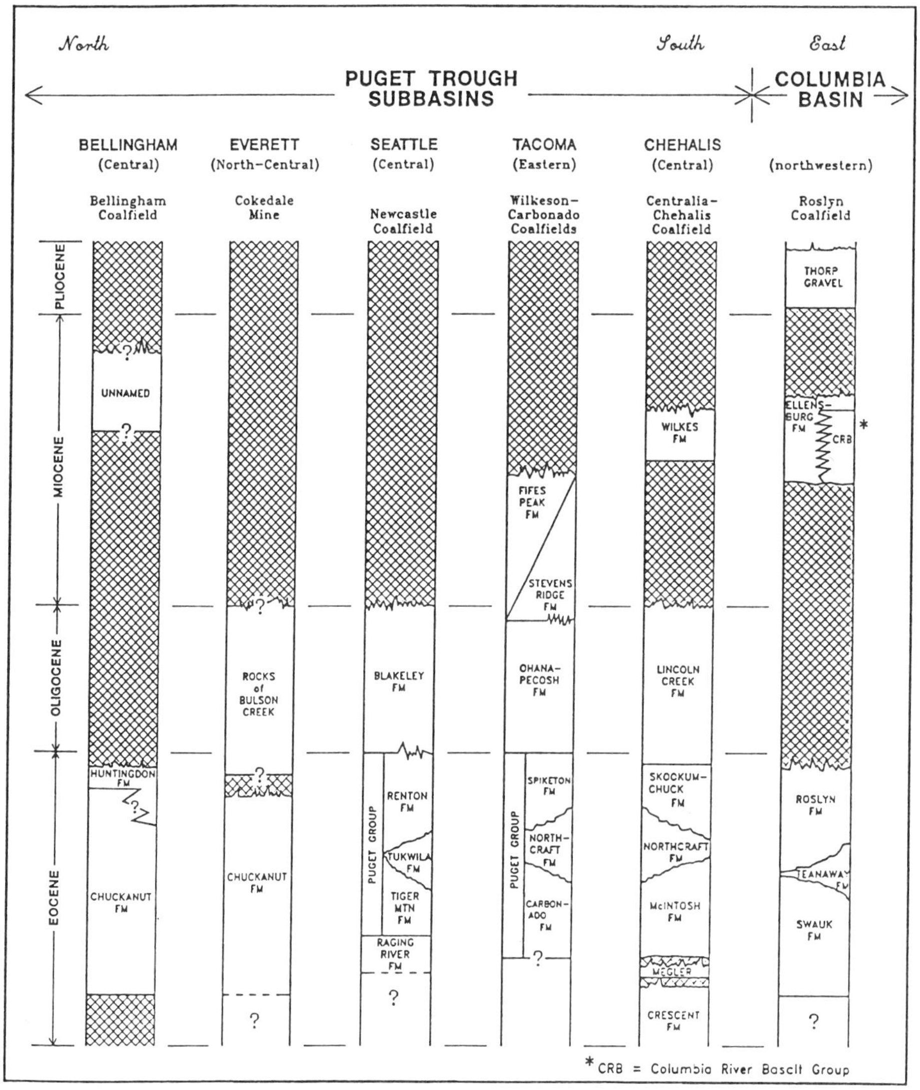

Figure 5. Correlation chart of Tertiary strata in western and central Washington. After Walsh and Lingley (1991).

Structure

Magnetic and gravity data, tectonic mapping, and outcrop mapping indicate that the northern and southern margins of several subbasins are controlled by west- and northwest-trending faults (Johnson, 1985). Based on sedimentological evidence, researchers have suggested that deposition in the Bellingham and Seattle subbasins was controlled by strike-slip faulting (Johnson, 1987). The southern boundary of the Seattle subbasin, which was active during deposition of the lower middle Eocene Raging River Formation and the Oligocene Blakeley Formation, may have had more than 3,500 ft (1,067 m) of displacement in Quaternary time (Walsh, 1986; Gower and others, 1985).

Small-scale structure superimposed on regional trends is common in most coal districts. Folds often are tight and asymmetrical, with thrust-faulted crests in some areas of steep, vertical, and overturned bedding. In general the intensity of deformation increases toward the Cascade front and, based on the geometries of most faults and folds, most structural stresses are compressional and probably related to post-Eocene episodes of subduction, oblique subduction, and strike-slip faulting.

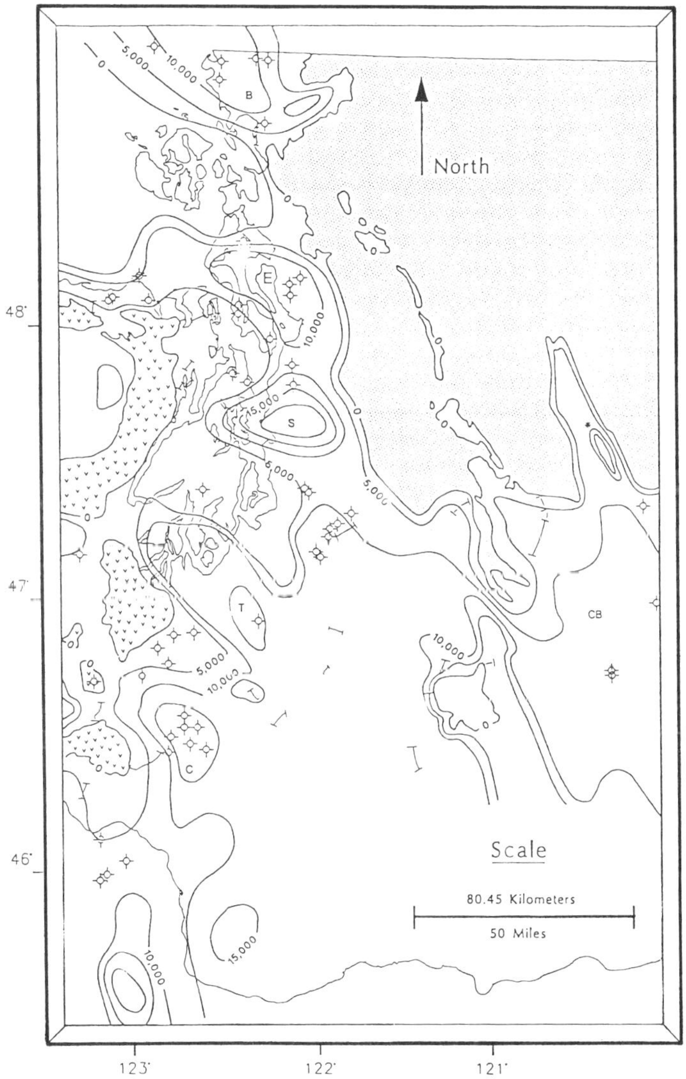

Figure 6. Isopach map of Eocene strata in the Puget trough, western Washington. Subbasins—B, Bellingham; E, Everett; S, Seattle; T, Tacoma; C, Chehalis. Contour interval 5,000 ft. After Walsh and Lingley (1991).

Structural deformation and the accompanying tectonic fracturing may provide enhanced porosity and permeability within coalbed reservoirs and improve the inherent capability of coalbeds to store and relinquish methane. However, tectonic stressing of the coal-bearing section, and the structural complexity that resulted, created obstacles for the early underground mining operations. Coalbeds were difficult to follow and to correlate, especially across faults, and the danger of explosions from coalbeds already gas-charged by rapid geothermal heating was compounded. These coal mines, clustered along the eastern flank of the Puget downwarp, provide a remarkable source of data that can be used to help define methane drilling prospects. More than 1,100 maps of approximately 150 mines and coal prospects are available (Walsh and Lingley, 1991). These maps illustrate structure with far more resolution than is generally available for petroleum exploration by providing exquisite details for determination of local structure, structural styles and trends, stratigraphic sequence, and coalbed thickness.

An isopach map of Ulatisian and Narizian rocks, exclusive of Crescent Formation basalt (economic basement in the southern Puget downwarp) has been prepared by Walsh and Lingley (1991) of the Washington Department of Natural Resources (Figure 6). The shape of the contours is controlled by gravity data, which correlate well with isopachous trends where well control is dense. The contours define five subbasins within the Puget trough that contain more 10,000 ft (3,050 m) of middle and upper Eocene section. The subbasins and adjacent areas on their eastern flanks have been extensively mined for coal since 1853, when the Bellingham Mine in Whatcom County was opened.

PROSPECTIVE COALBED METHANE TRENDS

Coal Rank and Thermal Maturation

Throughout most of the southern Puget downwarp, a correlation exists between trends of increasing coal rank and increasing present-day geothermal gradients. According to Walsh and Lingley (1991), drillstem tests and wireline-log temperature data from conventional wildcat wells suggest that correlation continues throughout the entire region. However, coal rank in the eastern Puget downwarp and the Cascade foothills generally is higher than what would be achieved solely by present-day heat flow. Igneous breaching of the coal basins superimposed a regional aureole of thermal maturation upon the coals that resulted in a trend of increased rank that is apparent in coal deposits located progressively closer to the crest of the

Cascade Range (Walsh and Lingley, 1991). Many coal deposits in the region reflect an anomalously advanced level of thermal maturity more typical of coal of pre-Tertiary age and therefore are anticipated to exhibit correspondingly high methane content.

Contact metamorphism also affects local coal rank (Vine, 1968; Gard, 1968; Walsh and Lingley, 1991). Vitrinite reflectance values in eastern Lewis and Pierce Counties have been observed to increase near the contacts of small sills, which selectively intrude the coal seams (Walsh and Lingley, 1991). In another instance, westward-increasing coal rank in central King County appears to have been caused by contact effects from Oligocene intrusives located close to several coal mining areas (Walsh and Phillips, 1985).

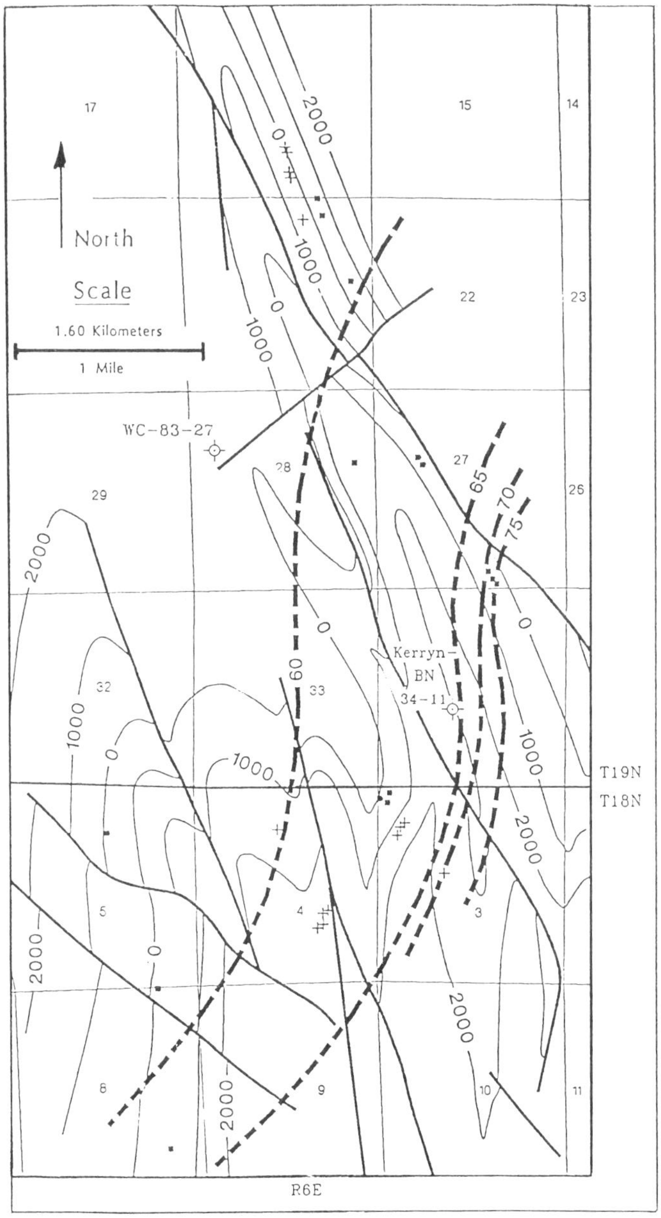

Figure 7. Isorank (% dry mineral-matter-free fixed carbon; heavy dashed lines) contours and structure contours (feet below sea level) for a coal seam in Wilkeson-Carbonado coal field, Pierce County, Washington. Isorank contour interval 5% dmmf fixed carbon. Structure contour interval 1,000 ft. After Walsh and Lingley (1991).

At Wilkeson-Carbonado coal field in Pierce County, site of the first coalbed methane pilot drilling program in Washington (see below), coal rank contours cross structure contours at a high angle and postdate folding (Figure 7). Although it is possible to explain the relationship of structure and coal rank in this area as the result of burial after folding, regional heating events are indicated as the most likely cause (Walsh and Lingley, 1991). Other geologic relationships in Pierce County, and elsewhere in western Washington, appear to bracket the present coal-maturation levels in the Wilkeson-Carbonado area as a result of a post-Oligocene thermal event. The timing of this event correlates with the emplacement of numerous lower Miocene intrusives and voluminous Miocene extrusives. Although present throughout the coal-bearing regions of western Washington, older intrusive rocks are less common and less voluminous than those of Miocene age and, as such, appear to have had less effect on coal maturation.

A similar relationship between rank and structure contours is present along the eastern flank of the Cascade Mountains. At Roslyn coal field in central Kittitas County, coal rank increases westward from high-volatile C to high-volatile A bituminous (Figure 8). Moderately folded Eocene rocks at Roslyn are conformably overlain by gently dipping, poorly indurated, Pliocene gravel, constraining the age of coal maturation as Miocene or Oligocene. Coal-rank trend surfaces generally cut structure in the Whatcom and eastern Lewis County areas as well.

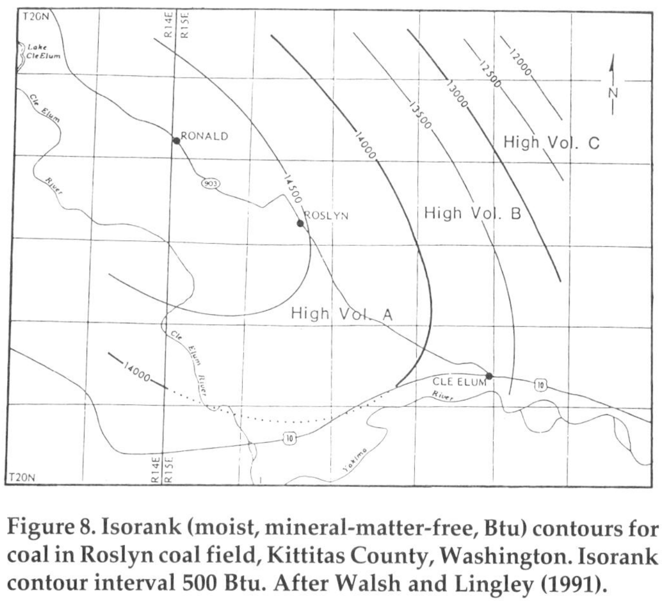

Figure 8. Isorank (moist, mineral-matter-free, Btu) contours for coal in Roslyn coal field, Kittitas County, Washington. Isorank contour interval 500 Btu. After Walsh and Lingley (1991).

Indications of Occluded Gas in Coalbeds

Mine Emissions

The most striking indicator of favorable coalbed methane potential in the central and western Washington coal districts is contained in the historical records and accounts of underground coal mines. The coal fields of Washington have long been recognized as gassy from the high frequency of methane-related mine accidents, as well as from U.S. Bureau of Mines direct measurements of methane emissions from mine ventilation (Ash, 1931; Choate and Johnson, 1980). As early mining operations proceeded from the surface to ever greater depths, they were increasingly troubled by explosive hazards from methane. Many mine accidents and much loss of life occurred as a result of the gassy nature of the coals. In some cases coal extraction in the deeper mines had to be slowed or suspended because of unsafe methane accumulations. The deepest mines in the state were approximately 2,000 ft (610 m) below the surface. Two surface coal mines currently operate in Washington, but all underground coal mining has been abandoned.

The western Washington coal fields were included in the U.S. Department of Energy (DOE) study (Choate and Johnson, 1980) of prospective coalbed methane areas in the United States. Based on the propensity for methane explosions during the coal mining era, the region was considered highly prospective for coalbed methane development. The study concluded, for example, that a potential of up to 395 BCF (11.2×10^9 m^3) of methane may be stored in coals within King County (including the Seattle metropolitan area), based on potential coal resources and gas-content values generally associated with bituminous coal rank. In addition, Choate and Johnson pointed out that their estimates of both coal resources and, consequently, coalbed methane resources, likely are conservative and may possibly be only a fraction of what actually may be present. The ubiquitous nature of the coal-bearing section throughout the region suggests that estimates based on coal resources that are identified only from active mining areas are substantially low.

Drilling Records

In addition to mining records and analyses of coal rank and quality, the DOE study also examined drilling records of conventional wildcat wells and numerous coal coreholes in the region. To date, no commercial production of oil or gas has been established in Washington. In many areas, especially those close to present and past volcanic trends, few reservoir-quality sandstones have been found by drilling (Walsh and Lingley, 1991). Although exceptions exist, the migration of thermally influenced fluids (usually water), probably associated with episodic emplacement of igneous bodies, has adversely affected the composition and porosity/permeability framework of arkosic sandstones in the Puget Group and its equivalents. An additional obstacle to charging clastic reservoir rocks with hydrocarbons is the timing of structural development with respect to generation of gas and oil—much of the thermal generation of hydrocarbons predated the formation of structural traps.

The same processes that rendered most clastic formations too tight to store hydrocarbons have created a greater ability and potential for coalbeds to generate, store, and produce methane gas. High paleogeothermal gradients associated with several episodes of regional igneous activity have elevated the coal rank (and the associated ability of the coals to generate and store gas) beyond that which normally would have been achieved in the same time by the heat and pressure of burial. The poor porosity and permeability of the encasing sandstone, siltstone, and carbonaceous shale create favorable conditions for maintaining high reservoir pressure and for retaining a significant proportion of the methane generated within the coalbeds. Subsequent structural deformation has acted both to bring deeper, higher pressure-regime coals to optimum target depths of 1,500 to 4,500 ft (460 to 1,370 m) and to enhance the inherent permeability (fractures or "cleats") with additional tectonic fracturing.

The existence of high reservoir pressure was confirmed in the Green River district of King County in 1911 during drilling of a corehole to delineate the trend of coalbeds and coal resources approximately 2 miles (3 km) south of an active mining area. The well encountered gas and water at depths of 900 to 1,000 ft (275 to 305 m). The flows increased as successive coalbeds were encountered until drilling was suspended at 1,411 ft (430.4 m) because of the vigorous flow. The well continued to flow large volumes of gas and water for years after abandonment, and, upon ignition of the gas, flares were reported as high as 25 to 30 ft (7.6 to 9 m). A small flow of gas still is evident today at the site of the "Flaming Geyser" well at the state park that bears its name. Eyewitness accounts, gas analyses, and the longevity and manner of gas production indicate that production probably came from coalbeds. Assuming conservative estimates for the gas flow rate over the nearly 80 years that the well has produced, a total gas volume of approximately 500,000 MCF (14.2×10^6 m^3) has been produced. Drilling records, sample logs, cores and core logs, hydrocarbon or mud logs, and electric logs from conventional wildcat wells and other modern logs contain numerous gas indications in the coalbeds.

Depth to Potential Coalbed Methane Targets

Because of the repetitive nature of clastic horizons in the sedimentary section and the complexity of structure in most of the mining districts, prospective gas-bearing coalbeds have been encountered from the surface to the maximum depths of drilling. Outcropping coalbeds prompted the initial surface exploitation of coal resources in western Washington in the middle 1800s. Subsequently, mining development followed the seams underground where gas hampered operations, and prohibitive economics eventually caused abandonment. Conventional drilling has not encountered the base of the coal-bearing section, nor is the base is exposed at the surface. Coal is widespread in the subsurface within the coal districts within target depth ranges for coalbed methane production testing. However, because of structural complexities and coal-rank variability within each district, depths to specific prospective coalbeds or coal-laden intervals will vary in each district.

Walsh and Lingley (1991) reported that depth to the 0.5% vitrinite reflectance contour (generally considered to be the initiation of peak oil generation) increases westward and eastward away from the crest of the Cascade Range (Figure 9). Exceptions occur near Tacoma and Chehalis, east of Tacoma near the Wilkeson-Carbonado mining district, and in central and western Whatcom County. Regional maturation of these coals was caused by the thermal effects accompanying the intrusion of numerous post-Eocene plutons of the Cascade magmatic arc, as discussed above.

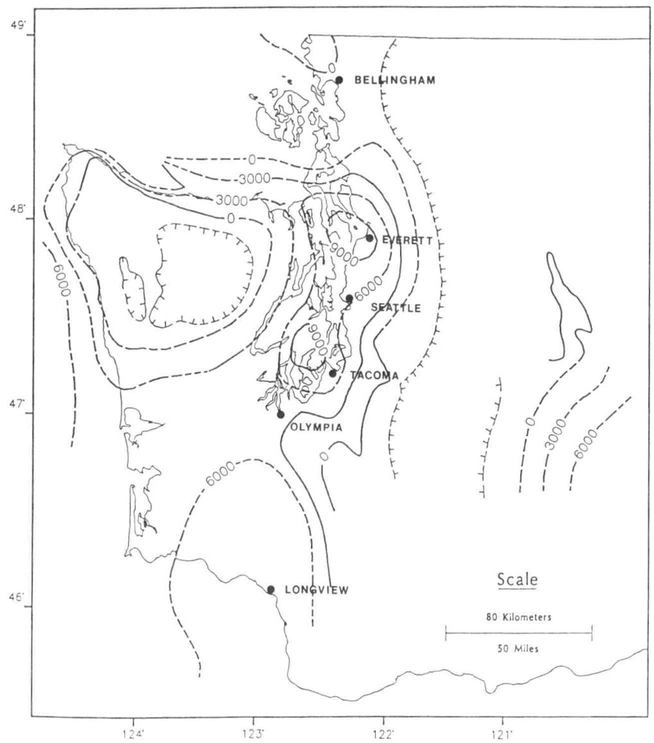

Figure 9. Depth to the 0.5% vitrinite reflectance contour in western Washington. Contour interval 3,000 ft. After Walsh and Lingley (1991).

Although some of the data from coreholes, conventional tests, coalbed methane wells, and other exploration activity described above remain proprietary, a significant volume of information on coalbed thickness and correlation, structure, water quality, and other data pertaining to coalbed methane production have been added to the public domain during the last decade. Of particular interest is the consistency of gas-desorption measurements and resultant gas-content values associated with the bituminous rank of coals sampled. Nearly all the samples meet (and many exceed) expected gas-content values for corresponding rank, depth, and pressure as predicted by widely used laboratory standards. As a result, thegas-content estimate of 400 SCF/T (12.5 cc/g) for bituminous coal, the basis for Choate and Johnson's (1980) gas-in-place assessment of 24 TCF (679×10^9 m^3) in western Washington, is rendered conservative by existing measurements (Table 1).

WASHINGTON DRILLING PROGRAM

Introduction

In September 1986, Sound Energy Development Company, a coalbed methane exploration firm, formed a partnership with Seattle Steam Energy Corporation, a nearly century-old district heating franchise that provides service to most of downtown Seattle, to explore for and develop natural gas resources. The resultant Carbon River Energy Partnership (CREP) conducted exploration for coalbed methane in Colorado and Washington in an attempt to secure for Seattle Steam up to 3 BCF (85×10^6 m^3) of natural gas annually for its operations. As part of that effort, the Partnership drilled five closely spaced coalbed methane exploration wells on prospective leasehold acreage in Pierce County, Washington.

However, because of inadequate capital, production testing and dewatering operations on the Champion project were halted prematurely. The Partnership eventually was forced to dissolve, prompting the abandonment of the Washington properties, which included the Champion acreage and wells, and the salvage of production equipment.

Project Objectives

The primary objectives of the CREP program in Washington were to obtain geological and engineering data with which to evaluate production potential, and to establish the region's first commercial production. The geological objectives of the drilling program in Pierce County were to penetrate Puget Group coalbeds and, given sufficient thickness of coal with suitable gas content, to complete closely spaced wells—on approximately 160-acre (65-ha) centers—in one or more seams. Sufficient individual coalbed thickness initially was defined as 6 ft (1.8 m) or greater. Minimum gas-content values of 250 SCF/T (7.8 cc/g) were anticipated to represent viable development prospects. Engineering objectives included design of effective completion procedures and stimulation treatments to establish producibility in multiple coalbeds, whose structural dips were anticipated to be in the range of 45° to 70°. Operational objectives included installing capable water-removal (pumping) facilities and maintaining consistent dewatering efforts to optimize reservoir management.

Project Area Selection

The CREP wells were to be the first attempt at an integrated coalbed methane development program in the state of Washington, but few specific coal degasification data existed in the region. Data were available for most coal

Table 1. Coal reserves and estimated gas in place for western Washington.

Coal Field/Region/District	County	Coal Reserves		Minimum Methane Resource at 50 CF/T		Maximum Methane Resource at 400 CF/T	
		(million short tons)	(million t)	(BCF)	(10^6 m^3)	(BCF)	(10^6 m^3)
Bellingham and Glacier coal fields	Whatcom	381	346	19	538	152	4,302
Cokedale; Cumberland-Day Creek, Mount Vernon-Big Lake-McMurray and Rock Creek areas	Skagit	507	460	25	708	203	5,745
Issaquah-Grand Ridge and Green River districts	King	987	895	49	1,387	395	11,178
Wilkeson-Carbonado coal field, Spiketon, Melmont, Fairfax-Montezuma and Ashford areas	Pierce	362	328	18	509	145	4,104
Roslyn coal field and Taneum and Manstash areas	Kittitas	282	256	14	396	113	3,198
Centralia-Chehalis district	Lewis-Thurston	3,694	3,351	185	5,236	1,478	41,827
Alpha-Cinnabar, Morton-Mineral Lake, Packwood and Carlton Pass areas	Lewis	47	43	2	57	19	538
Kelso-Castle Rock area	Cowlitz, Lewis	149	135	7	198	60	1,698
Totals							
Western Washington		6,409	5,814	300	8,490	2,600	73,580
Western Washington (Choate and Lent, 1977)		60,000*	54,432	3,000	84,900	24,000	679,200

* Total coal resource.

mining districts, generated by long-abandoned underground mining operations, recent coal exploration test wells, published coal analyses, and surface mapping. However, the need to avoid local mined-out areas, together with the availability of leasable acreage, dictated that lands adjacent to mining areas were most suitable for project development. Based on generally favorable geologic and coal quality attributes, as well as other operational and logistical factors, several prospective coal districts were high graded for their exploration and production potential. After analyzing these areas, CREP decided to target leasehold acreage near the Wilkeson-Carbonado mining district in Pierce County.

The project location was determined primarily by the following factors:

1. Availability, based on previous drilling and mining activity in the general area, of geologic data suggesting the widespread distribution of thick, multiple horizons of bituminous coal;
2. Documented presence of the largest known volume of high-quality coking coal in the western coastal states, which suggested that other coalbeds in the stratigraphic section were thermally mature and likely to be gassy;
3. Mining records and other published data that indicated an unusually high number of methane-related mine accidents in the Wilkeson-Carbonado district;
4. Proximity to gas delivery lines and end-user markets in the Seattle-Tacoma area approximately 6 to 12 miles (10 to 20 km) to the west, plus the potential for onsite gas utilization;
5. Logistical ease of access to potential drilling sites in the rugged, heavily timbered region;
6. Availability of large, unleased tracts of prospective mineral rights.

Geologic Setting

The Pierce County project area is traversed by a regional structure known as the Carbon River anticline (Gard, 1968), which extends along a north-plunging trend a distance of more than 30 miles (48 km) from its southern terminus in the vicinity of the Nisqually River drainage to just beyond the White River drainage (Figure 10). Erosional breaching of the anticline provides access beneath post-Eocene volcanics and other strata into a sequence of interbedded arkosic sandstone, siltstone, and coal of the Puget Group along a 5- to 6-mile-wide (8- to 10-km) band of outcroppings parallel to the crest of the structure. Superimposed on the anticline are numerous smaller-scale folds and faults at various angles to the anticlinal axes. In most places, however, bedrock exposures are covered by Pleistocene and Holocene volcanic and glacial deposits and dense vegetation, which makes even surface studies extremely difficult.

The target stratigraphic section selected for the project consists of the lower known part of the Puget Group anticipated to be present in the vicinity of sec. 20, T18N R6E. The Puget Group in this area of Pierce County is subdivided into the coal-bearing Carbonado Formation and younger Spiketon Formation, which are separated

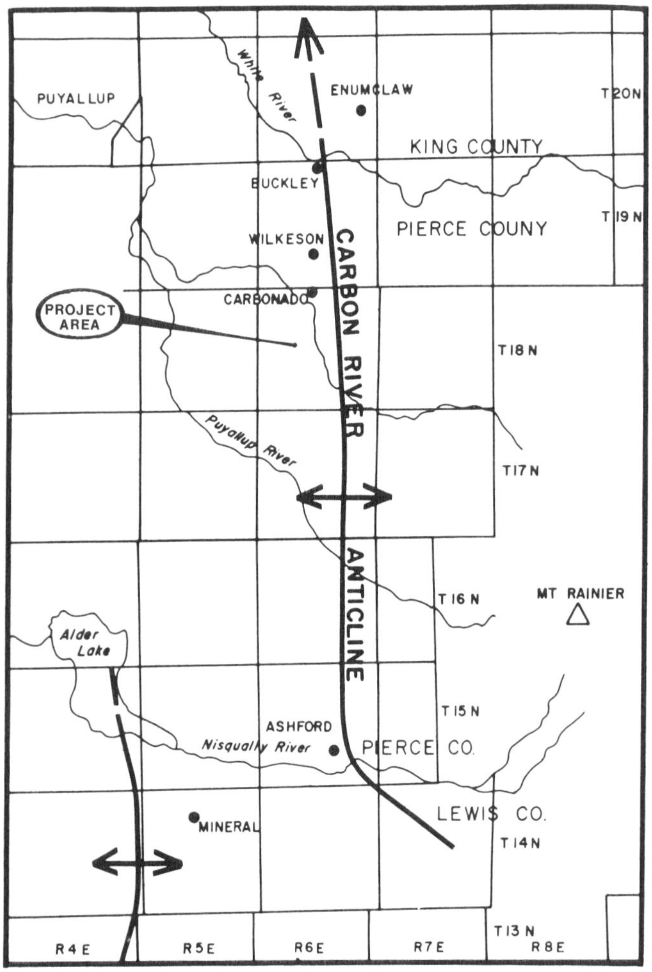

Figure 10. Regional structure of the Wilkeson-Carbonado district, Washington, and location of Champion drill sites, sec. 20, T18N R6E, Pierce County. Structure after Gard (1968).

stratigraphically by volcanic deposits of the Northcraft Formation, all of which are of late Eocene age. Subsurface information relevant to the project included mine data from the Wilkeson-Carbonado and Melmont districts, which are located 5 miles (8 km) north and 2 miles (3.2 km) east, respectively, of the drill-site area.

Additional subsurface stratigraphic data were derived from geophysical logs of the 1,800-ft-deep (549-m) Utah International C–83–1 exploratory coal corehole in the northeastern corner of the drill-site section. These log data indicated a sequence of interbedded sandstone, siltstone, and coal dipping 60° to 70° east, which was correlated as being stratigraphically equivalent to part of the Wilkeson coal series of Daniels (1914) and Gard (1968).

Despite the paucity of outcrop and subsurface data locally in the project area, the crest of a small (less than several kilometers long) asymmetrical anticline trending north-northwest was anticipated to be present in the proposed drill-site section. Its existence was suspected based on evaluation of a dipmeter log from the nearby corehole, as well as the interpretation of trends of other structural features in the area. The first well drilled in the five-well program, Carbon River 1–20, was located structurally low on the eastern flank of the postulated anticline. The four wells drilled subsequently, Carbon River 2–20, 3–29, 4–20, and 5–20, were located updip west of the 1–20 well closer to and along the anticlinal axial.

The wells were drilled in structurally high locations to take advantage of the possibility of encountering conditions favorable to methane production potential. These conditions, evident in mining records, included (1) greater degree of fracturing near the axial planes of folds (enhanced reservoir permeability); (2) lower hydrostatic head in updip locations (facilitates reservoir dewatering); and (3) slight overpressuring adjacent to fault planes, anticlinal axes, and plunging noses of folds.

Drilling Program

Well design consisted of drilling approximately 300 ft (92 m) of 12$^{1}/_{4}$-in. (31-cm) hole, setting 8$^{5}/_{8}$-in. (22-cm) casing, and cementing it to the surface. A 7$^{7}/_{8}$-in. (20-cm) hole was drilled below the 8$^{5}/_{8}$-in. casing to an average depth of 3,650 ft (1,113 m) into which 4$^{1}/_{4}$-in. (10.8-cm) production casing was set to total depth and cemented to approximately 500 ft (152 m) above the shallowest potentially productive coalbed. A "stiff" three-point packed-hole bottom assembly was used in an attempt to control wellbore deviation while drilling with optimum weight-on-bit into the steeply dipping stratigraphic section.

Drilling Results

Drilling the five wells confirmed the presence of an asymmetrical anticline with a north-northwest trend and northerly plunge in the drill-site section, as shown in Figure 11. Even though the surface locations of the four wells drilled along the axial trend are west of the anticlinal axis, all the coalbeds that subsequently were completed for production are situated on the eastern flank of the structure. The axial trend of the anticline actually may consist of a high-angle reverse fault with minor west-side-up displacement.

The drill-site area is intruded in places by igneous material, primarily along bedding planes. However, cursory petrographic analysis of coal samples in contact with the intrusives shows no obvious signs of thermal metamorphism, such as a baked zone of ash and coke.

The four wells drilled subsequent to the 1–20 well were located at average intervals of 1,300 ft (396 m), which was estimated to represent approximately 160-acre (65-ha) drainage centers in the steeply dipping reservoir system. This drainage pattern was selected as the most appropriate starting point to initiate reservoir testing. All four wells were situated near, and on trend with, the anticipated position of the anticlinal crest to take advantage of potentially more favorable reservoir characteristics that were anticipated to be present in such a structural position, as discussed previously.

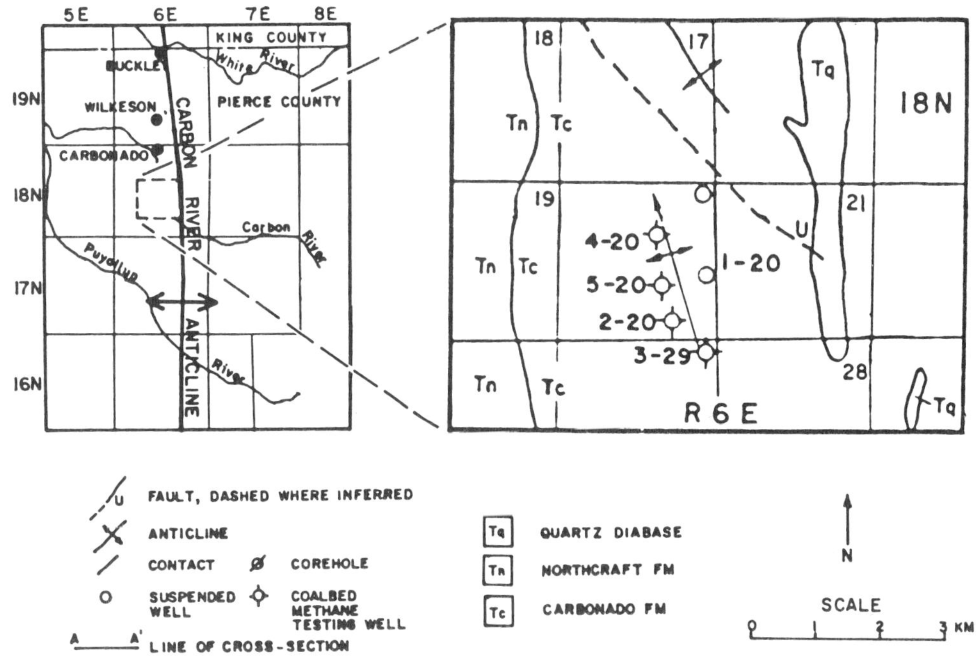

Figure 11. Local geology and structure and well locations of Champion drill-site area, secs. 20 and 29, T18N R6E, Pierce County.

Completion Program

The four test wells were drilled along the axial trend of the anticline to an average depth of 3,650 ft (1,113 m), and each penetrated a net thickness of more than 90 ft (27 m) of bituminous coal, consisting of beds several inches to more than 20 ft (6 m) thick (Figure 12). The coalbeds that were completed for production, although correlative with the No. 2, No. 2 1/2, No. 3 and, possibly, the No. 6 coalbeds of the Melmont Series (Beikman and others, 1961), are respectively designated herein as the Carbon River, Beaver, Middle, Big Red, and Rainier coalbeds (Figures 13 and 14). Several seams that were penetrated in the 3–29 well below the Rainier (No. 6?) represent coalbeds that were not encoun-tered in nearby mining districts. The 3–29 well was not completed in the newly discovered coalbeds because they were too deep to be penetrated in offsetting wells along the trend of the north-plunging anticline.

Casing opposite the entire wellbore thickness of each targeted coalbed was perforated with 4 jet spf. Following limited attempts to swab and test the wells without artificial stimulation, it became apparent that they would require such treatment to overcome effective and/or inherent permeability problems in the coal reservoir. Therefore, the completed zones were stimulated with water and sand proppant. Treatment sizes ranged from 8,500 gal (32.2 m^3) of fluid with 7,500 lb (3,402 kg) of sand to 60,000 gal (227 m^3) of fluid and 75,000 lb (34,020 kg) of sand. The fracture treatments were pumped down 2 3/8-in. (6-cm) tubing between a retrievable bridge plug and packer because all the perforating and breakdown tests already had been completed.

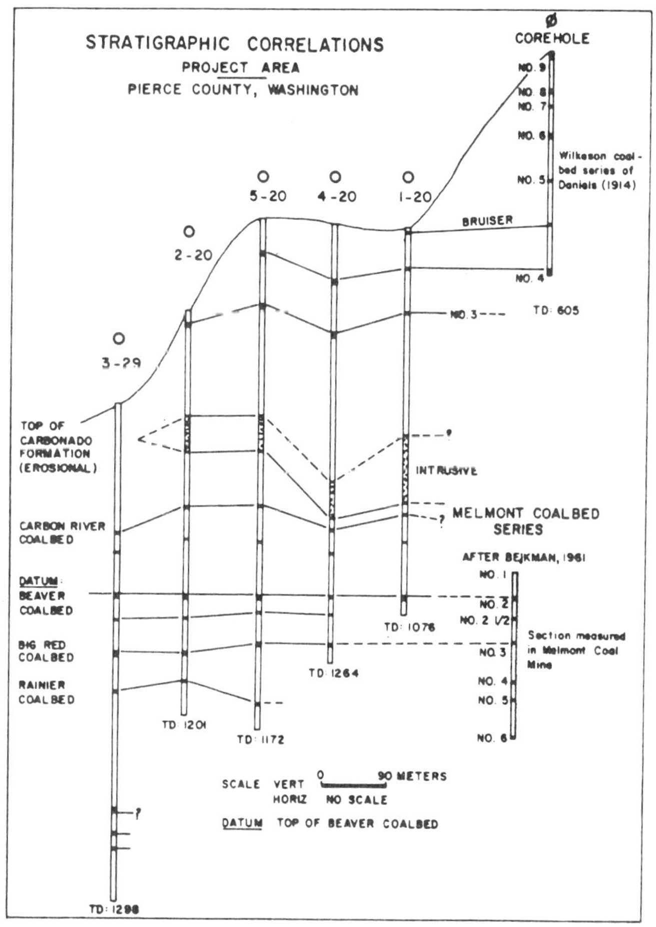

Figure 12. Stratigraphic correlations in the Champion project area, Pierce County. See figure 14 for well locations.

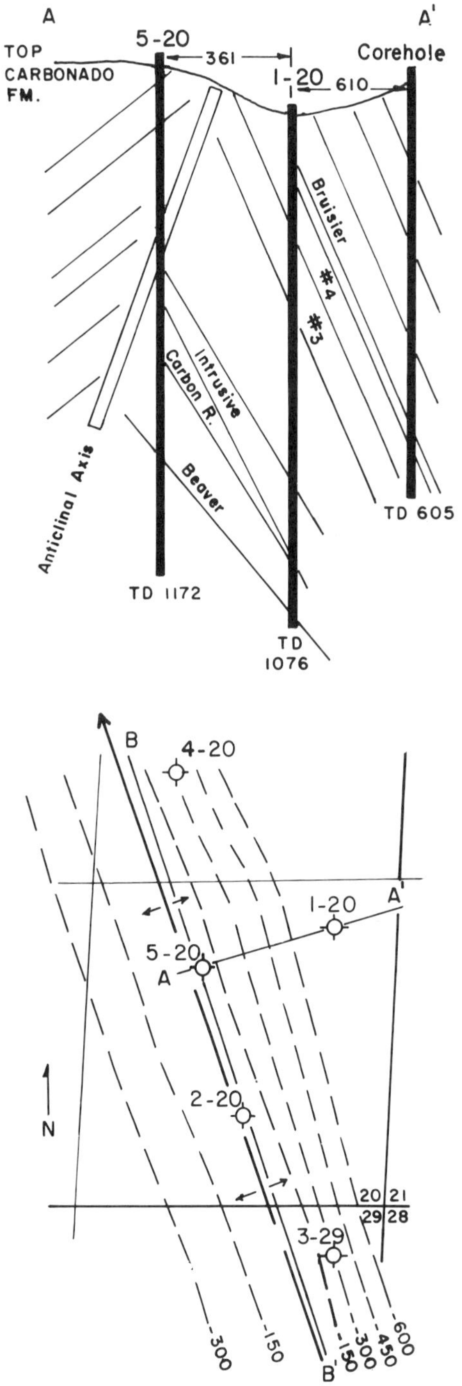

Figure 13. West-to-east log sections through wells Carbon River 5–20 and 1–20 and structure contours on top of the Beaver coalbed. Depths in meters. Structure contour interval 150 m.

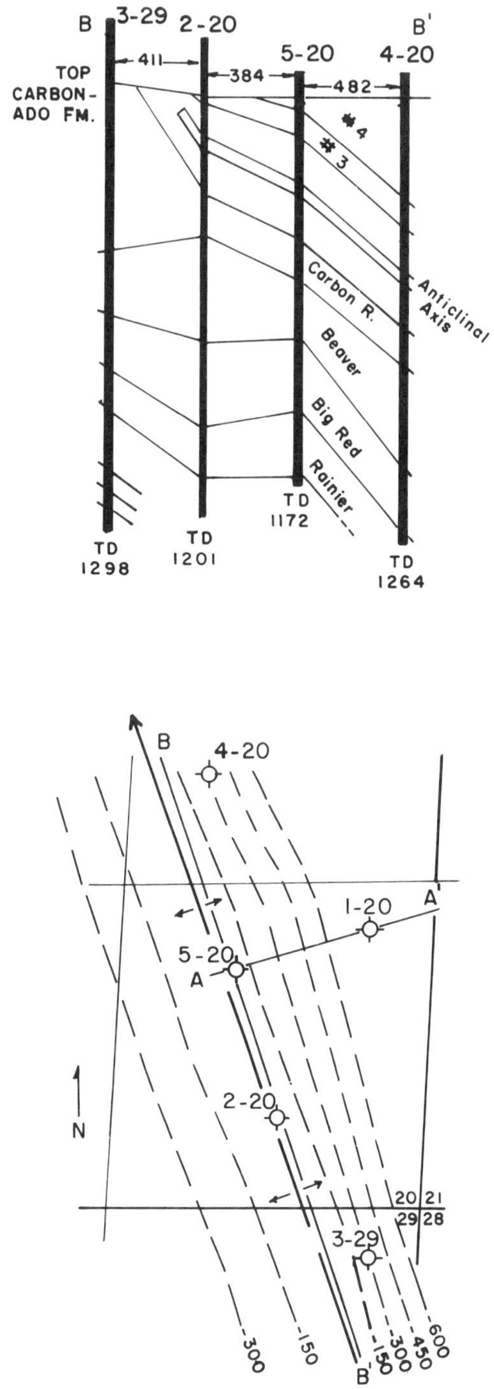

Figure 14. Southeast-to-northwest log sections through wells Carbon River 3–29, 2–20, 5–20, and 4–20 and structure contours on top of the Beaver coalbed. Depths in meters. Structure contour interval 150 m.

Final completion consisted of anchoring 2 3/8-in. (6-cm) tubing just above the top set of perforations and connecting 1 3/4-in. × 2-in. (4.4-cm × 5-cm) tubing barrel pump via a string of 7/8-in. (2.2-cm) sucker rods to a 160-size pumping unit driven by electric motors. Because of the remoteness of the well sites, power for the motors was supplied by diesel generators.

Coal, Gas, and Water Analyses

Measurements of vitrinite reflectance of samples from six of the completed intervals exhibited a range in values from 0.84% R_o to 1.18% R_o (Table 2). The values agree with prior coal-rank classifications of the Melmont coalbeds as high-volatile A to medium-volatile bituminous—a level of

Table 2. Vitrinite reflectance values, Pierce County, Washington.

Well	Location	Depth (ft)	Depth (m)	R_o (%)
CR 1–20	NE SE sec. 20, T18N R6E	350	107	0.88
		2,960	903	1.00
CR 2–20	SE SE sec. 20, T18N R6E	1,470	448	0.99
CR 3–29	NE NE sec. 29, T18N R6E	3,600	1,098	1.03
CR 4–20	SW NE sec. 20, T18N R6E	1,180	360	0.84
CR 5–20	NW SE sec. 20, T18N R6E	400	122	0.86

maturity capable of generating a significant volume of methane.

The gas content of the completed coalbeds averaged 423 SCF/T (13.2 cc/g), based on desorption measurements made while drilling (Table 3). Analyses of the coal gas indicated pipeline quality. Gas samples consisted of slightly less than 99% methane, approximately 1% nitrogen, and less than 1% carbon dioxide. Calculations of heating values ranged from approximately 950 to 1,000 Btu/CF (35,396 to 37,259 kJ/m^3).

Table 4. Representative water analyses, Pierce County, Washington.

Constituent	Value Range		
Total suspended solids (mg/L)	1.4	–	25.0
Total dissolved solids (mg/L)	1,330.0	–	1,660.0
Conductivity (µmhos/cm @ 25°C)	2,100.0	–	2,400.0
Alkalinity (mg/L as $CaCO_3$)	1,130.0	–	1,220.0
Total organic carbon (mg/L)	2.2	–	33.0
Dissolved oxygen (mg/L)	1.1	–	4.8
Turbidity (NTU)	1.2	–	20.0
pH	8.09	–	8.48
Nitrate and nitrite (mg/L)	<0.01	–	0.039
Fluoride (mg/L)	5.6	–	6.13
Magnesium (mg/L)	0.48	–	1.58
Ammonia nitrogen (mg/L)	0.482	–	723
Sodium (mg/L)	440.0	–	647.0
Potassium (mg/L)	2.9	–	12.9
Chloride (mg/L)	<1.0	–	55.7
Calcium (mg/L)	1.2	–	3.31

Table 3. Desorption values, Pierce County, Washington.

	Coalbed									
	Carbon River		Beaver		Middle		Big Red		Rainier	
Well	(SCF/T)	(cc/g)	(SCF/T)	(cc/g)	(SCF/T)	(cc/g)	(SCF/T)	(cc/g)	(SCF/T)	(cc/g)
CR 2–20	217	6.8	310	9.7	166	5.2	342	10.7	416	13.0
	236	7.4	550	17.2						
CR 3–29	224	7.0	339	10.6	ID	–	393	12.3	284	8.9
			467	14.6			553	17.3		
CR 4–20	ID	–	435	13.6	ID	–	320	10.0	NP	–
CR 5–20	ID	–	563	17.6	ID	–	499	15.6	550	17.2
			390	12.2			563	17.6		

ID indicates insufficient data for desorption analysis.
NP indicates coalbed was not penetrated.

The results of numerous analyses of water produced during limited pump testing of the four completed wells (Table 4) demonstrated that water quality was sufficient to meet state and federal requirements for surface discharge into the Carbon River, which drains the Carbon Glacier on nearby Mount Rainier. A temporary surface discharge permit was granted by the Washington Department of Ecology while application was made to the Environmental Protection Agency.

Currently, hundreds of abandoned underground coal mines in western Washington, including many in Pierce County and some along the Carbon River, discharge water into natural drainages. Although these mine waters are thoroughly oxidized, none contribute unacceptable levels of undesirable constituents. No indications of toxic water associated with underground coalbeds have been reported in Washington. Many streams and rivers that emanate from melting Cascade glaciers contain extremely high dissolved and suspended solids in their natural state.

Production Potential

The Carbon River wells underwent intermittent production testing on pump over a five-month period during which the produced gas was flowed to atmosphere. On average, initial gas-water ratios were on the order of 1:8. This value is representative of similar initial results in a majority of other coalbed methane projects that eventually established economic levels of production, as additional offset wells were drilled to dewater the reservoir and liberate increasing volumes of gas. It was not possible to document changes in the gas-water ratio during the sporadic test interval because frequent mechanical and logistical problems (parted rods, generator breakdown, lack of readily available parts and service) prevented simultaneous multiwell pumping for any adequate length of time. Consequently, an accurate assessment of production potential was not obtained prior to the exhaustion of the limited capital available for the project.

The reservoir system in the immediate drill-site area appears to be normally pressured to slightly underpressured; therefore, future dewatering operations in the project area will be required prior to achieving economic rates of gas flow. Such pressure conditions are typical of most other productive coalbed methane development programs in the United States. The specific reservoir conditions at the Carbon River drill-site area may not, however, be indicative of the production potential of the 60 billion short tons (54.4×10^9 t) of prospective coal elsewhere in the region. Considering the geologic complexity of the coal basins, it is likely that different reservoir conditions, such as overpressuring, will be encountered in other locations.

The average gas-content value of 423 SCF/T (13.2 cc/g) of coalbeds in the project area represents approximately 18 BCF/mi^2 (197×10^6 m^3/km^2) of in-place methane resources, given approximately 40 net ft (12 m) of completable coal and other standard conditions. This value agrees closely with the maximum value of 400 SCF/T (12.5 cc/g) used in the DOE study, which suggested up to 24 TCF (679×10^9 m^3) of coalbed methane resources in western Washington (Choate and Johnson, 1980).

Computer projections of potential peak-production rates range from 200 to 800 MCFD (5,660 to 22,640 m^3/day) per well in a worst-case/best-case scenario, based on data from the drilling and preliminary testing programs. Assuming an average production rate of 500 MCFD (14,150 m^3/day) for 125 wells that could be drilled on 160-acre (65-ha) centers on approximately 20,000 prospective acres (8,100 ha) within just the one township of the project area, the resulting production of 62,500 MCFD (1.77×10^6 m^3/day) represents significant potential in a state where all gas supplies currently are imported (Table 5).

Considering the fact that the project township represents only a fraction of the prospective coal-bearing acreage in western Washington, the potential for producing coalbed methane in this region should not be discounted solely on the incomplete results of the present study. The encouraging results of the Pierce County wells justify additional effort to explore for and develop coalbed methane resources in the region.

Table 5. Assumptions for computer-generated coalbed methane production forecast.

Parameter	Expected Value
Input	
Permeability, md	5
Porosity, %	2
Coal thickness, ft (m)	40 (12.2)
Initial pressure, psi (kPa)	1,350 (9,308)
Temperature, °R	520
Desorption, SCF/T (cc/g)	420 (13.1)
Diffusion coefficient, psi $^{-1}$ (kPa^{-1})	1.5×10^{-4} (2.18×10^{-5})
Density, g/cc	1.6
Initial gas saturation, %	5
BHF pressure, psi (kPa)	60 (414)
Skin factor	–1
Non-Darcy term	1×10^{-3}
Radius of investigation, ft (m)	1,489 (454)
Results	
Original gas in place, BCF/160 acres (m^3/km^2)	6.9 (18.8×10^6)
IP gas, MCFD (m^3)	520 (14,716)

CONCLUSIONS

A thick, widely distributed sequence of Eocene coal-bearing strata in western and central Washington represents a potential source for development of coalbed methane resources. The most prospective trends for exploration are defined by fairways of bituminous-rank coal that occur within coal-bearing strata on both the eastern and western flanks of the Cascade Range. A thermal aureole, created by several episodic, post-middle Eocene igneous heating events, has resulted in the creation of predictable coal-rank trends and the thermal generation of methane in a manner much faster than might be expected by burial-induced coalification alone. Based on the inferred presence of up to 60 billion short tons (54.4×10^9 t) of coal in the region and based on the assumption that 400 SCF/T (12.5 cc/g) represents a reasonable average gas-content value for the coal, coalbed methane resources have been estimated at 24 TCF (679×10^9 m^3) (Choate and Johnson, 1980).

One of the first attempts to produce coalbed methane in western Washington was a five-well pilot program drilled into multiple coalbeds of the Puget Group. The project location was selected based on the documented presence of multiple horizons of gassy, coking-quality, bituminous coal as determined from historical mining data, coal analyses, and limited corehole information. Based on preliminary results, the cumulative thickness of coal encountered in each well and the gas contents measured by desorption tests confirmed initial exploration expectations. The formal documentation of gas content and in-place gas resource values renders conservative the resource estimates of the DOE study (Choate and Johnson, 1980). Given the accumulated data, computer projections indicate the potential for an average per-well production rate of 500 MCFD (14,150 m^3/day) from an in-place resource base of more than 18 BCF/mi^2 (197×10^6 m^3/km^2).

The majority of the potential coalbed methane resources in Washington are situated close to the densely populated and rapidly growing area around Puget Sound to the west. A significant indigenous source of clean-burning natural gas may be available for many years in this import-dependent region. If even conservative assumptions are made regarding only the most logistically and economically viable prospects, the resulting potential natural gas volume represents a resource base worthy of development.

ACKNOWLEDGMENTS

In view of the exploratory nature of coalbed methane resource identification in Washington, it was necessary for us to depend on many regional geologic and coal studies. We relied heavily on the work of many other authors, but not all could be referenced here. As a result, we may have neglected to trace the origin of concepts that we have assimilated over the years. To the originators of those concepts that we failed to develop in a satisfactory manner, we offer humble apologies. All errors, regardless of form, reflect our shortcomings, and we accept sole responsibility.

Of special note is the ongoing coal-related work of geologists of the Washington Department of Natural Resources' Division of Geology and Earth Resources in Lacey. The coal-maturation research of Timothy Walsh, William Lingley, and William Phillips contributed to our work in a significant way. Hank Schasse of WDNR has done extensive field work in the project area as well. Cathy Wyckoff reviewed and edited the manuscript. We are grateful to all for their many helpful suggestions.

REFERENCES

Armentrout, J.M., and Suek, D.H., 1985, Hydrocarbon exploration in western Washington and Oregon: AAPG Bulletin, v. 69, p. 627–643.

Ash, S.H., 1931, Explosions in Washington coal mines: U.S. Bureau of Mines Technical Paper 507, 52 p.

Beikman, H.M., Gower, H.D., and Dana, T.A.M., 1961, Coal reserves of Washington: Washington Division of Mines and Geology Bulletin 47, 115 p.

Buckovic, W.A., 1974, The Cenozoic stratigraphy and structure of a portion of the west Mount Rainier area, Pierce County, WA: University of Washington M.Sc. Thesis.

Buckovic, W.A., 1979, The Eocene deltaic system of west-central Washington, *in* Armentrout, J.M., Cole, R.M., and Ter Best, H., Jr., eds., Cenozoic paleogeography of the western United States: SEPM Pacific Coast Paleogeography Symposium 3, p. 147–163.

Byrer, C.W., Mroz, T.H., and Covatch, G.L., 1987, Coalbed methane production potential in U.S. basins: Journal of Petroleum Geology, v. 39, no. 7, p. 821–834.

Choate, R., and Lent, J., 1977, Steeply dipping coalbeds suitable for in-situ gasification—A resource definition, *in* Proceedings of the third annual Underground Coal Conversion Symposium—Fallen Leaf Lake, California, June 6–9, 1977: U.S. Dept. of Energy.

Choate, R., and Johnson, C.A., 1980, Geologic overview, coal and coalbed methane resources of the western Washington coal region: Lakewood, Colorado, TRW Energy Systems Group report for U.S. Department of Energy, Morgantown Energy Technology Center, Contract DE–AC21–78MC08089.

Culbertson, W.C., 1964, Geology and coal resources of the coal-bearing rocks of Alabama: USGS Bulletin 1182–B, 79 p.

Culver, H.E., 1919, The coal fields of southwestern Washington: Washington Geologic Survey Bulletin 19, 155 p.

Daniels, J.Y., 1914, The coal fields of Pierce County Washington: Washington Geological Survey Bulletin 10, 146 p.

Engebretson, D.C., Cox, Allen, and Gordon, R.G., 1985, Relative motions between oceanic and continental plates in the Pacific basin: GSA Special Paper 206, 59 p.

Evans, J.C., 1989, Facies analysis and sandstone diagenesis of the Carbonado formation, western Washington: Colorado State University M.Sc. Thesis, 173 p.

Fassett, J.E., and Hinds, J.S., 1971, Geology and fuel resources of the Fruitland Formation and Kirtland Shale of the San Juan basin, New Mexico and Colorado: USGS Professional Paper 676, 76 p.

Gard, L.M., Jr., 1968, Bedrock geology of the Lake Tapps quadrangle, Pierce County, Washington: USGS Professional Paper 388–B, 33 p.

Gower, H.D., Yount, J.C., and Crosson, R.S., 1985, Seismotectonic map of the Puget Sound region, Washington: USGS Miscellaneous Investigations Series Map I–1613, 15 p.

Gwilliam, W.J, Stanley, W.D., Latham, G.V., and Westhusing, J.K., 1990, Deep seismic surveys of a dormant subduction zone in the Pacific Northwest, United States (abs.): AAPG Bulletin, v. 74, no. 5, p. 667–668.

Hagen, R.A., 1987, The geology and petrology of the Northcraft Formation, Lewis County, Washington: University of Oregon M.Sc. Thesis.

Hammond, P.E., 1979, A tectonic model for evolution of the Cascade Range, *in* Armentrout, J.M., Cole, R.M., and Ter Best, H., eds., Cenozoic paleogeography of the western United States: SEPM Pacific Coast Paleogeography Symposium 3, p. 219–238.

Heller, P.L., Taber, R.W., and Suczek, C.A., 1987, Paleogeographic evolution of the United States Pacific Northwest during the Paleogene: Canadian Journal of Earth Science, v. 24, p. 1652–1667.

Hull, D.A., Beaulieu, J.D., and Rau, W.W., 1983, Correlation of Cenozoic stratigraphic units of western Oregon and Washington: Oregon Department of Geology and Mineral Resources Oil and Gas Investigation 7, 90 p.

Johnson, S.Y., 1982, Stratigraphy, sedimentology, and tectonic setting of the Chuckanut Formation, northwest Washington: University of Washington Ph.D. thesis, 221 p.

Johnson, S.Y., 1985, Eocene strike-slip faulting and nonmarine basin formation in Washington, in Strike-slip deformation, basin formation and sedimentation: SEPM Special Publication 37, p. 283–302.

Johnson, S.Y., 1988, Eocene strike-slip tectonics and sedimentation in central and western Washington (abs.): Northwest Petroleum Association 1988 Meeting.

Kelso, B.S., and Wicks, D.E., 1988, A geologic analysis of the Fruitland Formation coal and coal-bed methane resources of the San Juan Basin, southwestern Colorado and northwestern New Mexico, *in* Fassett, J.E., ed., Geology and coal-bed methane resources of the northern San Juan Basin, Colorado and New Mexico: Rocky Mountain Association of Geologists Guidebook, p. 69–90.

Kim, A.G., 1978, Experimental studies on the origin and accumulation of coal bed gas: U.S. Bureau of Mines Report of Investigations 8317.

Landes, H., 1914, The mineral resources of Washington, with statistics for 1912: Washington Geological Survey Bulletin 11, 53 p.

Larson, V.E., 1985, Preliminary report on the coalbed methane deposits of western Washington: Quarterly Review of Methane from Coal Seams Technology, v. 3, no. 2, p. 3–6.

Mackin, J.H., and Cary, A.S., 1965, Origin of Cascade landscapes: Washington Department of Conversation Information Circular 41, 35 p.

Mooney, W.D, and Weaver, C.S., 1989, Regional crustal structure and tectonics of the Pacific coastal states, California, Oregon, and Washington, *in* Pakiser, L.C., and Mooney, W.D., eds., Geophysical framework of the continental United States: GSA Memoir 172, p. 129–161.

Mroz, T.H., Ryan, J.G., and Byrer, C.W., 1983, Methane recovery from coalbeds—A potential energy source: U.S. Department of Energy, Morgantown Energy Technology Center Rept. DOE/METC/83-76.

Robertson, R.L., and Rightmire, C.T., 1986, Commercial development of coalbed methane: Chemical Engineer Progress, February, p.48–53.

Snavely, P.D., and others, 1958, Geology and coal resources of the Centralia-Chehalis District, Lewis and Thurston Counties, Washington: USGS Bulletin 1053.

Snavely, P.D., Jr., Wagner, W.C., and Lander, D.L., 1980, Interpretation of the Cenozoic geologic history, central Oregon continental margin—Cross-section summary: GSA Bulletin, v. 91, pt. 1, p. 143–146.

Stremel, K., 1988, Coming out of the closet: Oil and Gas Investor, v. 8, p. 38–44.

Vine, J.D., 1968, Depositional environment of coal-bearing Paleogene strata, western Washington (abs.): GSA Special Paper 101, p. 230.

Vine, J.D., 11969, Geology and coal resources of the Cumberland, Hobart, and Maple Valley quadrangles, King County, Washington: USGS Professional Paper 624, 67 p.

Walsh, T.J., 1986, Age determination along the Olympic-Wallowa lineament near Seattle, Washington (abs.): GSA Abstracts with Programs, v. 18, no. 2, p. 195.

Walsh, T.J., and Lingley, W.S., 1991, Coal maturation and the potential for natural gas accumulation in western Washington State: Washington Division of Geology and Earth Resources, Open-File Rept. 91-2.

Walsh, T.J., and Phillips, W.M., 1983, Rank of Eocene coals in western Washington and central Washington state—A reflection of plutonism?: Washington Division of Geology and Earth Resources Open-File Rept. 83-16, 21 p.

Wells, R.E., Engebretson, D.C., Snavely, P.D., and Coe, R.S., 1984, Cenozoic plate motions and the volcano-tectonic evolution of western Oregon and Washington: Tectonics, v. 3, no. 2, p. 275–294.

Coalbed Gas Accumulations in the Paleocene Fort Union Formation, Powder River Basin, Wyoming

Ben E. Law[1]
Dudley D. Rice[1]
Romeo M. Flores[1]

1. U.S. Geological Survey, Denver, Colorado

ABSTRACT

Thick, low-rank coalbeds in the Paleocene Fort Union Formation in the Powder River basin, Wyoming, are potentially important sources and reservoirs of gas. Coal gas in the Powder River basin is of a bacterial origin, and structural and stratigraphic trapping aspects of the gas accumulation are of primary importance. Recent gas discoveries in the Wyodak-Anderson coal, at depths of 150 to 475 ft (45 to 145 m) and near outcrops on the east side of the basin, show that gas can economically be produced from these lignitic and subbituminous coals under favorable trapping and hydrodynamic conditions.

The search for coalbed gas in the Powder River basin may be improved by recognizing the significance of compaction folds. These folds, some with as much as 250 ft (75 m) of structural relief, developed as a result of contrasting lithologies and compaction potential related to different depositional features. Compaction folds indicate the presence of early-formed structural and/or stratigraphic traps that have provided optimal conditions for the entrapment of early-formed bacterial gas in coalbeds. Gas production, and usually minimal water production, is facilitated near the eastern edge of the basin where compaction structures lie above the regional water table and where dewatering by surface coal mines has affected the coal reservoirs.

INTRODUCTION

The Powder River basin of northeastern Wyoming is one of the more important energy-producing regions in the United States. Historically, the basin has produced large amounts of oil, uranium oxide, coal, and some conventional natural gas. Recent discoveries of coal gas in the Paleocene Fort Union Formation indicate the potential of a new source of gas in the basin. Estimates of in-place coalbed gas resources in the Powder River basin are as high as 39 TCF (1.1×10^{12} m^3) (Choate and others, 1984).

Interest in coal gas in the Powder River basin has been spurred by the occurrence of thick coalbeds at shallow drilling depths, successful production of gas from higher rank coalbeds in the San Juan basin, New Mexico and Colorado, and in the Black Warrior basin, Alabama, and tax credits for production. However, in order to realize the potential of this resource, exploration and exploitation programs must be developed that incorporate the geologic and geochemical attributes of coal gas accumulations in the basin. As a consequence of the uncertainties regarding the nature of such gas accumulations in basins with few data, coalbed gas-producing areas such as the San Juan basin often are used analogs. These comparisons should, however, be tempered with the realization that although some similarities exist among coal gas accumulations in different basins, significant differences also exist that may be unique and, therefore, not applicable to other coal-bearing regions. With that concern in mind, the purposes of this report are to (1) provide information on geologic and geochemical characteristics of coal and coal-derived hydrocarbons from the Powder River basin; (2) investigate the conditions under which gas may accumulate in coalbeds; and (3) speculate on the economic coalbed gas potential of the Powder River basin.

GEOLOGIC SETTING

The Powder River basin of northeastern Wyoming is elongated north-south and is structurally asymmetrical, with its axis along the western side (Figure 1). The basin extends into southeastern Montana; however, this study is restricted to the Wyoming part of the basin. In Wyoming the basin is bounded on the west by the Bighorn uplift and Casper arch, on the east by the Black Hills uplift, and on the south by the Laramie and Hartville uplifts. Structural and sedimentological data indicate that the basin and adjacent structural features evolved primarily as a result of compressive deformation during the Laramide orogeny (Late Cretaceous to Early Tertiary) (Whipkey and others, in press; Flores and others, 1989).

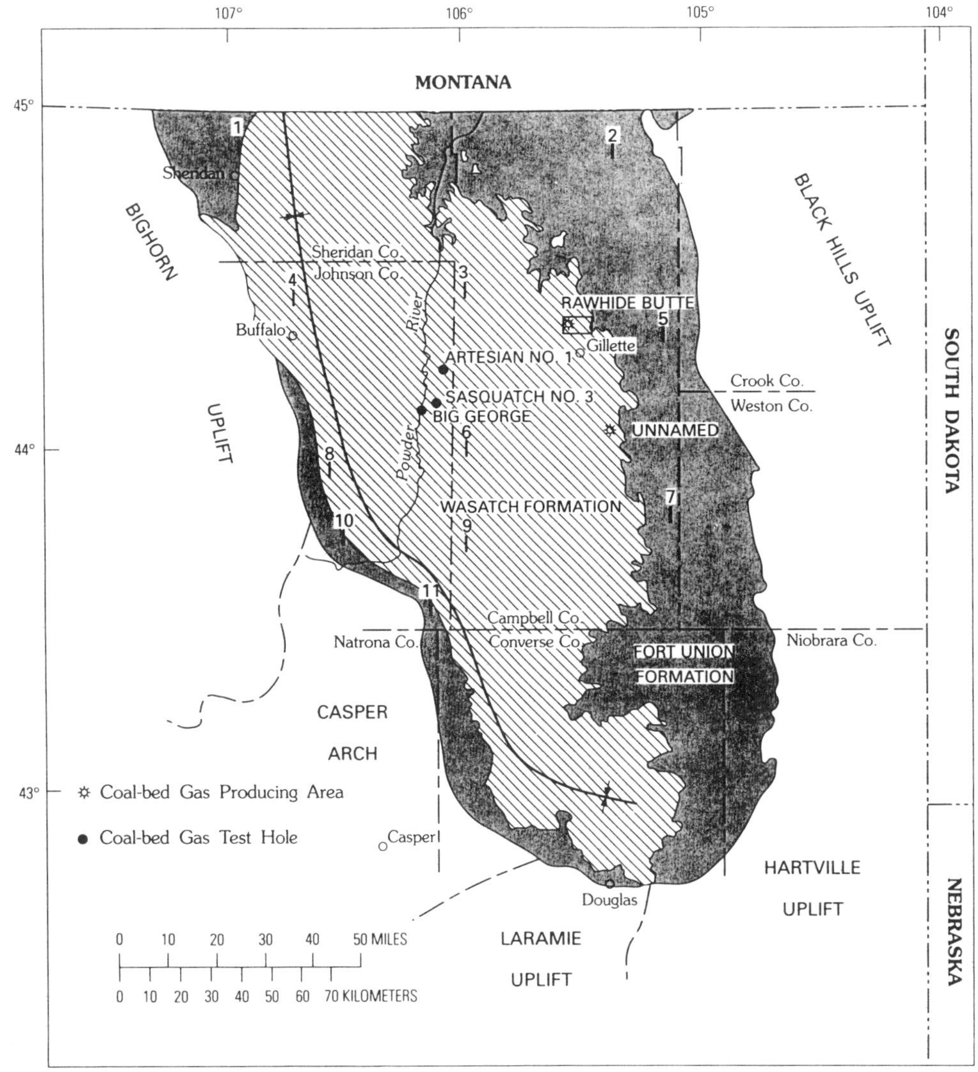

Figure 1. Powder River basin, Wyoming, and adjacent uplifts. Outcrop of Fort Union Formation shaded; outcrop of Wasatch Formation ruled. Circled numbers refer to approximate locations of coal sections shown in Figure 3. Rectangular area north of Gillette is shown in detail in Figure 4.

Tertiary rocks ranging in age from Paleocene to Miocene are as thick as 8,000 ft (2,440 m) (Love, 1989). The source terranes for nearly all Tertiary rocks were the Bighorn, Black Hills, and Hartville uplifts (Figure 1). The major coal-bearing rocks include the Paleocene Fort Union Formation and the Eocene Wasatch Formation. This paper focuses on the coal-bearing part of the Fort Union Formation.

FORT UNION FORMATION

The Fort Union Formation, which is as thick as 6,200 ft (1,890 m), is divided into three members, which in ascending order are the Tullock, Lebo Shale, and Tongue River (Figure 2). The Tullock and Tongue River Members are distinguished from the Lebo Shale by larger proportions of sandstone and coaly lithologies, whereas the Lebo Shale Member consists of shale, mudstone, siltstone, and minor coal.

The Tullock Member conformably overlies the Upper Cretaceous Lance Formation (Figure 2) and ranges in thickness from 370 to 1,400 ft (110 to 440 m) (Brown, in press). It consists of sandstone, siltstone, mudstone, coal, and carbonaceous shale that were deposited in fluvial environments as channels, flood-plain-crevasse splays, and mires. Flores and Ethridge (1985) interpreted the Tullock to have been deposited by a northerly flowing trunk-tributary fluvial system. Similarly, Brown (in press) concluded that the Tullock was deposited in a fluvial system. The sediment source was the Bighorn, Laramie, and Hartville uplifts.

The Lebo Shale Member conformably overlies the Tullock Member and ranges in thickness up to 3,000 ft (910 m). It consists of sandstone, siltstone, mudstone, coal, and carbonaceous shale that were deposited in fluvial and lacustrine environments. Flores and Ethridge (1985) interpreted the Lebo to have been deposited by a north-flowing fluvial system, in which crevasse-delta systems drained into a lacustrine setting juxtaposed to the Bighorn Mountains. They indicated that the large amount of mudstone in the Lebo probably was derived from adjacent highlands where Upper Cretaceous marine shales, such as the Pierre, Cody, and Mowry, were eroded, transported, and redeposited in the Powder River basin. These interpretations are supported by the presence of glauconite in Fort Union channel sandstones (Whipkey and others, in press).

The Tongue River Member conformably overlies the Lebo Shale Member (Figure 2) and is as thick as 1,750 ft (530 m). It consists of sandstone, conglomerate, siltstone, mudstone, limestone, coal, and carbonaceous shale that were deposited in fluvial and lacustrine environments. Flores and Ethridge (1985) and Flores (1986) interpreted the Tongue River as having been deposited by northerly flowing trunk-tributary fluvial systems flanked by alluvial fans along the western basin margin. Alternatively, Ayers (1986) interpreted the Tongue River to have been deposited in a deltaic-lacustrine system. The fluvial system is composed of a continuum of braided to meandering, and anastomosing rivers, with anastomosing rivers more commonly developed in the trunk fluvial systems (Flores, 1986). Sediment in this fluvial system was derived from the nearby Casper arch and surrounding uplifts.

The Tongue River Member contains 8 to 10 coalbeds ranging in thickness from 5 ft to over 200 ft (1.5 to 61 m); several coalbeds are less than 5 ft (1.5 m) thick. Some of the thicker, more laterally continuous coalbeds are commonly 20 to 90 ft (6 to 27 m) thick. Attempts to correlate the coalbeds through the basin have been only partly successful because of their extreme convergence and divergence over short distances, rapid facies changes, and limited subsurface data. Figure 3 shows an interpretation of the stratigraphic distribution and correlation of coalbeds in the Tongue River and the overlying Wasatch Formation in the Wyoming part of the basin. A discussion of coalbed correlations is beyond the scope of this paper; however, some of the coalbeds have been correlated through large areas of the basin, and their names are widely used—Roland, Smith, Anderson-Dietz, Wyodak-Anderson, Big George-Sussex.

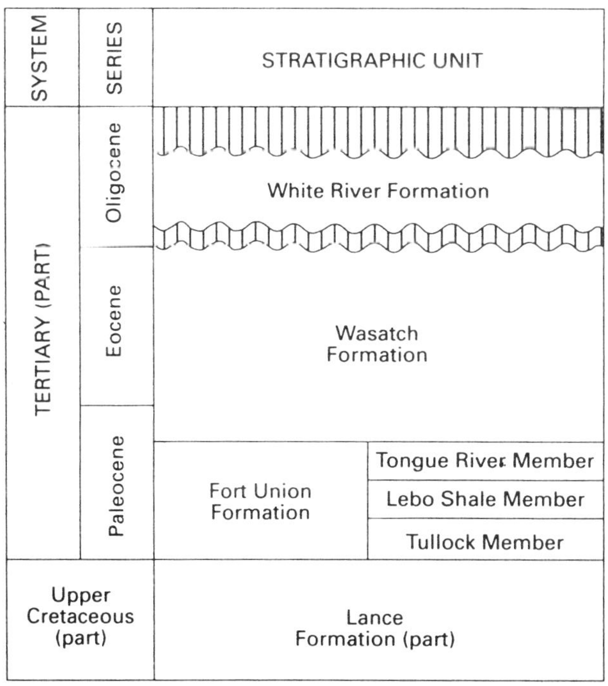

Figure 2. Generalized stratigraphic column of Upper Cretaceous (part) and Tertiary rocks in Powder River basin. Modified from Nuccio (1990).

Thick coalbeds, such as the Wyodak-Anderson, Big George-Sussex, and Anderson-Dietz, are interpreted to have formed in mires associated with meandering and anastomosing rivers. The Wyodak-Anderson coalbed, which commonly consists of an upper and lower bench, was interpreted by Warwick and Stanton (1988) to have been deposited in restricted parts of a fluvial flood plain. They concluded that the coal benches were separated by contemporaneous anastomosing channels that caused the seams to converge and diverge and in some areas form "wants," or areas of no coal. Petrologic investigations of the Wyodak-Anderson coals suggest that they developed in raised peat mires or ombrogenous (rain-fed) peat swamps. Based on the characteristics of Fort Union macrofloras, Nichols and others (1989) suggested that annual precipitation during late Paleocene time was as much as 90 in. (230 cm). This period of heavy rainfall coincides with deposition of the thick Tongue River coalbeds. Although some thick seams, such as the Anderson-Dietz coal, may not have completely developed into raised mires (Moore, in press), their low ash content suggests raised topographies during the early stages of the mires. The Tongue River trunk-tributary fluvial system is analogous to the modern Mahakam trunk-tributary system in the Kutai basin in Kalimantan Timor, Borneo (Flores, 1986; Moore, in press).

COAL GAS ACCUMULATIONS

For many years it has been known, or suspected, that coals in the Powder River basin contain gas. Early reports by Olive (1957), and later by Hobbs (1978), noted gas emanations from coalbeds or closely associated sandstones. To date, economic recovery of gas from coalbeds in the basin has been limited to two small fields near the eastern outcrop of Paleocene and Eocene rocks at depths of 150 to 475 ft (45 to 145 m) (Figure 1). Attempts to produce gas from thick coalbeds at depths of 1,200 to 1,800 ft (365 to 550 m) in the deeper part of the basin (Figure 1) have, as yet, been unsuccessful; large volumes of water with low volumes of gas have been reported.

Gas production from the east side of the basin is from the Wyodak-Anderson or equivalent coalbeds in the Fort Union Formation at shallow depths. These gas accumulations are closely associated with areas that are structurally deformed or that lie above the water table. Initial gas production ranges from less than 100 MCFD (2,832 m^3/day) to nearly

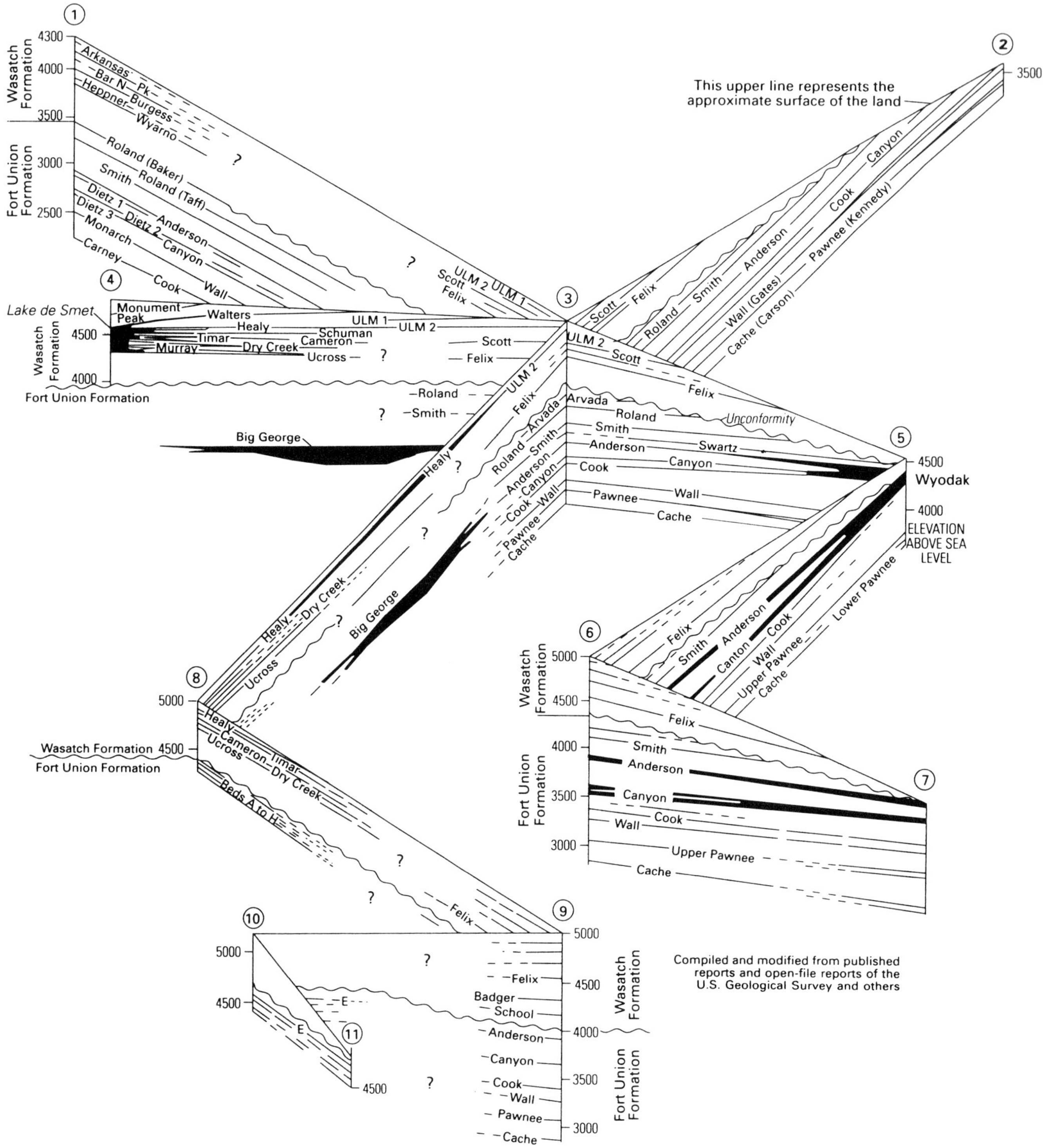

Figure 3. Coal correlation diagram for Fort Union and Wasatch Formations, Powder River basin, Wyoming. Locations of coal sections shown on Figure 1. Elevations in feet. From Glass (1989).

1,000 MCFD (28,320 m^3/day), with water production of 0 to 700 BPD (0 to 111 m^3/day).

In Rawhide Butte field (Figure 1) north of Gillette, gas is produced from both the upper and/or lower coalbeds in the Wyodak-Anderson coal at depths of 150 to 475 ft (45 to 145 m). Cumulative gas production from 11 wells, through January 1991, was about 487,000 MCF (13.8 × 10^6 m^3). The wells producing the most gas and least amount of water are located in relatively higher structural positions, probably reflecting proximity to the top of the water table. Some wells produce from different intervals within the same coal (C.W. Peck, 1990, pers. comm.). This apparent heterogeneity within the coals may be caused by variations in coal quality, seals, and/or cleating. Mining activity nearby possibly may have lowered the water level in the coal and enhanced gas production. Pressure surveys indicate that the coalbeds are underpressured, with pressure gradients ranging from 0.26 to 0.29 psi/ft (5.9 to 6.6 kPa/m); normal hydrostatic gradient in this area is about 0.43 psi/ft (9.73 kPa/m). In contrast, coal reservoirs in the San Juan basin are both overpressured and underpressured (Kaiser and Swartz, 1988).

Gas production from an unnamed field south of Gillette (Figure 1) is from the Wyodak-Anderson coal at depths of 169 to 340 ft (52 to 104 m). Initial production from two wells, which are located near coal mines, ranges from 30 to 108 MCFD (849 to 3,056 m^3/day) with no water. Structure on the producing coalbed is characterized by gentle folding, and the productive intervals in the coal lie above the water table (Dan Coolidge, 1990, pers. comm.). As in Rawhide Butte field, mining activity nearby may have lowered the water level.

COAL COMPOSITION

Tertiary coals in the Powder River basin have low ash and low sulfur contents, and for this reason they are used as fuel for electrical generating facilities throughout the United States. The coals are composed of an average 88% huminite (vitrinite precursor), 5% liptinite, and 7% inertinite (U.S. Geological Survey National Coal Resources Data System). Based on this composition, the coals can be classified as humic, and although they are typical of other Rocky Mountain coals (Meissner, 1984), their composition varies considerably. For example, Warwick and Stanton (1988) observed that the upper and lower parts of the Wyodak-Anderson coal contain large amounts of woody remains (humotellinite), whereas the middle part contains relatively large amounts of humic gels (eugelinite). They attributed these lithofacies to different chemical and biological environments of peat deposition. Moore and others (1990) concluded that maceral and palynological variations in the Smith and Anderson coals reflect different plant communities in different parts of the peat swamp.

COAL RANK

The rank of coal in the Fort Union varies from lignite to subbituminous B, which is equivalent to mean random vitrinite reflectance (R_m) values of 0.28 to 0.45%. Coal samples from three wells (Big George, Sasquatch No. 3, and Artesian No. 1) in the central part of the basin (Figure 1), at depths of 80 to 1,860 ft (24 to 567 m) have R_m values ranging from 0.34% to 0.43%. Within the general area of these wells, R_m values of Upper Cretaceous Cody Shale, at depths of 8,600 to 9,100 ft (2,620 to 2,775 m), range from 0.54% to 0.61% (Nuccio, 1990), indicating a low thermal maturity gradient. Compared to other Rocky Mountain basins, where exploration and production of coalbed gas has taken place, the rank of Tertiary coals in the Powder River basin is low. The low levels of thermal maturity probably result from shallow burial depth and/or low thermal gradient. By comparison, the major gas production in the San Juan basin comes from subbituminous A to low-volatile bituminous coals (0.6% to 1.54% R_m), with highest production rates coming from high-volatile bituminous A coal (0.8% to 1.0% R_m).

The low rank of Tertiary coals in the Powder River basin implies that gas content also should be low and, indeed, direct measurements of desorbed gas from Tertiary coalbeds in the basin confirm that relation. Boreck and Weaver (1984) reported coal-gas desorption values ranging from 56 to 74 CF/T (1.75 to 2.3 cc/g) for the Anderson coal (also known as the Big George coal) in the deeper part of the basin. Choate and others (1984) reported desorption values of less than 1 to 71 CF/T (0.03 to 2.2 cc/g) from various coalbeds in the Fort Union and Wasatch Formations. In contrast, coal-gas desorption values for high-volatile bituminous B to low-volatile bituminous Fruitland Formation coal in the San Juan basin range from 70 to 537 CF/T (2.2 to 16.8 cc/g) (Kelso and others, 1988).

GAS GENERATION, STORAGE, AND MIGRATION

Methane, carbon dioxide, and water are the main byproducts of the coalification process as larger amounts of volatile matter are lost (Jüntgen and Karweil, 1966a, 1966b; Karweil, 1969; Jüntgen and Klein, 1975). The major stage of thermal methane generation by devolatilization is believed to begin at ranks of high-volatile bituminous (0.73% R_m; Meissner, 1984) and higher, although recent work in the San Juan basin has shown that the threshhold of thermogenic gas generation for Fruitland coal occurs at about 0.6% R_m (Law and others, 1990). The main stages of water and carbon dioxide generation precede thermal methane generation (Jüntgen and Karweil, 1966a, 1966b; Karweil, 1969; Jüntgen and Klein, 1975) and are interpreted to have occurred in the thick, low-rank coals of the Powder River basin.

In relatively low rank coals significant amounts of gas can be produced through the decomposition of organic matter by anaerobic bacteria (Rice and Claypool, 1981) prior to the main stage of thermogenic gas production. This gas is referred to as "bacterial gas" to emphasize its origin. Although the amounts of bacterial gas generated are lower than those by thermal generation, the fact that coal resources in the Powder River basin are so large—approximately 1.3 trillion short tons (1.18×10^{12} t) (Choate and others, 1984)—has resulted in a large potential coalbed gas resource. The estimated in-place coalbed gas resources in the basin range from 6 to 39 TCF (0.170×10^{12} to 1.104×10^{12} m^3) (Choate and others, 1984).

Most of the bacterial gas generated from low-rank coals probably has left the coal and either was lost to the atmosphere or migrated into adjacent sandstone reservoirs. A large volume of gas, however, remains in the coal, where some is stored in fractures (cleat system) but most is adsorbed onto the internal surfaces of micropores (Kim, 1977). Gas adsorption in coal depends primarily on temperature, pressure, rank, and moisture; the amount of adsorbed gas increases with increasing pressure and decreases with increasing temperature. At a given temperature and pressure, gas adsorption increases with increasing rank. Moisture also affects the gas adsorption capacity of coal (Joubert and others, 1973), which is markedly reduced by even small amounts of moisture. In low-rank coals, like those in the Powder River basin, the adsorption capacity to water is high because of the relatively high oxygen content. Water is adsorbed into the micropores at the sites of hydrophylic functional groups, such as the hydroxyl (OH^-), carboxyl ($COOH^-$), methoxyl (OCH_3^-), and carbonyl (C=O) groups. With increasing coalification, these functional groups are lost, and water-adsorption capacity is reduced (Stach and others, 1982).

In high-rank coals, like those in the San Juan basin, where gas adsorption is an important aspect of gas production, it commonly is necessary to reduce pressure to achieve economic rates of gas flow to the wellbore. By comparison, in the Powder River basin, gas adsorption probably is not a major obstacle to gas production; the low pressure in the coals (less than hydrostatic), their high moisture content (20% to 30%), and low rank (implying relatively high water-adsorption capacity) indicate that gas adsorption is not significant. Therefore, gas flow from micropores to cleats is relatively unimpeded, insuring replenishment of gas produced from the cleat system.

GAS COMPOSITION AND ORIGIN

Natural gas desorbed and produced from coalbeds and adjacent sandstones in the Powder River basin is composed of methane with minor amounts of carbon dioxide, at depths ranging from a few hundred feet to more than 1,800 ft (550 m). In all gas samples, methane accounts for more than 99% of the hydrocarbon fraction, with ethane and heavier hydrocarbons always less than 1%. Carbon dioxide content is, however, variable. It may amount to as much as 10% in desorbed coal-gas samples, less than 3% in produced coal-gas samples, and less than 1% of gas produced from adjacent sandstone reservoirs.

Methane, which is dominant in all gas samples, is isotopically light ($\delta^{13}C_1$ values of –54‰ to –58‰) and depleted in deuterium (δD_1 values of –307‰ to –354‰). In addition, the fact that gas in sandstone reservoirs is nearly identical in composition to that in the coalbeds is evidence that both have been generated from a common source, the coalbeds. Based on chemical and isotopic compositions and low rank of the coal, the gases are interpreted to be bacterial in origin and to have been generated by the *fermentation* pathway, whereby methyl groups in the coal are converted to methane (Whiticar and others, 1986).

The other bacterial-gas pathway is *carbon dioxide reduction*, which results in mostly methane that is even lighter isotopically ($\delta^{13}C_1$ values lighter than –60‰) and enriched in deuterium (δD_1 values of –150‰ to 250‰) (Whiticar and others, 1986). Gases of thermogenic origin can be distinguished from bacterial gases by their heavier isotopic composition and larger proportion of heavier hydrocarbons (Rice and Claypool, 1981). Although fermentation-derived methane can overlap in carbon-isotope composition with thermogenic gases, it is depleted in deuterium relative to thermogenic methane.

Gas derived from low-rank coal in the Powder River basin is the only documented occurrence of ancient bacterial gas generated by the fermentation pathway, as opposed to all other reported ancient occurrences, which are the result of carbon dioxide reduction (Schoell, 1988). Schoell (1988) speculated that burial depth and age of sediment are the main factors regulating the relative significance of these two pathways. That is, young, fresh, near-surface (tens of feet) sediments can generate bacterial gas by either pathway, but most of this early-formed gas probably is lost to the atmosphere and usually does not represent a major component of ancient accumulations. In contrast, older, altered, more deeply buried (hundreds of feet) sediments generate bacterial gas mainly by carbon dioxide reduction, which is more likely to become trapped and thus form a major component of ancient accumulations.

Five factors are considered critical to the entrapment of early-formed methane-rich gas by the fermentation pathway. First, coal is characterized by relatively low matrix porosity and permeability (Jones and others, 1988), compared to most other rock types. Second, gas is stored or retained within the coal mostly by adsorption or molecular attraction onto the large internal surface areas characteristic of coal (Van Krevelen, 1981). Third, the Powder River basin was characterized by high rates of subsidence during buildup of the thick coal deposits, which for beds as thick as 200 ft (61 m), originally were as thick as 720 ft (220 m) (Kent, 1986) before dewatering and compaction. These rapid rates of subsidence, accompanied by rapid siliciclastic deposition, are believed to have continued after the time of peat accumulation and therefore were critical in inhibiting the loss of early-formed gas. Fourth, the coalbeds are commonly interbedded with shale, which, when dewatered and compacted, formed effective seals. Finally, the development of compaction folds (described below) provided the primary trapping mechanism in several fields, including those where sandstone is the reservoir and in Rawhide Butte field where coal is the reservoir.

COAL CLEAT CHARACTERISTICS

Although a large proportion of gas in coal is adsorbed, substantial recoverable gas is believed to occur as free gas within the cleat system. Cleats also provide permeability pathways for movement of gas and water to the wellbore. From observations at the Eagle Butte Mine, located a short distance east of Rawhide Butte field (Figure 4), face cleats in the Wyodak-Anderson coalbed (0.28% to 0.30% R_m) are normal to bedding, strike east-northeast, and have an average spacing of 3.3 in. (8.4 cm) in the lower bench and 4.9 in. (12.5 cm) in the upper bench. In contrast, producing coalbeds in the San Juan basin (>0.6% R_m) have a cleat spacing of 1.4 in. (3.6 cm) or less. Cleat spacing is highly variable and seems to be dependent, in part, on coal composition—more closely spaced cleats are associated with bright banded coal, whereas wider spaced cleats are associated with dull banded coal and shaly coal bands. Henkle and others (1978) determined that the dominant (face) cleat direction in the Wyodak-Anderson coal near Gillette ranges from N35°W to N85°W with subordinate (butt) cleat directions ranging from N25°E to N35°E. They concluded that the coal cleat orientations were the same as for joints in overlying noncoal lithologies. Measured cleat directions in Wyodak-Anderson coal at the Wyodak Mine, about 6 miles (10 km) east of Gillette, range from north to east-southeast (Glass, 1975). Neither of these studies addressed cleat spacing. Because stress associated with compaction folds probably has developed locally, the directions of face and butt cleats, as well as cleat spacing, are anticipated to vary widely over small areas.

areas produce large volumes of water with little or no gas. The structural configuration (Figure 4) at the base of the Wyodak-Anderson coal, in the vicinity of Gillette, is surprisingly complex. Several small folds and "wants" are present, yet structure on the Muddy Sandstone in this area shows a west-southwest, gently dipping, homoclinal slope (Law, 1976), typical of the eastern flank of the basin. Law (1976) concluded that structural features associated with coalbeds in the Powder River basin are penecontemporaneous compaction structures that formed in response to facies changes associated with specific depositional environments. Such compaction structures also have been recognized by Choate and others (1984) and Warwick and Stanton (1988) and most likely occur throughout the basin. Compaction structures are mappable on the surface where they occur at shallow depths, and some can even be identified on airphotos.

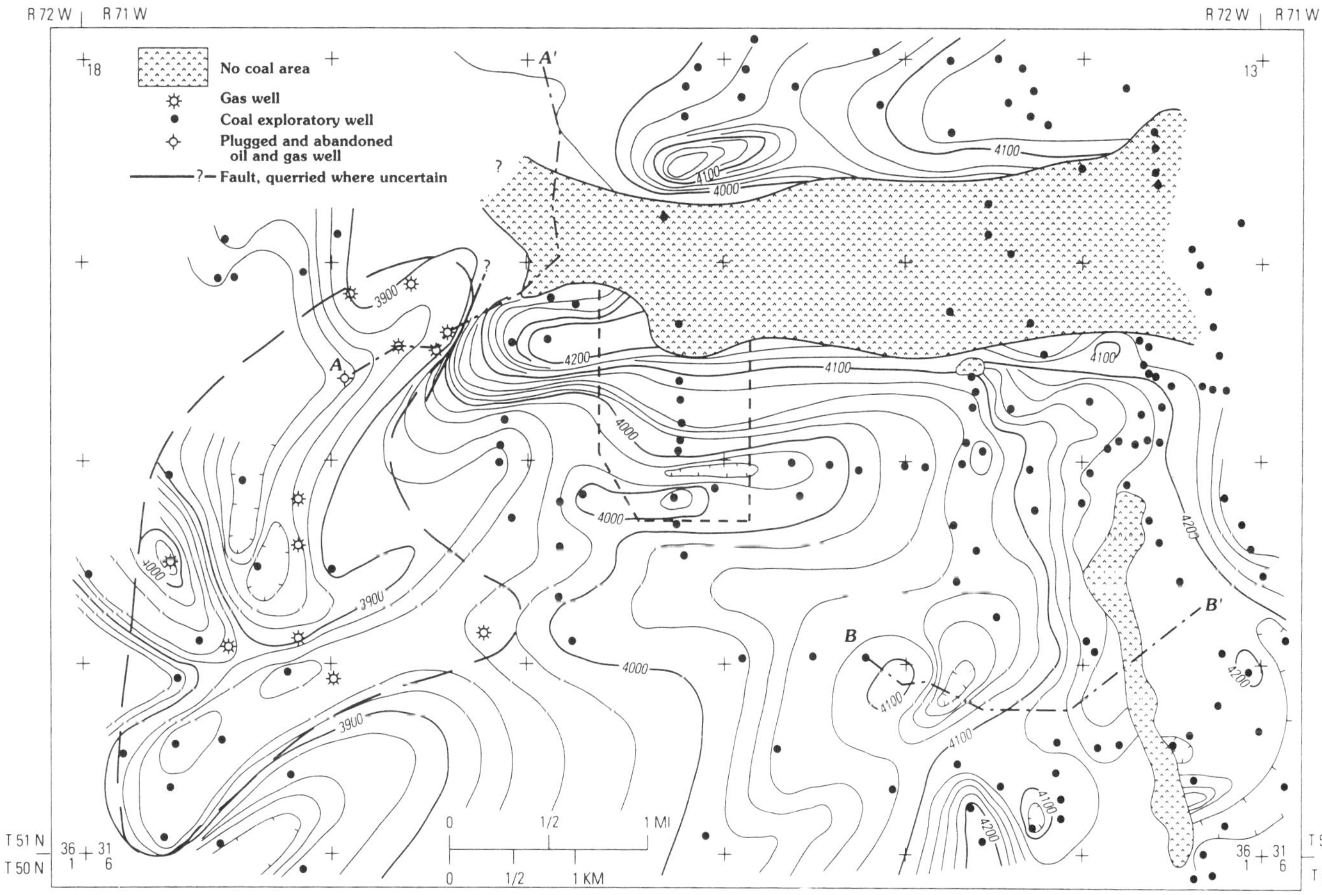

Figure 4. Structure contour map drawn at base of the Wyodak-Anderson coal. Contour interval 20 ft; structural datum, mean sea level. Approximate boundaries of Rawhide Butte field and Eagle Butte Mine are delineated by long and short dashed lines, respectively. Location of map area shown in Figure 1. Cross sections A–A' and B–B' are shown on Figure 5.

COMPACTION STRUCTURES AND GAS TRAPS

Gas production in Rawhide Butte field is associated with structurally deformed areas. Wells located on or near structurally high features often are gas productive with little or no water, whereas wells located in structurally low

Compaction structures, some of which exhibit as much as 250 ft (75 m) of structural relief, began to form shortly after peat accumulation as a consequence of contrasts in compaction potential between different lithologies. Examples of these structures associated with the Wyodak-Anderson coal in the Gillette area (Figures 4 and 5) are of three types—(1) one in which both coal benches form an anticline; (2) one in which the lower coal bench undergoes a facies change into noncoal lithology and the upper bench arches over the noncoal lithologic equivalent of the lower coal; and (3) one in which both coal benches undergo facies changes into noncoal lithologies. Because compaction structures are penecontemporaneous with deposition, the thickness and quality (that is, shaliness or ash content) of coalbeds associated with the structures also are affected. In many cases coalbeds thin or undergo facies changes into noncoal lithologies within the structurally deformed areas, such as the thinning that occurs in the upper bench of the Wyodak-Anderson coal, as shown in cross section A-A' of Figure 5.

terminating in relatively undeformed overlying and underlying strata.

Within the complexity of these compaction structures, five potential sites may evolve for trapping hydrocarbons. These trap types all assume the presence of an effective seal, such as carbonaceous shale or mudstone. The first type (Figure 6a) is a structural trap wherein gas, because of its buoyancy, has migrated within a laterally continuous coalbed to the crest of an anticline. Some of the anticlinal folds in Rawhide Butte field are believed to be of this type. The potentially hazardous occurrence of coal-derived gas inside homes in the Rawhide Village residential area (Jones and Taucher, 1989), located between Rawhide Butte field and the Eagle Butte Mine (secs. 20 and 21, T51N R71W,

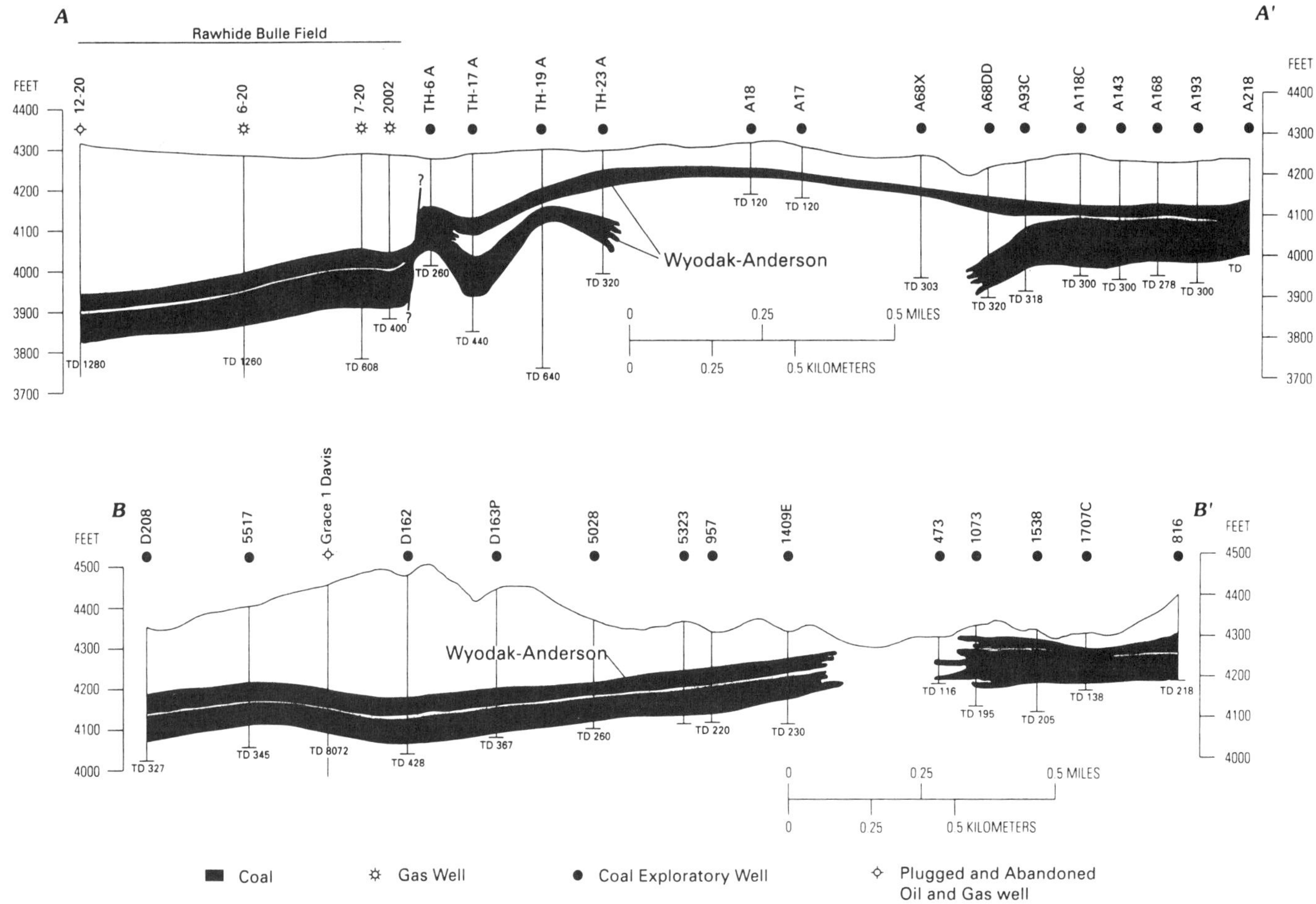

Figure 5. Cross sections A–A' and B–B', showing compaction structures associated with the Wyodak-Anderson coal. Location of sections shown on Figure 4.

Additional structural features that may be associated with compaction folds are faults and stress fractures. Along the edges of the large east-west-oriented want area (Figure 4), for example, Law (1976) described *compaction faults* associated with the folds. We suggest that these faults develop near the edges of underlying, relatively uncompactable lithologies in response to high stress. Within the high-stress areas, fractures also may develop, thereby enhancing permeability of the coal reservoir. Compaction faults only affect the more severely deformed rocks,

Figure 4), may, in part, be attributable to location of the subdivision near the crest of one such prominent anticline. Stress fractures on the steeply dipping southern and western flanks of the anticline, together with the inferred fault on the west, may have provided pathways for gas to migrate out of the coal and into overlying rocks and/or to the surface.

The second type is a structural-stratigraphic trap (Figure 6b) in which gas may accumulate near the updip boundary of a facies change between coal and a noncoal

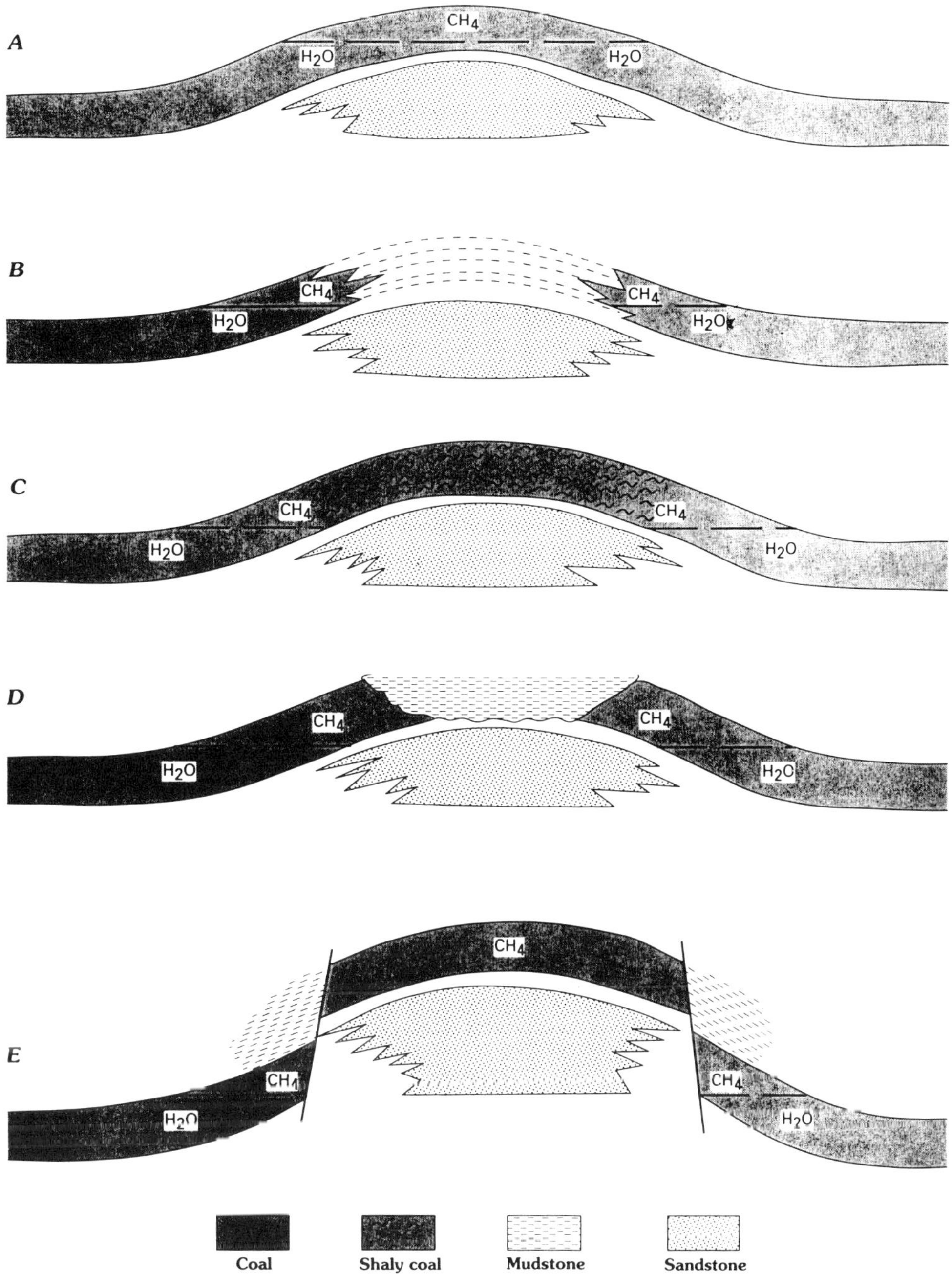

Figure 6. Diagrammatic sections of gas traps that may be associated with compaction folding.

lithologic seal. Traps of this type may occur along the flank of an anticline, such as shown in cross-section B-B' (Figure 6b). The third type is also a structural-stratigraphic trap but is different from the second type in that the trap occurs within a coal that has undergone a coal quality facies change into a shaly coal (Figure 6c). This type of trap would most likely occur on the flanks of compaction structures. A fourth type is a structural-truncation trap (Figure 6d) in which a coal, located on or near the crest of an anticline has been eroded and sealed by a mudstone or shale. This type of trap has not been documented but is one that could occur within a fluvial depositional system like that in the Tongue River Member. Finally, a fifth type, referred to as a compaction-fault trap (Figure 6e), may provide a gas trap. In this type of trap, gas may accumulate on the downthrown side of compaction faults where the faulting has juxtaposed a mudstone or shale against an updip, fault-truncated coal. Gas may also accumulate in the structurally higher, upthrown side of faults. Part of the gas production in the Rawhide Butte field may come from this type of trap, as shown in cross-section A-A' (Figure 5). However, the potential for gas production along this line of section quite likely extends to the upthrown side of the fault, but because no production tests were conducted in wells located on the upthrown side of the fault, there remains some uncertainty.

The entrapment of early-formed bacterial gas requires the existence of an early-formed trap. Otherwise, a large portion of the bacterial gas will be lost to the atmosphere. Since compaction folds begin to form shortly after deposition, they provide optimal sites for the entrapment of bacterial gas. Indeed, the identification of bacterial gas derived from the fermentation process lends some support to the early formation and entrapment of gas in the Powder River basin.

HYDROLOGY

One of the largest obstacles to economic recovery of coalbed gas is water. The presence of water in coal inhibits desorption of gas from the coal and flow to the wellbore. Consequently, it commonly is necessary to dewater coalbeds to a point at which gas begins to desorb from the matrix. The period of time necessary to accomplish sufficient dewatering is highly variable and in some cases is never reached. The success of dewatering efforts depends on development of the cleat system and on the source of water, which may be from recharge at the outcrop, from adjacent aquifers, or from inherent moisture in the coal.

Coalbeds contain large amounts of inherent moisture. Volumetrically, peat contains a large amount of water, possibly as much as 80% to 90% (Allardice and Evans, 1978). As coalification proceeds, moisture content decreases, and at coal ranks of high-volatile A bituminous (0.8% to 1.0% R_m) moisture content is less than 5%. Coals also are good aquifers; although matrix permeability is low, the cleat system often provides an excellent pathway for water, providing that the cleats remain open.

Potentiometric mapping by Daddow (1986) indicates that recharge of the Wyodak-Anderson coal occurs along its eastern outcrop and that regional flow is toward the northwest (Martin and others, 1988). Based on the isotopic composition of water samples from the Wyodak-Anderson coal, Martin and others (1988) concluded that at least some of the water near outcrops was of meteoric origin. Some coalbeds associated with larger compaction structures near the eastern side of the basin have sufficient structural relief to be above the potentiometric surface, facilitating the accumulation and production of gas. In addition, the water table in coalbeds near active open-pit coal mining may be lowered. Maps of water-level change indicate cones of depression near active mines. According to Martin and others (1988), annual measurements of ground-water levels in the Wyodak-Anderson coal, since 1980, have declined by as much as 80 ft (24 m) in wells near active mines. At distances of 1 to 3 miles (1.6 to 4.8 km) from mining, water levels have declined less than 10 ft (3 m) and at distances greater than 3 miles (4.8 km), water levels have been unaffected.

In the central and western basin, where thick Fort Union coals occur at depths greater than 1,200 ft (370 m), the coals produce large volumes of water—as much as 4,000 BPD (636 m^3/day). Water produced from coal in the deeper part of the basin in the Sasquatch No. 3 well (Figure 1) is mainly a sodium bicarbonate type, which based on Decker and others' (1987) interpretations in the San Juan basin, is the result of coalification. In contrast, waters in overlying Wasatch aquifers are distinctly different; they have a mixed character (calcium, magnesium, bicarbonate, and sulfate) and are quite different from coal-derived water. These data, in conjunction with the long distance from outcrops and the discontinuous nature of coalbeds, may indicate that the source of water in these deeply buried coalbeds is inherent moisture in the coalbeds. Therefore, dewatering efforts in the deeper part of the basin should be possible. However, because of the great thickness of individual coalbeds and the large drainage area, dewatering may take a long time to accomplish. Closely spaced drilling on compaction or tectonic structures in the deeper part of the basin may expedite the dewatering process and thus improve the chances of encountering economically recoverable gas accumulations.

SUMMARY

Gas generated in thick, low-rank coalbeds of the Fort Union Formation in the Powder River basin is of bacterial origin. Although the volume of gas generated by bacterial degradation of organic matter is small compared to that generated by thermal degradation, these thick coalbeds nevertheless have given rise to a large coal-gas resource. Because of the low volume of gas generated by bacterial processes, economically recoverable gas accumulations in the Powder River basin have, to date, required structural or stratigraphic traps with effective seals, or conditions under which gas can accumulate above the regional water table. The area that best meets these requirements occurs at shallow depths along the eastern edge of outcropping coalbeds. The development of early-formed compaction structures associated with coalbeds has provided optimal conditions for entrapment of early-formed bacterial gas. In deeper areas of the basin, gas is present; however, the large amount of water encountered in the thick, low-rank, high-moisture coals may render dewatering uneconomical. Insufficient data presently are available to determine the feasibility of gas production from coals in the deeper part of the basin.

ACKNOWLEDGMENTS

We are grateful to the following individuals for samples, subsurface data, and discussions related to coalbed gas—Chuck Peck, Martens and Peck Operating Company, Denver, Colorado; Owen Stone, Betop Inc., Carrollton, Texas; Bruce Taylor, AMAX Coal Company, Gillette, Wyoming; Dan Coolidge, DCD Inc., Gillette, Wyoming; Bill Siruta, Gilmore Oil and Gas Exploration, Midland, Texas; Jerry Cuzella, NCRA, Denver, Colorado; and Richard Jones, Geological Survey of Wyoming, Laramie, Wyoming.

REFERENCES

Allardice, D.J., and Evans, D.G., 1978, Moisture in coal, *in* Karr, C., Jr., ed., Analytical methods for coal and coal products: New York, Academic Press, v. 1, p. 247–261.

Ayers, W.B., Jr., 1986, Lacustrine and fluvial-deltaic depositional systems, Fort Union Formation (Paleocene), Powder River basin, Wyoming and Montana: AAPG Bulletin, v. 70, no. 11, p. 1651–1673.

Boreck, D.L., and Weaver, J.N., 1984, Coalbed methane study of the Anderson coal deposit, Johnson County, Wyoming—A preliminary report: USGS Open-File Rept. 84–831, 16 p.

Brown, J.L., (in press), Sedimentology and depositional history of the Paleocene Tullock Member of the Fort Union Formation, Powder River basin, Wyoming and Montana: USGS Bulletin 1917.

Choate, R., Johnson, C.A., and McCord, J.P., 1984, Geologic overview, coal deposits, and potential for methane recovery from coalbeds—Powder River basin, *in* Rightmire, C.T., Eddy, G.E., and Kirr, J.N., eds., Coalbed methane resources of the United States: AAPG Studies in Geology, Series 17, p. 335–351.

Daddow, P.B., 1986, Potentiometric surface map of the Wyodak-Anderson coal bed, Powder River structural basin, Wyoming, 1973–84: USGS Water Resources Investigations Rept. 85–4305.

Decker, D., Klusman, R., and Horner, D.M., 1987, Geochemical techniques applied to the identification and disposal of connate coal water, *in* The 1987 Coalbed Methane Symposium Proceedings—University of Alabama, Tuscaloosa, November 16–19, 1987: School of Mines and Energy Development, Gas Research Institute, and U.S. Mine Safety and Health Administration, p. 229–242.

Flores, R.M., 1986, Styles of coal deposition in Tertiary alluvial deposits, Powder River basin, Montana and Wyoming, *in* Lyons, P.C., and Rice, C.L., eds., Paleoenvironmental and tectonic controls in coal-forming basins in the United States: GSA Special Paper 210, p. 79–104.

Flores, R.M., and Ethridge, F.G., 1985, Evolution of intermontane fluvial systems of Tertiary Powder River basin, Montana and Wyoming, *in* Flores, R.M., and Kaplan, S.S., eds., Cenozoic paleogeography of west-central United States: SEPM Rocky Mountain Section, Rocky Mountain Paleogeography Symposium 3, p. 107–126.

Glass, G.B., 1975, Analyses and measured sections of 54 Wyoming coal samples (collected in 1974): Geological Survey of Wyoming Rept. of Investigations 11, 219 p.

Glass, G.B., 1989, Seam analyses and description of coal fields–Wyoming, *in* 1989 Keystone coal industry manual: Chicago, McClean Hunter Publications Co., p. 611–636.

Henkle, W.R., Jr., Muhm, J.R., and DeBuyl, M.H.F., 1978, Cleat orientation in some subbituminous coals of the Powder River and Hanna basins, Wyoming, *in* Hodgson, H.E., ed., Proceedings of the Second Symposium on the Geology of Rocky Mountain Coal—1977: Colorado Geological Survey Resource Series 4, p. 129–141.

Hobbs, R.G., 1978, Methane occurrences, hazards, and potential resources—Recluse geologic analysis area, northern Campbell County, Wyoming: USGS Open-File Rept. 78–401, 20 p.

Jones, A.H., Bell, G.J., and Schraufnagel, R.A., 1988, A review of the physical and mechanical properties of coal with implications for coalbed methane well completion and production, *in* Fassett, J.E. ed., Geology and coal-bed methane resources of the northern San Juan basin, Colorado and New Mexico: Rocky Mountain Association of Geologists Guidebook, p.169–181.

Jones, R.W., and Taucher, P.J., 1989, Coal geology, geophysical logs, and lithologic descriptions from a drilling program at the Rawhide Village subdivision, Campbell County, Wyoming: Geological Survey of Wyoming Open-File Rept. 89–2, 59 p.

Joubert, J.I., Grein, C.T., and Bienstock, Daniel, 1973, Sorption of methane in moist coal: Fuel, v. 52, no. 4, p. 181–185.

Jüntgen, H., and Karweil, Joachim, 1966a, Gasbildung und Gasspeicherung in Steinkohlenflözen, part I—Gasbildung: Erdol und Kohle, Erdgas, Petrochemie, v. 19, p. 251–258.

Jüntgen, H., and Karweil, Joachim, 1966b, Gasbildung und Gasspeicherung in Steinkohlenflözen, part II—Gasspeicherung: Erdol und Kohle, Erdgas, Petrochemie, v. 19, p. 339–344.

Jüntgen, H., and Klein, J., 1975, Entstehung von Erdgas aus Kohligen Sedimenten: Erdol and Kohle, Erdgas, Petrochemie, v. 28, no.2, p. 65–73.

Kaiser, W.R., and Swartz, T.E., 1988, Hydrology of the Fruitland Formation and coalbed methane producibility, *in* Ayers, W.B., Jr., Kaiser, W.R., Swartz, T.E., Zellers, S.D., and Scanlon, A.H., Geologic evaluation of critical production parameters for coalbed methane resources, part 1—San Juan basin: Gas Research Institute, Annual Rept. GRI–88/0332.1 (August 1987–July 1988), p. 61–81.

Karweil, J., 1969, Aktuelle Probleme der Geochemie der Kohle, *in* Schenk, P.A., and Havernaar, I., eds., Advances in organic geochemistry, 1968: New York, Pergamon Press, p. 59–84.

Kelso, B.S., Wicks, D.E., and Kuuskraa, V.A., 1988, A geologic assessment of natural gas from coal seams in the Fruitland Formation, San Juan basin: Gas Research Institute, Topical Rept. GRI–88/034 [Final Geologic Report (March 1988)], 56 p.

Kent, B.H., 1986, Evolution of thick coal deposits in the Powder River basin, northeastern Wyoming, *in* Lyons, P.C., and Rice, C.L., eds., Paleoenvironmental and tectonic controls in coal-forming basins in the United States: GSA Special Paper 210, p. 105–122.

Kim, A.G., 1977, Estimating methane content of bituminous coalbeds from adsorption data: U.S. Bureau of Mines Rept. of Investigations 8245, 22 p.

Law, B.E., 1976, Large-scale compaction structures in the coal-bearing Fort Union and Wasatch Formations, northeast Powder River basin, Wyoming, *in* Landon, R.B., Curry, W.H., III, and Runge, J.S., eds., Geology and energy resources of the Powder River: Wyoming Geological Association, 28th Annual Field Conference Guidebook, p. 221–229.

Law, B.E., Anders, D.E., and Michael, G.E., 1990, Use of Rock-Eval pyrolysis and vitrinite reflectance data in characterizing type and maturity of organic matter in coal, Upper Cretaceous Fruitland Formation, San Juan basin, New Mexico and Colorado (abs.): AAPG Bulletin, v. 74, no. 8, p. 1333.

Love, J.D., 1989, Geology of the Powder River basin, northeastern Wyoming and southeastern Montana, *in* Barrs, D.L., and others, eds., Basins of the Rocky Mountain region: GSA, Geology of North America, v. D–2, p. 204–217.

Martin, L.J., Naftz, D.L., Lowham, H.W., and Rankl, J.C., 1988, Cumulative potential hydrologic impacts of surface mining in the eastern Powder River structural basin, northeastern Wyoming: USGS Water-Resources Investigations Rept. 88–4046, 201 p.

Meissner, F.F., 1984, Cretaceous and lower Tertiary coals as sources for gas accumulations in the Rocky Mountain area, *in* Woodward, Jane, Meissner, F.F., and Clayton, J.L., eds., Hydrocarbon source rocks of the greater Rocky Mountain region: Rocky Mountain Association of Geologists Symposium Guidebook, p. 401–431.

Moore, T.A., (in press), The effects of clastic sedimentation on organic facies development within a Tertiary subbituminous coalbed, Powder River basin, Montana, U.S.A.: International Journal of Coal Geology.

Moore, T.A., Stanton, R.W., Pocknall, D.T., and Flores, R.M., 1990, Maceral and palynomorph facies from two Tertiary peat-forming environments in the Powder River basin, U.S.A.: International Journal of Coal Geology, v. 15, no. 4, p. 293–316.

Nichols, D.J., Wolfe, J.A., and Pocknall, D.T., 1989, Latest Cretaceous and early Tertiary history of vegetation in the Powder River basin, Montana and Wyoming, *in* Tertiary and Cretaceous coals in the Rocky Mountains: American Geophysical Union, Field Trip Guidebook T132, p. 28–33.

Nuccio, V.F., 1990, Determination of source-rock thermal maturity by direct measurements and predictive modeling–Application to Upper Cretaceous Cody Shale, Powder River basin, Wyoming, *in* Nuccio, V.F., and Barker, C.E., eds., Application of thermal maturity studies to energy exploration: SEPM Rocky Mountain Section, p. 167–175.

Olive, W.W., 1957, The Spotted Horse coalfield, Sheridan and Campbell Counties, Wyoming: USGS Bulletin 1050, 83 p.

Rice, D.D., and Claypool, G.E., 1981, Generation, accumulation, and resource potential of biogenic gas: AAPG Bulletin, v. 65, no. 1, p. 5–25.

Schoell, Martin, 1988, Multiple origins of methane in the Earth, *in* Schoell, M., ed., Origin of methane in the earth: Chemical Geology, v. 71, nos. 1–3, p. 1–10.

Stach, E., Mackowsky, M.T., Taylor, G.H., Chandra, D., Teichmüller, M., and Teichmüller, R., 1982, Stach's text book of coal petrology, 3d ed.: Stuttgart, Gebrüder Borntraeger, 535 p.

Van Krevelen, D.W., 1981, Coal—Typology, chemistry, physics, constitution, 2d printing: Amsterdam, Elsevier, 514 p.

Warwick, P.D., and Stanton, R.W., 1988, Depositional models for two Tertiary coal-bearing sequences in the Powder River basin, U.S.A.: Journal of the Geological Society of London, v. 145, p. 613–620.

Whipkey, C.E., Cavaroc, V.V., and Flores, R.M., (in press), Uplift of the Bighorn Mountains–A sandstone provenance study: USGS Bulletin 1917–D.

Whiticar, M.J., Faber, E., and Schoell, M., 1986, Biogenic methane formation in marine and freshwater environments—CO_2 reduction *vs.* acetate fermentation, isotope evidence: Geochimica et Cosmochimica Acta, v. 50, no. 5, p. 693–709.

Evaluation of Coalbed Methane Potential of Recluse Muddy Field, Campbell County, Wyoming

Jill D. Henderson[1]

1. Apache Corporation, Denver, Colorado

ABSTRACT

In the Recluse Muddy field (T56N R74W) of Campbell County, Wyoming, the Tongue River Member of the Fort Union Formation consists of thick sequences of subbituminous coal, sandstone, siltstone, and shale from the surface to a depth of 1,000 ft (305 m). These units were correlated through nearby Oedekoven, Ruff Draw, and Chan fields, located in T55&56N R73W, which produce water-free gas from sandstones at depths of 300 to 600 ft (92 to 183 m). Stratigraphic correlations of coals were made with available well logs, which consist of induction logs with a spontaneous potential curve and an occasional porosity log with a gamma ray curve. Correlation and recognition of potential reservoir units were complicated by the fact that much of the coal-sandstone sequence lies behind surface casing in many old wells and therefore was not logged.

The comparison of log responses of various lithologies provided a qualitative method for evaluating productive potential. For example, resistivity values across productive sandstones in nearby fields are, in most cases, significantly higher than those across coalbeds or nonproductive sandstone intervals. A maximum resistivity for water-filled sands in this area was calculated by the formula $R_o = (1/\phi^{2.00}) R_w$, using actual values from nearby wells.

Other conventional methods of prospect generation were used to evaluate the field, including structural mapping of several coalbeds and isopaching of strata between coals.

During 1989 and 1990, cased-hole logs were run in a number of old wellbores in the study area by various operators in an effort to identify gas-bearing zones. An unsuccessful recompletion was attempted by Apache Corporation in a depleted Muddy producer in section 15, T56N R74W, and dry holes were drilled by other operators in sections 16 and 26. Five Fort Union gas producers were completed and subsequently shut in, waiting on pipeline. One well in section 29 was perforated in four coalbeds, and the others in sections 25 and 36 were perforated in sandstones on the upthrown side of a normal fault. The evaluation of methane potential from shallow sandstones in T56N R74W by the analytical methods used here, and their results, are applicable to other similar depleted Muddy fields in this area of the Powder River basin.

INTRODUCTION AND OBJECTIVES

In the northern Powder River basin of Wyoming, the Tongue River Member of the Fort Union Formation contains thick sequences of fine- to medium-grained calcareous sandstone and siltstone, shaly siltstone, shale, and carbonaceous shale. These lithologies interfinger with and are interbedded with subbituminous coal (Choate and others, 1978). Several depositional models have been proposed for these sequences, which accumulated during rapid basin subsidence associated with the Laramide orogeny during the Paleocene. The models represent deposition in (1) a flood-plain environment dominated by fluvial channel facies and interchannel peat swamps with open sea to the east and northeast (Flores, 1980); (2) a fluvial-deltaic system adjacent to a large lake with swamps forming in an interdeltaic setting (Ayers, 1986); or (3) a lacustrine-swamp ecosystem analogous to the modern Okefenokee Swamp of the southeastern United States (McClurg, 1988). The Lebo Shale Member of the Fort Union, below the Tongue River Member, is believed to represent lacustrine deposition.

Whatever their origin, these sedimentary sequences recently have become the targets of methane exploration and production from depths above 1,000 ft (305 m) in northern Campbell County. Coals are assumed to be the methane source; because they also are aquifers, however, they require dewatering during the early stages of methane production, while the sandstones produce water-free gas.

The objective of this study was to apply conventional prospecting methods generally used in deeper plays to evaluate shallow coal-gas potential from sandstones in Recluse field, a depleted Muddy Formation oil and gas field operated by Apache Corporation, and in the surrounding area of T56N R74W.

DATA SOURCES

Well Logs

Evaluation of methane potential in the Recluse area is complicated by lack of modern well log data. Many wells here were drilled in the 1960s and 1970s by oil and gas exploration companies to test the Muddy Formation. In most of these older wells, induction and spontaneous potential logs were run from total depth to the base of surface casing, at depths of 300 to 800 ft (92 to 244 m); porosity and gamma ray logs were run only across deep zones of interest. Exceptions include occasional porosity or resistivity logs with gamma ray run to the base of surface casing or through casing to the ground surface.

During the course of this study, various operators evaluating shallow Fort Union Formation gas potential ran at least eight cased-hole logs in old wellbores to define gas-bearing zones. Most used Schlumberger's Thermal Decay Time (TDT[2]) cased-hole log, which analyzes the "decay of a burst of fast neutrons in the downhole environment" (Schlumberger, 1989) to recognize the presence of gas in formations behind pipe. Although the log is incapable of determining the quantity of gas present, it has been used successfully in old wellbores to identify gas zones that later were confirmed by crossover on FDC/CNL[3] open-hole logs run after the well was offset drilled. A recompletion for shallow methane was attempted in one well, and seven other shallow wells were drilled and logged in 1989 and 1990 (Table 1). Some information from these wells is held confidential by the operators.

Published data, including gas shows and gamma ray logs, also are available from shallow water wells drilled in the study area (Table 2). These wells were routinely checked for the presence of methane using a handheld detector (methanometer) as a safety precaution. The quantity of the gas was, however, not measured.

Recompletion Attempt

In July 1989 Apache Corporation attempted a recompletion in the Fort Union through 9 5/8-in. (24.4-cm) surface casing in the U.S. Fagerness No. 3 well (sec. 15, T56N R74W), a depleted Muddy producer. A cased-hole neutron density log showed crossover in sands below the Canyon coal at 456 to 500 ft (139 to 152 m) depth and Cook coal at 675 to 688 ft (206 to 210 m) depth. After perforating, swab tests recovered water, but no gas shows were reported. This may not have been a definitive test, however, because conditions behind the casing were poorly known. For this reason, old wells in this area usually are offset drilled or twinned rather than recompleted.

REGIONAL RELATIONSHIPS

Three Fort Union gas fields produce near the study area (Figure 1)—Oedekoven field (secs. 18, 19, and 30, T55N R73W), and Ruff Draw and Chan fields (secs. 33 and 34, T56N R73W, respectively). Oedekoven field produces from sandstones below the Anderson and Canyon coals, and Ruff Draw and Chan fields produce from below the Upper and Lower Cook coals (Figure 2). The coalbed terminology used in this paper was established by McKay (1973) and currently is used in completion reports and scout cards for wells in these fields. Producing intervals are recognized on open-hole GR-FDC-CNL logs on the basis of crossover of the neutron and density curves. In many cases this is the only log run; however, some resistivity logs are also available.

From analysis of logs in these fields, several conclusions can be offered that are applicable to the Recluse area.

1. *Correlation of coals.* Coalbeds were correlated regionally throughout the study area and into the producing fields (Figure 2; Kent and others, 1980). Their correlation is difficult not only because of lateral facies changes but also because of variation in the degree of compaction, which results in great variation in thickness of both the strata between the coalbeds and the coalbeds themselves. For example, in section 26 in the study area, the Upper and Lower Pawnee coalbeds can be as close to one another as 14 ft (4.3 m), but in section 9 an interval as thick as 112 ft (34 m) separates the two coals. Similarly, between sections 25 and 10, respectively, the interval between the Cook and Wall coals varies from 14 ft (4.3 m) to as much as 98 ft (30 m).

The coalbeds themselves also can split or coalesce. The Cook is a thick coalbed throughout the study area but splits into two thinner coalbeds (Upper and Lower Cook) in Ruff Draw and Chan fields.

2. *Understanding lithologies.* A better understanding of the various lithologies was gained by analyzing their log responses from wells in these fields (Table 3). Although sandstones appear "dirty" on the gamma ray log, for example, it is questionable whether this reflects actual clay plugging or the presence of potassium in the K-feldspar in arkosic sandstone.

3. *Comparison of resistivities.* Meaningful comparisons can be made between modern and older resistivity logs, particularly in Fort Union wells directly offsetting tests of the Muddy Formation. Results of these comparisons are applicable where resistivity logs are the only data sources. In most cases, the comparisons show that very high resistivity readings across Fort Union producing intervals also can be seen on older logs (Table 4).

An example from the study area is the induction log from the 36–8 State well, SE/4 NW/4 sec. 36, which showed a maximum resistivity of 80 ohm-m at the top of the sand below the Wall coal. An offset well, State 6–36, has 3 ft (0.9 m) of crossover on the FDC/CNL open-hole log in this interval and reported an IP of 138 MCFD (3,905 m^3/day).

4. *Resistivity of water-filled sands.* Fresh water in Fort Union sandstone reservoirs is believed to mask productive

---

2. TDT, Thermal Decay Time log (mark of Schlumberger).

3. FDC/CNL, Formation Density Log (Compensated) and Compensated Neutron Log (marks of Schlumberger).

Table 1. Summary of recent Fort Union Formation and Muddy Formation drilling activity in T56N R74W, Recluse field area, Campbell County, Wyoming.

Operator	Well Name	Location	Formation	Logs[1]	Interval (ft)	Interval (m)	Comments/Status
Hondo	Greenough Govt 9	NE SE sec. 8	Muddy	TDT	–	–	Tight hole; not redrilled yet.
Hondo	Greenough Govt 3	NW SE sec. 9	Muddy	TDT	–	–	Tight hole; not redrilled yet.
Hondo	Greenough Govt 5	NW SW sec. 9	Muddy	TDT	–	–	Tight hole; not redrilled yet.
Hondo	Greenough Govt 4	NW SW sec. 10	Muddy	TDT	–	–	Tight hole; not redrilled yet.
Hondo	Greenough Govt 6	SE SW sec. 10	Muddy	–	–	–	Tight hole; not redrilled yet.
Apache	U.S. Fagerness 3	NW SW sec. 15	Ft. Union	Cased hole Neutron density	504–540 674–688	154–165 206–210	Neutron-density crossover, but recompletion unsuccessful. P&A.
Apache	U.S. Fagerness 4	SE NW sec. 15	Ft. Union	TDT neutron	–	–	No potential. P&A.
Lowe	State 32–16	SW NE sec. 16	Ft. Union	FDC/CNL	–	–	No sand development. P&A.
Apache	U.S. Cochise 4	SE SW sec. 23	Ft. Union	Cased hole neutron	–	–	Log not definitive.
Hondo	1–A Burnett	NW NW sec. 25	Muddy	TDT	–	–	
Hondo	1–25 Tfu Burnett	NW NW sec. 25	Ft. Union	FDC/CNL	457–464	139–142	IP 600 MCFD from sandstone below Canyon coal; shut in.
Hondo	4 Burnett	NW SW sec. 25	Muddy	TDT	–	–	
Hondo	3–25 Tfu Burnett	NW SW sec. 25	Ft. Union	FDC/CNL	468–488	143–149	IP 500 MCFD from sandstone below Canyon coal; shut in.
Hondo	Apache 1–26	NE NE sec. 26	Ft. Union	FDC/CNL	–	–	Drilled in fault zone. P&A.
REI	Floodgate 1–29	SE SE sec. 29	Ft. Union	FDC/CNL	–	–	Perfed in coals, not sands.
Diasu	State 6–36	SE NW sec. 36	Ft. Union	FDC/CNL	701–703	214–215	IP 138 MCFD from sandstone below Wall coal; shut in.
Diasu	State 7–36	SW NE sec. 36	Ft. Union	Neutron density	374–378 382–386	114–115 116–118	IP 156 MCFD from sandstone below Canyon coal; shut in.

1. Wells drilled to the Muddy Formation had cased-hole TDT logs run over Fort Union interval only. TDT, Thermal Decay Time; FDC, Formation Density Log (Compensated); CNL, Compensated Neutron Log (marks of Schlumberger).

intervals. However, as shown in Table 4, the resistivity, R_t, of fresh water-bearing sands generally is lower than that of coals or producing sandstones. A maximum R_t can be calculated for water-filled sands in this area by the formula,

$$R_o = (1/\phi^{2.00})\ R_w$$

where R_o is resistivity of a water-saturated nonshaly formation rock, and ϕ is the average porosity of non-producing sandstones in nearby fields. Using actual values (ϕ of 40%, and R_w of 5.68 ohm-m at 68°F (20°C), obtained from the Floodgate 1–29 well (SE/4 SE/4 sec. 29) yields an R_o of 35 ohm-m, the maximum R_t of a water-filled sand in the study area. Therefore, any R_t greater than this value could indicate gas. Of course, R_o would decrease with a decrease in R_w or an increase in porosity. Although an estimation, this technique could be a useful tool for identifying productive sandstones in old wellbores, which could warrant offset drilling. For example, a dual-induction laterolog run in the 4 Burnett well (NW/4 NW/4 sec. 25) showed a maximum resistivity of 40 ohm-m in sandstone below the Canyon coalbed. The well was offset by the 3–25

Table 2. Summary of methane occurrences in unnamed water wells drilled in T56N R74W, Recluse field area, Campbell County, Wyoming. Data summarized from Cathcart and others (1983).

Location	Total Depth (ft)	(m)	Comments
NE NW sec. 1	244.0	74.4	Water level surging with methane discharge.
SE NW sec. 13	90.0	27.4	No methane detected.
SE NW sec. 13	56.5	17.2	No methane detected.
NW SW sec. 17	320.0	97.6	Water level surging with methane discharge.
NE NW sec. 23	147.0	44.8	No methane detected.
SW SE sec. 26	228.0	69.5	No methane detected.
NW SE sec. 30	201.0	61.3	No methane detected.
NE SE sec. 31	295.0	89.9	No methane detected.

Tfu Burnett well, which had 28 ft (8.5 m) of crossover on the FDC/CNL open-hole log in this interval and reported an IP of 500 MCFD (14,150 m^3/day).

High-resistivity sands also occur lower in the section. For example, in the Greenough Government No. 3 well (NW/4 SE/4 sec. 9), a sandstone between the Upper and Lower Pawnee coals has a resistivity of 100 ohm-m (deep induction) between 870 and 880 ft (265 and 268 m). A TDT log run in this well showed the presence of gas in this interval, which is considered potentially productive.

Structure and Isopach Mapping and Interpretation

A structure map on the top of the Lower Pawnee coal (Figure 3) is presented to show a general representation of regional structure at shallow depths. The Lower Pawnee coal is one of the lowest coals in the stratigraphic section and was chosen because it lies at a level below the surface casing in most wells. The known logged depth of this coalbed varies from 420 to 1,010 ft (128 to 308 m) below ground surface.

A northwest-dipping normal fault strikes N60°E in the southeastern corner of the map area. In May 1990, a Fort Union well drilled into the fault zone in NE/4 NE/4 section 26. When the log was correlated with those from surrounding wells, it was seen that the Cook and Wall coals, as well as the sandstone productive in the Chan field, are faulted out. By constructing an isopach of the interval

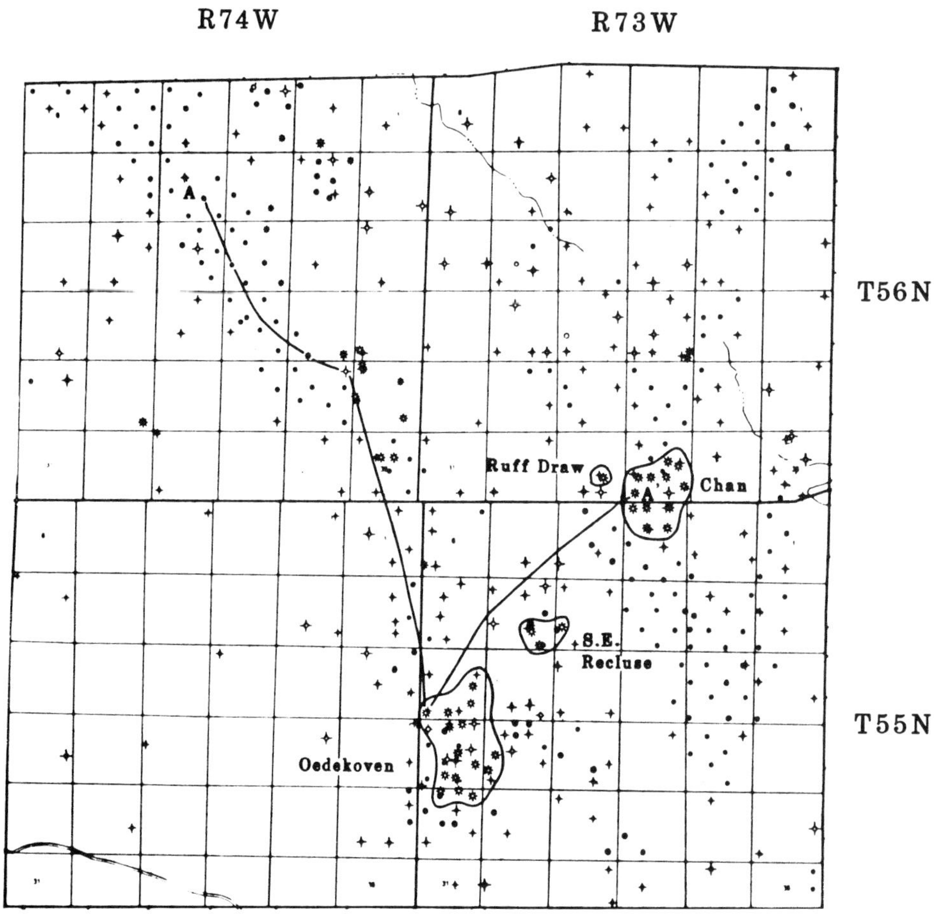

Figure 1. Well location map of Recluse area and nearby fields, Campbell County, Wyoming. See Figure 2 for logs and correlation of Fort Union Formation coal seams along line of section A–A'.

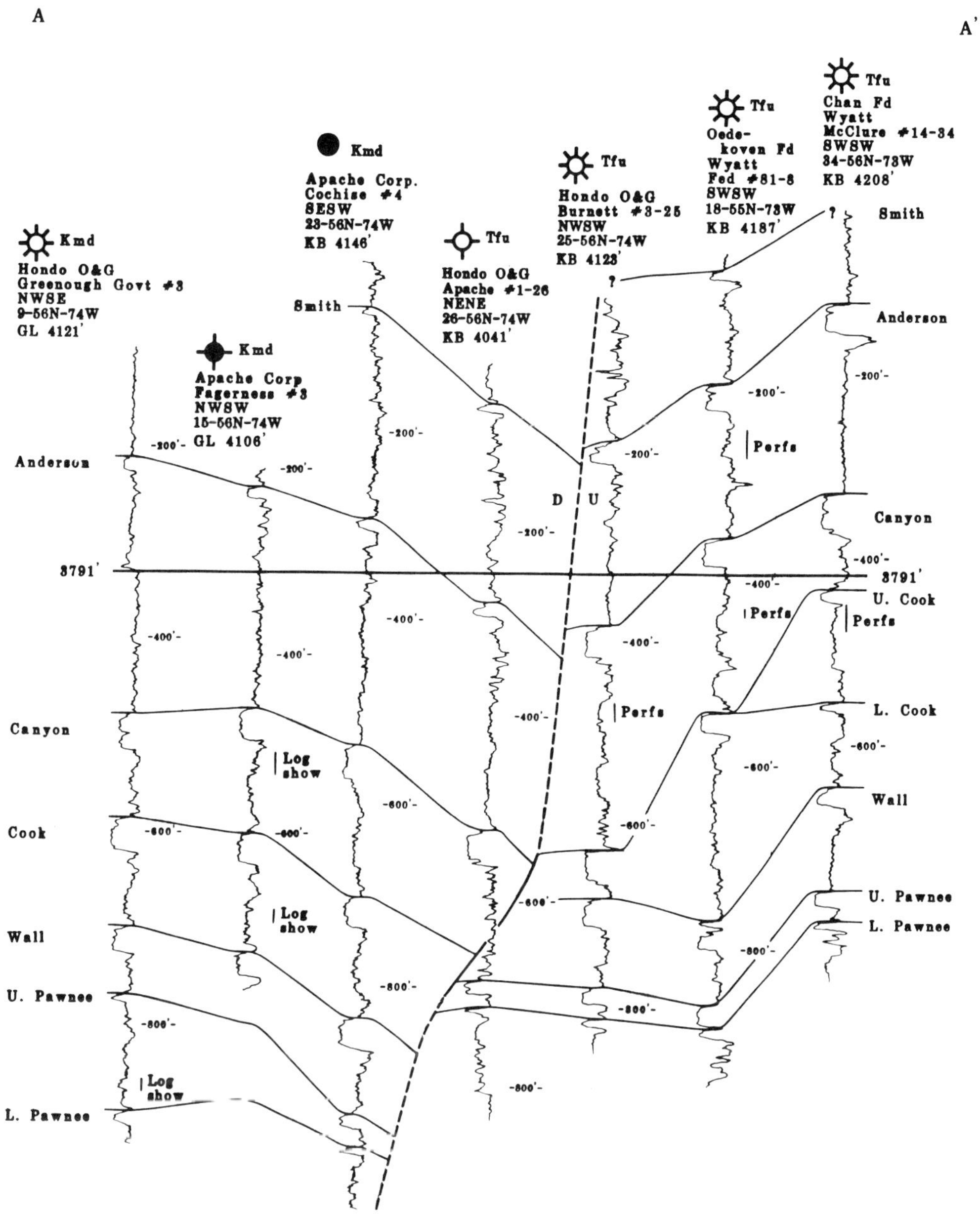

Figure 2. Cross-section showing gamma ray correlation of Fort Union Formation coal seams from wells in Recluse study area with analog fields. See Figure 1 for location of line of section A–A'.

between the base of the Canyon coal and the top of the Upper Pawnee coal, it was determined that approximately 230 ft (70 m) of section is missing at this location. Although other well logs in the area were examined to total depths greater than 7,000 ft (2,135 m), the presence of the fault was not noted.

Olive (1957) described an exposure of the Anderson coal in the footwall of this fault and indicated it on a structure map of the Smith coal, which lies above the Anderson. Kent and Berlage(1980) also mapped the fault on a surface geologic map of the Recluse area.

A second structure map (Figure 4) was drawn on top of the Cook coal, which produces in Chan field. Much of the data are unavailable because the Cook lies behind surface casing in many wells and was not logged. In such cases, values were interpolated from an isopach of the interval between the top of the Cook and the top of the Lower Pawnee coals. The known logged depth of the Cook coal throughout the study area varies from 288 to 715 ft below ground surface.

An isopach map of the gross sandstone interval between the Cook and Wall coalbeds is shown in Figure 5. Initially, an attempt was made to quantify net sand (minus siltstone) using a resistivity cutoff in conjunction with the spontaneous potential and gamma ray, if available. However, cutoffs appear to be inconsistent unless the gamma ray

Table 3. Principal log characteristics of lithologies (after White Dog Exploration Company).

Lithology	Gamma Ray	Resistivity	Spontaneous Potential*	Neutron Porosity	Density Porosity
Sandstone	Low	High	Negative	High	High
Producing sandstone	Low	Highest	Negative	Low	High
Siltstone	Moderate	Moderate	Negative	Moderate	Moderate
Shale	Very high	Very low	No deflection	High	Low
Coal	Very low	Very high	Negative	Very high	Very high

* If salinity of mud filtrate exceeds that of formation water, SP deflection is positive (to the right).

Table 4. Formation resistivities of Fort Union coal and sandstones in Oedekoven, Ruff Draw, and Chan fields, Campbell County, Wyoming. Original Muddy Formation wells are paired with more recent Fort Union "twins." (NR = Not run)

Well Name & Location	Objective	Coal		Producing Sand or Equivalent		Nonproducing Sand	
		GR (API)	R_t, deep (ohm-m)	GR (API)	R_t, deep (ohm-m)	GR (API)	R_t, deep (ohm-m)
Federal 81–8	Ft. Union	10–20	50–60	80–120	25–60	80–125	15–50
Oedekoven 1–D	Muddy	20–40	60	60–80	28–34	60–80	24–48
SW SW 18, T55N R73W							
Pal 82–C	Ft. Union	10–20	60	70–100	100–210	60–80	30–40
Oedekoven 5X–C	Muddy	NR	50	NR	35–120	NR	20–40
NE NW 19, T55N R73W							
Oedekoven 83–1	Ft. Union	10	50–60	70–110	23–130	60–100	20–40
Oedekoven B–2	Muddy	25–40	50	40–70	38–130	60–90	15–40
SW SE 19, T55N R73W							
Oedekoven 82–H	Ft. Union	10–20	50–60	80–110	110–130	80–115	35
Oedekoven 10–19	Muddy	NR	NR	NR	NR	NR	NR
NW SE 19, T55N R73W							
Oedekoven 45–1	Ft. Union	10	40–50	70–110	40–120	60–95	15–45
Oedekoven 1	Muddy	NR	60	NR	25–150	NR	15–33
NE NW 30, T55N R73W							
Lester 33–33	Ft. Union	10	NR	70–80	NR	60–80	NR
Lester 10–33	Muddy	NR	40–50	NR	28–50	NR	30–40
NW SE 33, T56N R73W							
Lester 13–34	Ft. Union	10	NR	60–90	NR	60–85	NR
Lester 1–34	Muddy	NR	50	NR	100–200	NR	30–45
NW SW 34, T56N R73W							
Oedekoven 42–34	Ft. Union	10–20	NR	70–100	NR	80–110	NR
Oedekoven 1–34	Muddy	NR	40	NR	28–140	NR	15–46
SE NE 34, T56N R73W							
Morse 44–34	Ft. Union	10–20	40–50	60–90	40–210	70–90	15–50
Morse 16–34	Muddy	NR	28–60	NR	370–390	NR	40–50
SE SE 34, T56N R73W							
Morse 33–34	Ft. Union	10–20	40–50	70–90	70–150	70–100	15–40
Morse 10–34	Muddy	20–40	25	60–80	60–200	60–80	23–40
NW SE 34, T56N R73W							

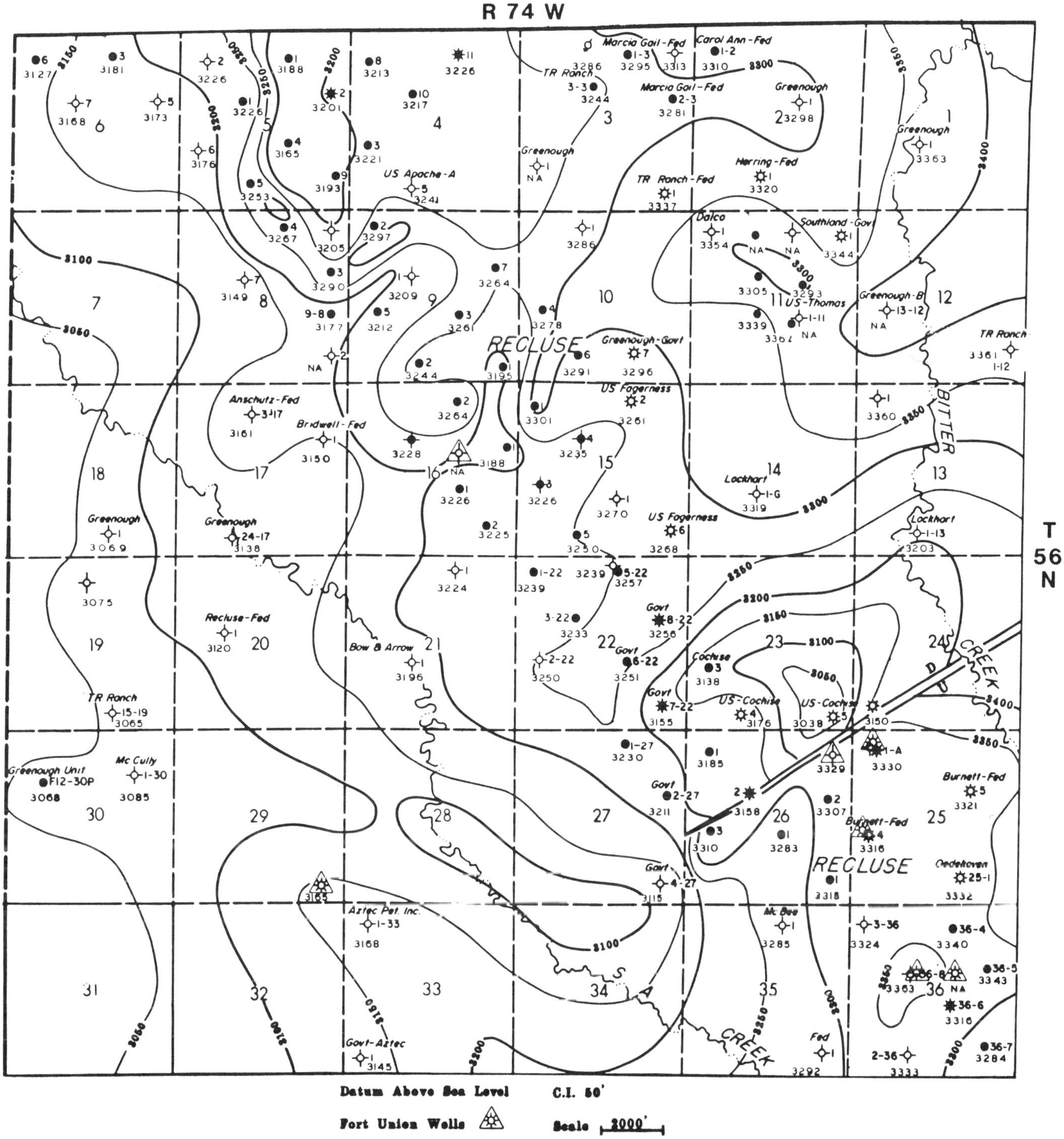

Figure 3. Structure contour map on top of Lower Pawnee coalbed, Fort Union Formation, Recluse field area, Campbell County, Wyoming. Contour interval 50 ft.

curves are normalized and resistivities are corrected for borehole size, drilling-fluid invasion, and other external effects. The SP varies from positive to negative deflection to almost no deflection at all, depending on the ratio of R_{mf} (mud-filtrate resistivity) to R_w (formation-water resistivity).

The nearby fields appear to demonstrate that conventional hydrocarbon accumulations are situated on structural highs (Kent and others, 1980), with stratigraphic variations accentuating the trapping mechanism.

In parts of the study area with close well control, small anticlinal noses plunge south and southwest (Figures 3 and 4). Productive potential may exist where thick sandstones are associated with these features. The structures flatten toward the center of the area and to the northeast

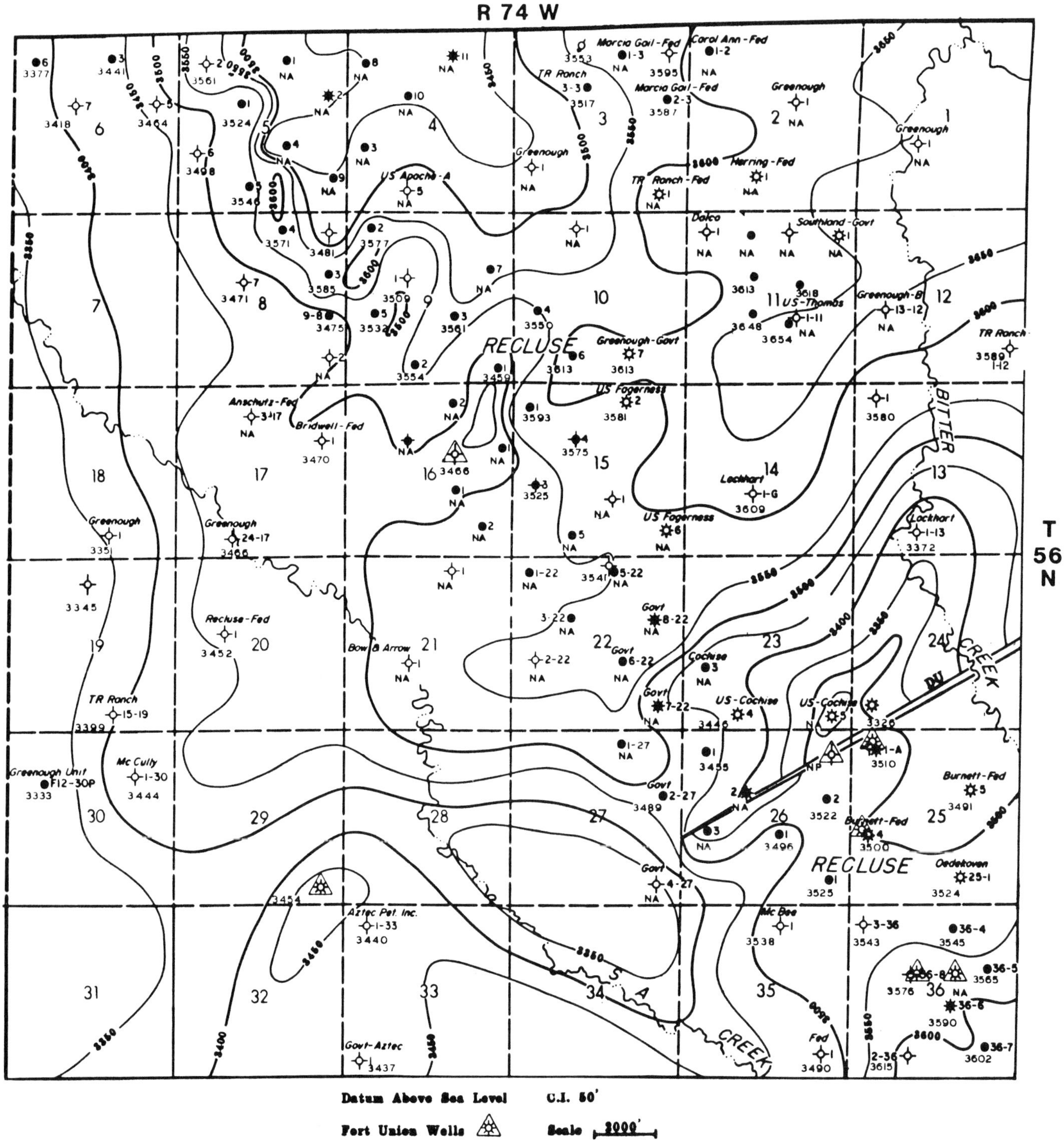

Figure 4. Structure contour map on top of Cook coalbed, Fort Union Formation, Recluse field area, Campbell County, Wyoming. Contour interval 50 ft.

and southwest, where subsurface control is sparse. The fault mapped in the southeastern corner of the map areas may act as a trap for hydrocarbons. A Fort Union well was drilled on the upthrown side of the fault (NW/4NW/4 sec. 25), offsetting an old location where shallow gas ignited and burned down the rig. This well was completed as a gas producer from sandstone below the Canyon coalbed.

CONCLUSIONS

A wide variety of data are available that, if accessed, can decrease the risks of drilling. These include information on methane occurrences in shallow water wells, cross-sections showing correlations of coals and sandstones (Kent and others, 1980), and structure and isopach maps of various

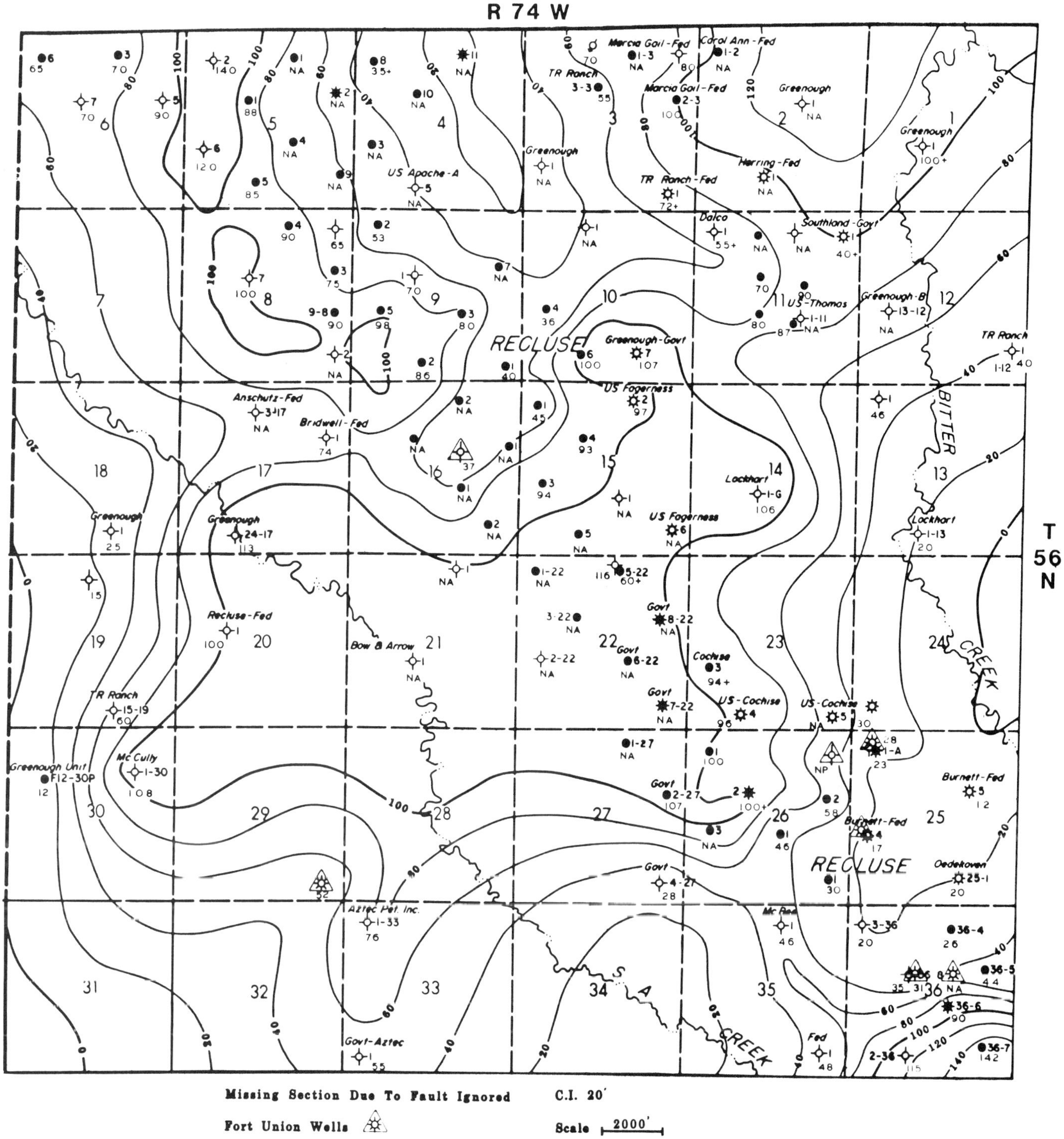

Figure 5. Gross isopach of sandstone between the Cook and Wall coalbeds, Recluse field area, Campbell County, Wyoming. Contour interval 20 ft.

coalbeds showing faults and other major structural features (Olive, 1957; Kent and others, 1978). Cased-hole logs can be used to define the presence, but not the quantity, of formation gas through casing in old wellbores. The study of dual-induction logs from nearby fields can provide guidelines for general resistivity cutoffs to be used in defining gross sand and net pay intervals for isopach mapping. Because Fort Union wells are relatively inexpensive to drill and complete, some operators have tended to simply "punch" holes rather than generate true prospects by the more conventional means used in deeper plays. However, increased emphasis on science can result in significant economic rewards.

ACKNOWLEDGMENTS

Sincere thanks go to Logan MacMillan, Consulting Geologist, previously with Apache Corporation, for his technical support and inspiration, without which this paper would not have been written. Thanks also are extended to Tom Ryer and Scott Uttley for each of their reviews and suggestions. Special thanks to Michel Henderson for drafting the maps and cross section. Finally, thanks to Apache Corporation for granting permission to prepare this paper.

REFERENCES

Ayers, W.B., Jr., 1986, Coal resources of the Tongue River Member, Fort Union Formation (Paleocene), Powder River Basin, Wyoming and Montana: Geological Survey of Wyoming Rept. of Investigations 35, p. 1–22.

Cathcart, J.D., Babcock, R.N., and Hobbs, R.G., 1983, Geophysical logs of water wells logged for coal occurrences in northern Campbell and eastern Sheridan Counties, Wyoming: USGS Open-File Rept. 83–411, p. 1–4, 52–60.

Choate, R., Johnson, C.A., and McCord, J.P., 1978, Powder River basin report—A study of Early Tertiary geology, coal and the potential for methane recovery from coalbeds in Montana and Wyoming: TRW Inc. for U.S. Department of Energy, p. 3–11, 3–12, 4–3, 4–4, 5–5, 5–6, 5–7.

Flores, R.M., 1980, Fluvial coal settings of the Tongue River Member of the Fort Union Formation in the Powder River-Clear Creek area, Wyoming, *in* Glass, G.B., ed., Guidebook to the coal geology of the Powder River Coal Basin, Wyoming: Geological Survey of Wyoming Public Information Circular 14, p. 71–96.

Kent, B.H., Berlage, L.J., 1980, Geologic map of the Recluse 1° by 1/2° quadrangle, Campbell and Crook Counties, Wyoming: USGS Coal Investigations Map C–81–D, scale 1:100,000.

Kent, B.H., Berlage L.J., and Boucher, E.M., 1980, Stratigraphic framework of coal beds underlying the western part of the Recluse 1° by 1/2° quadrangle, Campbell County, Wyoming: USGS Coal Investigations Map C–81–C, scale 1:100,000.

Kent, B.H., and Munson, B.E., 1978, Structure contour maps of the Canyon and associated coal beds, western half of the Recluse 1° by 1/2° quadrangle, Campbell County, Wyoming: USGS Coal Investigations Map C–81–A, scale 1:100,000.

McClurg, J.E., 1988, Peat forming wetlands and the thick Powder River Basin coals, *in* Diedrich, R.P., Dyka, M.A.K., and Miller, W.R., eds., Eastern Powder River basin–Black Hills: Wyoming Geological Association, 39th Annual Field Conference Guidebook, p. 229–236.

McKay, E.J., 1973, Preliminary geologic map of the Rawhide School quadrangle, Campbell County, Wyoming: USGS Open-File Rept., scale 1:24,000.

Olive, W.W., 1957, The spotted Horse coalfield, Sheridan and Campbell Counties, Wyoming: USGS Bulletin 1050, p. 58–59.

Schlumberger, 1989, Cased hole log interpretation, principles and applications: Schlumberger, p. 3–16 to 3–33.

White Dog Exploration Company, [n.d.], Coal geophysical well-log interpretation manual: White Dog Exploration Co., p. 24–25.

Coalbed Methane Resource and Producibility Potential of the Rock Springs Formation, Great Divide Basin, Wyoming

Bruce S. Kelso[1]
Woodruff G. Leel, Jr.[2]
David L. Carr[3]

1. ICF Resources Inc., Lakewood, Colorado
2. Triton Oil and Gas Corporation, Dallas, Texas
3. Geological Consultant, Denver, Colorado

ABSTRACT

Recent exploration activities for coalbed methane in the Rock Springs Formation (Mesaverde Group) in Sweetwater County, Wyoming, indicate favorable geologic and reservoir conditions for commercial development. Triton Oil and Gas Corporation drilled three wells between October 1989 and February 1990 in the Great Divide basin, part of the greater Green River basin, to test the potential of coals in the Rock Springs Formation. Data were collected from these wells to characterize the coals for their production potential. Analysis of desorbed gas-content data and adsorption isotherm data indicated that within the depth range of 3,400 to 4,000 ft (1,037 to 1,220 m), gas contents of the coal seams uncharacteristically decrease with depth; above and below this range, gas content remained fairly constant.

Gas-in-place resource estimates ranging from 4 to 30 BCF/mi^2 (43.7 × 10^6 to 327.6 × 10^6 m^3/km^2) were determined for Triton's acreage. One of Triton's wells, the 2 UPRC–1, was characterized by history matching a 140-day production test using the COMETPC 3D reservoir simulator. A 25-year gas-recovery prediction of approximately 1.0 BCF (28.3 × 10^6 m^3), 61% of the gas in place, is estimated for this well from 14.5 ft (4.4 m) of coal. This paper presents the geologic analysis of Triton's Great Divide coalbed methane prospect and discusses production potential for the area based on reservoir simulation results.

INTRODUCTION

Recent development of coalbed methane in the Rock Springs Formation, Sweetwater County, Wyoming, indicates favorable conditions for commercial development. A number of operators, including Triton Oil and Gas Corporation, Buttonwood Petroleum Inc., Eagle Oil and Gas Company, and Rocky Mountain Energy Company, have evaluated the potential for coalbed methane around the Rock Springs uplift. Triton's property, referred to as the Great Divide Prospect, is located approximately 30 miles (48.3 km) north-northeast of the town of Rock Springs on the northeastern limb of the Rock Springs uplift (Figure 1), which marks the structural western boundary of the Great Divide basin, a component of the Greater Green River coal region.

The geological assessment of the property to determine gas-in-place resources included compilation of coal seam thicknesses and depths, rank assessment, and gas-content data analysis. In-depth discussions of the property's geological characteristics, gas content of the coals, including adsorption isotherm samples, and gas-in-place resources are presented. An example of reservoir characterization through production history matching, including a simulated production forecast, is included for the 2 UPRC–1 well. This paper presents some of the first publicly available data specific to coalbed methane potential of the Mesaverde Group in the Green River basin.

GEOLOGY AND RESERVOIR CHARACTERISTICS

The geologic analysis of the Great Divide Prospect was based on subsurface data collected from geophysical logs of 46 exploration wells in the area. A generalized Upper Cretaceous stratigraphic section is shown in Figure 2. The target coals are found in the Rock Springs Formation of the Mesaverde Group. In most cases, gamma ray-bulk density and gamma ray-sonic type logs were used to identify coal seam thickness and depth, and tops of the Rock Springs and Blair Formations. A maximum bulk density value of 1.75 g/cc was applied in determining coal thickness. In several instances, gamma ray/SP-resistivity logs were used, with caution, for data in areas where wells had not been logged with density tools.

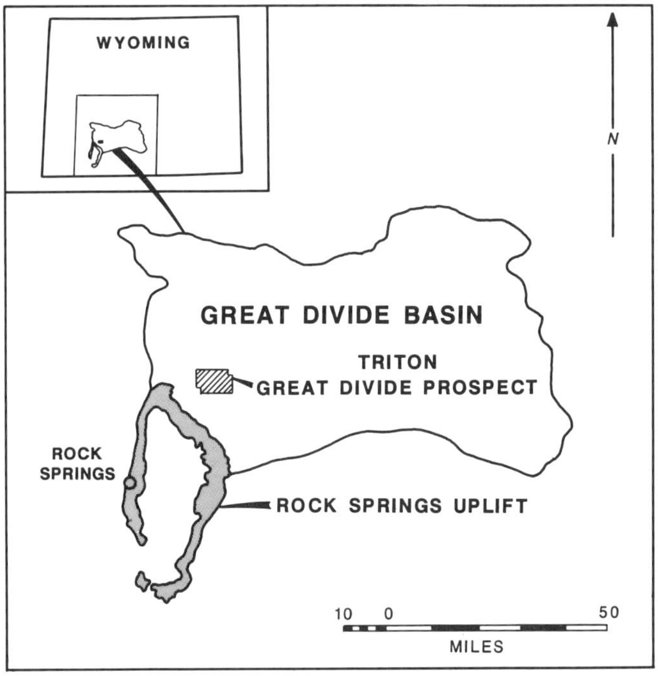

Figure 1. Location Triton Oil and Gas Corporation's Great Divide Prospect, Great Divide basin, Wyoming.

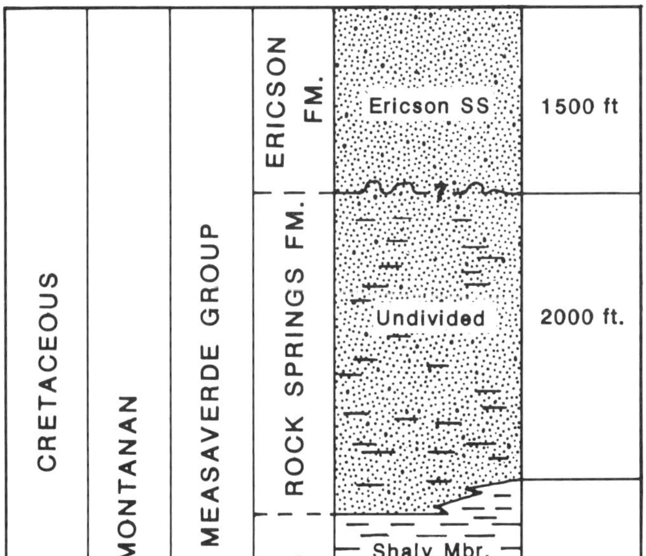

Figure 2. Generalized stratigraphic column of Mesaverde Group strata in Great Divide basin, Wyoming.

Geometry and Structure

From investigations of coal-seam geometry and continuity in the area of the Rock Springs uplift (Levey, 1985), lateral continuity of more than 30 miles (48 km) for individual coal seams can be demonstrated. However, well control in the prospect area is insufficient to prove or disprove lateral continuity of the Rock Springs Formation coal seams, although the three wells drilled by Triton all exhibit similar lithologic units and coal types, as described by Levey. In addition, the lack of well control in the region does not permit detailed structural analysis of the prospect. A structure map drawn on the top of the Rock Springs Formation (Figure 3) shows a regional northwest-southeast structural strike and a northeastward dip of about 500 ft/mile (95 m/km). As more wells are drilled in the area, identification of subtler tectonic and depositional features should become more apparent; these will play an important role in exploration and development of the prospect.

Coal Thickness

The Rock Springs Formation has been interpreted to be a series of wave-dominated deltaic deposits (Levey, 1985), whose thickness across the prospect ranges from 1,420 ft (433 m) to 1,985 ft (605 m) in the northwest. Coal is distributed throughout most of the interval. Thicker coal seams commonly are located in the middle to lower part of the formation, as seen in the 2 UPRC–1 well (Figure 4).

A net-coal isopach map was constructed for coals 5 ft (1.5 m) or more thick (Figure 5). Net coal thickness in Triton's three wells, including all seams greater than 5 ft (1.5 m), ranges from 21 to 68 ft (6.4 to 20.7 m). The mean thickness of individual seams greater than 5 ft (1.5 m) is 7.5 ft (2.3 m) and, on average, six such "thick" seams occur in the Rock Springs Formation at the prospect. Many thinner coals that occur in the Rock Springs Formation were not included in this analysis because of well completion considerations. For comparison, if all seams thicker than 2 ft (0.6 m) were included, net coal thickness observed in the three wells would range from 74 to 110 ft (22.6 to 33.6 m), an increase of 30 to 60% over that by the 5-ft (1.5-m) cutoff.

Coal Rank

Mesaverde Group coals in this area of the Greater Green River coal region are bituminous in rank. Based on 26 proximate analyses and four vitrinite reflectance analyses of coal samples from the 2 UPRC–1 well, the coals are ranked as high-volatile B bituminous at shallow depths and as high-volatile A bituminous at depths below 4,000 ft (1,220 m). Vitrinite reflectance values measured on four samples ranged from 0.66 to 0.74%. Ash content of less than 5% to more than 25% were determined both from sample analysis and from interpretation of high-resolution bulk-density logs and log-derived parameters based on Fruitland Formation coals of the San Juan basin (Mavor and others, 1990).

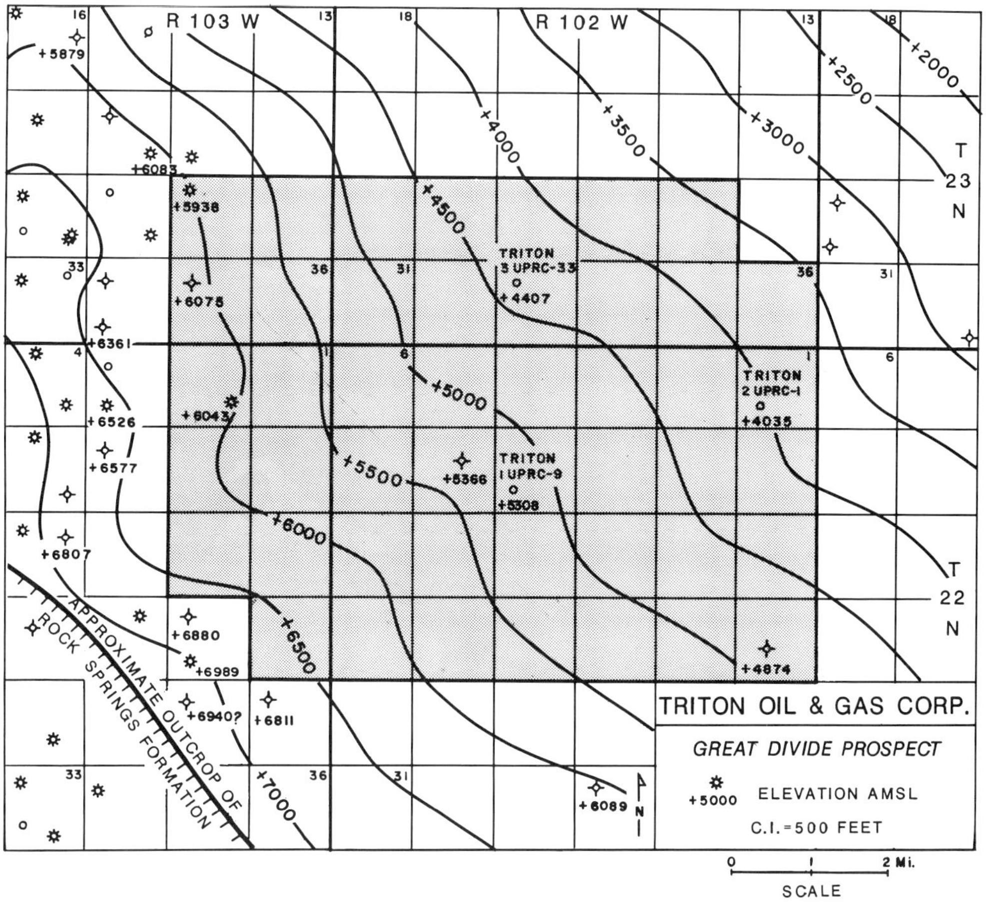

Figure 3. Structure map of Triton Great Divide Prospect and vicinity, drawn on top of Rock Springs Formation.

Desorption Gas-Content Data

Gas-content data were collected from three Triton wells—1 UPRC–9, 2 UPRC–1, and 3 UPRC–33. Gas-content data determination from core samples followed the U.S. Bureau of Mines' (Diamond and Levine, 1981) and the Smith and Williams' (1981) procedures. Data collected from drill cuttings included a four-part handling procedure. First, the cuttings from the coal seams of interest were collected from the mud return system, then sealed in canisters. The amount of desorbed gas was measured, and finally, the data were analyzed to estimate the gas content and correct for "lost gas." Results from the cutting samples lost-gas corrections by both the Bureau of Mines' and Smith and Williams' methods were reasonable, and based on those methods, the average lost-gas estimate constituted 15.5% and 20% of the total gas-content value, respectively. Conventional core and cutting samples were collected from 1 UPRC–9, but only cutting samples were collected from 2 UPRC–1 and 3 UPRC–33 (Table 1). Marked variations in desorption results were observed among the three wells.

From the 1 UPRC-9 well, located near the center of the prospect area (sec. 9, T22N R102W), 33 samples for desorption were collected from conventional core. Results were disappointing. The average gas content (noncorrected ash basis) was 11 SCF/T (0.34 cc/g) and ranged from 1 to 50 SCF/T (0.03 to 1.6 cc/g). Although low, these values are considered to be representative of the reservoir at this location and not the result of sampling or handling errors. Given the proximity of the well to the Rock Springs Formation outcrop (approximately 5 miles), perhaps the shallower coal depths had facilitated natural desorption and leakoff into either adjacent strata or the atmosphere. Based on these gas-content results, an updip limit for resource evaluation can be established, as is discussed in the gas resource section.

Cuttings desorption samples were collected from the 2 UPRC–1 well, located in sec. 1, T22N R102W. Average gas content (noncorrected ash basis) for 26 samples is 211 SCF/T (6.6 cc/g), with a range of 30 to 435 SCF/T (0.9 to 13.6 cc/g). Proximate analyses were run on all the samples, and for comparative purposes, average gas content on an ash-free basis is 260 SCF/T (8.1 cc/g).

Fifteen cuttings desorption samples were collected from the 3 UPRC–33 well, located in sec. 33, T32N R102W. The average gas-content value is 270 SCF/T (8.4 cc/g) noncorrected for ash, with values ranging from less than 1 to 541 SCF/T (0.03 to 16.9 cc/g).

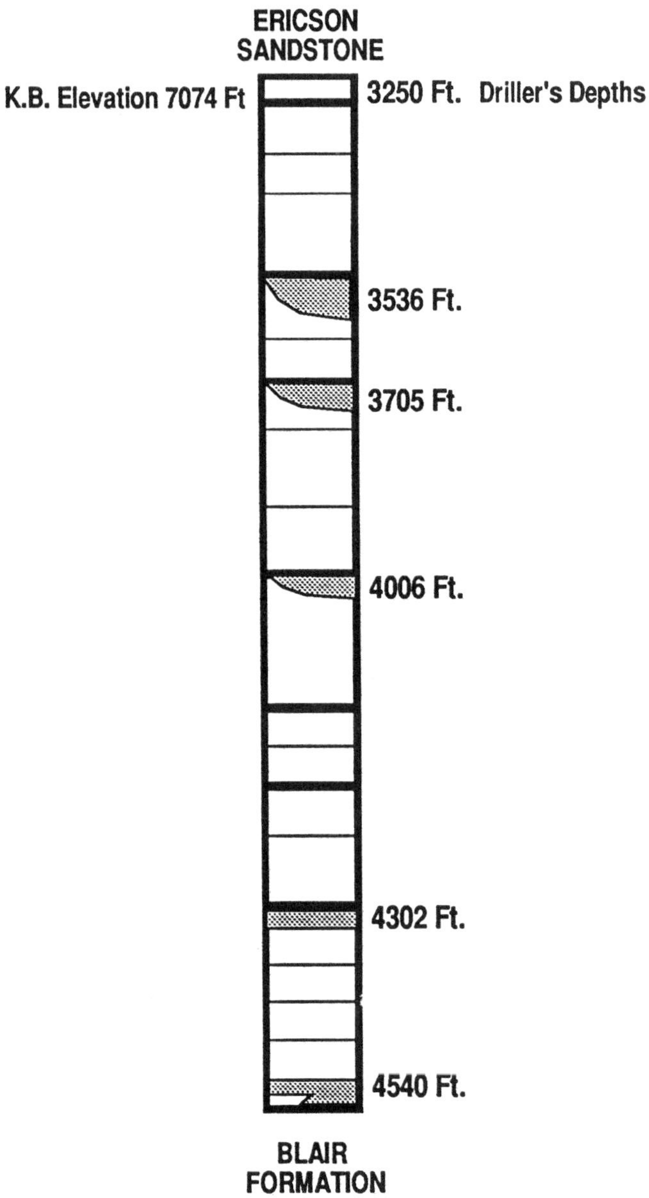

Figure 4. Generalized section of Rock Springs Formation, showing positions of major coal seams and associated sandstones, 2 UPRC–1 well.

Sorption Isotherm Data

Sorption isotherm analyses were conducted on four coal cuttings samples collected from depths of 3,397 to 4,258 ft (1,036 to 1,299 m) in the 2 UPRC–1 well. From the isotherm curves shown in Figure 6, an interesting observation regarding the Langmuir constants is that, with the exception of isotherm sample U12, Langmuir volumes (V_L) decrease with depth. This is contrary to what one would expect in most coal-bearing formations and is opposite to the generalized trends of Eddy and others (1982), who found gas contents to increase with depth and coal rank. To confirm this observation for the isotherm data, gas-content data from similar depth intervals from the 2 UPRC–1 and 3 UPRC–33 wells were reviewed.

As a result of the measured gas-content data and the adsorption isotherms, a relatively complex gas-content vs depth relationship was developed for use in the resource evaluation of the prospect. Figure 7 illustrates the decreasing gas-storage capacity of coal samples collected from the 2 UPRC–1 well, with the exception of sample U12. In addition, 21 desorbed gas-content samples collected from the 2 UPRC–1 and 3 UPRC–33 wells, over corresponding depth intervals to those of the adsorption isotherm samples, demonstrate a similar decreasing gas-content vs depth relationship.[4]

GAS RESOURCE ESTIMATION

Estimation of gas resource in place (*GIP*, in units of CF) for Rock Springs coal seams within the Triton prospect was calculated by the equation, $GIP = GC \times h \times A \times \rho$, where *CG* is gas content (CF/T); *h* is net-coal thickness (ft); *A* is drillable area (acres); and ρ is coal density (tons/acre-ft). Coal volumes were determined for each section and partial section within the property boundaries using net coal thickness from Figure 5, drillable area (number of acres in the section or partial section), and coal density (assumed to be 1,800 tons/acre-ft).

Estimated gas-content values for each section were derived in a two-step procedure. For each section, a midpoint depth of the coal interval first was determined. The second step, estimation of coal seam gas contents was based on a threefold subdivision of midpoint depth . In areas of midpoint depths between 2,500 and 3,400 ft (762.5 and 1,037 m), gas contents were estimated to range from 100 to 200 SCF/T (3.1 to 6.2 cc/g), based on data collected from the 1 UPRC–9 well. For midpoint depths between 3,400 and 4,000 ft (1,037 and 1,220 m), the decreasing gas content vs depth relationship from Figure 7 was used. At midpoint depths below 4,000 ft (1,220 m), a gas content of 275 SCF/T (8.6 cc/g) was assumed for lack of sufficient data necessary to develop a statistical gas-content relationship. It is the authors' opinion that an increasing gas-content trend may exist at depths below 4,000 ft (1,220 m), based on gas-content analyses from other western coal basins.

The combination of coal volume values and estimated coal-gas contents yielded gas-in-place estimations on a BCF/mi^2 basis. Resource estimates of less than 4 BCF/mi^2 to 30 BCF/mi^2 (43.7×10^6 to 327.6×10^6 m^3/km^2) were derived for the prospect area, which was divided into three areas (Figure 8). Most of the resource, approximately 273 BCF ($7{,}726 \times 10^6$ m^3) is contained within Area 1, where Rock Springs coalbeds lie at optimum depths of approximately 3,400 to 4,400 ft (1,037 to 1,220 m) and have high gas contents, approximately 200 to 400 SCF/T (6.2 to

4. The interpretation and development of the decreasing gas-content vs depth relationship presented in Figure 7 differs slightly from the material presented by Young and others (1991). At the time their paper was prepared, only preliminary desorption results were available for samples from the 3 UPRC–33 well. The finalized data used in this paper results in a slightly different interpretation, although the decreasing gas-content vs depth relationship still prevails.

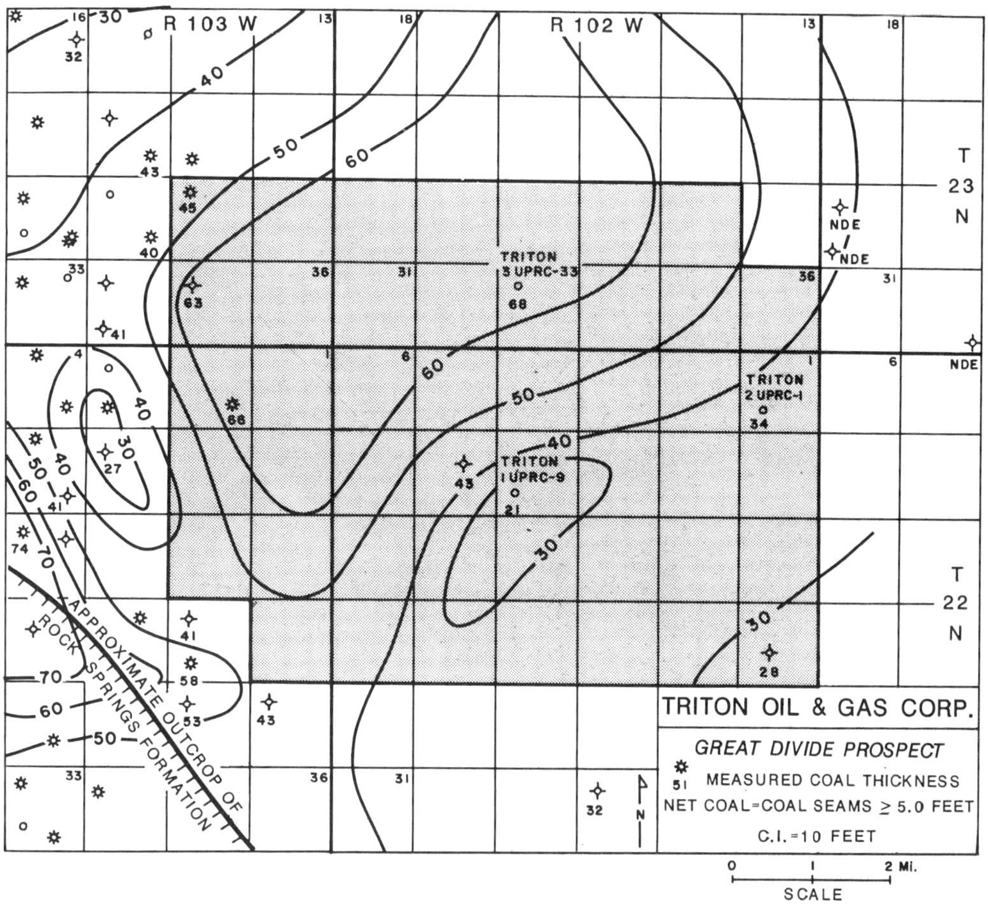

Figure 5. Net-coal thickness of Rock Springs Formation.

12.5 cc/g). Area 2 is transitional between the relatively high gas contents observed in the 3 UPRC–33 well and low gas contents, less than 200 SCF/T (6.2 cc/g), seen updip in the 1 UPRC–9 well. Area 2 contains an estimated resource of 56.6 BCF ($1{,}602 \times 10^6$ m^3). Coalbeds occurring at shallower depths updip from the 1 UPRC–9 well (Area 3) probably have negligible gas contents, and thus no gas in-place evaluation was performed because the area is considered to have limited potential for Rock Springs Formation coal-gas production.

RESERVOIR SIMULATION

History Match of 2 UPRC–1 Well

A history match was performed on the 2 UPRC–1 well using COMETPC 3D, a reservoir simulation model de-

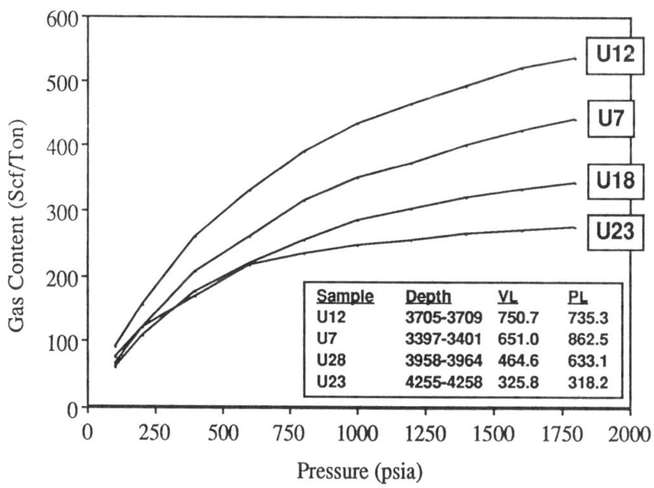

Figure 6. Sorption isotherms and Langmuir constants for four coal cuttings samples from 2 UPRC–1 well.

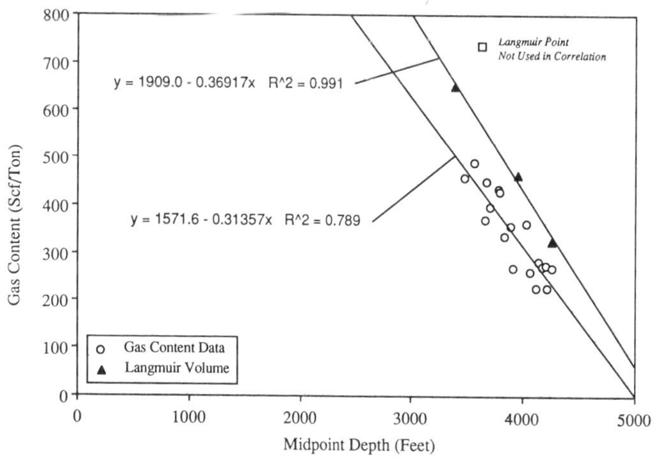

Figure 7. Gas content and sorption isotherm vs sample depth, 2 UPRC–1 and 3 UPRC–33 wells. Langmuir volumes derived from isotherm data for 2 UPRC–1 well (Figure 6).

Table 1. Desorption gas-content data for the 1 UPRC–9, 2 UPRC–2, and 3–UPRC–33 wells, Great Divide Prospect.

Sample Type	Average Depth (ft)	(m)	Gas Content (naf) (SCF/T)	(cc/g)
Well 1 UPRC–9				
Core	1,769.5	539.7	26.25	0.82
Core	1,782.5	543.7	5.66	0.18
Core	1,797.5	548.2	22.86	0.71
Core	1,804.5	550.4	52.04	1.62
Core	1,819.4	554.9	15.87	0.49
Core	1,825.5	556.8	43.42	1.36
Core	1,827.5	557.4	22.32	0.70
Core	2,183.5	665.9	1.79	0.06
Core	2,184.5	666.3	4.32	0.13
Core	2,186.5	666.9	17.57	0.55
Core	2,186.5	666.9	6.50	0.20
Core	2,187.5	667.2	3.22	0.10
Core	2,292.0	699.1	1.79	0.05
Core	2,295.5	700.1	1.86	0.05
Core	2,313.9	705.7	4.05	0.13
Core	2,316.8	706.6	5.84	0.18
Core	2,321.2	708.0	5.92	0.18
Core	2,324.8	709.1	4.51	0.14
Core	2,325.5	709.3	2.89	0.09
Core	2,329.7	710.6	6.79	0.21
Core	2,330.5	710.8	1.11	0.03
Core	2,332.5	711.4	4.78	0.15
Core	2,336.9	712.8	5.92	0.18
Core	2,365.4	721.4	6.63	0.21
Core	2,367.6	722.1	7.96	0.25
Core	2,479.0	756.1	2.12	0.07
Core	2,515.8	767.3	12.57	0.39
Core	2,516.5	767.5	9.48	0.30
Core	2,517.5	767.8	5.82	0.18
Core	2,518.5	768.1	11.97	0.37
Core	2,535.5	773.3	9.17	0.29
Core	2,587.7	789.2	4.94	0.15
Core	2,589.8	789.9	21.18	0.66
Well 2 UPRC–1				
Cuttings	3,036.5	926.1	34.16	1.07
Cuttings	3,126.0	953.4	34.23	1.07
Cuttings	3,162.5	964.6	35.07	1.09
Cuttings	3,162.5	964.6	29.90	0.93
Well 2 UPRC–1, continued				
Cuttings	3,345.0	1,020.2	117.75	3.67
Cuttings	3,345.0	1,020.2	127.90	3.99
Cuttings	3,399.0	1,036.7	155.94	4.87
Cuttings	3,479.0	1,061.1	80.68	2.52
Cuttings	3,539.5	1,079.5	161.10	5.03
Cuttings	3,629.0	1,106.8	225.18	7.03
Cuttings	3,667.5	1,118.6	371.68	11.60
Cuttings	3,707.0	1,130.6	396.42	12.37
Cuttings	3,786.0	1,154.7	434.98	13.58
Cuttings	3,835.0	1,169.7	336.60	10.51
Cuttings	3,917.0	1,194.7	269.90	8.42
Cuttings	3,961.0	1,208.1	203.01	6.34
Cuttings	4,037.0	1,231.3	363.64	11.35
Cuttings	4,065.0	1,239.8	262.23	8.18
Cuttings	4,119.0	1,256.3	226.21	7.06
Cuttings	4,146.0	1,264.5	281.76	8.79
Cuttings	4,173.0	1,272.8	270.76	8.45
Cuttings	4,219.0	1,286.8	226.63	7.07
Cuttings	4,256.0	1,298.1	268.56	8.38
Cuttings	4,296.5	1,310.4	337.32	10.53
Cuttings	4,372.0	1,333.5	74.59	2.33
Cuttings	4,397.5	1,341.2	150.70	4.70
Well 3 UPRC–33				
Cuttings	519.5	158.4	809.1	34.75
Cuttings	1,459.5	445.1	1	0.03
Cuttings	1,769.5	539.7	1	0.03
Cuttings	2,819.0	859.8	87.5	2.73
Cuttings	2,838.0	865.6	79.5	2.48
Cuttings	3,061.5	933.8	233.8	7.30
Cuttings	3,119.5	951.4	183.7	5.73
Cuttings	3,210.0	979.1	58.2	1.82
Cuttings	3,393.0	1,034.9	205.2	6.40
Cuttings	3,475.0	1,059.9	591.2	18.45
Cuttings	3,566.0	1,087.6	475.9	14.85
Cuttings	3,674.0	1,120.6	373.4	11.65
Cuttings	3,790.0	1,156.0	386.3	12.05
Cuttings	3,891.0	1,186.8	290.3	9.06
Cuttings	4,206.0	1,282.8	369.2	11.52

veloped specifically for coalbed methane wells. The COMETPC 3D model has been documented and validated for use with this type of reservoir. For additional information about the model, refer to Sawyer and others (1990) and Paul and others (1990). A gas and water production history of 140 days was available from the 2 UPRC–1 well, and actual water rates were used for the history match. The completed interval in the well includes two coal seams at depths of 3,658.0 to 3,665.3 ft (1,115.7 to 1,117.9 m) and 3,698.0 to 3,705.2 ft (1,127.9 to 1,130.1 m), for a combined thickness of 14.5 ft (4.4 m). In addition to the two coal seams, a 28-ft-thick (8.5-m) water-bearing sandstone at 3,705.2 to 3,733.2 ft (1,130.1 to 1.138.6 m) directly underlies the deeper coal. In the history match, the two coals were modeled as a single layer. The sandstone appears to be in hydraulic communication with the coal and may be contributing to water production in the well.

The reservoir properties used in the history-matching exercise are shown in Table 2. Variables that were adjusted included (1) gas-water relative permeability curves, (2) porosity of the two coal layers, and (3) vertical permeability between the coal and sandstone. The simulated gas and water production, along with the observed historical rates, are presented in Figure 9. Greater detail of the history matching is presented in a paper by Young and others (1991).

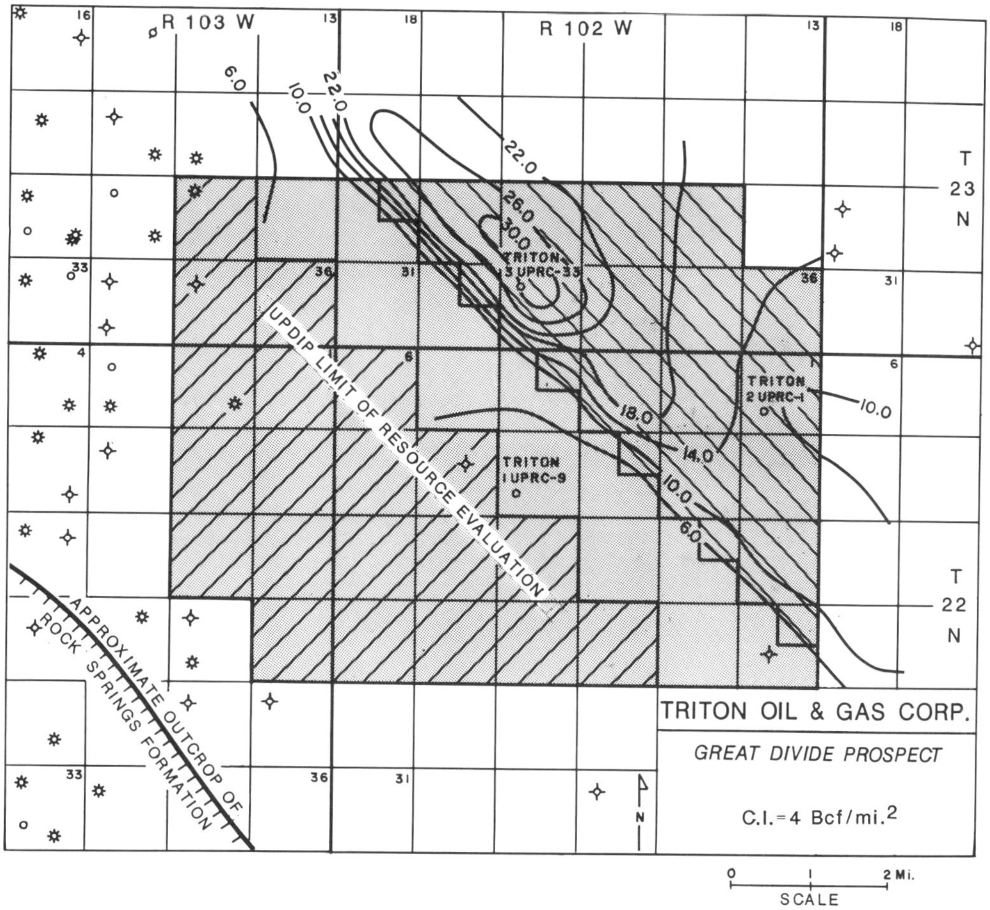

Figure 8. Gas-in-place resources of Triton Great Divide Prospect. Area 1, high gas-in-place (left-to-right diagonal hatchuring); Area 2, transitional (no pattern); Area 3, low gas-in-place (right-to-left diagonal hatchuring). Contour interval 4 BCF/mi^2.

Production Forecast

The reservoir characteristics determined in the history-matching exercise are used to predict a 25-year recovery for a well on 160-acre (65-ha) spacing. Because all the wells drilled in the prospect area might not be confined, well performances and recoveries could be lower. For a confined well with 14.5 ft (4.4 m) of net-coal pay and original gas in place of 1.63 BCF (46.1×10^6 m^3), 25-year predicted recoveries are 1 BCF (28.3×10^6 m^3) of gas and 250,000 bbl (39,750 m^3) of water. A peak gas rate of 240 MCFD (6,792 m^3/day) occurs in year two, and the well produces at a final rate of 40 MCFD (1,132 m^3/day) (Figure 10). Readers should note that this prediction is only for 14.5 ft (4.4 m) of coal and that the 2 UPRC–1 well contains more than 36 ft (11 m) of net coal in seams greater than 5 ft (1.5 m). Provided that the other coal seams have reservoir characteristics similar to those history matched, a substantial upside exists to the production potential from wells in this area.

Table 2. Reservoir properties for the 2 UPRC–1 well.

Parameter	Coal	Sandstone
Horizontal permeability, md	10.5	5.0
Vertical permeability, md	0.25	0.25
Porosity, %	12	10
Initial water saturation, %	100	100
Thickness, ft (m)	14.5 (4.4)	28.0 (8.5)
Gas content, SCF/T (cc/g)	348 (10.9)	0 (0)

SUMMARY

Geological conditions at Triton's Great Divide Prospect in the Great Divide basin favor a substantial in-place coal-gas resource. Net-coal thicknesses of 21 to 68 ft (6.4 to 20.7 m), together with gas-content measurements from canister desorption samples and sorption isotherms, are the basis for the resource estimation. Values of 4 BCF/mi^2 to more than 30 BCF/mi^2 (43.7×10^6 to 327.6×10^6 m^3/km^2) have been estimated within coal seams 5 ft (1.5 m) or more thick. Coal seams 5 ft thick and greater and the associated gas constitute 30% to 60% of the total coal-gas resource potential of the Rock Springs Formation on the prospect. Based on Triton's early testing, many favorable factors exist for economic recovery of this large methane resource.

Measured gas-content values from drill cuttings and core samples range from 1 to 541 SCF/T (0.03 to 16.9 cc/g). Samples from the 1 UPRC–9 well, located updip from the

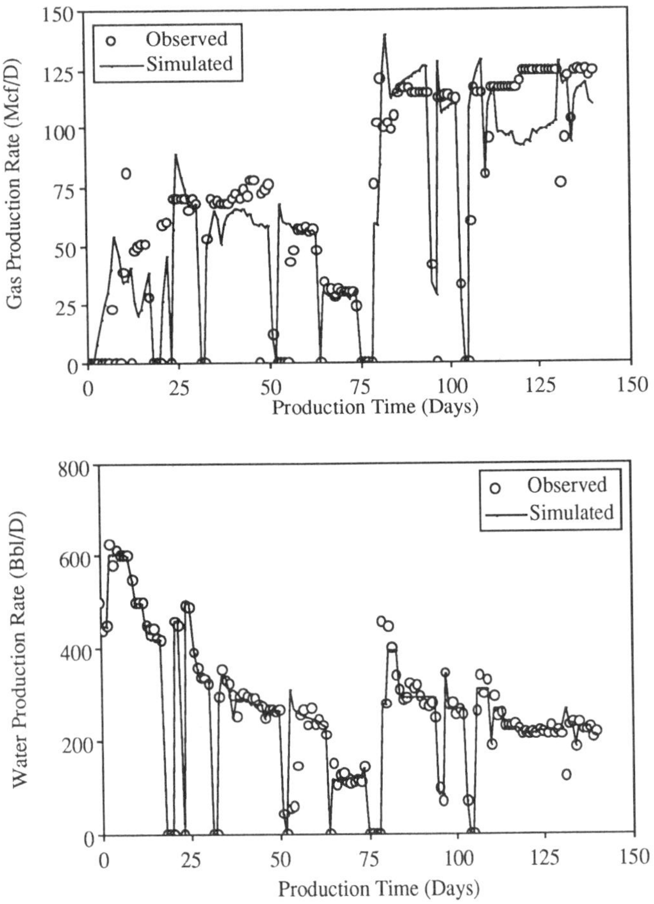

Figure 9. Historical and simulated gas and water production for the 2 UPRC–1 well, Triton Great Divide Prospect.

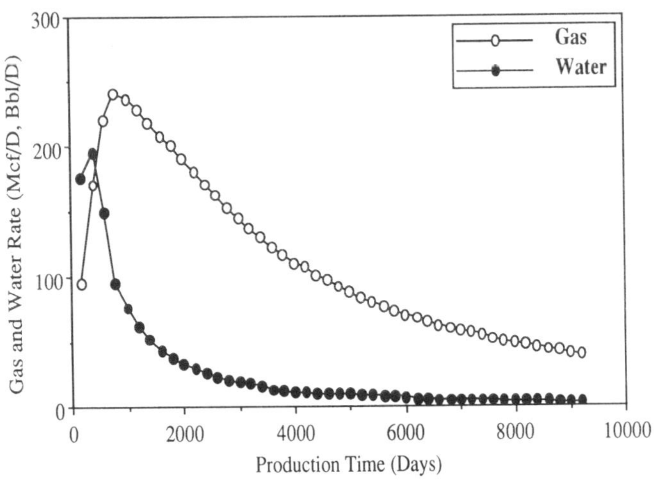

Figure 10. Production forecast for a confined 160-acre (65-ha) well.

two other Triton wells, show lower gas contents and may possibly be explained by the process of natural desorption and leakoff into more permeable zones or to the atmosphere. A relationship of decreasing gas content with depth, contrary to findings in other basins, was established with data from the 2 UPRC–1 and 3 UPRC–33 wells between depths of 3,200 ft and 4,000 ft (976 m and 1,220 m). This finding is supported by sorption isotherm data that reflect a similar trend. Additional gas content data from future wells in this area will undoubtedly help to define this relationship.

Production test data from the 2 UPRC–1 well were history matched to quantify coal reservoir characteristics, which then were used in a simulated 25-year recovery forecast by the COMETPC 3D model. A 25-year gas-recovery estimate of 1.0 BCF (28.3×10^6 m^3) was predicted for a confined well on a 160-acre (65-ha) spacing unit. This recovery represents approximately 61% of the gas contained in 14.5 ft (4.4 m) of coal and appears to represent favorable commercial development of the resource. Completion of additional coal seams in each well should add significantly to the reserves determined in this investigation.

ACKNOWLEDGMENTS

The authors wish to thank Triton Oil and Gas Corporation for permission to release the data and findings presented in this paper. In addition, Resource Enterprises Inc. was responsible for collecting much of the field data from Triton's three wells, and we wish to thank REI for its efforts with the data interpretation.

REFERENCES

Diamond, W.P., and Levine, J.R., 1981, Direct method determination of the gas content of coal—Procedures and results: U.S. Burea of Mines Rept. of Investigations 8515, 36 p.

Eddy, G.E., Rightmire, C.T., and Byrer, C., 1982, Relationship of methane content of coal, rank, and depth, *in* Proceedings of the SPE/DOE Unconventional Gas Recovery Symposium, Pittsburgh, Pa., 1982: SPE/DOE Paper 10800, p. 177–122.

Levey, R.A., 1985, Depositional model for understanding geometry of Cretaceous coals—Major coal seams, Rock Springs Formation, Green River Basin, Wyoming: AAPG Bulletin, v. 69, no. 9, p. 1359–1380.

Mavor, M.J., Close, J.C., and McBane, R.A., 1990, Formation evaluation of exploration coalbed methane wells: CIM/SPE Paper 90–101 [presented at the 1990 Petroleum Society of CIM/Society of Petroleum Engineers Technical Meeting, Calgary, Alberta, June 10–13].

Paul, G.W., Sawyer, W.K., and Dean, R.H., 1990, Validation of 3–D coalbed methane simulators: SPE Paper 20733 [presented at the 1990 Technical Conference and Exhibition, New Orleans, La., September 23–26].

Sawyer, W.K., Paul, G.W., and Schraufnagel, R.A., 1990, Development and application of a 3–D coalbed simulator: CIM/SPE Paper 90–119 [presented at the 1990 Petroleum Society of CIM/Society of Petroleum Engineers Technical Meeting, Calgary, Alberta, June 10–13].

Smith, D.M., and Williams, F.L., 1981, A new method for determining methane content of coal, *in* Proceedings, 16th Intersociety Energy Conversion Engineering Conference—August 9–14, 1981, Atlanta, Ga.: American Society of Mechanical Engineers, p. 1267–1272.

Young, G.B., McElhiney, J.M., Dhir, R., Mavor, M.J., and Anbouba, I.K., 1991, Coalbed methane production potential of the Rock Springs Formation, Great Divide Basin, Sweetwater County, Wyoming: SPE Paper 21487 [presented at the 1991 Gas Technology Symposium, Houston, Texas, January 23–25].

Characteristics and Development of Fluvial Sandstone and Coalbed Reservoirs of Upper Cretaceous Mesaverde Group, Grand Valley Field, Colorado

Kurt M. Reinecke[1]
Dudley D. Rice[2]
Ronald C. Johnson[2]

1. Barrett Resources Corporation, Denver, Colorado
2. U.S. Geological Survey, Denver, Colorado

ABSTRACT

Two major gas reservoirs are being developed in the Upper Cretaceous Mesaverde Group in the area of Parachute, Colorado. Gas is produced by dual-completion techniques from Cameo coalbeds and lenticular, fluvial sandstone reservoirs in the Williams Fork Formation. Production rates from sandstone reservoirs are a function of depth, porosity, and degree of natural fracturing. These characteristics are highly variable from well to well, but reservoir quality appears best developed on the northeastern flank of the Crystal Creek anticline. A good mud log is the best qualitative indicator of the potential of a sandstone reservoir. Gas productivity from low-volatile bituminous Cameo coalbeds is less predictable, and no mappable parameters have been identified to explain the variations.

Gases from coal and overlying (non-Cameo) sandstone reservoirs generally can be distinguished by chemical composition. Gases from Cameo coals are chemically drier and contain a larger proportion of carbon dioxide compared to gases from the sandstones. However, the fact that many gases are mixtures of the two types suggests some contribution to the sandstone reservoirs by expelled coal-generated gases. The greater wetness of coal gases in the area, compared to other areas of production, indicates that, as a result of different maceral and/or chemical composition, the coal may be richer in hydrogen.

Wells produce relatively low volumes of sodium chloride-type water, which usually is associated with sandstone reservoirs. The occurrence of such water in coalbeds, instead of sodium bicarbonate type, indicates either contamination by treatment fluids or a past hydrologic communication with the sandstones.

Although ultimate reservoir performance can be predicted, coal reservoirs cannot conclusively be distinguished from sandstone reservoirs on the basis of short-term decline characteristics. However, sandstone reservoirs are prone to brief periods of flush production after shut in, whereas coal reservoirs generally are not.

INTRODUCTION

This paper is a progress report of the geology, development history, and characteristics of sandstone and coal reservoirs of the Upper Cretaceous Mesaverde Group that are undergoing extensive development drilling in Grand Valley field of the Piceance basin, western Colorado (Figure 1). Grand Valley field, as defined here, covers parts of T6–7S R96–97W west of the town of Parachute (Figure 2). For convenience, the boundary between Grand Valley field and Parachute field to the east is the R96–95W boundary. Field boundaries generally are defined by the operator and by field development history, rather than by geologic features.

STRATIGRAPHY OF MESAVERDE GROUP

The geology and characteristics of Mesaverde Group reservoirs in the Piceance basin have been documented by Johnson (1989) and by Myal and others (1989). In the area of Grand Valley field, the Mesaverde Group consists, in ascending order, of the Iles and Williams Fork Formations (Figure 3). The Iles Formation intertongues with underlying marine Mancos Shale (Upper Cretaceous) and includes three regressive marine cycles separated by tongues of Mancos Shale—in ascending order, the Corcoran, Cozzette, and Rollins Sandstone Members. In Grand Valley field the tongue of Mancos Shale between the Cozzette and Rollins Sandstones is well developed, but that between the Cozzette

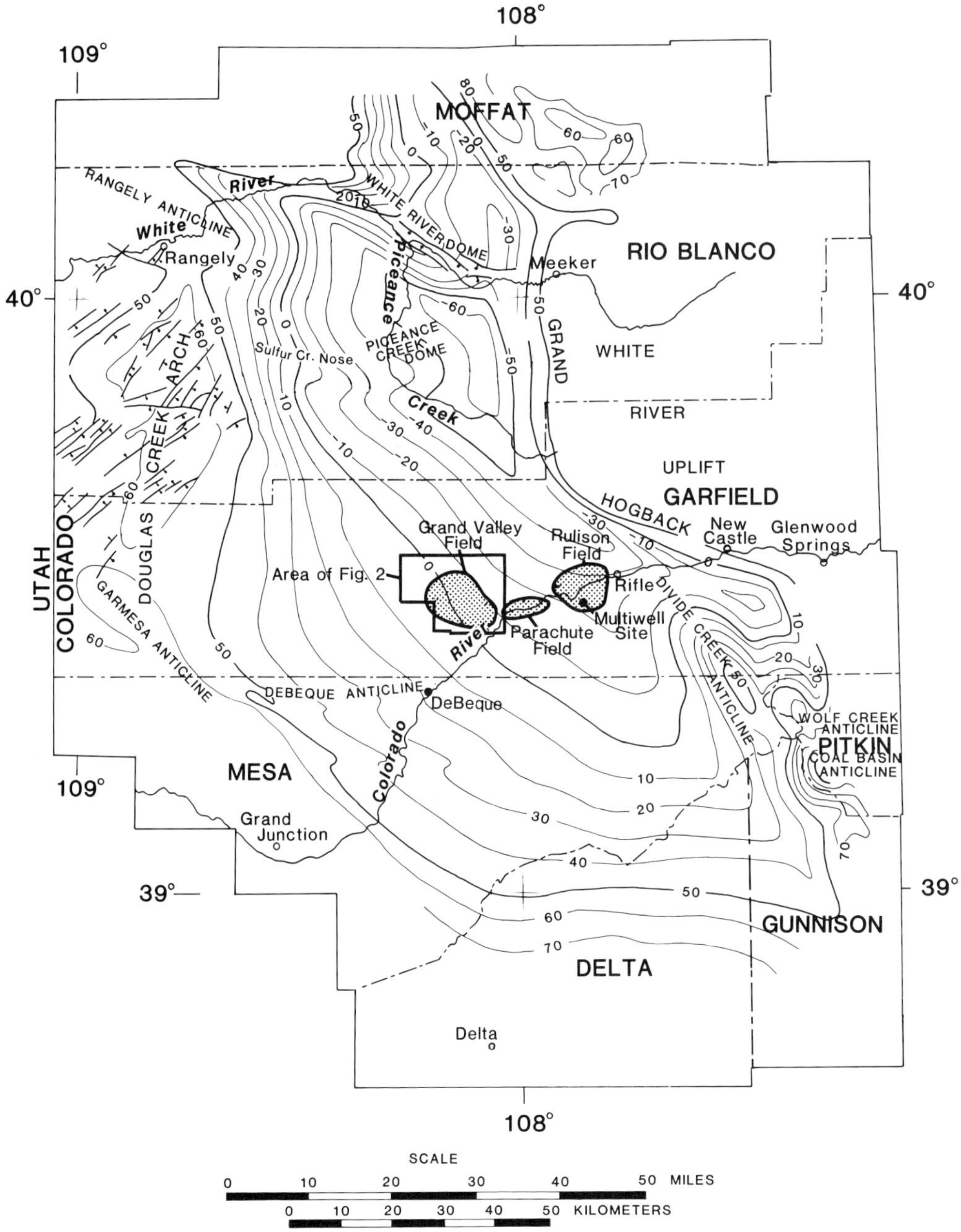

Figure 1. Map of Piceance basin, Colorado, showing location of Grand Valley, Parachute, and Rulison gas fields. Modified from Johnson and Rice (1990). Structure contours on top of the Rollins Sandstone or equivalent Trout Creek Sandstone Members of the Iles Formation. Contour interval 1,000 ft (305 m), numbers shown are in hundreds of feet.

and the Corcoran Sandstones represents its landward limit and is so poorly developed that the Cozzette-Corcoran boundary is difficult to recognize.

The Corcoran and Cozzette Sandstones are stratigraphically complex units deposited in diverse deltaic and lower coastal-plain environments. Distributary-channel, beach, and paludal coal deposits have all been recognized by their electric-log responses through this interval. In contrast, the Rollins Sandstone consists of a single coarsening-upward cycle capped by a littoral sandstone. This unit is continuous throughout much of the basin, although it is clearly time transgressive, rising stratigraphically toward the east and southeast.

The Cameo coal zone of the Williams Fork Formation overlies the Rollins Sandstone (Figure 3) and reflects a paludal depositional environment landward of the prograding Rollins paleoshoreline. Cameo coalbeds in Grand Valley field typically are thickest near the base of the zone and become progressively thinner toward the top (Figure 4). The Cameo zone is about 470 ft (143.4 m) thick and contains 50 to 70 net ft (15.2 to 21.4 m) of coal. The lower 200 ft (61 m) of section generally contains more laterally persistent seams than the upper part, together with very fine-grained sandstone, siltstone, or shale. Lenticular sandstones of probable crevasse-splay or distributary-channel origin commonly are interbedded with upper Cameo coals (Lorenz, 1988).

An isopach map of Cameo coals greater than 4 ft thick in the field shows a highly irregular northwest-trending thin flanked by thicker coal development (Figure 5). However, this pattern is most clearly defined where well control is most concentrated, in the valley of Parachute Creek. North-northeast isopach trends apparent in the vicinity of Grand Valley field change to more northwesterly trends east of Rulison field (Tremain, 1983). A 4-ft (1.2-m) cutoff was selected to minimize errors associated with estimating thin-seam thickness from geophysical logs.

The Cameo zone thickens eastward from 470 ft (143.4 m) in Grand Valley field to nearly 1,000 ft (305 m) in the 1–36 Arco-Exxon well (NE/4 sec. 36, T6S R93W, Figure 6). This increase results from the stacking of coal zones related to two or possibly three regressive marine cycles above the Rollins Sandstone farther east. Total coal thickness in the Cameo zone also increases as a result of this stacking, and in many wells in the eastern part of the basin net coal thickness exceeds 100 ft (30.5 m). East of the Arco-Exxon well, two of these regressive marine cycles divide the Cameo coal zone into three coal zones—in ascending order, Wheeler (Fairfield), South Canyon, and Coal Ridge. In the 1–16 Snyder-Barton-Porter well (SW/4 sec. 16, T6S R90W, Cactus Valley field), the interval from the top of the uppermost coal in the Coal Ridge coal zone to the top of the Rollins Sandstone is 1,320 ft (402.6 m) (Figure 6). The two regressive cycles resemble the Rollins in that each generally consists of a single coarsening-upward cycle capped by a littoral sandstone. Collins (1976) informally referred to these two units as the "middle" and "upper" sandstones. As with the Rollins, these regressive marine cycles appear to rise stratigraphically toward the southeast (Johnson, 1989).

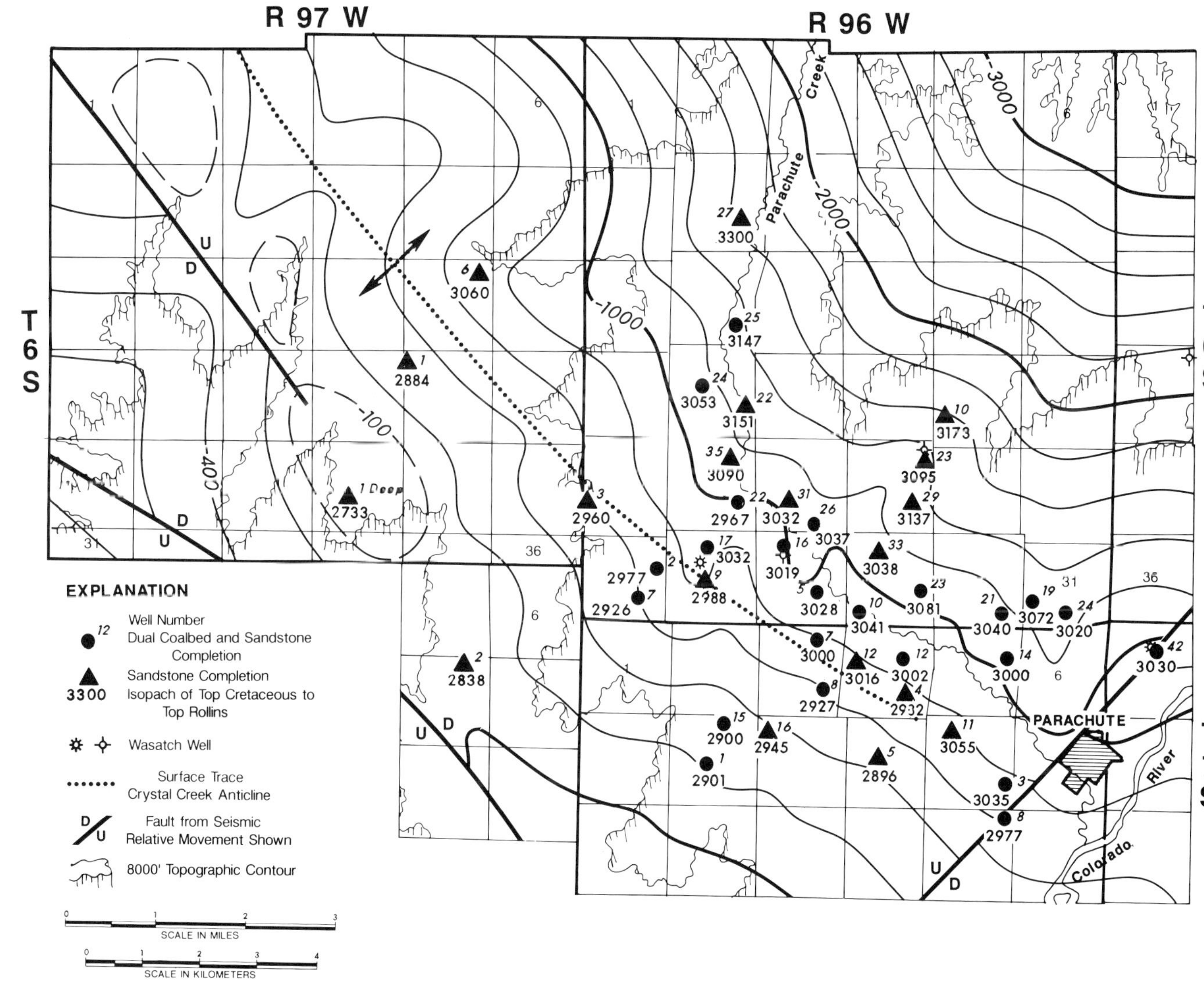

Figure 2. Structure map on top of the Rollins Sandstone Member of the Iles Formation, Grand Valley field. Contour interval 200 ft (61 m). Well locations also show thickness of Williams Fork Formation.

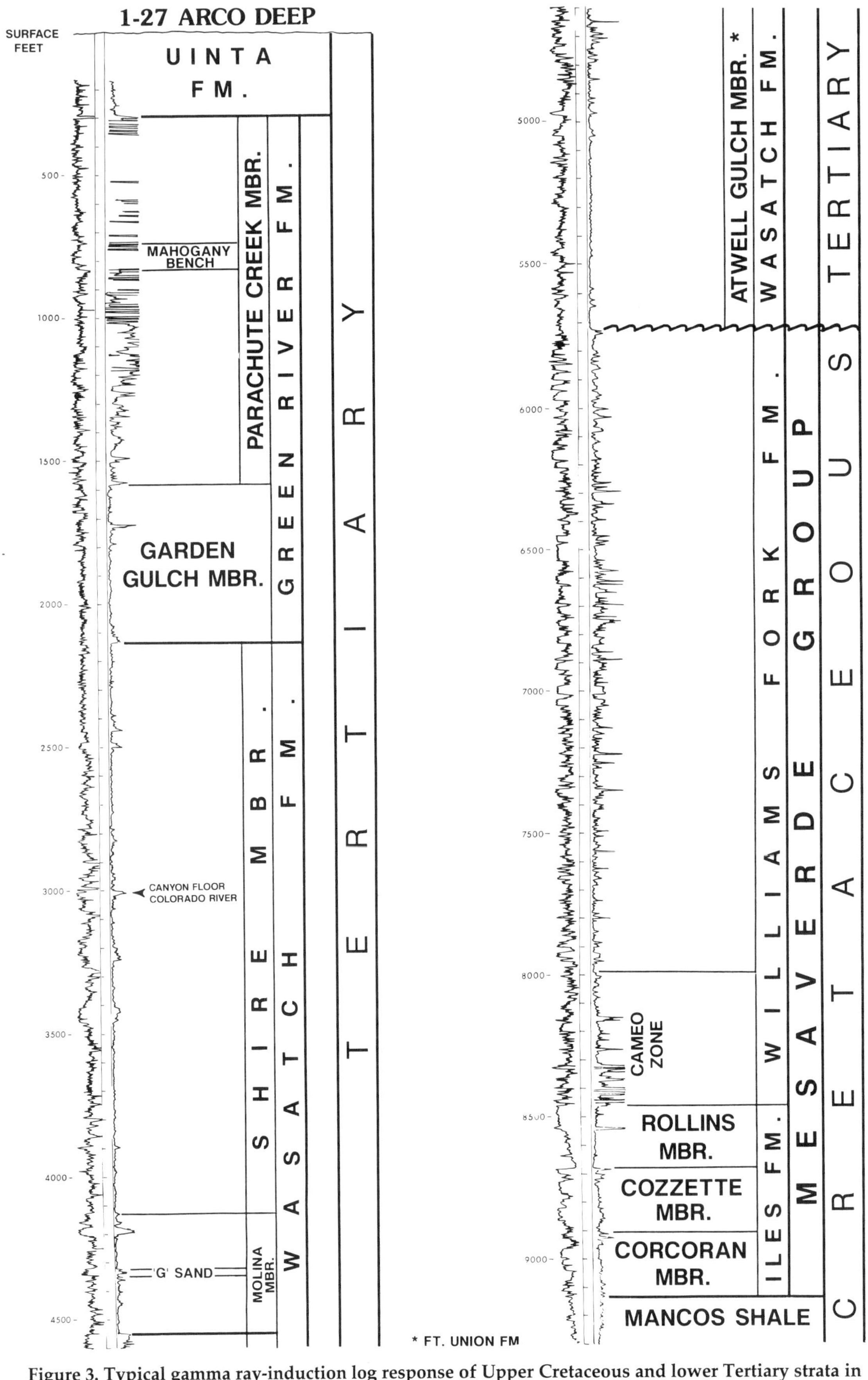

Figure 3. Typical gamma ray-induction log response of Upper Cretaceous and lower Tertiary strata in Grand Valley field. Log is from the Barrett 1–27 Arco Deep well (sec. 27, T6S R97W).

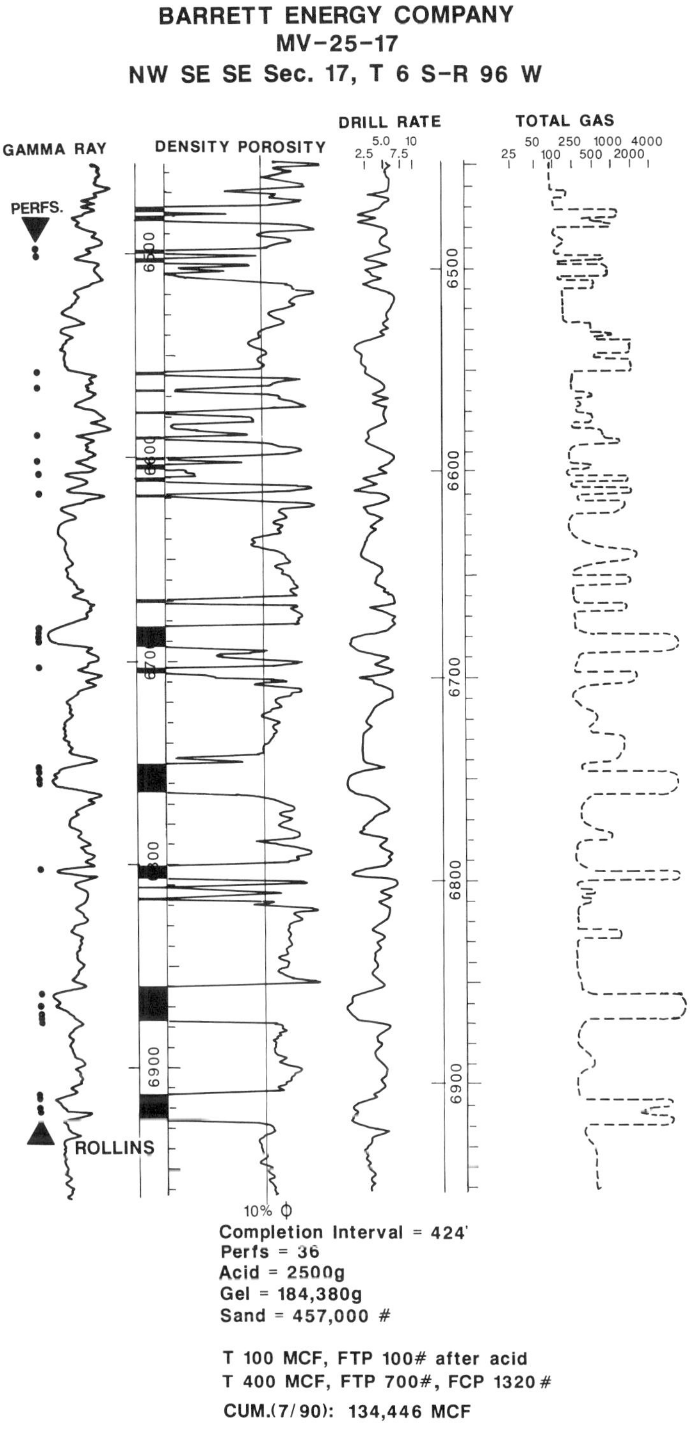

Figure 4. Typical electric and mud-log responses, and completion information for Cameo coal zone, Williams Fork Formation, Grand Valley field. Information is from the Barrett MV-25-17 well (sec. 17, T6S R96W). FTP, flowing tubing pressure; FCP, flowing casing pressure.

Above the Cameo coal zone in Grand Valley field is a 2,200- to 2,800-ft-thick (671 to 854 m) section of mostly fluvial rocks constituting the upper Williams Fork Formation. These sandstones were deposited as accretion point bars and associated strata bounded by flood-plain shale (Figure 7). Although they typically appear on electric logs as a single homogeneous unit, they are, in many cases, a complex amalgamation of several fluvial sandstones. High-resolution logging and cores can help distinguish them.

STRUCTURE

Grand Valley field is located on the northeastern flank of the northwest-trending Crystal Creek anticline (Figure 2). This structure, which is subtly expressed at the surface in the Green River Formation (Eocene), appears to be better developed in the Mesaverde, based on well control and seismic data; however, the crest of the fold at the Rollins level lies southwest of its surface trace. The Mesaverde Group may thin slightly across the fold (Figure 2), but critical well control is lacking on the southwestern flank to confirm this. Seismic data indicate that the southwestern flank is truncated by a northeast-dipping reverse fault. Sedimentary rocks older than the Chinle Formation (Upper Triassic) are largely missing on the upthrown block. The 1-27 Arco Deep well (SW/4 sec. 27, T6S R97W, Figure 2) either drilled directly into Middle Proterozoic (1,100 Ma) metamorphic basement after penetrating the Chinle Formation or drilled through a greatly thinned section (≤270 ft [82.4 m]) of Permo-Pennsylvanian sandstone and shale before penetrating crystalline basement. No Mississippian carbonates were recognized in the well. This relationship suggests that the Crystal Creek anticline may have been a horst during Late Paleozoic time, as was the Rulison anticline to the east (DeVoto and others, 1986; Waechter and Johnson, 1986).

Detailed structural mapping on top of the Rollins Member in Grand Valley field indicates an irregular homoclinal dip northeastward (Figure 2). The irregularities may be caused by small, undetected northeast-trending faults or by the Rollins' stratigraphic rise to the southeast. Extensive natural fracturing has been documented to the east in Rulison field, especially at the U.S. Department of Energy MWX test site (Figure 1) (Finley and Lorenz, 1988). In Rulison field a northwest-trending zone of probable enhanced natural fracturing approximately follows the crest of the Rulison anticline, a Pennsylvanian-age horst that apparently was reactivated during the Laramide orogeny. Within this zone, which has been recognized indirectly from well production histories (Peterson, 1984; Mercer and Frohne, 1986), some wells completed in the Mesaverde have produced 300 percent more gas than wells beyond the zone. Based on well tests and performance, natural fracturing also is developed in Grand Valley field but not to the degree indicated in Rulison field. The density of natural fractures probably is greater on the northeastern flank of the Crystal Creek anticline than along its crest. Although this apparent relationship may be attributable to the lack of well control, it may be the result of compactional folding in repeated sandstone-shale sequences of the Williams Fork Formation (Gay, 1989).

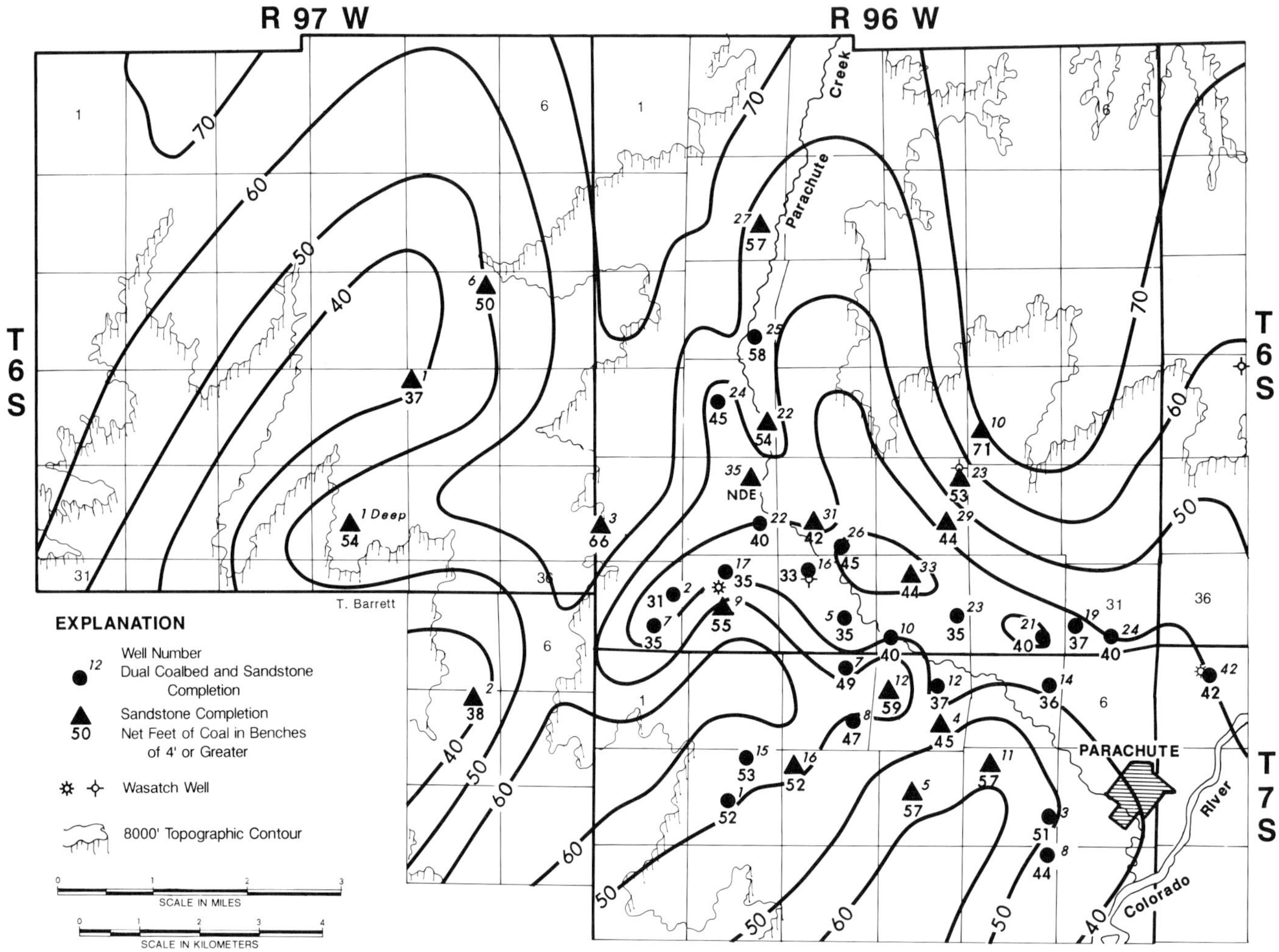

Figure 5. Isopach map of net Cameo coal in seams greater than 4 ft (1.2 m) thick, Grand Valley field. Contour interval 10 ft (3 m).

EXTENT OF GAS ACCUMULATION

Gas accumulations in low-permeability (tight) Mesaverde reservoirs in the Piceance basin are similar to accumulations in tight reservoirs in other Rocky Mountain basins in that they occur in the structurally deepest part of the basin, cross lithologic boundaries, and lack conventional seals. In the basin-center gas-trap model, Masters (1979) and Brown and others (1986) proposed that gas generated deep within a sedimentary basin migrates upward until it is trapped by water-saturated, tight reservoirs. Masters also noted that permeability to gas in extremely tight rocks is only about 30% that of water at 40% water saturation, a typical saturation level for Cretaceous tight reservoirs. In Grand Valley field, sandstones in the upper one-third of the Williams Fork Formation appear to be only sporadically gas saturated, whereas all sandstones in the lower two-thirds appear to be gas saturated. (Gas has been tested but never produced from upper Williams Fork sandstones north of Grand Valley field.)

A cross section from De Beque to Rulison field (Figure 8) shows that, based on mud-log gas shows, the top of the gas-saturated section rises stratigraphically toward the deep basinal trough. Factors that may have influenced the extent of the gas pocket in these tight reservoirs include porosity and permeability changes, local structure, variation in reservoir pressure, proximity to outcrop, and lateral continuity of sandstones. The Rollins Sandstone, near the base of the Mesaverde Group, is unique in that it probably contains movable water throughout most of the basin. It appears to have acted as a major conduit for fluid movement because of its lateral persistence (Johnson and others, 1987). Localized traps might exist in the Rollins either where it steps up stratigraphically or in structurally deep parts of the basin.

DEVELOPMENT HISTORY

In 1984 the Crystal Creek A–2 well (now the MV–1–23) was drilled to 9,877 ft (3,012 m) in the NW/4 sec. 23, T6S R97W on the axis of the Crystal Creek anticline. This well has been designated as the discovery well for Grand Valley field. Crystal Creek A–2 was the first in a four-well wildcat program that targeted tight sandstone reservoirs in the Iles and Williams Fork Formations and conventional sandstone reservoirs in the Paleocene-Eocene Wasatch Formation

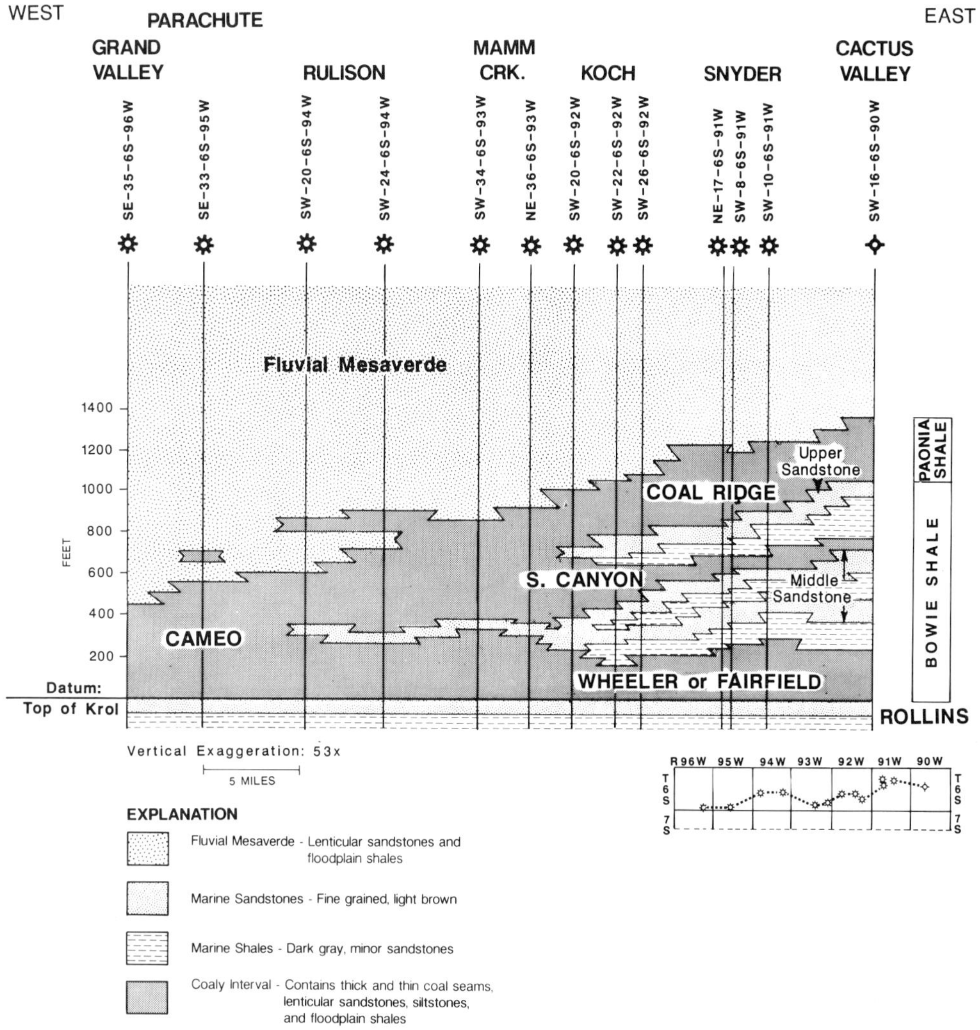

Figure 6. Cross section from east end of Grand Valley field to southeast of New Castle, showing stratigraphic changes in the coal interval.

Figure 7. Accretion point-bar deposit in the Williams Fork Formation in NW/4 NE/4 sec. 5, T9S R100W. Scale shown at upper left is approximately 20 ft (6 m). Note lenticularity of sandstone.

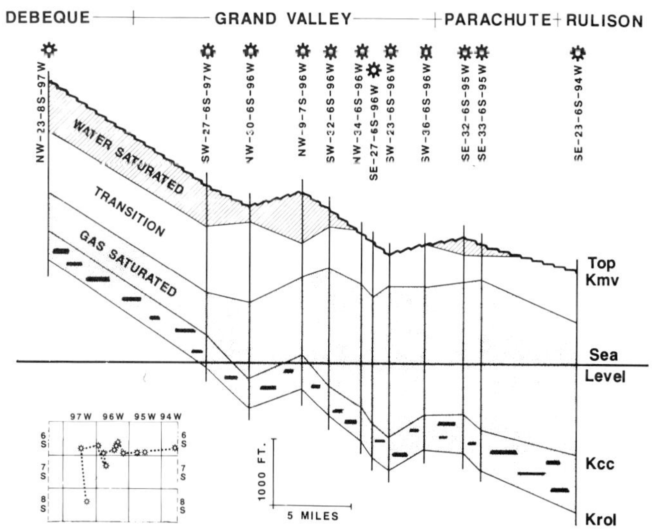

Figure 8. Cross section from De Beque to center of Rulison field. Note relative position of gas-saturated sandstones above the Cameo coal zone (Kcc) in Williams Fork Formation based on mud-log shows.

(Figure 3). The potential for gas in the Cameo coal zone was recognized early but not considered a primary objective because of anticipated high water-production rates. Testing later demonstrated that water rates were minimal, and the coalbeds would be favorable economic objectives.

Rulison field, on the other hand, had undergone extensive developmental drilling between 1979 and 1982 and became the site of the U.S. Department of Energy Multiwell Experiment (MWX), at which extensive geologic and engineering studies were conducted on three closely spaced wells (Northrop, 1990b). The field produces from most of the reservoirs that were expected to be found at Grand Valley. Most wells at Rulison, however, reached total depth above the Cameo coal zone, and the coalbeds were not a producing horizon in 1984.

All four Grand Valley wildcats became gas discoveries, but the MV–4–3 well (SE/4 sec. 3, T7S R96W) was the greatest commercial success. Although the Cozzette and Corcoran Sandstones were tested, initial results indicated noneconomical production rates. The better reservoirs lay within the inferred distributary-channel facies (Cozzette Sandstone in well MV–4–3), which occurs erratically.

During 1985, drilling activity was focused on an offset to the MV–4–3 and on the 15,531-ft (4,737-m) Arco Deep wildcat, which tested the potential of the Dakota Sandstone (Lower Cretaceous) and Morrison and Entrada Formations (Jurassic) on the Crystal Creek anticline. Development of Iles and Williams Fork reservoirs in Grand Valley field continued at a slow pace during 1986, when only three wells were drilled, all on the northeastern flank of the anticline, where thicker, more numerous high-porosity sandstone reservoirs were developed. These wells were located in the valley of Parachute Creek, where drilling depths are approximately 3,000 ft (915 m) less than on adjacent plateaus (Figure 2). During 1985 and 1986 most of the activity occurred east of Grand Valley field in Parachute field (Figure 1), where shallow Wasatch Formation reservoirs were being developed at depths of 1,200 to 1,800 ft (366 to 549 m).

After pipeline connections were secured in late 1986, field activity accelerated during 1987 and 1988, spurred also by the opening of new gas markets, acquisition of new leases, and drilling obligations on old leases. In 1988 Cameo coalbeds were first production tested in wells MV–15–8 (NE/4 sec. 8, T7S R96W) and MV–25–17 (SE/4 sec. 17, T6S R96W). Before that, only sandstone reservoirs, or commingled coal and sandstone reservoirs, within the Cameo interval were perforated and hydraulically fractured. Short-term production testing indicated that Cameo coals could be commercial reservoirs, even though they were deeper than most, if not all, coal reservoirs developed elsewhere. Because of the encouraging test results and because of the availability of Section 29 tax credits and the technology to produce gas simultaneously from coals and tight sandstones, a coalbed methane development program was begun in late 1989.

Currently in Grand Valley field 41 wells produce from the Mesaverde, 23 of which produce from coalbeds. Cumulative production through November 1990 from all Mesaverde zones was 8.73 BCF (247×10^6 m^3)—6.67 BCF (188.7×10^6 m^3) from tight sandstones and 2.06 BCF (58.3×10^6 m^3) from coalbeds—with a total deliverability of about 15,000 MCFD (424,500 m^3/day).

WELL SPACING

Well spacing in Grand Valley field currently is unregulated by the Colorado Oil and Gas Conservation Commission, but both sandstone and coal reservoirs are being developed on 160-acre (65-ha) spacing. In Parachute field the Commission has determined spacing in the Mesaverde to be three wells per section (640 acres); in Rulison field, spacing ranges from 640 acres to 320 acres (259 to 130 ha).

Although Mesaverde fluvial sandstone reservoirs are discontinuous, sandstone-rich packages 200 to 400 ft (61 to 122 m) thick do tend to correlate over wide areas. Barrett Resources Corporation has attempted to determine the net thickness of individual sandstones that are unique to a single well on 160-acre (65-ha) spacing. When two wells are compared, only about 33% of the sandstone beds (based on thickness) appear to correlate. Among three wells, the percentage of sandstones that correlate between at least two of the wells varies from 49% to 63%. Obviously, as a well becomes surrounded by offset wells, the number of sandstones that can be correlated will increase. Knutson and others (1971) estimated only 5% of the sandstones extend 1,200 ft (366 m) from the wellbore. Well spacing becomes important when one considers that, according to Lorenz and others (1985), the average meander-belt width for the fluvial section of the Williams Fork Formation is 1,500 ft (458 m) (Figure 9). In general, sandstones in the upper Williams Fork appear to be thicker and more widespread than those in the lower part.

A laterally continuous reservoir may not be present even though a given sandstone appears to correlate between wells, based on its log character and relative stratigraphic position. In one example, a sandstone bed with good porosity and gas shows and consistent thickness in one well was tight and nonproductive in an offset well. Heterogeneities, such as shale drapes, fractures, or other permeability barriers, may have prevented effective fluid communication. Another example of reservoir compartmentalization has been documented in reservoirs of the Frio Formation (Oligocene) in southern Texas (Jackson and Ambrose, 1989). Although labor intensive, quantitative correlations may be necessary to determine the economic feasibility of decreasing well spacing to 80 acres (32.4 ha).

COAL RESERVOIR PROPERTIES

The rank of Cameo coals in Grand Valley field is low-volatile bituminous (Johnson and Nuccio, 1986), with mean

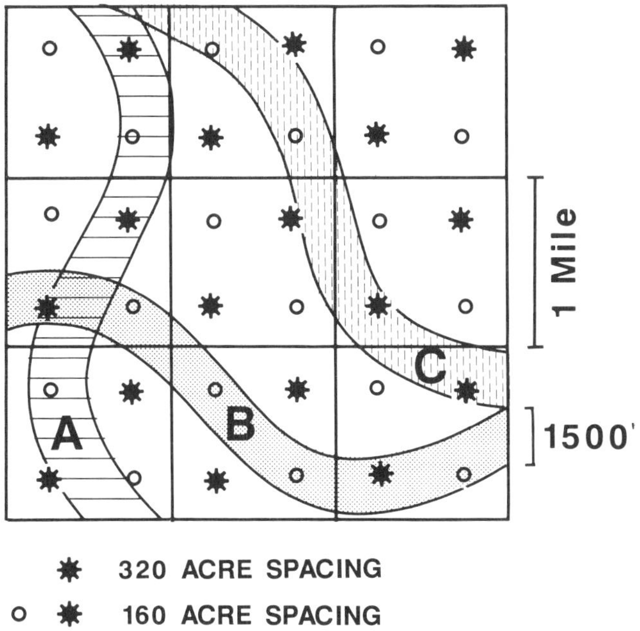

Figure 9. Idealized well spacing to increase the probability of encountering fluvial channel sandstone units (A, B, C). A typical well will encounter approximately 30 discrete sandstones within the Williams Fork Formation.

random vitrinite reflectance (R_m) values of 1.54% to 1.90%. Proximate analyses yielded the following results:

Component	Average (%)	Range (%)
Ash	13.0	5.1–27.3
Fixed carbon	69.0	53.5–77.1
Moisture	3.0	1.4–6.8
Volatile matter	15.7	13.2–16.9

Ash content of the thick, basal coalbeds (<200 ft above the top of the Rollins Member) average 14.3% (range 8.8% to 27.3%), whereas ash contents of the thin, upper coals average 11.1% (range 5.1% to 21.6%). If the high-end value is removed, the average ash content for upper coals becomes 8.9%. These observations may be skewed because of the limited data set and because of changes in the Rollins' stratigraphic position.

Limited coring of the basal coalbeds revealed face-cleat development at a spacing of about 0.1 to 0.3 in. (2.5 to 7.6 mm). Butt cleats were poorly developed to nonexistent in the basal beds. No oriented cores were taken to determine orientation of the cleating system.

Thin coals cored from the upper Cameo zone generally disintegrated when removed from the core barrel, suggesting better development of both face and butt cleats than in the thick, lower coals. Besides compositional variations, other possible explanations for this difference in cleat development are that (1) thinner coal seams fracture more readily than thicker seams (Price, 1966); and (2) coals interbedded with sandstone fracture more readily than those interbedded with shale (Ayers, 1989; Donaldson, 1979). Core disintegration also may have been aided by escaping gas. Production testing indicates that most of the coalbed gas comes from the upper thinner seams rather than the thick basal coal seams.

Gas desorption values for core samples from depths of 6,572 to 6,976 ft (2,004 to 2,128 m) range from 234 to 604 SCF/T (7.3 to 18.8 cc/g) by the Smith and Williams (1981) method. No corrections were made for ash content. An adsorption/desorption isotherm generated from a coal sample at depth 6,586 to 6,587 ft (2,008.7 to 2,009 m) in well MV–24–20 is shown in Figure 10. Note that at high pressures the curves have not appreciably leveled out, as is normally observed. This suggests that reservoir pressure need not be lowered much to initiate desorption.

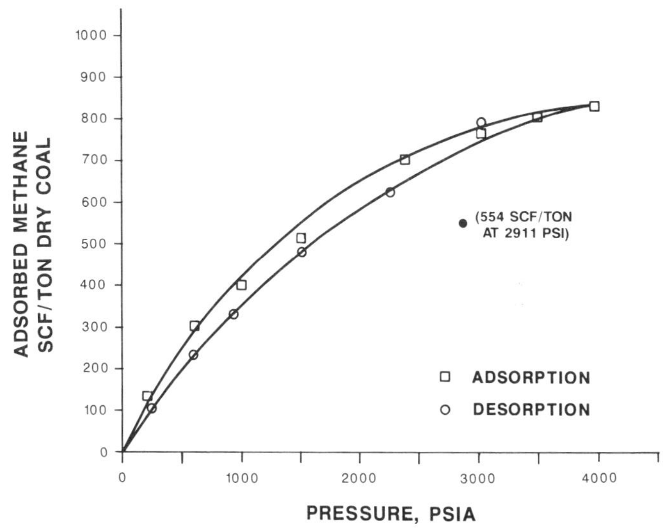

Figure 10. Methane adsorption vs pressure. Measured at 200°F (93.3°C) from a Cameo coal seam at depth 6,586 to 6,587 ft (2,008.7 to 2,009 m) in well MV–24–20 (NW/4 sec. 20, T6S R96W). Estimated ash content is 11.66%. Position of the data point, 554 SCF/T (17.3 cc/g) desorption gas content at 2,911-psi (20,071-kPa) reservoir pressure, below adsorption isotherm indicates that coal reservoirs should require dewatering to initiate gas flow.

When the gas content, 554 SCF/T (17.3 cc/g), measured by desorption and the reservoir pressure, 2,911 psi (20,071 kPa), are plotted on this graph, the data point lies below the adsorption isotherm, indicating that dewatering would be required to initiate gas flow. However, upon actual completion of the coal reservoir, virtually water-free gas was immediately produced, indicating a gas-saturated cleat system. Furthermore, reservoir pressure data suggest that the coal reservoirs are normally pressured. Therefore, the most likely explanation for why gas contents do not plot on the curve is that the measured amount of desorbed gas is erroneous. The error probably arises from the comparatively long trip time (2 to 4 hours) of the core to the surface from an average depth of 6,500 ft (1,982.5 m). Lost-gas ratios (correction factors) for these measurements using the Smith and Williams (1981) technique are at or near their maximum values and were not designed to account for long trip times. Alternatively, gas-content estimates from drill cuttings for these depths are at best only qualitative because of their large surface areas and long circulation times. The magnitude of gas content suggested by the adsorption curve has been verified by pressure-core data from the 1 SHCT well (NW/4 sec. 34, T6S R94W), where a gas content of 765 SCF/T (23.9 cc/g), uncorrected for ash, was reported from Cameo coal at 7,090 to 7,100 ft (2,162 to 2,166 m) TVD (F.R. Myal, 1990, pers. comm.).

Cameo coal reservoirs in Grand Valley field produce little associated water. Although this is beneficial for water disposal, it is disadvantageous in that the reservoirs lack the well developed natural fracture permeability associated with prolific water and gas flows on the Divide Creek anticline in eastern Piceance basin and from areas of the northern San Juan basin. Well test data from Parachute field indicate that in situ coal permeability ranges from 0.02 to 0.2 md. The absence of abundant open fractures at Grand Valley field, compared to highly fractured reservoirs at Divide Creek, probably resulted from the coals' deeper burial and from their nonfaulted nature.

SANDSTONE RESERVOIR PROPERTIES

A typical density-neutron log response for low-permeability sandstone reservoirs is shown in Figure 11. Porosities measured from density logs are about 10% to 12%, but measured core porosities from the MV-8-4 well (SE/4 sec. 4, T7S R96W) are only 7% to 9%. Natural permeability measured at reservoir conditions—about 3,000-psi (20,685-kPa) confining pressure—is on the order of 5 to 30 μd, unless natural fracturing, which can cause a 1:100 permeability anisotropy, is present (Lorenz, 1990). These permeability values are similar to those found in cores from the MWX site (Sattler, 1989).

The mud log provides a good qualitative assessment of sandstone reservoir quality in a given well (Figure 11), although absolute values for total gas vary from well to well because of differences in gas-detection devices and drilling environments. Features that better performing sandstone reservoirs share include distinct gas kicks throughout the porosity zone and good drilling breaks (<2 minutes/ft). In addition, these reservoirs cause a higher background gas in the drilling mud after penetration than before penetration. The upper one-third of the Williams Fork Formation is considered to have low reserve potential. Gas kicks from these sandstones may increase slowly to a peak and then diminish slowly, suggesting gas and water in solution within the reservoir; or the gas show occurs only in a restricted part of a thick homogeneous sandstone, suggesting permeability barriers or distinct gas-water contacts. Gas-water transition zones such as these are typical of low-permeability basin-center gas accumulations (Masters, 1979).

Although some density-neutron logs of the Rollins Sandstone show gas effect or gas crossover, the Rollins probably is incapable of production. Evaluation of the formation with mud logs is difficult because the drilling mud contains high levels of background gas derived from the overlying Cameo coals. Gas kicks from the Rollins itself are rare. What testing has been done in Grand Valley field indicates that the Rollins contains movable water.

Gas shows from thin intervals of the stratigraphic section that apparently contain no reservoirs must be explained. Detailed logging of drilling rate and examination of high-resolution electric logs have revealed that some gas shows originated from thin (<1 ft [30 cm]) coalbeds. The remainder appear to have come from carbonaceous shale or thin sandstones representing the pinch-out of a lenticular reservoir not fully encountered by the wellbore. If these shows had come from natural fractures, background gas probably would have remained at a high level after the show because of high permeability. This was not the case, however.

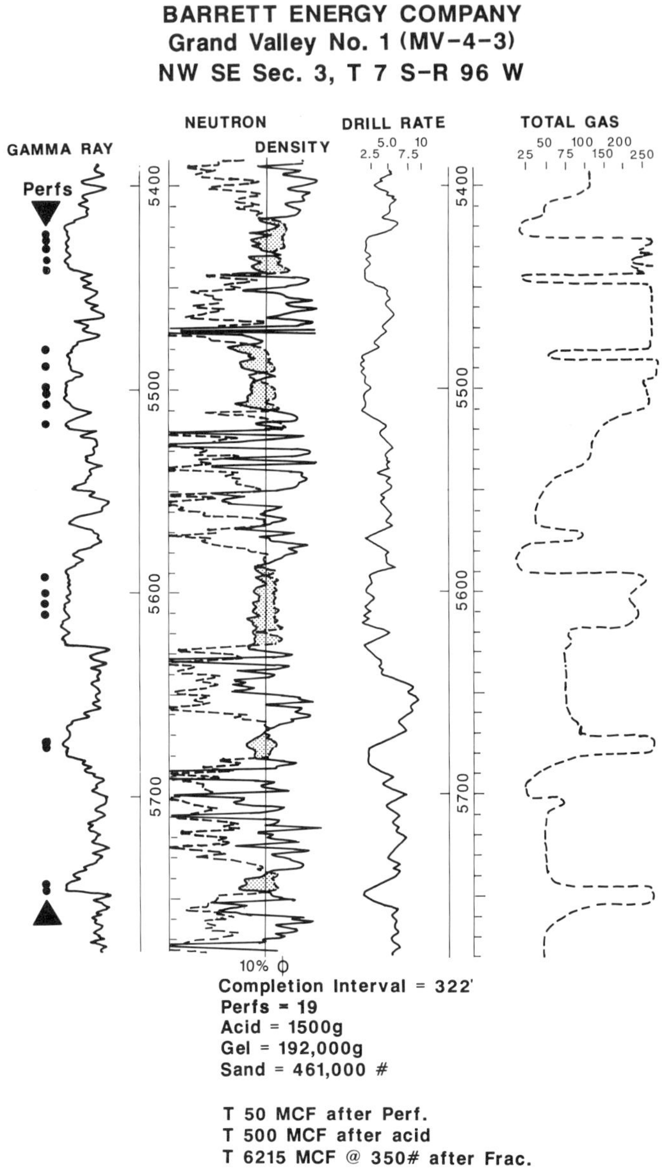

Figure 11. Typical electric and mud-log response of sandstone in Williams Fork Formation, with completion information from the Barrett MV-4-3 well, Grand Valley field (sec. 3, T7S R96W).

DRILLING AND COMPLETION TECHNIQUES

Important markers encountered during drilling in Grand Valley field include, from youngest to oldest (1) sandstones of the Molina Member of the Wasatch Formation; (2) a coaly zone in the lower Wasatch Formation (upper Atwell Gulch Member), which sometimes is a lost-circulation zone; (3) top of the Cretaceous, which is marked by the loss of variegated shale and siltstone and the appearance of "salt-and-pepper" sandstone and dark-gray shale; (4) onset of sustained gas shows in the lower two-thirds of the Williams Fork Formation; (5) Cameo coals (<1 minute/ft drilling rate); and (6) light brown, fine-grained Rollins Sandstone, the most distinctive Mesaverde marker bed in the field. No distinctive sample or electric log markers have been recognized in the Williams Fork Formation.

Drilling depths to the Rollins Sandstone in Grand Valley field range from as shallow as 5,900 ft (1,800 m) below the canyon floors to as deep as 8,900 ft (2,714 m) below the top of the Roan Cliffs. In the valley of Parachute Creek, where most development activity has occurred, the depth to the top of the Rollins averages 6,500 ft (1,982 m). Surface casing (8 5/8 in.) (21.9 cm) normally is set to 300 ft (91.5 m), and a 7 7/8-in. (20-cm) hole then is drilled to total depth. Drilling fluid consists of a fresh-water (2% KCl) polymer gel-mud system that inhibits swelling of shale in the Wasatch. One-person mud-logging units are used from the base of surface casing to total depth. Typically, drilling is concluded in 19 days after drilling 150 to 200 ft (46 to 61 m) below the top of the Rollins.

At total depth, an open-hole log suite consisting of dual induction, gamma ray, SP, dual caliper, density, and neutron logs is run back to surface casing. In Grand Valley field a short-axis kit is used to provide valid porosity data. The need arose when it was discovered that asymmetrical borehole spalling had occurred (Figure 12). This spalling, which is related to the present stress field (Blümling and others, 1983; Lacy and Smith, 1989), causes the borehole to elongate as much as 9 to 15 in. (22.9 to 38 cm). Porosity-tool pads tend to become aligned and ride in the rough long axis, thereby giving erroneous porosity readings. Ideally, the short-axis kit forces the density pads to follow the smoother, short axis (8 to 10 in. [20 to 25 cm]) for more reliable readings. After logging is completed, 5 1/2-in. (14-cm) casing is run, and any Mesaverde zones considered to be productive are covered with cement. Two-stage cementing sometimes is necessary if the interval is sufficiently long—>3,000 ft (915 m)—or if Wasatch gas potential exists.

When Rulison field was developed between 1979 and 1982, various drilling techniques were used. Setting casing to the top of the Mesaverde and continuing drilling with air allowed penetration of the fluvial interval in only three or four days—two to three times as fast as with water-based drilling. Another advantage of air is minimal fluid contact with the Mesaverde sandstones. Research at the MWX site suggests that natural fractures are extremely sensitive to blockage by fluids (Northrop, 1990a). The disadvantage of air is that the drilling assembly often would become stuck in the Cameo coals.

Completion technology for coalbeds and fluvial sandstone reservoirs in the Mesaverde is an ever-evolving process. Early well completions in Rulison field were small, <100,000 lb (45,359 kg), sand-propped hydraulic fracture treatments in individual sandstones or in multiple-sandstone intervals as long as 300 ft (91.5 m). No evidence was found that larger fracture treatments were used for longer intervals. During the period of major development of Rulison field, 1979 to 1982, the practice of completing in long multiple-pay intervals was carried to an extreme with limited-entry techniques. The length of completion intervals increased to more than 1,100 ft (336 m), with fracture-

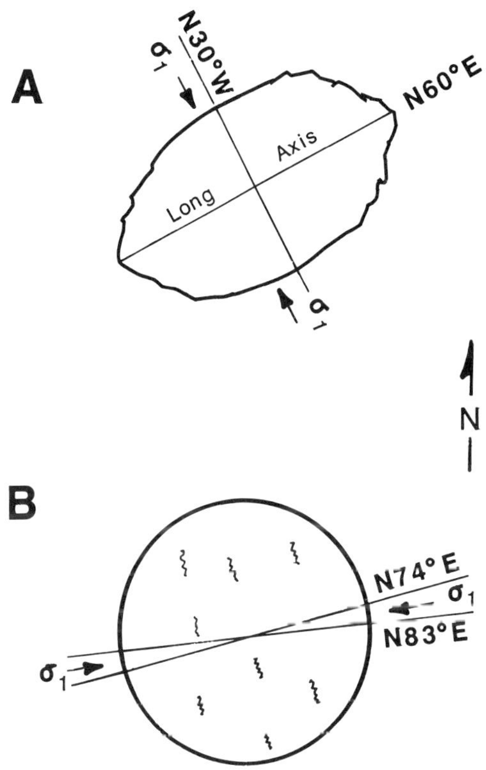

Figure 12. Orientation of horizontal compressive stress. *A.* Horizontal section through asymmetrical wellbore of well GV–22–29 (SE/4 sec. 29, T6S R96W). Blümling and others (1983) suggest that maximum principal compressive stress (σ_1) will be oriented parallel to the short axis of the hole. However, according to Lacy and Smith (1989), orientation of σ_1 parallel to the long axis also is valid. The degree of natural fracturing probably influences which axis will represent the σ_1 orientation. *B.* Acoustic-velocity anisotropy in Williams Fork Formation sandstone in well MV–8–4 (SE/4 sec. 4, T7S R96W) indicates σ_1 orientation between N74°E and N83°E. Experiments assume that as the core is relieved of in situ stress, microcracks will open with their long axes perpendicular to σ_1 orientation. Acoustic velocity tool records slower travel times as it measures data perpendicular to microcrack orientation. Data gathered and interpreted by N.R. Warpinski and J.C. Lorenz, Sandia National Laboratories.

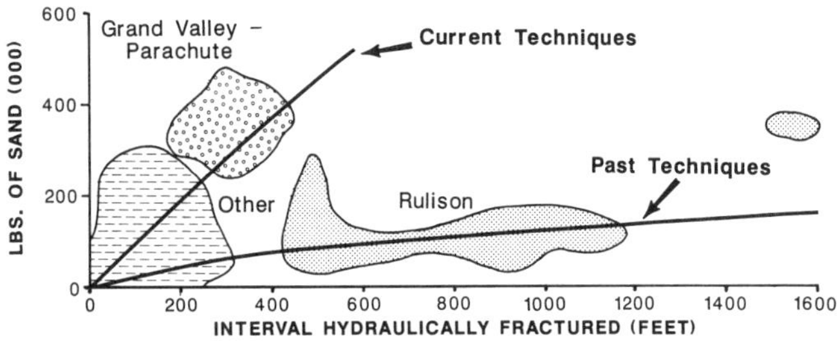

Figure 13. Length of hydraulically fractured treatment interval vs weight of sand proppant. Patterns isolate general differences in past and current completion techniques applied to sandstones in the Williams Fork Formation.

treatment sizes generally below 150,000 lb (68,038 kg) of sand (Figure 13).

In Grand Valley field the limited-entry technique has been modified so that treatment size is proportional to the length of the interval treated and to total thickness of reservoir sandstone within the interval. At the same time, the number of perforations is restricted to 20, and the length of the interval to be treated is limited to 450 ft (137 m), preferably less than 350 ft (107 m) (Figure 11). Wherever possible, a minimum 100-ft (30.5-m) barren buffer zone is left between completed intervals. Figure 14 compares the relative stratigraphic positions of present and anticipated completion zones for three Grand Valley wells.

For fluvial sandstone reservoirs, 1,000 gal (3.8 m^3) of acid are injected to clean up perforations. Natural flow rates average 225 MCFD (6,368 m^3/day) at 145-psi (1,000-kPa) flowing tubing pressure, although a natural rate as high as 1,706 MCFD (48,280 m^3/day) has been recorded from a naturally fractured reservoir. Current treatment size in Grand Valley field averages 327,000 lb (148,324 kg) of sand and 110,000 gal (416.4 m^3) of gelled 2% KCl water.

Limited-entry completion methods for coal reservoirs differ little from those for sandstones. Coalbeds within a designated completion interval of no more than 450 ft (137 m) are perforated with approximately 20 holes (Figure 4). More perforations are allowed in the thick basal coalbeds than in the thinner upper coalbeds. Because the number of coalbeds typically exceeds the number of perforations allowed, preference is given to those seams closest to other thin seams. In thicker seams, perforations are either grouped near the center of the seam or distributed evenly across it; neither technique appears to affect completion results. In most cases, fracture half-lengths are believed to be relatively short, and fracture apertures are wide.

Following acidization to clean up perforations, natural flow rates from Cameo coalbeds average 100 MCFD (2,830 m^3/day) at 100-psi (690-kPa) flowing tubing pressure. An average 362,000 lb (164,200 kg) of sand and 152,000 gal (575.4 m^3) of gelled 2% KCl water are pumped

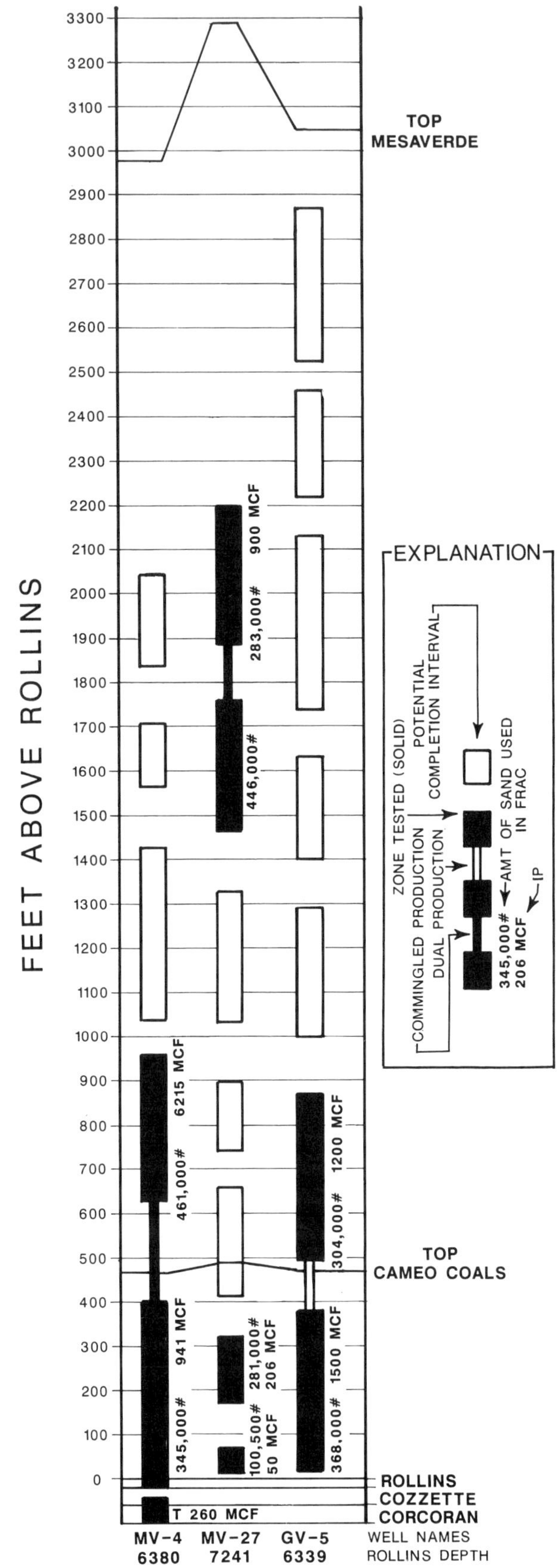

Figure 14 (above right). Completion intervals, stimulation used, and production data for three Grand Valley field wells producing from coal and sandstone reservoirs within the Williams Fork Formation, normalized to top of the Rollins Sandstone.

down 5 1/2-in. (14-cm) casing to stimulate the coalbeds. In Grand Valley field, a Cameo coalbed completion is allowed a short cleanup time (48 hours) before a packer and brass knockout plug are set in the casing above the completion. The sandstone interval above the Cameo then is treated down the casing, and tubing is immediately lowered downhole to knock out the brass plug. Coal gas is produced up the tubing and sandstone gas up the casing. These two flow streams are metered separately for purposes of the Section 29 tax credit and then combined in the gathering system at approximately 700 psi (4,826 kPa).

Other completion techniques are being investigated for both sandstone and coal reservoirs as development continues outside Grand Valley field. These techniques involve smaller treated interval sizes, higher numbers of perforations, reduced fluid viscosity, and larger treatments per foot of interval; all are done with the objective of increasing fracture half-length. It is believed that flow rates and/or ultimately recoverable reserves will increase as more coal surface area is exposed to the induced fracture.

GAS COMPOSITION

Gases from low-volatile bituminous coals in the Cameo zone and from sandstone reservoirs above the Cameo zone generally can be distinguished by their chemical composition. Coal gases are chemically drier (90% to 93% methane, 1% to 3% ethane) and contain relatively large amounts of carbon dioxide (3% to 6%) compared to gases from sandstones, which are chemically wetter (89% to 92% methane, 3% to 6% ethane) with less carbon dioxide (1% to 4%). Furthermore, gases from coals in the lower Cameo zone generally are chemically drier with larger amounts of carbon dioxide than are gases from upper Cameo coalbeds. Differences in composition are better illustrated in Figure 15, a plot of ethane content vs carbon dioxide content for gas samples from Grand Valley and Parachute fields. Although two end members (coal and sandstone) are present, a spectrum of compositions exists, including a transitional area with characteristics of both coal and sandstone gases.

Analyses of gases desorbed from whole core show that the percentage of ethane increased and percentage of methane and carbon dioxide decreased with increasing desorption time (D. Dowden, 1990, pers. comm.). This phenomenon may be attributable either to different sorptive characteristics of the gases or to relative permeability effects among them that, because of their different molecular sizes and weights, result in dissimilar desorption rates (Harpalani, 1989). Gas samples collected from desorbed coal cuttings show considerable variation in composition but generally contain higher methane and lower carbon dioxide than gases from coalbeds (Figure 15). Most of the cuttings analyses also were normalized to 1% nitrogen because of air contamination. The difference in composition of gases from coal cuttings compared with coalbeds may be the result of the larger surface area of cuttings, which tends

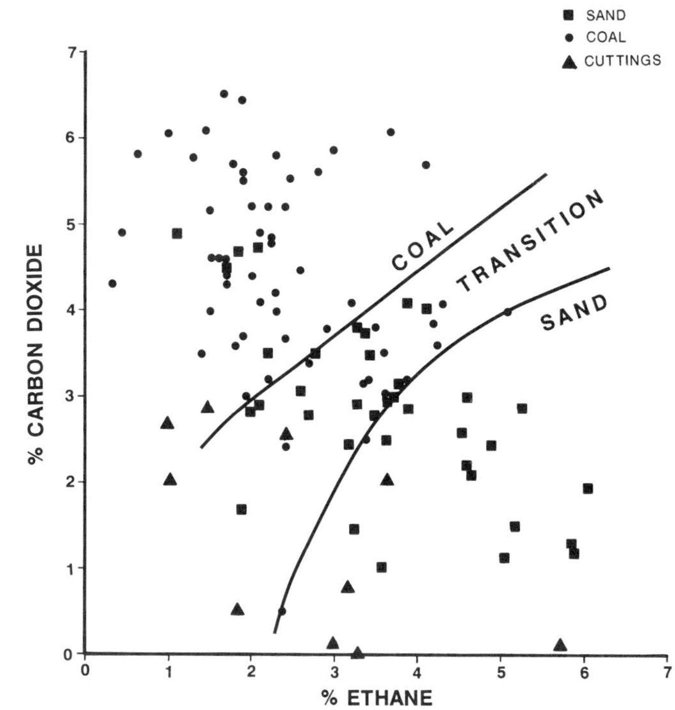

Figure 15. Comparison of gas composition based on carbon dioxide and ethane contents of gas samples desorbed from Cameo coal cuttings and produced from Cameo coalbeds and from sandstones above the Cameo in Grand Valley and Parachute fields.

to increase desorption rates, and/or to errors introduced by normalization. Although analyses of cuttings gases are helpful, they do not accurately represent the composition of gases produced from coalbeds. Produced gas is similar to that obtained from desorption of whole core.

In the De Beque area (secs. 23 and 29, T8S R97W) the coalbeds are of lower rank (high-volatile A to medium-volatile bituminous), and gas composition is different. Gases desorbed from coal cuttings and from well completion tests generally are chemically wetter (79% to 88% methane, 3% to 16% ethane) and contain less carbon dioxide (0.2% to 5.4%) than gases from the higher rank coals in Grand Valley field. Thus, chemical distinction of coalbed and sandstone gases should be attempted only in areas of similar coal rank.

Carbon isotopic composition of the methane fraction ($\delta^{13}C_1$) of gases desorbed and produced from coal and sandstone reservoirs does not appear to be a distinguishing characteristic. $\delta^{13}C_1$ values for desorbed gas from coal cuttings are variable (–29.6‰ to –37‰); values for gases from coal, sandstone, and commingled sandstone and coal reservoirs cluster in the range of –35.4‰ to –37.8‰.

The gases desorbed and produced from the coalbeds are inferred to be thermogenic and to have been generated within the coalbeds. However, coal-derived gases from Grand Valley field are chemically wetter than those produced either from high- to medium-volatile bituminous coal in the San Juan basin (Rice, Clayton, and Pawlewicz,

1989) or from medium- and low-volatile bituminous coal in the Black Warrior basin (Rice, Epsman, and Mancini, 1989). Chemically wet gases desorbed from coalbeds have been reported by Hanson (1990) from several Rocky Mountain basins and by Tremain and Toomey (1983) from coals of the same rank as those in Grand Valley field but from elsewhere in the Piceance basin. The differences in gas composition probably result from the coals' chemical and/or maceral composition—those with higher hydrogen content (exinite/liptinite) generate chemically wetter hydrocarbon gas and less carbon dioxide than coals that are more oxygen (vitrinite) rich. Collins (1976) reported higher amounts of exinite in the southeastern Piceance basin, where he suggested marine influence on coal deposition. We suggest that hydrogen-rich organic matter was deposited in the area of Grand Valley field and that it is more prevalent in the upper Cameo zone than in the lower. Although carbon isotopic composition of methane is not useful in distinguishing gases among the various reservoirs, it is helpful in determining if the gases are thermogenic and if they were generated at relatively high levels of thermal maturity. As indicated by isotopic composition and by coal rank, the gases in Grand Valley field were generated at higher levels of thermal maturity than those in the San Juan basin.

Because of compositional differences, most of the gases produced from sandstone reservoirs above the Cameo coal zone are interpreted to have been generated not from the coals but from dispersed type III kerogen abundant in the interbedded nonmarine shale. Similar conclusions were reached by Rice, Clayton, and Pawlewicz (1989) and by Johnson and Rice (1990) for gases produced from sandstone and coal reservoirs in the San Juan and Piceance basins, respectively. In Grand Valley field, gas compositions for Cameo sandstone reservoirs plot in the transitional zone of Figure 15 and appear to contain components derived and expelled from both nonmarine shale and coal. Evidence for this comes from analyses of three gas samples collected from unstimulated Cameo sandstone reservoirs where no coals were open. Their compositions plot within the transitional zone, two toward the coal boundary and one toward the sandstone boundary. This is not surprising, considering that most Cameo sandstones are bounded by coaly intervals and in some cases are separated from underlying coalbeds by less than 2 ft (0.6 m) of low-permeability rock. An upward limit of coal-derived gas must exist because fewer than one-half of the gases from sandstones above the Cameo plot in the transition zone with gases derived directly from coalbeds (Figure 15).

In reality, the composition of gases from Cameo sandstones probably varies from reservoir to reservoir. Because completion techniques in Grand Valley field target either Cameo sandstones and coals commingled together or only the coals, the data set for Cameo sandstones alone is limited. Most produced gases from commingled Cameo sandstone and coal plot within the transitional zone.

WATER CHARACTERISTICS

The amount and composition of water produced from Cameo coalbeds in Grand Valley field are different from that of waters produced from coals in the San Juan and Black Warrior basins and from other coals in the Piceance basin. Cameo coals generally produce less than 4 BWPD (0.6 m^3/day), whereas Fruitland coals in the San Juan basin commonly produce more than 100 BWPD (15.9 m^3/day) during the early months of production (Decker and others, 1988). Two reasons may explain this. First, the Cameo cleat system, which holds much of the water, probably is not as well developed because of the greater burial depth of the coal. Second, during coalification smaller volumes of water may have been produced from more hydrogen-rich coals that generate chemically wetter gases than from more oxygen-rich coals that generate methane-rich gas.

Water composition also is different. From their analyses of coal- and sandstone-derived waters in the eastern Piceance and San Juan basins, Decker and others (1987) concluded that coalbed waters are a sodium bicarbonate type resulting from coalification. In contrast, sandstones contain sodium chloride-type waters, which result from selective retention and expulsion of certain ions. Waters produced from coal, sandstone, and commingled coal and sandstone reservoirs in Grand Valley field plot as sodium-chloride waters characteristic of sandstones (Figure 16). One explanation for this anomaly is that the produced water was contaminated by treatment fluids that contain KCl. Such contamination can, however, be evaluated by periodic sampling and analysis. Another possibility is that a natural hydrologic communication existed between the coalbeds and adjacent sandstones during coalification.

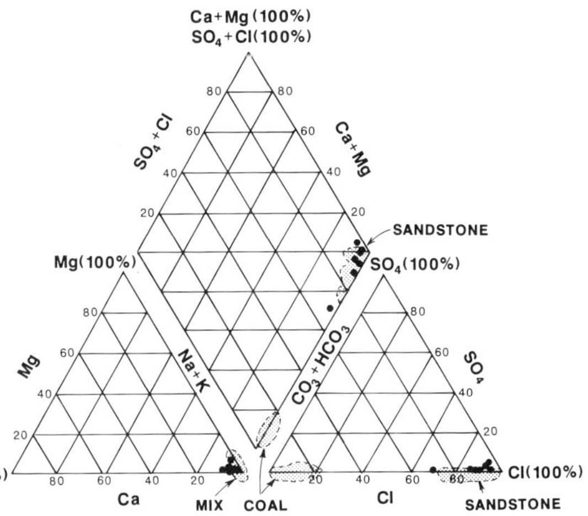

Figure 16. Piper (trilinear) diagram of produced-water composition from coalbed completions in Grand Valley field. Data point that lies closest to the coal region of diagram is from well completion at De Beque (NW/4 sec. 23, T8S R97W).

PRODUCTION ANALYSIS

Analysis of production histories is another possible method of distinguishing gas production from coal and sandstone reservoirs. Production from tight sandstones, such as those above the Cameo coals, typically exhibits rapid initial decline and then slow decline over an extended period of time. Such decline behavior has been documented nearby in Rulison field, where production histories of tight Mesaverde sandstones span 10 to 30 years. Production from water-saturated coalbeds, such as those in the San Juan basin, commonly is steady or even increases for a time (Decker and others, 1988). This type of production behavior results from draining the water-saturated cleat system, followed by diffusion and desorption of gas from the coal matrix. Because Grand Valley field produces little or no water and because it lacks a well developed fracture system, production histories of Cameo coal reservoirs were expected to differ from those of water-saturated, highly fractured coal reservoirs in the San Juan basin.

Analysis of short-term production data shows that gas production from Grand Valley sandstones declines rapidly at first, as expected, and then begins a long period of slow decline (Figure 17a). Production from coal reservoirs, however, is characterized by initial rapid decline, followed by a flatter decline curve than that for sandstone reservoirs (Figure 17b). However, results are highly variable, and longer term production histories are needed to determine any meaningful trends that can be used to distinguish coal from sandstone reservoirs. Prediction of ultimate reservoir performance using decline curves is possible.

Finally note that sandstone reservoirs tend to yield 15 days of flush production following a shut-in period. That is, initial flow rates are significantly higher when a well is brought back into production than were rates prior to shut in (Figure 17a, cumulative 55,000 MCF [1.56×10^6 m^3]). The flush production results from the tight sandstone near the wellbore repressurizing with time. In contrast, flow rates from water-free coal reservoirs are comparable before and after a shut in. In water-saturated coals, such as in the San Juan basin, gas flow rates decrease and water rates increase after a period of shut in (Decker and others, 1988).

SUMMARY

Grand Valley field, on the northeastern flank of Crystal Creek anticline, produces gas from sandstone and coalbed reservoirs in the Mesaverde Group, with lesser amounts of gas from shallower sandstones in the Wasatch Formation. The Cameo coal zone contains low-volatile bituminous coals at drilling depths of 6,000 to 6,500 ft (1,830 to 1,982 m) and with average thicknesses of 50 to 70 net ft (15.2 to 21.4 m) per well. Desorbed gas contents from whole cores range from 234 to 604 SCF/T (7.3 to 18.8 cc/g), but adsorption isotherms and pressure-core data indicate that in situ gas contents are much higher. The fact that butt cleats are better developed in upper Cameo coals than in the lower seams may result from variations in coal composition, seam thickness, or interbedding relationships with sandstones. The better cleat development in the upper coal seams also is reflected by significantly higher gas flow rates from the upper seams. Overall lower gas production rates and minimal water production suggest that cleating generally is more poorly developed here than in the northern San Juan and eastern Piceance basins. This may be the result of burial depth and the relatively unfaulted nature of the Cameo interval in Grand Valley field.

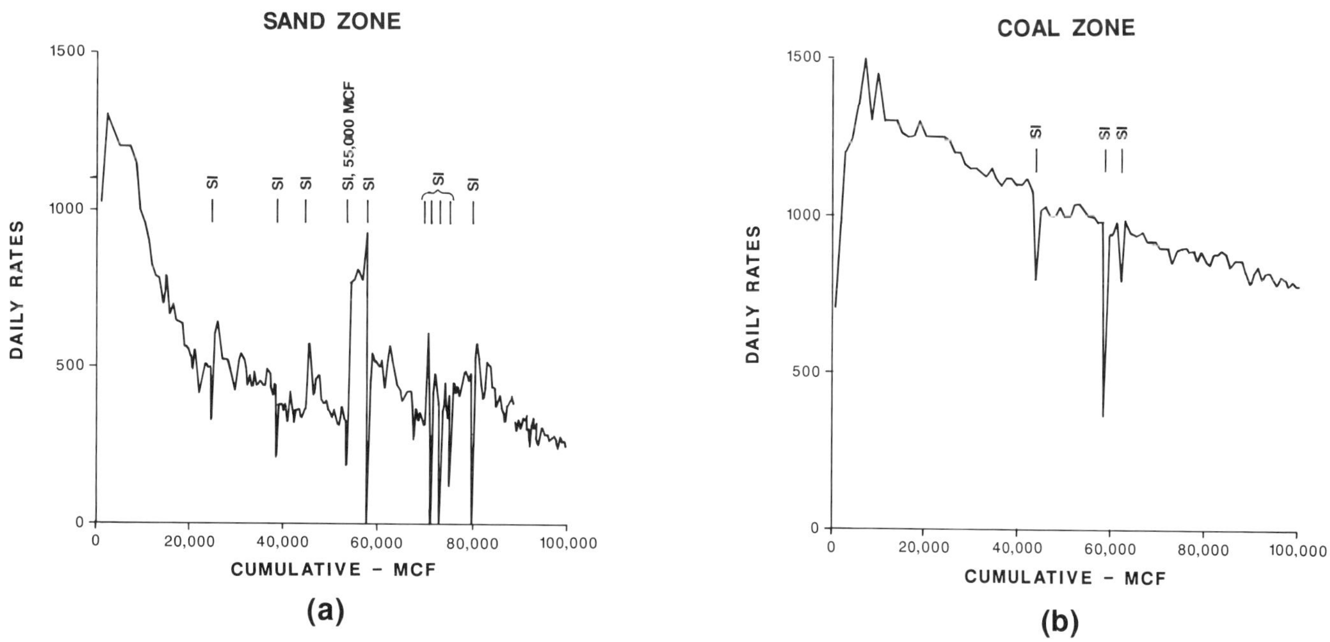

Figure 17. Typical daily flow rates vs cumulative gas produced from (*a*) sandstone reservoir above the Cameo coal zone, and (*b*) Cameo coal reservoir, Williams Fork Formation. For the sandstone, note flush production spike after a prolonged shut-in period at cumulative production of approximately 55,000 MCF (1.56×10^6 m^3). Shorter shut-in periods did not show as pronounced a rate increase. Coal reservoirs generally show no evidence of flush production after a shut-in period.

Tight sandstone reservoirs above the Cameo are produced from the same wellbore as the coalbeds by dual-completion techniques. These sandstones exhibit a highly variable depositional nature and require at least 160-acre (65-ha) well spacing for optimum development. Only the lower two-thirds of the sandstones of the Williams Fork Formation in Grand Valley field appear to contain economically recoverable gas reserves. Upper Williams Fork sandstones contain mostly water because of the basin-center nature of the gas trap. A good mud log is the best tool to qualitatively assess the potential of individual sandstone reservoirs within a given well.

In Grand Valley field, limited-entry hydraulic fracturing has been applied to both coal and overlying sandstone reservoirs. This technique specifies a maximum 450-ft-long (137-m) treatment interval and a maximum of 20 perforations. The amount of sand proppant and treatment fluids are directly related to interval length and to thickness of reservoir identified in the completion interval.

Chemical composition of reservoir gases shows that chemically drier Cameo coal gases can be distinguished from chemically wetter gases from sandstone reservoirs above the Cameo zone. A complete compositional spectrum between the two suggests that both coal and interbedded shale were sources for gases in the Cameo sandstones and to a lesser degree in the sandstones above the Cameo zone. The influence of coal-derived gas in sandstones above the Cameo is more erratic and decreases rapidly above the top of the Cameo. Chemical composition of the gas also changes with desorption time and coal rank.

The small amount of water produced from the coals is a sodium-chloride type rather than the more typical sodium-bicarbonate type. Treatment fluids may have skewed the data, or the coalbeds may naturally contain sodium chloride-type water because of past hydrologic communication with sandstones.

Decline curves are highly variable, and no firm trends can be recognized to distinguish production from coal reservoirs from that of sandstone reservoirs, although individual reservoir performance can be predicted. Sandstone reservoirs, however, tend to yield flush production after a shut-in period, whereas coal reservoirs show no increased production rates after shut in. As more production information is obtained, some mappable parameter may be found to high grade underdeveloped coal areas for future drilling.

ACKNOWLEDGMENTS

The authors extend their thanks to William Barrett, President of Barrett Resources Corporation, for his encouragement and permission to publish this work. The paper benefited from the work of and discussions with Terry Barrett, Terry Dobkins, Merle Evers, and Frank Farnham. Ben Law and Charles Spencer of the U.S. Geological Survey, Denver, Colorado, reviewed the manuscript and offered helpful suggestions. Thanks also are extended to Judy Whittemore for typing the manuscript and to Karen Connally for drafting the illustrations.

REFERENCES

Ayers, W.B., Jr., 1989, Geologic evaluation of critical production parameters for coalbed methane resources: Quarterly Review of Methane from Coal Seams Technology, v. 6, nos. 3 and 4, p. 52–55.

Blümling, P., Fuchs, K., and Schneider, T., 1983, Orientation of the stress field from breakouts in a crystalline well in a seismic active area: Amsterdam, Elsevier Science Publishers, Physics of the Earth and Planetary Interiors, no. 33, p. 250–254.

Brown, C.A., Smagala, T.M., and Haefele, G.R., 1986, Southern Piceance basin model—Cozzette, Corcoran and Rollins sandstones, *in* Spencer, C.W., and Mast, R.F., eds., Geology of tight gas reservoirs: AAPG Studies in Geology, Series 24, p. 207–219.

Collins, B.A., 1976, Coal deposits of the Carbondale, Grand Hogback, and southern Danforth Hills coal fields, eastern Piceance basin: Colorado School of Mines Quarterly, v. 71, no. 1, 138 p.

Decker, A.D., Jeu, S.J., Cooper, J.D., and Wicks, D.E., 1988, Geology, geochemistry, reservoir engineering, and completion methods at the Cedar Hill field, San Juan County, New Mexico—A field study of classic coal degasification behavior, *in* Fassett, J.E., ed., Geology and coal-bed methane resources of the northern San Juan basin, Colorado and New Mexico: Rocky Mountain Association of Geologists Guidebook, p. 221–235.

Decker, A.D., Klusman, R., and Horner, D.M., 1987, Geochemical techniques applied to the identification and disposal of connate coal water, *in* The 1987 Coalbed Methane Symposium Proceedings—University of Alabama, Tuscaloosa, November 16–19, 1987: School of Mines and Energy Development, Gas Research Institute, and U.S. Mine Safety and Health Administration, p. 229–242.

DeVoto, R.H., Bartleson, B.L., Schenk, C.J., and Waechter, N.B., 1986, Late Paleozoic stratigraphy and syndepositional tectonism, northwestern Colorado, *in* Stone, D.S., and Johnson, K.S., eds., New interpretations of northwest Colorado geology: Rocky Mountain Association of Geologists Symposium Guidebook, p. 37–49.

Donaldson, A.C., 1979, Origin of coal seam discontinuities, *in* Donaldson, A.C., Presley, M.W., and Renton, J.J., eds., Carboniferous coal guidebook, volume 1—Carboniferous coal short course: West Virginia Geological and Economic Survey Bulletin B–37–1, p. 102–132.

Finley, S.J., and Lorenz, J.C., 1988, Characterization of natural fractures in Mesaverde core from the multiwell experiment: Sandia National Laboratories Rept. SAND 88–1800, 90 p.

Gay, S.P., Jr., 1989, Gravitational compaction, a neglected mechanism in structural and stratigraphic studies—New evidence from mid-continent, USA: AAPG Bulletin, v. 73, no. 5, p. 641–657.

Hanson, W.B., 1990, Chemistry of Western Interior USA coal-bed gases based upon desorption of subsurface coal samples (abs.): AAPG Bulletin, v. 74, no. 8, p. 1326.

Harpalani, Satya, 1989, Permeability changes resulting from gas desorption: Gas Research Institute Final Rept. GRI–89/0219 (May 1988–June 1989), 88 p.

Jackson, M.L., and Ambrose, W.A., 1989, Influence of reservoir heterogeneity on gas-resource potential for geologically based infill drilling, Brooks and I–92 reservoirs, Frio formation, south Texas: Gulf Coast Association of Geological Societies Transactions, v. 39, p. 127–140.

Johnson, R.C., 1989, Geologic history and hydrocarbon potential of Late Cretaceous-age, low permeability reservoirs, Piceance basin, western Colorado: USGS Bulletin 1787–E, 51 p.

Johnson, R.C., Crovelli, R.A., Spencer, C.W., and Mast, R.F., 1987, An assessment of gas resources in low-permeability sandstones of the Upper Cretaceous Mesaverde Group, Piceance basin, Colorado: USGS Open-File Rept. 87–357, 165 p.

Johnson, R.C., and Nuccio, V.F., 1986, Structural and thermal history of the Piceance Creek basin, western Colorado, in relation to hydrocarbon occurrence in the Mesaverde group, *in* Spencer, C.W., and Mast, R.F., eds, Geology of tight gas reservoirs: AAPG Studies in Geology, Series 24, p. 165–205.

Johnson, R.C., and Rice, D.D., 1990, Occurrence and geochemistry of natural gases, Piceance basin, northwest Colorado: AAPG Bulletin, v. 74, no. 6, p. 805–829.

Knutson, C.F., Maxwell, E.L., and Millheim, K., 1971, Sandstone continuity in the Mesaverde Formation, Rulison field area, Colorado: Journal of Petroleum Technology, v. 23, p. 911–919.

Lacy, L.L., and Smith, M.B., 1989, Fracture azimuth and geometry determination, *in* Gidley, J.L., Holditch, S.A., Nierode, D.E., and Veatch, R.W., eds., Recent advances in hydraulic fracturing: Society of Petroleum Engineers, p. 341–356.

Lorenz, J.C., 1988, Geology, Multiwell Experiment final report, part II—The Paludal interval of the Mesaverde Formation: Sandia National Laboratories Rept. SAND 88–1008, 21 p.

Lorenz, J.C., 1990, Geology, Multiwell Experiment final report, part IV—The fluvial interval of the Mesaverde Formation: Sandia National Laboratories Rept. SAND 89–2612/A, p. 3.1–3.54.

Lorenz, J.C., Heinze, D.M., Clark, J.A., and Searls, C.A., 1985, Determination of widths of meander-belt sandstone reservoirs from vertical downhole data, Mesaverde Group, Piceance Creek basin, Colorado: AAPG Bulletin, v. 69, no. 5, p. 710–721.

Masters, J.A., 1979, Deep basin gas trap western Canada: AAPG Bulletin, v. 63, no. 2, p. 152–181.

Mercer, J.C., and Frohne, K.H., 1986, Production characterization of tight lenticular gas sands in the Rulison area of western Colorado: SPE Paper 15248, 6 p.

Myal, F.R., Price, E.H., Hill, R.E., Kukal, G.C., Abadie, P.A., and Riecken, C.C., 1989, Geologic and production characteristics of the tight Mesaverde group—Piceance basin, Colorado: Report prepared for U.S. Department of Energy, Morgantown Energy Technology Center, Contract DE AC21–88MC24120, 123 p.

Northrop, D.A., 1990a, Background and summary, Multiwell Experiment final report, part IV—The fluvial interval of the Mesaverde Formation: Sandia National Laboratories Rept. SAND 89–2612/A, p. 1.0–1.27.

Northrop, D.A., 1990b, Enhanced gas recovery bibliography Sandia National Laboratories December 1975 through December 1989: Sandia National Laboratories Rept. SAND 90–0759 (April), 32 p.

Peterson, R.E., 1984, Geological and production characteristics of the nonmarine part of the Mesaverde group, Rulison field area, Piceance basin, Colorado: SPE/DOE/GRI Paper 12835, 10 p.

Price, N.J., 1966, Fault and joint development in brittle and semi-brittle rock: New York, Pergamon Press, 176 p.

Rice, D.D., Clayton, J.L., and Pawlewicz, M.J., 1989, Characterization of coal-derived hydrocarbons and source-rock potential of coal beds, San Juan basin, New Mexico and Colorado, U.S.A.: International Journal of Coal Geology, v. 13, p. 597–626.

Rice, D.D., Epsman, M.L., and Mancini, E.A., 1989, Origin of conventional and coalbed gases in the Black Warrior basin region, northwestern Alabama (abs.), in The 1989 Coalbed Methane Symposium, Proceedings—University of Alabama, Tuscaloosa, April 17–20, 1989: School of Mines and Energy Development, Gas Research Institute, U.S. Mine Safety and Health Administration, and Geological Survey of Alabama, p. 321.

Sattler, A.R., 1989, Core analysis in a low permeability sandstone reservoir—Results from the multiwell experiment: Sandia National Laboratories Rept. SAND 89–0710, 64 p.

Smith, D.M., and Williams, F.L., 1981, A new method for determining methane content of coal, *in* Proceedings, 16th Intersociety Energy Conversion Engineering Conference—August 9–14, 1981, Atlanta, Ga.: American Society of Mechanical Engineers, p. 1267–1272.

Tremain, C.M., 1983, Coal bed methane potential of the Piceance basin, Colorado: Colorado Geological Survey Open-File Rept. 82–1, 49 p.

Tremain, C.M., and Toomey, J.J., 1983, Coal bed methane desorption data: Colorado Geological Survey Open-File Rept. 81–4, 514 p.

Waechter, N.B., and Johnson, W.E., 1986, Pennsylvanian-Permian paleostructure and stratigraphy as interpreted from seismic data in the Piceance basin, northwest Colorado, *in* Stone, D.S., and Johnson, K.S., eds., New interpretations of northwest Colorado geology: Rocky Mountain Association of Geologists Symposium Guidebook, p. 51–64.

Geologic Controls on Coalbed Methane Occurrence and Producibility in the Fruitland Formation, Cedar Hill Field and COAL Site, San Juan Basin, Colorado and New Mexico*

William A. Ambrose[1]
Walter B. Ayers, Jr.[2]

1 Bureau of Economic Geology, The University of Texas at Austin, Austin, Texas
2 Formerly with Bureau of Economic Geology, The University of Texas at Austin, Austin, Texas. Now with Taurus Exploration Inc., Birmingham, Alabama.

ABSTRACT

Cedar Hill field contains some of the most productive coalbed methane wells in the San Juan basin. To determine geological controls on occurrence and producibility of coalbed methane at Cedar Hill field and the COAL site (Completion Optimization and Assessment Laboratory) we mapped structure, coal occurrence, net sandstone thickness, and production from coalbed methane wells.

The primary targets for coalbed methane production in Cedar Hill field are thick, basal Fruitland Formation coals (coal group A) that formed in a coastal plain setting. Group A coalbeds are the thickest, up to 35 ft (11 m), and most continuous. In contrast, at the COAL site the thickest coalbeds are more than 30 ft (9 m) thick and occur in the middle Fruitland (coal group C).

Fruitland coalbeds are complex reservoirs as a result of coalbed-sandstone relationships and minor structural deformation. Coals formed in fluvial and delta-plain settings; compaction-induced folds are present where coals are interbedded with channel-fill sandstones. Coalbeds may be more intensely fractured and thus more permeable where folded. The northwest-trending hingeline of the basin and a syncline that crosses Cedar Hill field contribute to reservoir complexity.

Maximum daily production rates of more than 1,500 MCF (42,450 m^3) per well in the field occur where coalbeds have been folded along the synclinal axis. Fractures associated with these folds may increase coal permeability and therefore enhance production. Other highly productive areas within Cedar Hill field coincide with greatest net-coal thickness.

INTRODUCTION

Cedar Hill field, located in northwestern San Juan basin (Figure 1), has produced coalbed methane since 1977. The field, operated by Amoco Production Company, produces predominantly from basal Fruitland Formation coalbeds at a depth of approximately 2,800 ft (854 m). These basal beds contain an estimated 313 million short tons (284 million t) of coal and 89 BCF (2.52 × 10^9 m^3) of gas in place (Decker and others, 1988). About 17.2 BCF (0.49 × 10^9 m^3) of coalbed methane and 737,000 bbl (117,183 m^3) of water were produced from 1976 through 1989 (Petzet, 1990).

The COAL site, approximately 2.5 miles (4 km) northwest of Cedar Hill field (Figure 1), is a field research laboratory operated jointly by Amoco Production Company, Gas Research Institute (GRI), and GRI's field contractor, Resource Enterprises Inc. Research here is designed to compare the effectiveness of open-hole cavity completions with cased-hole, hydraulically fractured completions, and to determine parameters that favor open-hole cavity completions.

Objectives

This area was chosen for study because Cedar Hill field has the longest history of coalbed methane production in the basin and thus has a large production database. The COAL site was included in this study to characterize reservoir conditions and to provide a framework for interpreting results from coalbed methane research at the site. The objectives of this study were to (1) identify Fruitland coal groups and depositional units in the Cedar Hill and COAL site areas; (2) define geologic controls on thickness, continuity, trends, and structural attitude of major coalbeds (reservoirs) within each coal group; (3) map Fruitland coalbed methane production; (4) characterize reservoir conditions; and (5) clarify geologic and hydrologic controls on coalbed methane occurrence and producibility.

* Published by permission of the Director, Bureau of Economic Geology, The University of Texas at Austin.

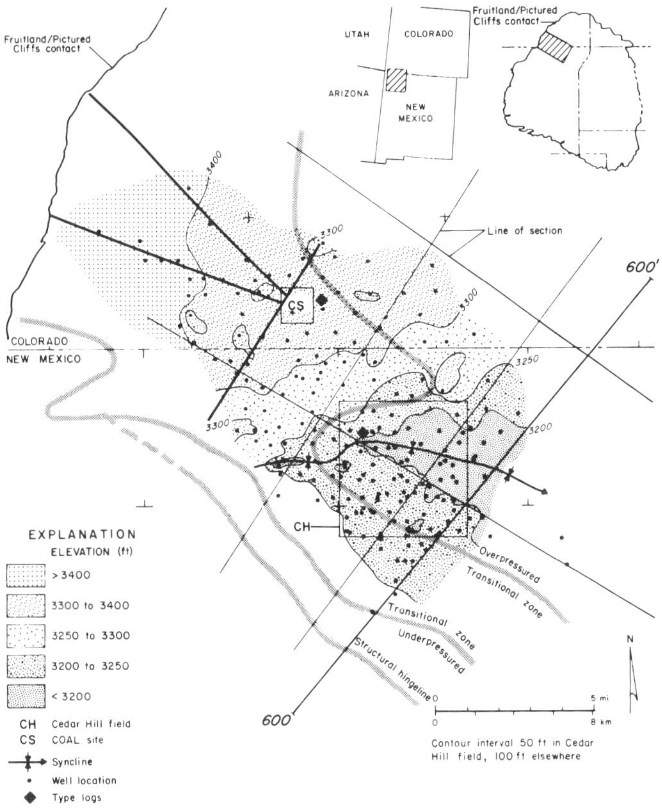

Figure 1. Structure map of the base of Fruitland coal group A at Cedar Hill field and the COAL site.

Methods

Data from 260 geophysical logs, including 22 from shallow boreholes less than 600 ft (180 m) deep near the Fruitland outcrop, were used in this study. Stratigraphic units below and above the Fruitland Formation were correlated in eight structural and stratigraphic cross sections as a framework for determining lateral extent of Fruitland coalbeds and sandstones. Marker beds that were correlated included the Huerfanito Bentonite Bed, five shale marker beds in the Lewis Shale and Pictured Cliffs Sandstone, tonsteins (ash layers) in Fruitland coalbeds, carbonaceous shales in the Fruitland, and a high-conductivity shale at the Fruitland-Kirtland contact. These marker beds previously were correlated in a regional study of the Fruitland Formation and Pictured Cliffs Sandstone (Ayers and others, 1991).

The Fruitland Formation was divided into three depositional subunits, each bounded by continuous coalbeds or by distinctive marker beds. Net-coal thickness, net-sandstone thickness, isopach, and structure maps were compiled for each subunit. Sandstone thickness was determined from resistivity and natural gamma ray curves; lithologic interpretations were confirmed with bulk density logs. Coal identification and thickness were determined from bulk density logs, and coal thickness was measured at a density of approximately 1.80 g/cc. We recorded the thickness of coalbeds thicker than 2 ft (0.6 m); partings thinner than 2 ft (0.6 m) were included as coal because of the limits of log resolution. For logs without density curves, other curves such as natural gamma ray, neutron, and resistivity were used; coals in these logs were identified by correlation with nearby logs having bulk-density curves. Gas and water production data for Fruitland wells were recorded by lithology (coal, sandstone, or both) and stratigraphic interval, and production maps were compiled for Cedar Hill field.

STRATIGRAPHIC AND STRUCTURAL SETTINGS

At Cedar Hill field and the COAL site, the Fruitland Formation is divided into three depositional subunits (lower, middle, and upper) (Ambrose and Ayers, 1991). Coalbeds in four informally designated groups (A, B, C, and D) bound these subunits (Figures 2 and 3). Where the coalbeds locally pinch out, subunit boundaries become shales or sandstones that occur at the stratigraphic horizon of the coal.

The lower subunit, which directly overlies the Pictured Cliffs Sandstone (Figures 2 and 3), consists of thin sandstone, mudstone, and two coal groups. Coalbeds in group A occur at the base of the Fruitland Formation, and those in group B are bounded by two sandstones, LF1 and LF2 (Figure 3). LF1 was mapped separately, whereas thinner, underlying sandstones in the package that includes coal group A were mapped collectively. The middle and upper subunits comprise most of the Fruitland Formation and contain coal groups C and D, respectively. Group C coalbeds overlie LF1 sandstones in Cedar Hill field but override the LF2 sandstone to the northeast (Figure 3).

The Fruitland-Kirtland contact is a regionally extensive, high-conductivity shale (Figure 2) inferred to be a condensed marine section deposited over older, Fruitland coastal plain deposits (Ayers and others, 1991). Foraminifera have been reported in this shale (Dilworth, 1960), whose log response (high natural gamma, high conductivity) and regional continuity are similar to that described for other Cretaceous shallow-water, condensed sections in the Western Interior (Leckie and others, 1990).

An east-plunging syncline crosses Cedar Hill field (Figure 1). Structural relief on the base of coal group A in Cedar Hill field is 70 ft (21 m). Up to 40 ft (12 m) of structural relief occurs across faults and where coalbeds override sandstones (Figure 3). A northwest-southeast-trending structural hingeline that bounds the southwestern margin of the basin floor lies 2 to 5 miles (3.2 to 8 km) south of Cedar Hill field and the COAL site (Figure 1). Southwest of the hingeline, strata dip gently northeastward at approximately 40 ft/mi (7.5 m/km); at Cedar Hill, strata are nearly flat lying, dipping less than 10 ft/mi (1.9 m/km), but are folded and faulted (Figure 3). The hingeline, which projects to the northwestern basin margin at La Plata Mine, is inferred to be a zone of northwest-southeast-trending normal faults. At La Plata Mine, three such faults displace Fruitland and Pictured Cliffs strata by more than 200 ft (61 m) in a zone more than 2,000 ft (610 m) wide. Fruitland strata are down-

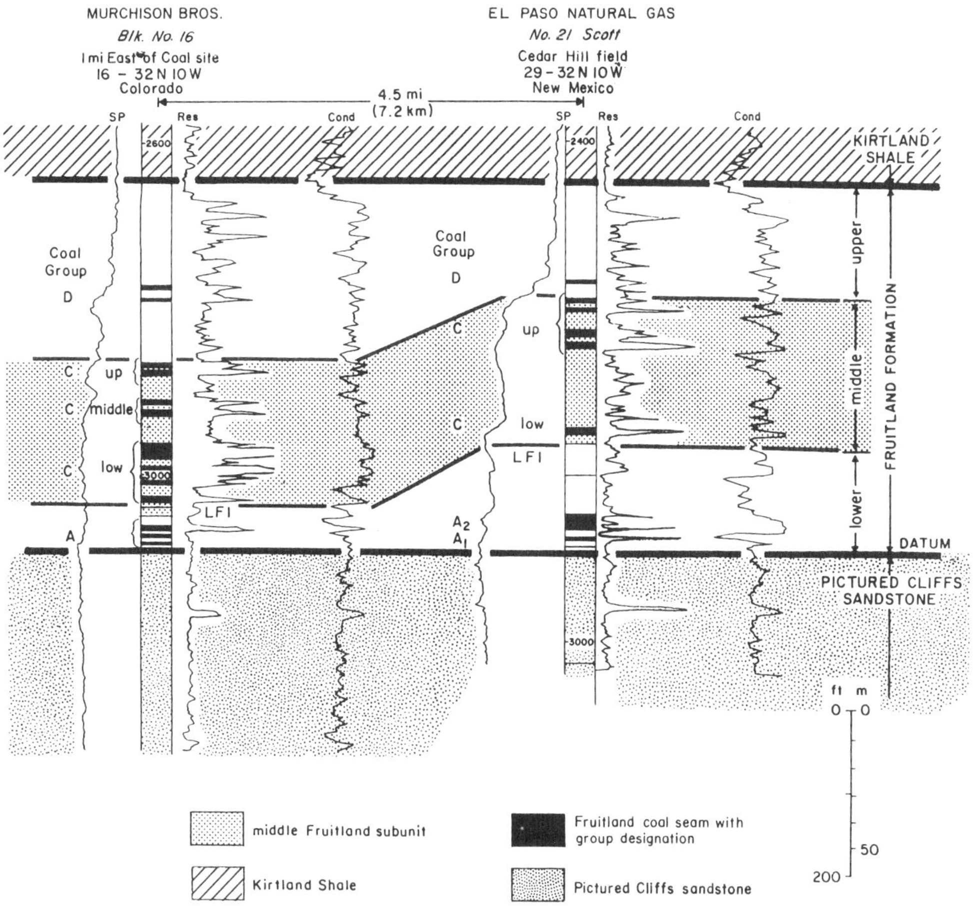

Figure 2. Type logs (located in Figure 1) showing Fruitland stratigraphic units and relations at Cedar Hill field and in the area of the COAL site. Coal group B (shown in Figure 3) is absent at Cedar Hill field and the COAL site.

dropped as much as 70 ft (21 m) to the northeast. Additional evidence for the hingeline includes (1) marked changes in coal-gas composition and vitrinite reflectance (Scott and others, this volume); (2) changes in coal thickness and trends (Ayers and others, 1991); (3) southwestward pinchout of retrogradational tongues of Pictured Cliffs Sandstone (Ambrose and Ayers, 1990); and (4) closely spaced contours in maps of potentiometric surface and pressure gradient (Kaiser and others, 1991).

DEPOSITIONAL SYSTEMS

Pictured Cliffs Sandstone

The Pictured Cliffs Sandstone consists of northwest-trending barrier-island and wave-dominated deltaic deposits that underlie and intertongue with the Fruitland Formation. Progradation of the Pictured Cliffs paleoshoreline, a function of complex interactions among sediment supply, basin subsidence, and eustasy, was intermittent, resulting in stillstands. The formation is 250 ft (76 m) thick in Cedar Hill field and intertongues basinward with the Lewis Shale. Isochronous marker beds in the Lewis Shale cross the proximal shelf and shoreface to terminate in Pictured Cliffs coastal sandstones (Figure 3). Similar marker beds occur in the Upper Cretaceous Point Lookout Sandstone in the San Juan basin (Wright, 1986).

Fruitland Formation

The Fruitland Formation is the continental facies (coal, mudstone, and sandstone) that overlies and intertongues with shoreline deposits of the Pictured Cliffs Sandstone (Figures 2 and 3). Net-sandstone thickness in the Fruitland Formation in the Cedar Hill field and COAL site areas ranges from 110 to 210 ft (33.5 to 64 m) (Figure 4a). Northeast-trending fluvial-deltaic sandstone belts, which constitute the Fruitland depositional framework, controlled the occurrence, trend, and thickness of the coalbeds.

Lower Subunit

Thin sandstones associated with coal group A, below LF1 (Figures 2 and 3), were mapped to help describe the depositional framework of basal Fruitland coalbeds. These northeast-trending sandstones locally are more than 40 ft (12 m) thick in Cedar Hill field but thin to only 10 ft (3 m) downdip to the northeast (Figure 4b). They are interpreted to be distributary-channel and crevasse-splay complexes deposited in a lower delta-plain setting.

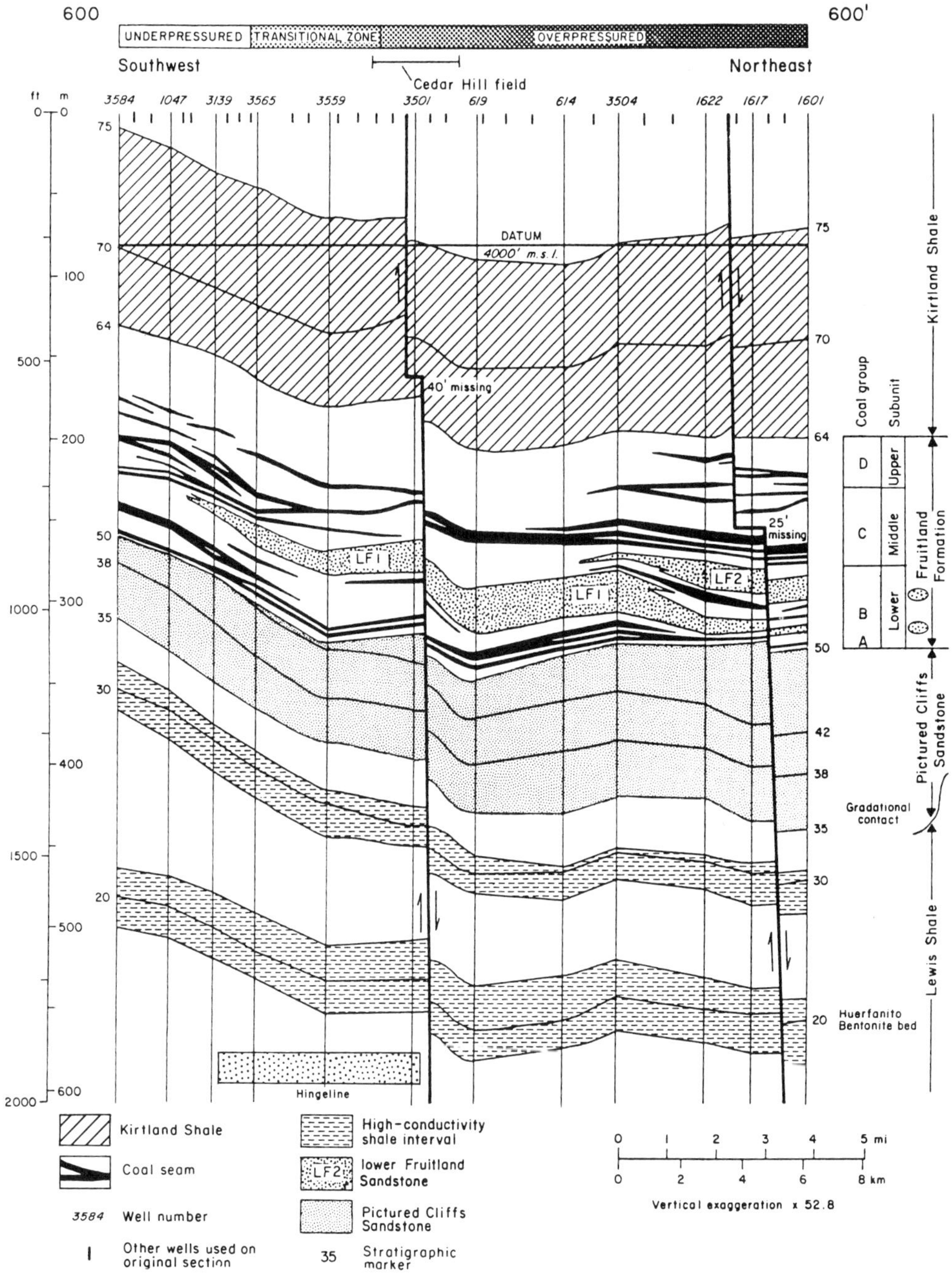

Figure 3. Structural dip section in northwestern San Juan basin (see Figure 1 for location). A transition from gently northeast-dipping strata to nearly flat-lying but folded and faulted strata coincide with the hingeline and southwestern margin of regionally overpressured Fruitland strata.

LF1 sandstone bodies, which exceed 30 ft (9 m) in thickness, are interpreted to be delta distributary channels, elongate to lobate in shape, northeast-trending, and northeast-bifurcating (Figure 5a). In a major depocenter that trends northeast through eastern Cedar Hill field, net-sandstone thickness exceeds 40 ft (12 m). The COAL site is situated between LF1 depositional axes.

LF2, a lenticular sandstone less than 20 ft (6 m) thick, occurs above coal group B but was not mapped because it occurs only locally at the margin of the study area (Figure 3).

Middle and Upper Subunits

The middle Fruitland subunit consists of strata from the base to the top of coal group C (Figures 2 and 3). Net-sandstone thickness varies from 40 to 60 ft (12 to 18 m) at

Figure 4 (right). ***a.*** **Sandstone isolith map of the Fruitland Formation.** ***b.*** **Sandstone isolith map of lower Fruitland subunit, below the LF1 sandstone. A depocenter at Cedar Hill field is defined by dip-elongate sandstone bodies locally more than 40 ft (12 m) thick.**

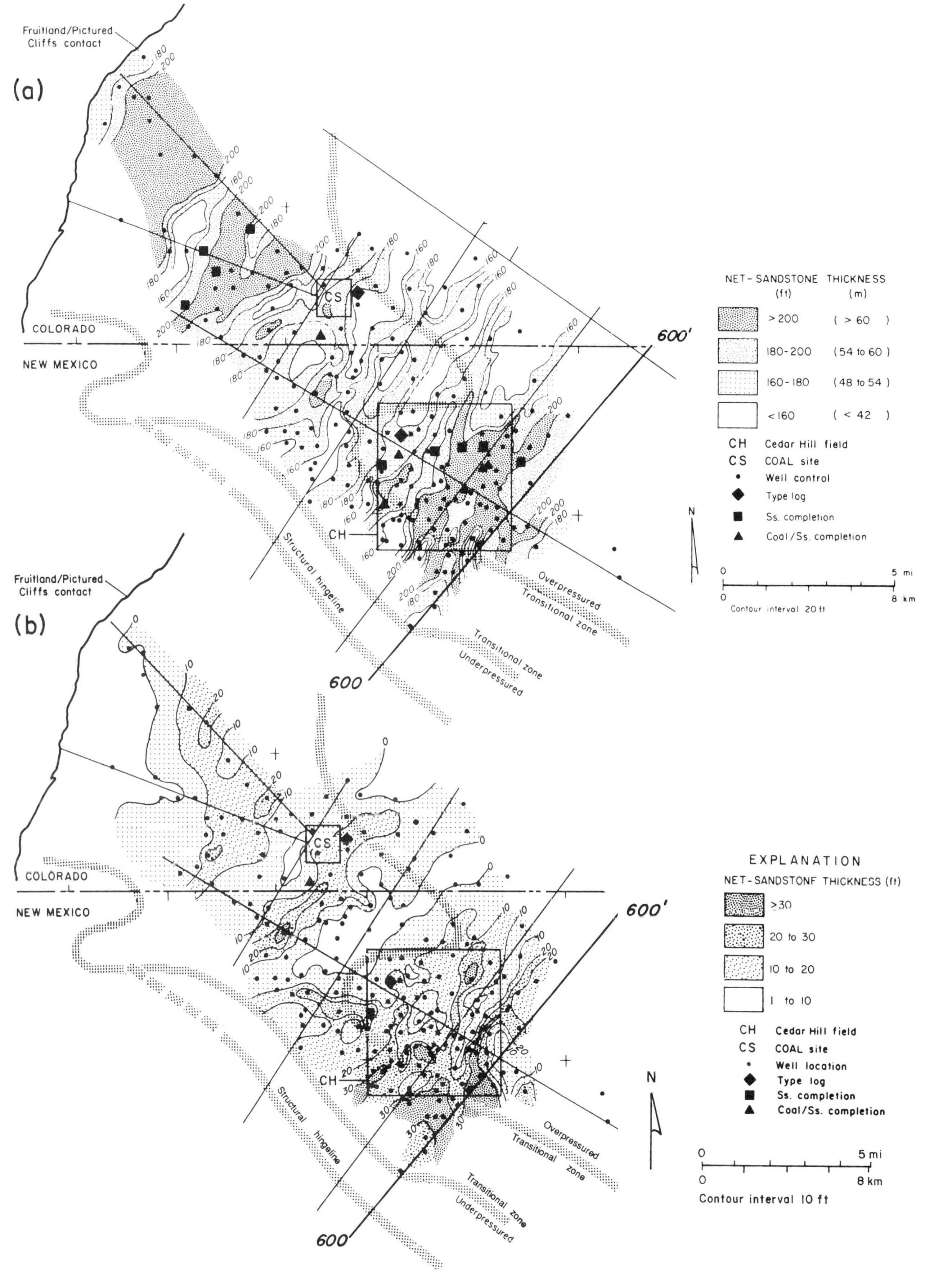
(a)
Fruitland/Pictured Cliffs contact
COLORADO
NEW MEXICO
CS
CH
600'
600
Structural hingeline
Overpressured
Transitional zone
Underpressured
NET-SANDSTONE THICKNESS
(ft) (m)
>200 (> 60)
180-200 (54 to 60)
160-180 (48 to 54)
<160 (< 42)
CH Cedar Hill field
CS COAL site
Well control
Type log
Ss. completion
Coal/Ss. completion
N
0 5 mi
0 8 km
Contour interval 20 ft
(b)
Fruitland/Pictured Cliffs contact
COLORADO
NEW MEXICO
CS
CH
600'
600
Structural hingeline
Overpressured
Transitional zone
Underpressured
EXPLANATION
NET-SANDSTONF THICKNESS (ft)
≥30
20 to 30
10 to 20
1 to 10
CH Cedar Hill field
CS COAL site
Well location
Type log
Ss. completion
Coal/Ss. completion
N
0 5 mi
0 8 km
Contour interval 10 ft

Figure 5. ***a.*** **Sandstone isolith map of lower Fruitland sandstone, LF1.** ***b.*** **Sandstone isolith map of the middle Fruitland subunit.**

Cedar Hill field to more than 80 ft (24 m) near the basin margin (Figure 5b). Framework sandstone bodies in the subunit are strongly dip elongate. The inferred depositional setting is transitional between upper coastal plain and lower alluvial plain. Because the Pictured Cliffs paleoshoreline prograded northeastward, a corresponding landward shift in depositional environments can be seen in successive Fruitland strata. As a result of net progradation, the paleoshoreline at this time was located farther basinward than it was during deposition of the lower subunit.

The upper subunit consists of all Fruitland strata above coal group C (Figure 2). This subunit contains the thickest individual sandstones, up to 50 ft (15 m) thick with net-sandstone thickness up to 130 ft (40 m). The upper subunit is interpreted to be the most proximal part of the Fruitland fluvial system at Cedar Hill field.

FRUITLAND FORMATION COAL

Depositional Setting

Previous researchers concluded that Fruitland coals accumulated in a back-barrier setting, proximal to and landward of high-energy, wave-dominated shorelines (Sears and others, 1941; Scruton, 1961; Fassett and Hinds, 1971). Erpenbeck (1979) and Flores and Erpenbeck (1981) suggested that Fruitland coalbeds in southwestern San Juan basin formed in two depositional settings—(1) thick, dip-elongate lower coals in an interdistributary setting, and (2) thin coals with abundant carbonaceous shale in a back-barrier setting. Ayers and Zellers (1988) reported that the thickest coals in the north-central basin occur landward of upper Pictured Cliffs sandstone tongues, which suggests that thick peats formed in persistent and recurring swamp environments landward of abandoned upper Pictured Cliffs shorelines. Based on palynology (Manfrino, 1984) and low sulfur content of Fruitland coals, the peat accumulated in fresh-water swamps isolated from marine influence. Coal occurrence and trends, together with the relationships between coalbeds and framework sandstones in the Cedar Hill field area, suggest that the coal formed in middle coastal-plain to lower alluvial-plain settings.

Coal Groups

The four coal groups (A, B, C, and D) at Cedar Hill field and the COAL site consist of one or more individual seams that vary in thickness and continuity as a result of depositional settings (Ambrose and Ayers, 1991). We will describe only the thickest groups (A and C) and relate coal characteristics to depositional setting. Coal group B occurs only locally and pinches out approximately 1 mile (1.6 km) northeast of Cedar Hill field (Figure 3); therefore, it was not evaluated. Coalbeds in group D were not mapped because they are relatively thin and discontinuous as a result of formation in an unstable flood-plain setting.

Net-coal thickness of the undivided Fruitland Formation in the study areas ranges from less than 40 ft (12 m) to more than 100 ft (30 m) (Figure 6a). Coal occurs in northeast-trending bodies, 60 to 80 ft (18 to 24 m) thick at Cedar Hill field and south of the COAL site. These dip-elongate bands merge downdip (northeastward), approximately 3 miles (5 km) northeast of Cedar Hill field, to form a strike-elongate (northwest-southeast) deposit of net thickness greater than 80 ft (24 m). The dip-elongate deposits at Cedar Hill field formed in flood-plain and interdistributary environments, whereas the strike-elongate deposits to the northeast accumulated in swamps landward (southwest) of the Pictured Cliffs paleoshoreline.

Coal Group A

Coal group A consists of as many as three coalbeds with net thickness as much as 35 ft (11 m) in northeast-trending belts (Figure 6b). At Cedar Hill field coal group A is the main target for coalbed methane wells. Net-coal thickness decreases northwestward from the field, and all the seams pinch out 1 to 3 miles (1.6 to 4.8 km) from the northwestern basin margin. The thickest coal in the group occurs in a 3-mile-wide (5-km) belt centered on Cedar Hill field. This belt coincides with an area in which net-sandstone thickness in the lower Fruitland exceeds 20 ft (6 m) (Figure 4b). This relationship suggests that the peat accumulated on a crevasse splay or on a small deltaic platform. However, within the belt of thick coal, net-coal thickness in group A locally is greatest, more than 30 ft (9 m), in sand-poor areas. This depositional setting is similar to that described by Levey (1985) for lower delta-plain coals in the Upper Cretaceous Rock Springs Formation in southwestern Wyoming. Peats formed on abandoned delta lobes, with greatest peat accumulation between splay channels and distributaries.

Coal Group C

Coal group C consists of lower, middle, and upper coalbeds (Figure 2). Net-coal thickness in lower C coalbeds is greatest, more than 30 ft (9 m), north of Cedar Hill field and around the COAL site, where the beds are the target for coalbed methane drilling (Figure 7a). These coals occur as dip elongate belts that merge downdip (northeastward) into a strike-elongate sheet north of Cedar Hill field. They overlie thick LF1 distributary sandstones (Figure 5a) over much of the area, but at Cedar Hill field they pinch out against them. Dip-elongate pods of lower C coal south of

Figure 6 (p. 234). ***a.*** **Coal isolith map of the Fruitland Formation, showing dip-elongate coal deposits broadening and merging downdip (northeastward).** ***b.*** **Coal isolith map of Fruitland coal group A, the thickest coal group at Cedar Hill field.**

Figure 7 (p. 235). ***a.*** **Coal isolith map of lower group C coalbeds. Coals are absent over thick LF1 sandstones (Figure 5a) in eastern Cedar Hill field.** ***b.*** **Coal isolith map of middle group C coalbeds. At the COAL site, net coal thickness is less than that of lower group C coalbeds. Coals thicken northwestward toward the Fruitland outcrop.**

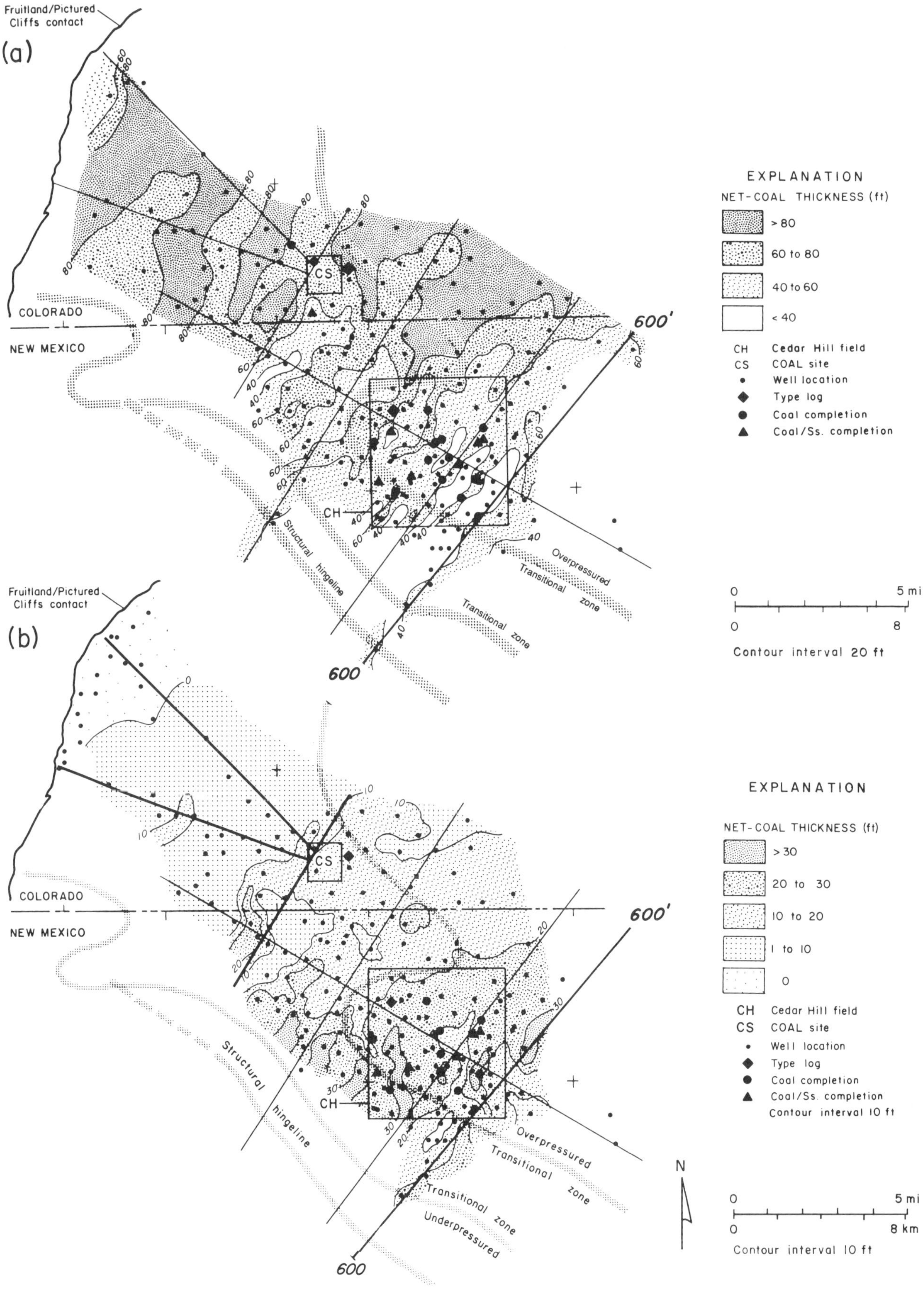
(a)
Fruitland/Pictured Cliffs contact
COLORADO
NEW MEXICO
CS
CH
600'
600
Structural hingeline
Overpressured
Transitional zone
Transitional zone
EXPLANATION
NET-COAL THICKNESS (ft)
> 80
60 to 80
40 to 60
< 40
CH Cedar Hill field
CS COAL site
Well location
Type log
Coal completion
Coal/Ss. completion
0 5 mi
0 8
Contour interval 20 ft
(b)
Fruitland/Pictured Cliffs contact
COLORADO
NEW MEXICO
CS
CH
600'
600
Structural hingeline
Overpressured
Transitional zone
Transitional zone
Underpressured
N
EXPLANATION
NET-COAL THICKNESS (ft)
> 30
20 to 30
10 to 20
1 to 10
0
CH Cedar Hill field
CS COAL site
Well location
Type log
Coal completion
Coal/Ss. completion
Contour interval 10 ft
0 5 mi
0 8 km
Contour interval 10 ft

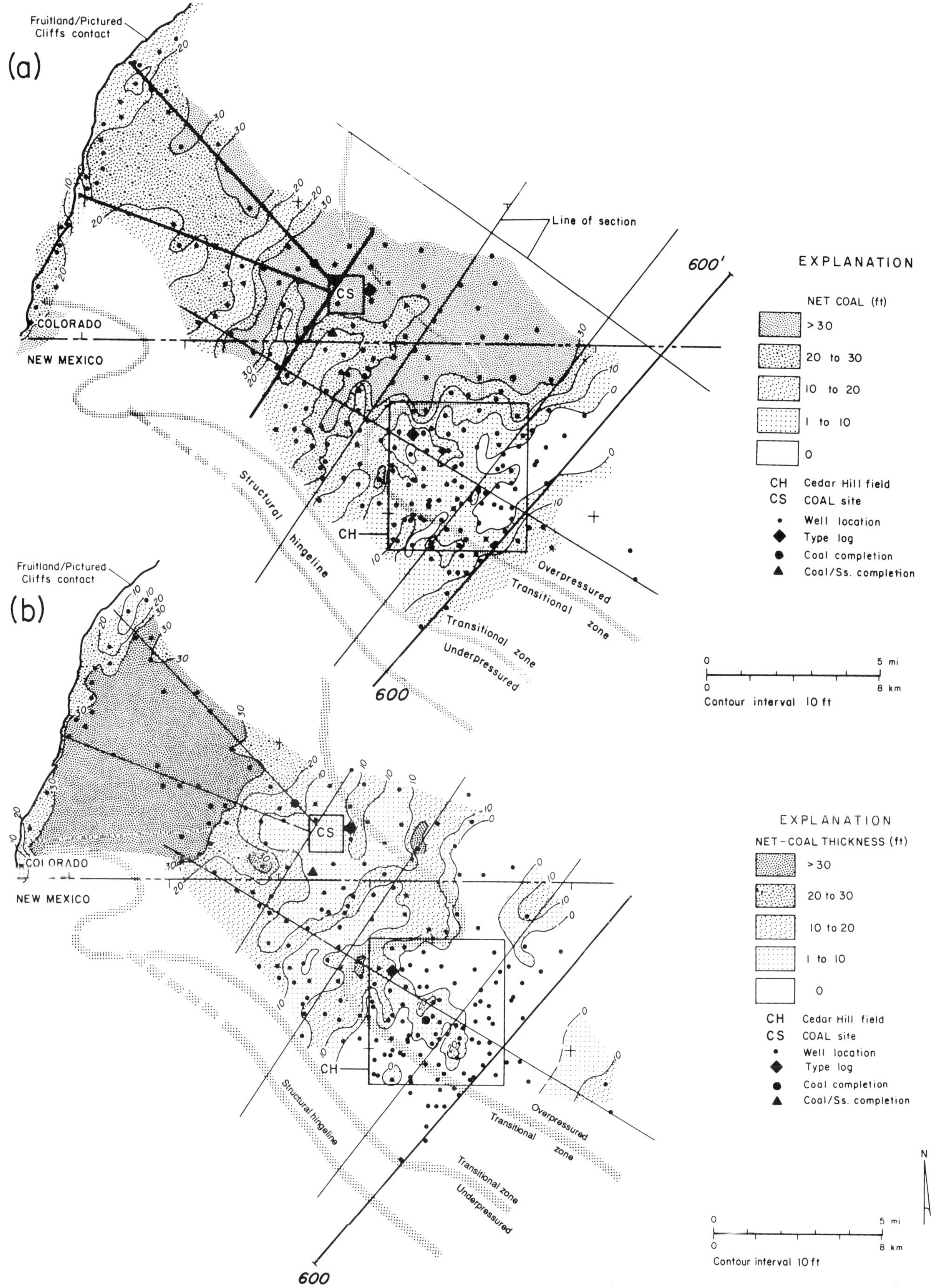
(a)
Fruitland/Pictured Cliffs contact
COLORADO
NEW MEXICO
CS
CH
Line of section
600'
600
Structural hingeline
Overpressured
Transitional zone
Transitional zone
Underpressured
EXPLANATION
NET COAL (ft)
>30
20 to 30
10 to 20
1 to 10
0
CH Cedar Hill field
CS COAL site
Well location
Type log
Coal completion
Coal/Ss. completion
0 5 mi
0 8 km
Contour interval 10 ft
(b)
Fruitland/Pictured Cliffs contact
COLORADO
NEW MEXICO
CS
CH
600
Structural hingeline
Overpressured
Transitional zone
Transitional zone
Underpressured
EXPLANATION
NET-COAL THICKNESS (ft)
>30
20 to 30
10 to 20
1 to 10
0
CH Cedar Hill field
CS COAL site
Well location
Type log
Coal completion
Coal/Ss. completion
N
0 5 mi
0 8 km
Contour interval 10 ft

the COAL site accumulated between distributary channel-fill sandstones, whereas widespread lower C coals north of the site formed in a lower delta-plain setting.

Middle and upper group C coalbeds (Figure 2) both are relatively thin and discontinuous; because of their similarities, only middle group C coalbeds will be described. Like the lower group C beds, middle C beds are absent in eastern Cedar Hill field (Figure 7b). Middle group C beds are present but less than 10 ft (3 m) thick at the COAL site, but they thicken to more than 40 ft (12 m) near the Fruitland outcrop at the basin margin.

DEPOSITIONAL CONTROLS ON COALBED CONTINUITY AND STRUCTURAL ATTITUDE

The occurrence, thickness, and extent of Fruitland coals were controlled by depositional environments. In the southwestern part of the study area, thick Fruitland coalbeds occur in northeast-trending belts between channel-fill sandstone complexes, whereas in the northeast, they occur in northwest-trending sheet deposits that formed landward of Pictured Cliffs barrier-strandplain deposits.

Coal depocenters migrated northwestward through time (Figure 8). Thick coalbeds in group A are the main coal reservoirs in Cedar Hill field, but they are thin at the COAL site (Figure 6b), where thick lower group C coals are the primary coalbed methane targets (Figures 7a and 8). Fruitland coalbeds vary considerably in thickness across the basin, and although they act as an integrated, regional reservoir, individual beds locally may be in poor hydraulic communication with one another. Stratigraphic studies therefore are required to delineate the limits of individual coal reservoirs and to evaluate well spacing and location.

The relationship between LF1 sandstones (Figure 5a) and lower group C coalbeds (Figures 7a and 8) typifies sandstone-body control on the extent, thickness, and trend of coalbeds. The thickness of lower group C coalbeds is inversely related to the thickness of underlying LF1 sandstones. At Cedar Hill field these seams are thin or absent over the thick, northeast-trending LF1 sandstones, whereas at the COAL site, lower group C coalbeds locally are more than 30 ft (9 m) thick, where LF1 sandstones are less than 10 ft (3 m) thick (Figure 5a).

Fruitland coalbeds commonly are interbedded with channel-fill sandstone complexes (Figure 9). Coalbeds overlying these sandstones formed in swamps that overspread abandoned-channel complexes, whereas other coalbeds are overlain by sandstones deposited during periods of swamps. Where they are interbedded with lenticular sandstone and adjacent mudstone, coalbeds commonly were subjected to differential compaction. Abrupt changes in structural attitude of the coals occur where they override or underlie the channel-fill sandstones, resulting in local relief as great as 60 ft (18 m). The significance of continuous coalbeds that override and underlie sandstones but are sharply folded at the facies boundaries is twofold—(1) as

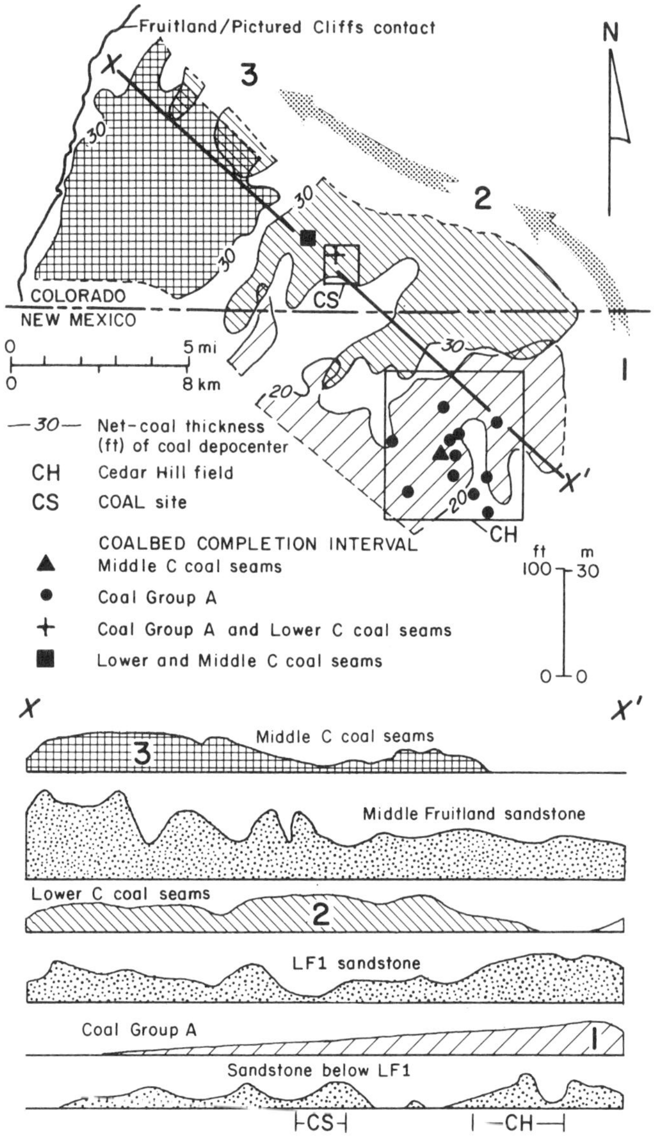

Figure 8. Fruitland coal depocenters (group A, lower group C, and middle group C) in Cedar Hill field area, defined by areas of greatest net coal thickness, and showing northwestward migration through time.

methane reservoirs, these coals are more extensive than previously believed; and (2) compactional folding of the coalbeds may induce fracturing (Donaldson, 1979; Houseknecht and Iannacchione, 1982) that enhances permeability.

FRUITLAND RESERVOIR CHARACTERISTICS AND PRODUCTION

Reservoir characteristics of coalbeds vary regionally and as well as locally. Assessment of coalbed reservoir quality requires knowledge of (1) gas occurrence and resources, (2) gas composition, (3) coal permeability, and (4) pressure regime and fracture-filling fluids (Kaiser and others, 1991). These parameters, in turn, are controlled by structural setting, regional cleat trends, and by coal occurrence, thickness, depth, and rank.

Coal Occurrence, Trends, and Thickness

Greatest thickness of Fruitland coalbeds occurs in the northern San Juan basin in northwest-southeast-trending belts of back-barrier and coastal-plain coal, where net thickness exceeds 50 ft (15 m) (Ayers and others, 1991). Extending southwestward from these belts are northeast-trending, dip-elongate belts of flood-plain coal deposits, 1 to 8 miles (1.6 to 12.8 km) wide where net-coal thickness is 30 to 70 ft (9 to 21 m). Cedar Hill field and the COAL site are located at the intersection of these two principal depositional trends. Discontinuous coalbeds (reservoirs) in this area occur where coastal plain coals split or pinch out landward (southwestward) into muddy flood-plain deposits or along strike (northwest-southeast), where they interfinger with fluvial complexes.

Coal Rank

The quantity of methane generated by and stored in coals varies directly with rank, or thermal maturity. Cedar Hill field is located at the transition between high-volatile B and A bituminous coal (vitrinite reflectance 0.78%) (Scott and others, this volume). Southwest of the field is an abrupt decrease in vitrinite reflectance values coincident with the structural hingeline (Ayers and others, 1991). Offset of the coals along the hingeline, together with southwestward pinchout of flood-plain coals, may have inhibited the southwestward migration of fluids derived from the more thermally mature northern part of the basin.

Formation Pressure

Methane storage capacity of coal also is pressure dependent. In the San Juan basin the Fruitland Formation is abnormally pressured relative to the fresh-water hydrostatic gradient of 0.433 psi/ft (9.8 kPa/km). The north-central area is overpressured, whereas much of the rest of the basin is underpressured (Kaiser and others, 1991). Because Fruitland overpressuring arises from hydrodynamics, most overpressured coalbed methane wells are water productive. The pressure gradient at Cedar Hill field, which is located near the southwestern margin of regional Fruitland overpressuring, is approximately

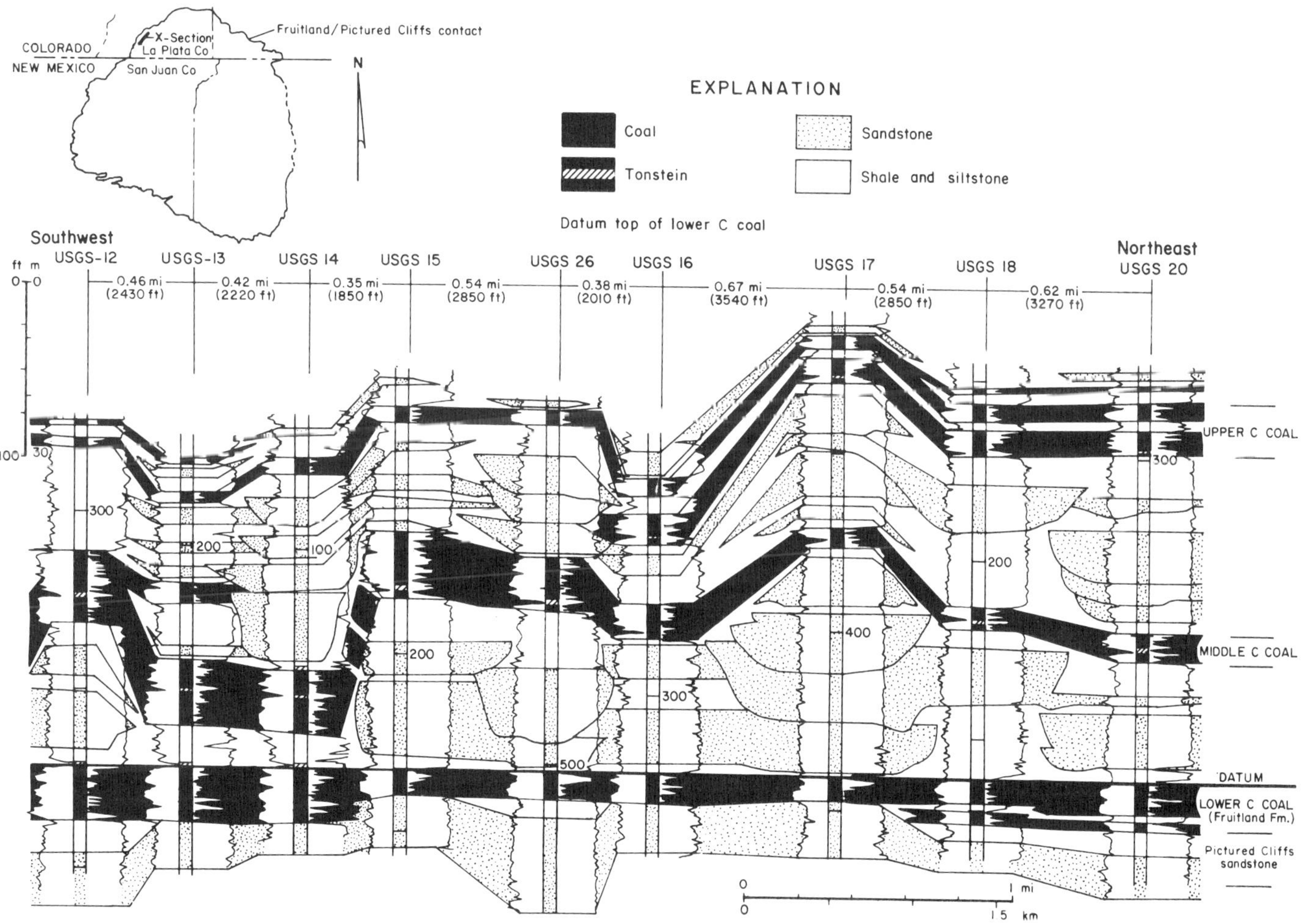

Figure 9. Fruitland coalbed and sandstone stratigraphic architecture, northwestern San Juan basin. Datum is top of lower group C coals. Geophysical logs from Roberts (1989).

0.50 psi/ft (11.3 kPa/km). Here, the southwestern boundary of overpressured strata conforms to a syncline (Figure 1). The transition between overpressuring and underpressuring is marked by pronounced steepening of the potentiometric surface and potential for upward fluid flow (Kaiser and others, 1991). High productivity from overpressured coalbeds near the transition zone may reflect the potential for upward flow and for free gas in conventional stratigraphic traps along the southwestward pinchout of the aquiferous coalbeds.

Coalbed Methane Resources

Cedar Hill field lies in a transitional area where in-place methane resource estimates range from 15 to 25 BCF/mi^2 (1.64×10^6 to 2.73×10^6 m^3/km^2), among the highest in the San Juan basin. This range of values reflects not only the presence of high coal rank and high formation pressure (Ayers and others, 1991), but also the change from locally thick, flood-plain coal deposits in the south to extensive, thick, delta-plain and back-barrier coal deposits in the north.

Controls on Coalbed Permeability

Permeability in coalbeds is controlled by the intensity and interconnectivity of fractures. The orientation of cleats (fractures) is determined by regional stresses present during coalification. At Cedar Hill, face cleats (dominant fracture) trend northeastward (Tremain and others, 1991). Folded coalbeds within the east-plunging syncline (Figure 1) may exhibit higher fracture intensity and thus enhanced permeability. Coalbed continuity is partly controlled by the basin's structural hingeline, which is inferred to be a complex system of northwest-trending normal faults. Individual faults within this complex offset the coals as much as 50 ft (15 m).

Gas Production Trends

The most productive coalbed methane wells in the San Juan basin are completed in overpressured coals along the southwestern margin of the overpressured region (Kaiser and others, 1991). This highly productive trend, which includes Cedar Hill field, occurs in the thick delta-plain and back-barrier coals. Within this belt, the highest production rates, more than 1,000 MCFD (28,300 m^3/day), come from local northeast-trending patterns that reflect the presence of thick, interdistributary coals.

Production in Cedar Hill field reflects geologic controls; coalbeds, channel-fill sandstones, and production patterns all trend northeastward (Figure 10). Individual wells typically produce 300 MCFD (8,490 m^3/day) to more than 1,000 MCFD (28,300 m^3/day). In 1980 the average coal-gas production per well was approximately 600 MCFD (16,980 m^3/day); by 1990, average well production had increased to 1,000 MCFD (28,300 m^3/day). Northeast-trending belts of high production through the southwestern part of the field coincide with basal Fruitland coalbeds (coal group A; Figure 6b) locally more than 30 ft (9 m) thick. In contrast, low-production trends of less than 500 MCFD (14,150 m^3/day) in the southeast part of the field coincide with thin net-coal in coal groups A and C (Figures 6b, 7a, and 7b). The most productive wells, with maximum production of more than 1,800 MCFD (50,940 m^3/day), lie on the southern flank of the synclinal axis in eastern Cedar Hill field (Figure 10), where folding may have caused fracture-enhanced permeability.

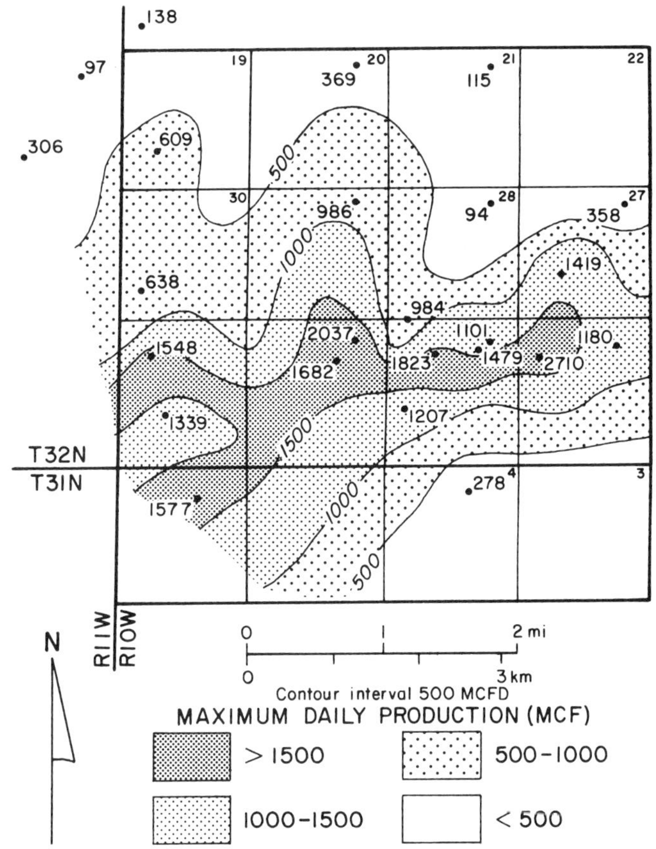

Figure 10. Fruitland coalbed methane production at Cedar Hill field. Values are average daily production (MCFD) in each well's most productive year. Modified from Yeh and others (1991).

Other Reservoir Characteristics

Gas Composition

At Cedar Hill field and throughout the northern, overpressured area of the basin, Fruitland coal gases are chemically "dry" ($C_1/C_{1\text{-}5}$ values greater than 0.94) (Scott and others, this volume). However, gases from the COAL site, 3 miles (4.8 km) southwest of the transitional zone of overpressured strata, are chemically "wet" ($C_1/C_{1\text{-}5}$ values less than 0.94). Carbon dioxide content of Fruitland coals also changes abruptly across the pressure transition. At Cedar Hill, carbon dioxide content ranges from less than 3% to more than more than 6%, whereas at the COAL site it is less than 1% (Scott and others, 1991). The fact that both gas wetness and carbon dioxide content mimic the pressure and structure contours is consistent with inferred structural control on formation pressure, fluid migration, and gas composition. Further geologic control on coal-gas composition is suggested by tongues of high carbon dioxide gas that extend southwestward from the overpressured area and coincide with thick flood-plain coals south of Cedar Hill field (Scott and others, this volume).

Produced Water

Overpressured wells in Cedar Hill field commonly produce water, requiring disposal and additional operating costs. Areas of highest water production, 20 to 152 BWPD (3.2 to 24.2 m^3/day) occur in the central and eastern areas along the syncline. Wells in the field are still dewatering—in 1980 the average water production per well was approximately 200 BWPD (31.8 m^3/day); by 1990, rates had declined to 20 BWPD (3.2 m^3/day) per well.

CONCLUSIONS

1. At Cedar Hill field and the COAL site, the thickest and most continuous coalbeds occur in the lower and middle Fruitland Formation. These coalbeds (coal groups A and C), which formed in a coastal plain setting, are of subregional extent and commonly override or underlie channel-fill sandstone bodies. Although the lower and middle coal groups are extensive, they are not present or equally well developed throughout the area. At Cedar Hill field the thickest and most extensive coals occur in group A, but they pinch out near the northwestern margin of the basin. At the COAL site, the thickest coalbeds are in coal group C, which are thin or absent at Cedar Hill field.

2. Increased fracture intensity, and therefore enhanced coalbed permeability, may occur where coals are folded. Tectonic folding is common at Cedar Hill field and the COAL site. In addition, many lower and middle Fruitland coals are compactionally folded where they underlie and overlie channel-fill sandstones complexes. At the COAL site, structural relief on the thickest coalbed is 50 ft (15 m).

3. Cedar Hill field and the COAL site are situated northeast of the structural hingeline of the San Juan basin. This hingeline may result from a northwest-southeast-trending fault complex that offsets overpressured, water-saturated coal reservoirs in the northern part of the basin from underpressured, predominantly gas-saturated coals in the south. Thermal maturity, chemical "wetness," and carbon dioxide content of Fruitland coals change abruptly across the hingeline.

4. Coal-gas occurrence and producibility in Cedar Hill field are controlled primarily by depositional setting. Productivity patterns trend dominantly northeastward and coincide with thick, northeast-trending coals that formed in interdistributary areas on a crevasse-splay platform. Other coalbed methane production trends in the area are controlled by structure; a high-productivity trend follows a syncline that crosses the field. Coal permeability may be enhanced where coalbeds locally are fractured because of differential compaction.

ACKNOWLEDGMENTS

This study was funded by the Gas Research Institute under Contract 5087–214–1544. The manuscript benefited from review by Tucker F. Hentz. Illustrations were drafted by Patrice Porter, Tari Weaver, Karel Cromie, and Maria Saenz, under the direction of Richard C. Dillon. Publication authorized by the director of Bureau of Economic Geology, The University of Texas at Austin.

REFERENCES

Ambrose, W.A., and Ayers, W.B., Jr., 1990, Barrier-strandplain deposits of the Pictured Cliffs Sandstone and coal occurrence in the Fruitland Formation, Northern San Juan basin, Colorado and New Mexico (abs.): AAPG Bulletin, v. 74, no. 8, p. 1313.

Ambrose, W.A., and Ayers, W.B., Jr., 1991, Geologic controls on coalbed occurrence, thickness, and continuity, Cedar Hill field and the COAL site, *in* Ayers, W.B., Jr., and others, Geologic and hydrologic controls on the occurrence and producibility of coalbed methane, Fruitland Formation, San Juan basin: Gas Research Institute, Topical Rept. GRI–91/0072 (August 1987–July 1990), p. 47–68.

Ayers, W.B., Jr., Ambrose, W.A., and Yeh, Joseph, 1991, Depositional and structural controls on coalbed methane occurrence and resources in the Fruitland Formation, San Juan basin, *in* Ayers, W.B., Jr., and others, Geologic and hydrologic controls on the occurrence and producibility of coalbed methane, Fruitland Formation, San Juan basin: Gas Research Institute, Topical Rept. GRI–91/0072 (August 1987–July 1990), p. 9–46.

Ayers, W.B., Jr., and Zellers, S.D., 1988, Sedimentologic and structural controls on the occurrence and producibility of coalbed methane, Fruitland Formation, northern San Juan and Rio Arriba Counties, New Mexico, *in* Ayers, W.B., Jr., and others, Geologic evaluation of critical production parameters for coalbed methane resources, part I—San Juan basin: The University of Texas at Austin, Bureau of Economic Geology, annual report prepared for the Gas Research Institute, under contract no. 5087–214–1544, p. 3–59.

Decker, A.D., Jeu, S.J., Cooper, J.D, and Wicks, D.E., 1988, Geology, geochemistry, reservoir engineering, and completion methods at the Cedar Hill field, San Juan County, New Mexico—A field study of classic coal degasification behavior, *in* Fassett, J.E., ed., Geology and coal-bed methane resources of the northern San Juan basin, Colorado and New Mexico: Rocky Mountain Association of Geologists Guidebook, p. 221–235.

Dilworth, O.L., 1960, Upper Cretaceous Farmington Sandstone of northeastern San Juan County, New Mexico New Mexico University, Master's thesis.

Donaldson, A.C., 1979, Origin of coal seam discontinuities, *in* Donaldson, A.C., Presley, M.W., and Renton, J.J., eds., Carboniferous coal guidebook, volume 1—Carboniferous coal short course: West Virginia Geological and Economic Survey Bulletin B–37–1, p. 102–132.

Erpenbeck, M.F., 1979, Stratigraphic relationships and depositional environments of the Upper Cretaceous Pictured Cliffs Sandstone and Fruitland Formation, southwestern San Juan basin, New Mexico: Texas Tech University, Master's thesis, 78 p.

Fassett, J.E., and Hinds, J.S., 1971, Geology and fuel resources of the Fruitland Formation and Kirtland Shale of the San Juan basin, New Mexico and Colorado: USGS Professional Paper 676, 76 p.

Flores, R.M., and Erpenbeck, M.F., 1981, Differentiation of delta front and barrier lithofacies of the Upper Cretaceous Pictured Cliffs Sandstone, southwestern San Juan basin, New Mexico: Mountain Geologist, v. 18, no. 2, p. 23–34.

Houseknecht, D.W., and Iannacchione, A.T., 1982, Anticipating facies-related coal mining problems on Hartshorne Formation, Arkoma basin: AAPG Bulletin, v. 66, no. 7, p. 923–946.

Kaiser, W.R., Ayers, W.B., Jr., Ambrose, W.A., Laubach, S.E., Scott, A.R., and Tremain, C.M., 1991, Geologic and hydrologic characterization of coalbed methane production, Fruitland Formation, San Juan basin, *in* Ayers, W.B., Jr., and others, Geologic and hydrologic controls on the occurrence and producibility of coalbed methane, Fruitland Formation, San Juan basin: Gas Research Institute, Topical Rept. GRI–91/0072 (August 1987–July 1990), p. 273–301.

Leckie, D.A., Singh, Chaitanya, Goodarzi, Fariborz, and Wall, J.H., 1990, Organic-rich, radioactive marine shale—A case study of a shallow-water condensed section, Cretaceous Shaftesbury Formation, Alberta, Canada: Journal of Sedimentary Petrology, v. 60, no. 1, p. 101–117.

Levey, R.A., 1985, Depositional model for understanding geometry of Cretaceous coals—Major coal seams, Rock Springs Formation, Green River basin, Wyoming: AAPG Bulletin, v. 69, no. 9, p. 1359–1380.

Manfrino, Carrie, 1984, Stratigraphy and palynology of the upper Lewis Shale, Pictured Cliffs Sandstone, and lower Fruitland Formation (Upper Cretaceous) near Durango, Colorado: Mountain Geologist, v. 21, no. 4, p. 115–132.

Petzet, G.A., 1990, Devon pressing Fruitland coal seam program: Oil & Gas Journal, v. 88, no. 45 (November 5), p. 28–30.

Roberts, L.N.R., 1989, Results of 1988 coal exploratory drilling in the Fruitland Formation, western part of the Southern Ute Indian Reservation, La Plata County, Colorado—Lithologic descriptions, preliminary correlations, and proximate analyses of coal samples: USGS Open-File Rept. 89–487, 221 p.

Scruton, P.C., 1961, Rocky Mountain Cretaceous stratigraphy and regressive sandstones, *in* Wiloth, G.J., and others, eds., Symposium on Late Cretaceous rocks, Wyoming and adjacent areas: Wyoming Geological Association, 16th Annual Field Conference Guidebook, p. 242–249.

Sears, J.D., Hunt, C.B., and Hendricks, T.A., 1941, Transgressive and regressive Cretaceous deposits in southern San Juan basin, New Mexico: USGS Professional Paper 193–F, p. 101–121.

Tremain, C.M., Laubach, S.E., and Whitehead, N.H., III, 1991, Coal fracture (cleat) patterns in Upper Cretaceous Fruitland Formation, San Juan basin, Colorado and New Mexico—Implications for coalbed methane exploration and development, *in* Ayers, W.B., Jr. and others, Geologic and hydrologic controls on the occurrence and producibility of coalbed methane, Fruitland Formation, San Juan basin: Gas Research Institute, Topical Rept. GRI–91/0072 (August 1987–July 1990), p. 97–117.

Wright, Robyn, 1986, Cycle stratigraphy as a paleogeographic tool—Point Lookout Sandstone, southeastern San Juan basin, New Mexico: GSA Bulletin, v. 97, no. 6, p. 661–673.

Evaluation of the Hydraulic Fracture Stimulation of the Colorado 32–7 No. 9 Well, San Juan Basin

Matthew J. Mavor[1]
Rahul Dhir[1]
John D. McLennan[2]
Jay C. Close[1]

1. Resource Enterprises Inc., Salt Lake City, Utah
2. Terra Tek Inc., Salt Lake City, Utah

ABSTRACT

Extensive formation evaluation and hydraulic fracture diagnostic data were collected from the Colorado 32–7 No. 9 well that was completed in Fruitland Formation coal zones in the northern San Juan basin. Thorough evaluation of these data, when combined with reservoir simulation, resulted in accurate reservoir characterization together with projections of gas and water production rates. A hydraulic fracture stimulation treatment effectively connected the well to the reservoir and resulted in a complex induced fracture geometry. The coal seams of interest produce gas at currently uneconomic levels because of the relatively low, 1.7-md absolute permeability of the cleat system.

INTRODUCTION

This paper presents the results of a detailed field study of a hydraulic fracture stimulation treatment performed in the Colorado 32–7 No. 9 Fruitland Formation (Upper Cretaceous) coalbed methane well. The well is located in sec. 4, T32N R7W in La Plata County, northern San Juan basin, Colorado. Data relevant to the fracture diagnostics also were collected from the Colorado 32–7 No. 4 well located 140 ft (42.7 m) north and 25 ft (7.6 m) west of the No. 9 well. Extensive stimulation diagnostics used to estimate induced fracture geometry included tiltmeter surface deformation data, microseismic data collected in the No. 4 well, tracer logs, and post-stimulation production testing. The cross-linked gel stimulation was evaluated by comparing observed pressure behavior to that estimated with a three-dimensional fracture-stimulation simulator. Reservoir properties were extensively quantified with core, log, and open-hole drillstem test data. The core analyses included fracture-orientation description and anelastic strain relaxation analysis for estimation of maximum horizontal stress orientation. Post-stimulation production performance was matched with a coalbed reservoir simulator.

The interpretation of induced fracture geometry was confirmed through agreement among the tiltmeter, microseismic, and well test data. A complex T-shaped fracture was induced with multiple vertical components extending in the direction of the coal face cleat trend (N11°E) and in the maximum horizontal strain direction (N45°W). Multiple horizontal components also were created. The well test-derived estimate of the effective fracture length controlling production was 168 ft (51.2 m) from well to tip (half-length). One single vertical fracture was controlling production. The effective fracture length predicted by the simulator was greater than that indicated by well tests, suggesting that complex fracture geometry had reduced the effectiveness of the stimulation treatment.

Simulation of well deliverability was based on estimates of reservoir properties obtained from extensive formation evaluation. The observed well performance was matched with a coalbed reservoir simulator, which confirmed that uneconomic levels of gas production resulted from low (1.7 md) absolute cleat permeability rather than failure of the stimulation treatment.

Funding for this cooperative field research study was provided by the Gas Research Institute as part of the Western Cretaceous Coal Seam Project (Mavor and Close, 1989). This effort represents the first detailed, publicly available diagnostic information from a Fruitland coalbed methane well.

This paper will review field operations, formation evaluation results, hydraulic fracture diagnostics, stimulation modeling results, and reservoir simulation results.

FIELD OPERATIONS

Field operations included efforts on both Colorado 32–7 No. 9 research well and the offset No. 4 well located 142 ft (43.3 m) to the northwest. This section tracks the chronology of operations performed on both wells. Further details are discussed by Mavor and Close (1989).

To accurately estimate the number, depth, and thickness of coal seams, the offset well No. 4 was logged with a cased-hole neutron/gamma log, background tracer log, and cement-bond/collar locater log. The cement-bond log was required to determine if bonding between cement and the formation was sufficient for seismic monitoring. Two such intervals were located.

The Colorado 32–7 No. 9 well was drilled with a 10.6-ppg (1,270-kg/m^3), fresh-water, bentonite and barite mud system using a 7 7/8-in. (20-cm) bit to core point. The main interval of interest to this project was cored during three core runs. At the conclusion of coring operations for the main interval, drillstem test tools equipped with surface recording, bottom-hole pressure gauges were run in the hole to test the coal interval between 2,881 and 2,945 ft (878.7 to 898.2 m). A total of 3 bbl (0.5 m^3) of mud filtrate-contaminated formation water was produced at rates decreasing from 40 to 10 BPD (6.4 to 1.6 m^3/day).

At the conclusion of coring operations, the well was drilled to a total depth of 3,196 ft (974.8 m) and logged with an extensive suite that included resistivity, density, natural and induced gamma ray, and other logs.

Following logging, 5 1/2-in.-outside-diameter (14-cm) casing was run and cemented to surface. The main interval was perforated underbalanced between 2,881 and 2,941 ft (878.7 and 897 m) with 4 jet shots/ft. Tubing equipped with a seating nipple and surface recording, bottom-hole electronic pressure gauge was run into the well to attempt to collect data during a prestimulation production and shut-in test. The well failed to produce, and the interval was acidized with 17 bbl (2.7 m^3) of 12% HCl and 3% HF in an attempt to initiate production. A rod pump then was installed with the surface recording, bottom-hole pressure transducer to perform a production test. Gas rates ranged from a low of 2 MSCFD (57 m^3/day) after acidizing to a high of 9 MSCFD (255 m^3/day). Water rates ranged between 10 and 20 BPD (1.6 and 3.2 m^3/day). Production was continued for four days and then followed by a three-day shut-in test. Data could not be accurately evaluated for estimation of reservoir properties because of wellbore-storage effects.

During the prestimulation testing, one tiltmeter was installed in each of fifteen 30-ft-deep (9.2-m) wells drilled around the location. A subsurface geophone was installed at 3,192 ft (973.6 m) in the No. 4 well prior to the treatment to record seismic activity during the stimulation operation.

The main interval was hydraulically fractured on May 26, 1989, according to the treatment schedule in Table 1. Significant problems arose with the pumping equipment and sand blenders during the treatment. The stage 4 sand proppant concentration actually varied from 0 to 4.3 ppg (0 to 520 kg/m^3). Cross-linker was not added to the gel during the final stage, and the treatment immediately screened out. Pumps failures necessitated shutdown during the initial pad treatment.

Table 1. Hydraulic-fracture treatment schedule for the Colorado 32–7 No. 9 well.

Stage	Fluid Volume (gal)	Sand Load (ppg)	Sand Size (mesh)	Injection Rate (BPM)	Surface Pressure (psig)
1	20,000	0.0	–	45	2,760
2	2,000	1.0	40/70	50	2,178
3	3,000	2.0	40/70	50	2,152
4	5,838	*	12/20	49	1,904
5	4,746	2.8	12/20	51	1,998
6	3,528	4.1	12/20	51	2,055
7	7,560	6.1	12/20	49	2,110
8	5,334	7.2	12/20	49	2,223
9	3,024	8.0	12/20	49	2,400
10	1,554	10.0	12/20	1.8	5,664

*Concentration ranged from 0 to 4.3 ppg due to sand-blending mechanical problems.

Following cleanout, a tracer log was run in both the No. 4 and No. 9 wells. No tracers was identified at the No. 4 well location, indicating that sand proppant and stimulation fluids had not intersected the offset well.

Tubing equipped with a surface recording, bottom-hole pressure transducer was run in the No. 9 well which was placed on production with a rod pump. During a 38-day production test, gas rate declined from a high of 207 to 119.5 MSCFD (5,858 to 3,382 m^3/day). Water production rate declined from an initial 200 BPD to a 80 BPD (31.8 to 12.7 m^3/day). Following the production test, a 31-day shut-in test was begun. The results of the shut-in test and other analyses are discussed in the following section.

FORMATION EVALUATION

Numerous evaluations and measurements were performed to quantify reservoir properties. The results from each of these analyses are discussed.

Thickness, Ash, and Gas Content

A total of 41.3 ft (12.6 m) of net coal reservoir was penetrated in the No. 9 well. Of this amount, 27.4 ft (8.4 m) is contained in two principal seams within the interval of interest, depth 2,881.05 to 2,945.15 ft (878.72 to 898.27 m), as illustrated in the density log in Figure 1. Evaluation of desorption tests indicated an average in situ gas content of 344 SCF/T (10.7 cc/g). The average ash content is 34.6%.

The productive intervals were selected as those with a density less than 1.75 g/cc (Close and others, 1990). A total of 24 separate continuous zones that met this criterion were identified in the No. 9 well. An area of 320 acres (130 ha) would contain 9.24 BSCF (261×10^6 m^3) of gas. Estimates of reservoir properties are summarized in Table 2.

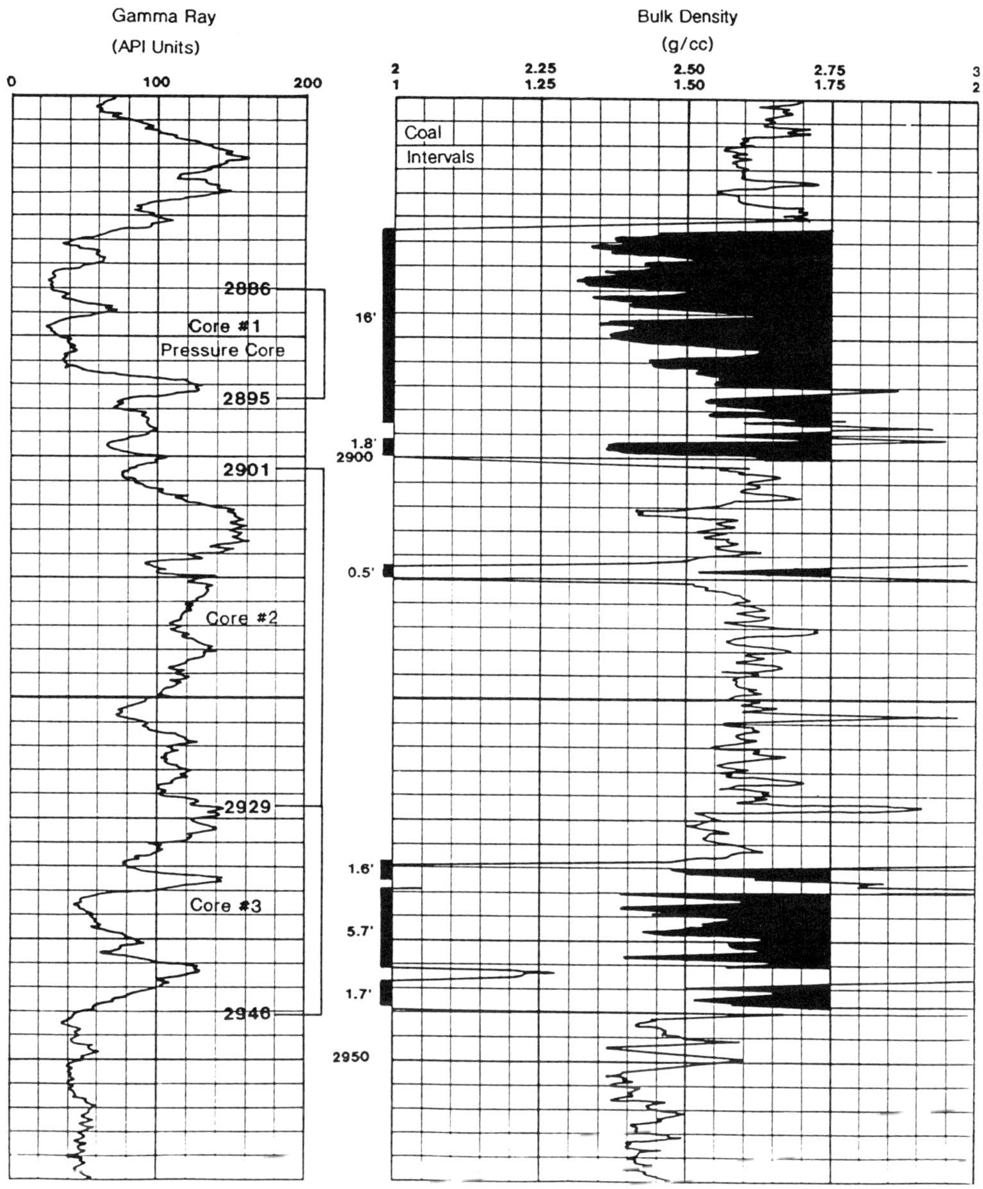

Figure 1. High-resolution density and gamma ray logs of the Colorado 32–7 No. 9 well, La Plata County, Colorado, showing identification of coal intervals based on density cutoff of 1.75 g/cc.

Petrology

A vitrinite reflectance measurement of 1.27% was obtained from a composite sample from the two seams illustrated in Figure 1. A rank of medium-volatile bituminous was confirmed through proximate analysis. Maceral composition was determined to be 83.6% vitrinite, 8.5% inertinite, and 7.9% liptinite.

Methane Storage Capacity

Sorption tests were conducted on a No. 9 well sample from 2,760 ft (841.8 m) with an ash content of 37.8% to determine methane storage capacity as a function of pressure at 110°F (43°C). The experiments were conducted with a coal moisture content of 1.3%. The estimated dry, ash-free Langmuir volume and pressure are 671.8 SCF/T (21 cc/g) and 287.6 psia (1,983 kPa), respectively.

Fracture Descriptions

The cores obtained from this project were oriented to allow detailed determination of natural fracture trends. The face cleat orientation (Terra Tek, 1989) in the interval of interest lies between due north and N20°E, with most face cleats observed between N10°E and N20°E (Figure 2).

Cleat dip varied from 90° to 35° with an average dip of 75.5°. Generally, the average cleat dip for other Fruitland formation coals is more toward vertical than at this location. We feel that the proximity of the No. 9 well to the Ignacio anticline may explain this difference.

Table 2. Estimated reservoir properties for the Colorado 32–7 No. 9 well.

Property	Value
* Gas gravity	0.6
* Average in situ gas content, SCF/T (cc/g)	344.3 (10.7)
Average ash-free gas content, SCF/T (cc/g)	526.0 (16.4)
* Average sorption time, hours	10.1
Dry ash-free Langmuir volume, SCF/T (cc/g)	671.8 (21)
* Langmuir volume (in situ), SCF/T (cc/g)	439.4 (13.7)
* Langmuir pressure, psia (kPa)	287.6 (1,983)
* Completed coal thickness, ft (m)	27.4 (8.4)
Total coal thickness, ft (m)	41.3 (12.6)
Average ash content, %	34.6
Average coal density, g/cc	1.494
Total gas in place (320 acres), BSCF (10^6 m^3)	9.24 (261.5)
Completed gas in place (320 acres), BSCF (10^6 m^3)	6.13 (173.5)
* Absolute permeability, md	1.7
* Fracture spacing, $in.^{-1}$ (cm^{-1})	10 (4)
* Initial reservoir pressure, psia (kPa)	1,620 (11,170)
* Pressure datum, ft (m)	2,830 (863)
Initial hydrostatic gradient, psi/ft (kPa/m)	0.572 (12.9)
* Cleat porosity, %	0.8
* Cleat water saturation, %	98
* Pore-volume compressibility, psi^{-1} (kPa^{-1})	100×10^{-6} (14.5×10^{-6})
* Temperature, °F (°C)	110 (43)
Vitrinite reflectance, %	1.27
Coal rank	Medium volatile
* Induced-fracture tip-to-tip length, ft (m)	336 (102)

* Required for simulator input.

The unstressed face and butt cleats range from 0.1 to 3 mm long and average 0.05 mm wide. Cleat frequency varies from 2.5/in. to 71/in. (1/cm to 28/cm) with an average frequency of 18/in. (7/cm). Cleat height is controlled by the thickness of vitrain lithotype layers that vary from 0.08 to 2 in. (0.2 to 5 cm).

Current Stress Orientation

A summary of anelastic strain measurements (Owen, 1989) is illustrated in Figure 2. The predominant direction of maximum horizontal stress lies between N80°W and N33°W, with an average direction of N55°W.

Well Test Analysis

Three well tests were performed in the interval of interest—an initial open-hole drillstem test (DST) produced water; a prestimulation test produced water and gas; and a post-stimulation test also produced water and gas. Reservoir properties obtained from the DST and post-stimulation test are summarized in Table 2.

The open-hole drillstem test was conducted in coal, sandstone, and shale intervals between 2,876 and 2,946 ft (877 and 898.5 m) (Figure 1). Coal thickness in this interval is 27.4 ft (8.4 m). The sandstone and shale in this well are impermeable, based on wireline formation-tester measurements. Water production rates were estimated from the change in bottom-hole pressure as the fluid level rose inside the drill string. Production rates decreased dramatically during the test from approximately 40 to 10 BPD (6.4 to 1.6 m^3/day). Estimated total recovery was 3 bbl (0.5 m^3).

Analysis of the pressure data was performed with a combination of diagnostic graph and semilog analysis techniques. The analysis indicated an average absolute permeability of 1.7 md. Initial reservoir pressure was 1,620 psia (11,170 kPa) at 2,820 ft (860 m), corresponding to a pressure gradient of 0.572 psi/ft (12.9 kPa/m) from the surface. The near-wellbore cleat permeability was severely reduced by the drilling and coring operations, as indicated by an estimated skin factor of 8.5, which corresponds to a flow efficiency of 49%.

Evaluation of the post-stimulation shut-in data was performed by history matching the measured pressure behavior with mathematical models that describe the pressure behavior of vertical, damaged, infinite- or finite-conductivity fractures (Cinco and others, 1978, 1981). The diagnostic graph in Figure 3 illustrates how the model simulations match the recorded pressure change and its derivative. The agreement between the simulation and actual data for reservoir behavior beyond the wellbore

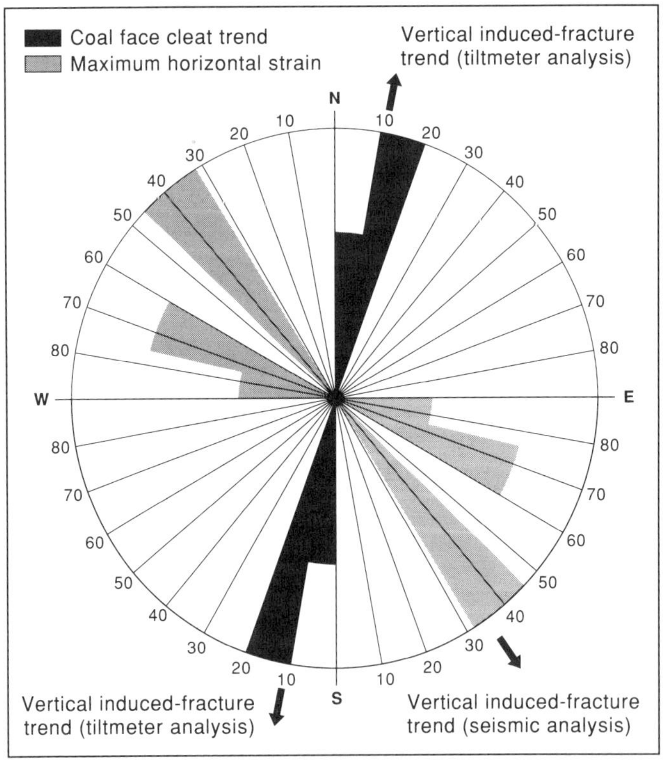

Figure 2. Face cleat and horizontal stress directions in Fruitland Formation coal in the Colorado 32–7 No. 9 well, and comparison with induced-fracture orientations inferred from tiltmeters and microseismic data.

effects and the agreement with DST-derived total permeability estimate validates the accuracy of the reservoir and fracture-property estimates.

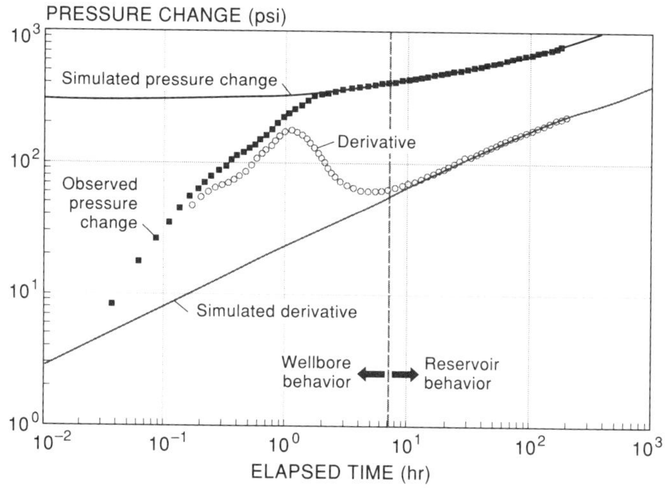

Figure 3. Well test pressure data for the Colorado 32–7 No. 9 well, showing history match with simulations.

Application of these models to coalbed methane reservoirs assumes that the cleat network is homogeneously distributed and in communication with a vertical, hydraulically induced fracture that is in communication with the wellbore. One primary, induced vertical fracture must dominate the pressure behavior. The "damaged-fracture" model assumes either a reduction in cleat permeability near the induced fracture or a reduction in near-wellbore conductivity of the induced fracture. Analysis with this model yields an estimate of the damaged-fracture skin factor. Estimates of reservoir and fracture properties include absolute permeability of 1.7 md (in agreement with DST analysis results), effectively infinite fracture conductivity, fracture length (tip to tip) of 336 ft (102.5 m), and damaged-fracture skin factor of 0.11. A damaged-fracture skin factor of this magnitude has little effect on production rates.

Although the mechanism for reduction of permeability near the induced fracture is unknown at this time, several hypotheses can be offered. Each assumes that near-fracture cleats have been plugged by one or more of the following:

1. Coal fines induced during stimulation and production.
2. Residual broken or unbroken cross-linked polymer.
3. Swelling of bentonite (ash) caused by fresh water-clay interaction.
4. Kaolinite fines release by fresh water-clay interaction.
5. Gas relative permeability reduction surrounding the fracture.
6. Swelling of the coal matrix due to adsorption of stimulation chemicals.

One can also hypothesize that the reduction in pressure near the fracture face is causing the highly compressible cleat system to close, thereby reducing cleat permeability near the hydraulic fracture. Another possibility is that conductivity of the induced fracture has been decreased near the well. This effect could result from proppant embedment, plugging by fines, or production of near-wellbore proppant.

FRACTURE DIAGNOSTICS

Tiltmeter data (Hunter Geophysics, 1989), microseismogram data from the offset No. 4 well (Worpel, 1989), and radioactive-tracer log data were collected during and after the hydraulic fracture treatment. Analysis of data collected by these techniques and the application of a hydraulic fracturing model are critical in interpretation of the induced-fracture geometry.

Tiltmeters

Based on analysis of tiltmeter data, the induced fracture is believed to be T-shaped with a vertical component in the direction N11°E (Figure 2). The vertical component extended 374 ft (114 m) tip to tip, with an average height of 80 ft (24.4 m), average width of 0.95 in. (2.4 cm), and dip of 84°SE. The horizontal component is inferred to be an ellipse whose major axis extends 93 ft (28.4 m) in the direction of N45°W with the minor axis extending approximately 50 ft (15 m) perpendicular to the major axis; the plane is inclined 6°NW (Figure 4). The estimated aperture of the horizontal component was 1.15 in. (2.9 cm) Note that the vertical component is believed to have propagated parallel to the face cleat trend and that the horizontal component dips in the direction of maximum horizontal strain.

The fact that no radioactive tracer was detected at the offset well confirms that the horizontal component length must be less than 142 ft (43 m) in the northerly direction.

Microseismicity

Based on microseismic data from the offset No. 4 well (Worpel, 1989), it was estimated that the vertical fracture propagated a distance of 300 to 400 ft (92 to 122 m) in the direction of S35°–45°E, which parallels the maximum horizontal stress direction (Figure 2). Although it was believed that the fracture propagated only southeastward from the wellbore, seismic activity due to propagation on the opposite side may have been was obscured. It also is possible that the seismic data were reacting more to horizontal rather than to vertical fracture components. Further interpretation suggests that multiple induced fractures propagated in each of the three coal seams between 2,881 and 2,900 ft (878.7 and 884.5 m), 2,909 and 2,910 ft (887.2 and 887.6 m), and 2,933 and 2,945 ft (894.6 and 898.2 m) (Figure 1). The estimated total fracture height was 150 ft (45.8 m).

Combining the tiltmeter and seismic data suggests that multiple vertical fracture components were created. These components propagated parallel to the face cleat orientation and to the maximum stress direction.

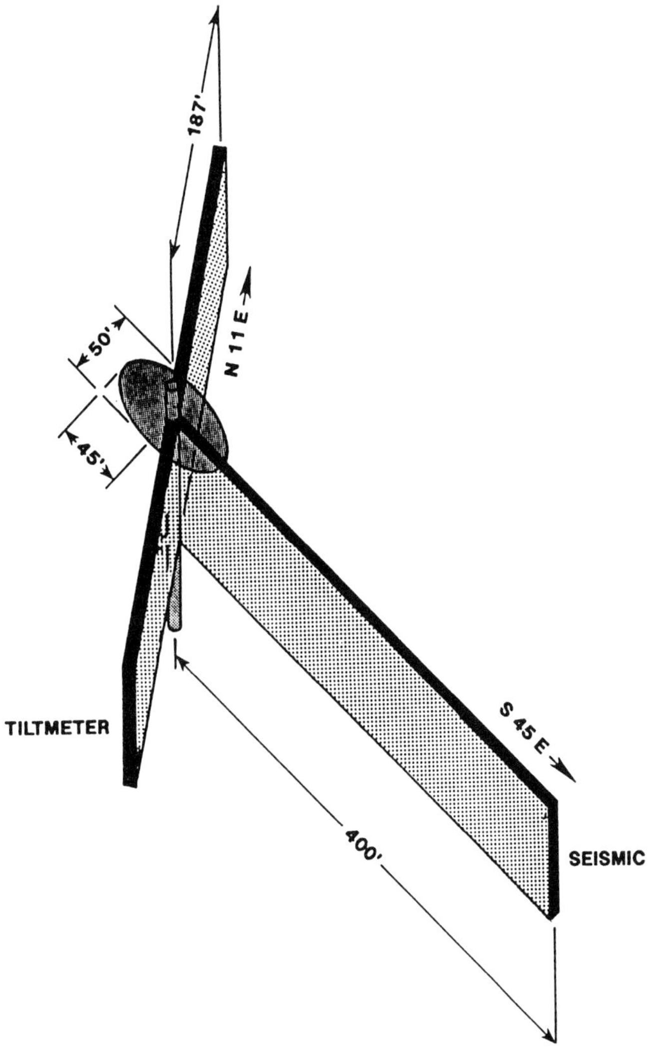

Figure 4. Inferred three-dimensional induced-fracture geometry in the Colorado 32–7 No. 9 well derived from tiltmeter and microseismic data.

Radioactive Tracer

Interpretation of post-stimulation tracer log data indicates that the induced fracture extended from 2,880 ft to 2,956 ft (878.4 to 901.6 m), for a height of 76 ft (23.2 m), which agrees with tiltmeter-derived estimates. Most of the treatment fluid exited the perforations opposite the upper seam between 2,881 and 2,900 ft (878.7 and 884.5 m). The scandium isotope tracer indicated that the entire induced fracture was filled with sand near the wellbore.

Well Test Analysis

Estimates of induced fracture length obtained from well test analysis agree closely with those from tiltmeter analysis. In general, well test-derived estimates tend to be less than those obtained from other sources, as is the case for this analysis. The well-test estimate is the "effective" fracture length—that which is propped and contributing to production. Estimates from the other sources are induced fracture lengths, the remote ends (from the well) of which may not effectively be filled with proppant and may not contribute to production.

Stimulation Modeling

To evaluate the effectiveness of the fracture stimulation, net pressure behavior during the treatment was estimated with a three-dimensional hydraulic fracture-propagation model. This model requires information about the mechanical properties of the coal and surrounding rocks and the vertical distribution of horizontal stresses. Mechanical properties for a discrete number of lithologic zones were determined from published sources (Jones and others, 1988), local experience, and logging evidence. Unfortunately, no measured information was available concerning the vertical distribution of in situ stresses. However, a reasonable stress profile was inferred from estimated mechanical properties and lithologies.

In addition to the actual treatment specifications, listed in Table 1, the simulation required several other input parameters. Bottom-hole pressures were not recorded during the stimulation and so had to be estimated from surface pressure data. These estimates required calculation of friction pressure, which was determined from pump-rate variations during the treatment, proppant-free fluid properties, and published relationships for the effects of sand content (Keck and others, 1989).

From an early shutdown period closure stress was determined to be 2,410 psi (16,617 kPa), corresponding to a fracture gradient of 0.83 psi/ft (18.8 kPa/m). Based on the closure stress and estimated friction-pressure losses, bottom-hole treatment pressure and excess treatment pressure were estimated. Excess treatment pressures are defined as the differences between bottom-hole treating (injection) pressures and closure stress. The estimated excess treatment pressures are illustrated in Figure 5 as a function of time during the treatment.

A fully three-dimensional planar hydraulic fracturing simulator was used to model vertical and horizontal fracture growth. Fracture geometry was estimated by varying input

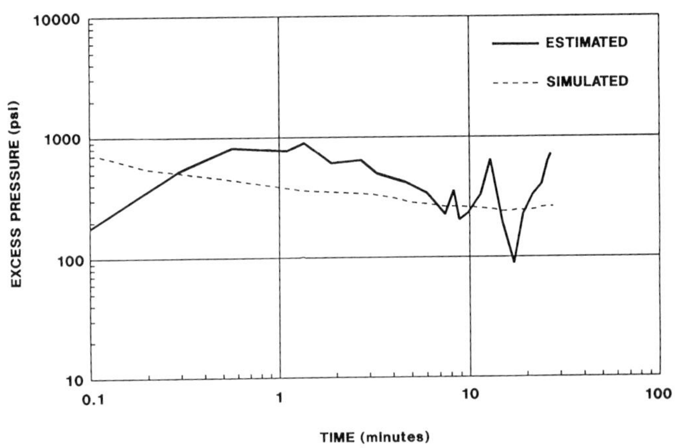

Figure 5. Comparison of estimated and simulated net treating pressure in the Colorado 32–7 No. 9 well.

parameters as required to obtain an approximate match between excess pressure computed by the model and that estimated from the treatment data. This procedure generally is "straightforward" except that the actual induced-fracture geometry was not restricted to a single plane, as determined by the other diagnostic tools. The simulator can account for neither the horizontal fracture component nor multiple vertical components. To account for differences in geometry between the computer model and the actual case, several first-order approximations and the following assumptions were required.

First, a vertical profile of the minimum horizontal principal stress was assumed. Fracture growth was controlled by the assumed stress profile, which was modified to obtain an approximate match with the excess pressure, as illustrated in Figure 5.

Second, because the fracture height predicted by the model was greater than the height determined from diagnostics, we assumed that the additional height represents the horizontal fracture component. This assumption is equivalent to assuming that propagation of the horizontal component of the fracture would create the same pressure response as for height growth.

Because of the assumed equivalence of the horizontal and vertical components on pressure behavior, predicted in-zone vertical fracture lengths and widths were believed to be approximately correct.

Numerous simulations were performed to obtain the approximate match illustrated in Figure 5. The computed propped-width profile (assuming minimal width change on shut in after screenout occurred) ranged from 0.5 in. (1.3 cm) at the wellbore to 0.38 in. (1 cm) at the tip. The essential vertical fracture geometry is summarized as follows:

Induced-Fracture Property	Value
Propped fracture half-length	379 ft (115.6 m)
Penetration of 12/20-mesh proppant (well to tip)	199 ft (60.7 m)
Length-weighted average fracture width	0.453 in. (1.15 cm)
Fracture conductivity	22.9 darcy-ft
Dimensionless fracture conductivity	35.5

The fracture segment from the tip of the 12/20-mesh proppant to the propped fracture length was filled with 40/70-mesh proppant. Fracture conductivity (permeability-width product) was estimated as a weighted average of the conductivity, noting the penetrations of 40/70 and 20/40 proppants and using published values for proppant conductivities (Gidley and others, 1989).

A dimensionless fracture conductivity of 35.5 would behave as effectively infinite, which is in agreement with the pressure behavior observed during the post-stimulation test analysis. Note that the 199-ft (60.7-m) penetration length of the 12/20-mesh proppant is in approximate agreement with the effective fracture penetration length of 168 ft (51.2 m) determined from well-test analysis. The part of the fracture propped with 40/70-mesh sand apparently does not have sufficient conductivity to affect productivity of the well.

One possible conclusion of this modeling exercise is that the current state of available technology does not properly explain the hydraulic fracturing behavior of coal seams.

RESERVOIR MODELING

Following the hydraulic fracturing and cleanout treatment, a 38-day production test was conducted on the No. 9 well. During this test the gas production rate declined from a high of 207 MSCFD (5,858 m^3/day) to 119.5 MSCFD (3,382 m^3/day), and water production declined from an initial rate of 200 BPD (31.8 m^3/day) to 80 BPD (12.7 m^3/day). Bottom-hole pressures recorded during the test then were history matched with the COMETPC–3D coalbed methane reservoir simulator (Sawyer and others, 1990). Results of the previous analyses (Table 2) were used to simulate the performance of the well.

Because of the coal's relatively low permeability and the short duration of the production test, no boundary effects were expected to influence production. A system area of 320 acres (130 ha) was considered adequate for the purpose of this study. Because the formation was assumed to be homogeneous, computational efficiency was achieved by simulating a representative quarter block of the 320-acre (130-ha) area and locating the fractured well in the northwest corner.

Gas and water flow in the cleat system is influenced by relative-permeability effects that are expected to vary with coal and reservoir characteristics. Because no laboratory permeability measurements were available for the No. 9 well, relative permeability data were obtained from the Northeast Blanco Unit 403 well (Mavor and others, 1990), located about 20 miles (32 km) to the southeast, as a starting point for this study. The data were adjusted during history matching to account for multiphase flow effects, and results are summarized in Table 3.

Table 3. Estimated relative-permeability relationships for the Colorado 32–7 No. 9 well.

Water Saturation (fraction)	Relative Permeability to Gas (fraction)	Relative Permeability to Water (fraction)
0.80	0.0455	0.0116
0.85	0.0417	0.0600
0.87	0.0340	0.1000
0.90	0.0264	0.1500
0.92	0.0239	0.1800
0.95	0.0053	0.3500
0.97	0.0044	0.5400
0.98	0.0020	0.7000
1.00	0.0000	1.0000

Compressibility and pore volume of the cleat system have a significant effect on the rate and decline of water production. At this time no direct measurements of cleat system pore volume and compressibility are available. However, previous experience with Fruitland coals indicated that cleat system porosity would fall in the range of 0.5 to 2%. Based on the history match, porosity was estimated to be 0.8%, and cleat compressibility was estimated to be 100×10^{-6} psia^{-1} (145×10^{-6} kPa^{-1}).

The early part of the test was characterized by higher gas production rates than predicted by desorption alone with bottom-hole and reservoir pressures above the desorption pressure. This implies the presence of free gas in the cleat system. Based on the reservoir history-match results, initial water saturation in the cleats was estimated to be 98%.

The consistent match between actual and simulated gas rates (Figure 6) not only confirmed the validity of the formation evaluation and stimulation diagnostic results but also verified that the noneconomic levels of gas production were the result of low absolute cleat permeability rather than failure of the stimulation treatment.

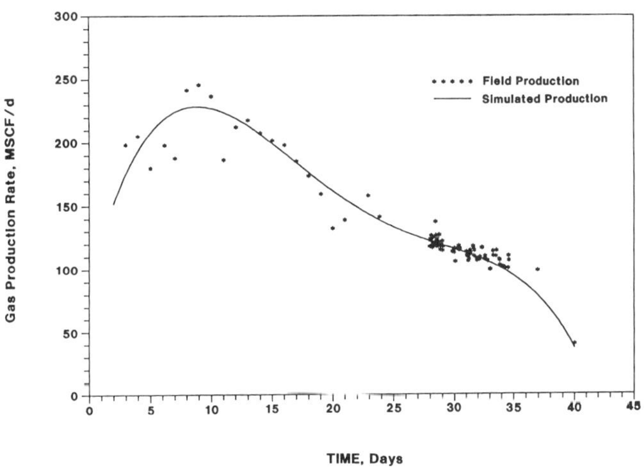

Figure 6. COMETPC–3D history match of production-test results in the Colorado 32–7 No. 9 well.

SUMMARY

Because of the combination of core, test, log, and stimulation diagnostic data, a complete evaluation of well performance of the Colorado 32–7 No. 9 well was possible. Noneconomic levels of gas production resulted from low (1.7 md) absolute cleat permeability rather than failure of the stimulation treatment. The induced-fracture geometry was complex—a T-shaped structure with multiple vertical components extending in the directions of both the face cleat trend and maximum horizontal stress trend. Multiple horizontal components also were created. Well-test analyses suggested that a single vertical fracture of length 336 ft (102 m) tip to tip was controlling production.

ACKNOWLEDGMENTS

The data collection efforts and subsequent analysis for this project was funded by the Gas Research Institute. We appreciate their support and permission to publish this material. We are grateful to Mobil Oil Corporation for access to the well and for the assistance provided to our efforts by numerous Mobil employees.

REFERENCES

Cinco-L., H., and Samaniego-V., F., 1981, Transient pressure analysis—Finite conductivity fracture case versus damaged fracture case: SPE Paper 10179 [presented at the 56th Annual Fall Technical Conference and Exhibition, San Antonio, Texas, October 5–7].

Cinco-L., H., Samaniego-V., F., and Dominguez-A., N., 1978, Transient pressure behavior for a well with a finite conductivity vertical fracture: Society of Petroleum Engineers Journal, v. 18, no. 4, p. 253–264.

Close, J.C., Mavor, M.J., and McBane, R.A. 1990, Importance, genesis and recognition of fracture permeability in Fruitland coalbed methane reservoirs of the northern San Juan basin, Colorado and New Mexico: CIM/SPE Paper 90–106 [presented at Petroleum Society of CIM and Society of Petroleum Engineers International Tech. Meeting, Calgary, Alberta, June 10–13].

Condon, S.M., 1988, Joint patterns on the northwest side of the San Juan Basin (Southern Ute Indian Reservation), southwest Colorado, *in* Fassett, J.E., ed., Geology and coal-bed methane resources of the northern San Juan Basin, Colorado and New Mexico: Rocky Mountain Association of Geologists Guidebook, p. 61–68.

Gidley, J., Holditch, S.A., Nierode, D.E., and Veatch, R.W., Jr., 1989, Recent advances in hydraulic fracturing: Society of Petroleum Engineers Monograph Series, v. 12.

Hunter Geophysics, 1989, Monitoring and interpretation of tilt response accompanying stimulation of May 26, 1989, Mobil well 32–7 No. 9, lower zone 2,881 to 2,941 feet: Sunnyvale, California, Hunter Geophysics.

Keck, R.G., Nehmer, W.L., and Strumolo, G.S., 1989, A new method for predicting friction pressure and rheology of proppant laden fracturing fluids: SPE Paper 19771 [presented at the 64th Annual Fall Technical Conference and Exhibition, San Antonio, Texas, October 8–11].

Mavor, M.J., and Close, J.C., 1989, Western Cretaceous Coal Seam Project, evaluation of the cooperative research well Colorado 32–7 No. 9 operated by Mobil Oil Corporation: Gas Research Institute, Topical Rept. GRI–90/0043, 21 p.

Mavor, M.J., Close, J.C., and McBane, R.A., 1990, Formation evaluation of exploration coalbed methane wells: CIM/SPE Paper 90-101 [presented at Petroleum Society of CIM and Society of Petroleum Engineers International Technical Meeting, Calgary, Alberta, Canada, June 10–13].

Owen, L., 1989, ASR measurements, Resource Enterprises, Inc. Project Number 5265 Well—Colorado 32–7 No. 9: Terra Tek Core Services Rept. TR 89–91 (March).

Sawyer, W.K., Paul, G.W., and Schraufnagel, R.A., 1990, Development and application of a 3D coalbed simulator: CIM/SPE Paper 90–119 [presented at Petroleum Society of CIM and Society of Petroleum Engineers International Technical Meeting, Calgary, Alberta, June 10–13].

Terra Tek Core Services, 1989, Fracture analysis of core no. 4 from Mobil Exploration and Production U.S. Inc.'s Colorado 32–7 No. 9 well, La Plata County, Colorado: Terra Tek Inc. Rept. TTCS File 5353 (September).

Worpel, C., 1989, Remote well monitoring of hydraulic fracture treatment, Mobil Colorado 32–7 No. 9: Teledyne Geotech Technical Rept. 89–3 (August).

Coalbed Methane—A Canadian Resource for the 1990s

Dennis J. Nikols[1]
Brian A. Rottenfusser[1]

1. Alberta Geological Survey, Edmonton, Alberta, Canada

ABSTRACT

Alberta holds in excess of 7.7 trillion short tons (7.0×10^{12} t) of coal, distributed over its three physiographic provinces (plains, foothills, mountains). These coals range in age from Jurassic to Tertiary and in rank from subbituminous through semianthracite. Promising geologic settings for coalbed methane recovery are found in all three regions. The Alberta petroleum industry recently has recognized this potential, and companies have begun testing samples and acquiring land positions. Resource estimates of the coal in nontraditional mining areas currently are being revised and improved. A resource evaluation study by the Alberta Geological Survey of the coalbed methane potential of the province has been heavily subscribed to by industry.

Alberta coals have not been extensively tested for gas content. Developing a coalbed methane resource estimate for such large, untested resources presents some interesting problems. Indirect evidence has been incorporated, such as relationships among depth, rank, and gas content, and between proximate analysis and gas content. These results will eventually be verified by direct data and field demonstrations. The 1990 published coalbed methane in-place resource estimate is 150 TCF (4.25×10^{12} m^3). The estimate from the procedures outlined in this paper is an order of magnitude larger—that is, 3,028 TCF (85.75×10^{12} m^3).

INTRODUCTION

Near-surface coal resources in Alberta have been known since 1793. By 1858 Sir James Hector of the Palliser Expedition noted that local coal already was in use by blacksmiths in Edmonton. These resources have been explored and exploited ever since. Measured and indicated reserves, as well as potential mining-related resources, have been well defined. These defined resources generally lie near the surface, within economic and practical mining depths. Deep resources have received considerably less attention. Although these coals have been penetrated by thousands of petroleum production and exploration wells over the years, they have not been thoroughly studied. These wells, together with geophysical logs and chip samples from them, comprise the data set available for this study. A program to study and redefine coal and coalbed methane resources in Alberta at greater depths is now under way at the Alberta Geological Survey. This paper presents an interim assessment and focal point for discussion until more definitive data become available.

GENERAL GEOLOGY

Alberta's coal deposits are concentrated in the southern half of the province. They are distributed over three physiographic regions—plains, foothills, and mountains (Figure 1). Most of the coal resources are found in the plains region of the province.

Sedimentation related to the formation of coal in Alberta was largely controlled by mountain-building processes in the west. Thick clastic wedges from the rising Cordillera prograded into the interior seaway during latest Jurassic and Cretaceous times. As this seaway periodically transgressed and regressed, peats developed and became buried in the intertonguing of marine and nonmarine sedimentation. In early Tertiary time, thick accumulations of peat developed on broad interior plains during periods of relative orogenic quiescence (Figure 2) (Mellon, 1967; Caldwell, 1984; Yurko, 1976).

COAL RESOURCES

Work of the Alberta Energy Resources Conservation Board (ERCB), Alberta Geological Survey, and Geological Survey of Canada have been combined to derive the coal resource estimates reported in this paper. The most recent summary of deep coal resources in western Canada was published by Williams and Murphy (1981), who found coal at depths of greater than 3,280 ft (1,000 m) in British Columbia, Alberta, and Saskatchewan. By far the largest resources are in Alberta.

Accurate measurements for shallow coal resources (Energy Resources Conservation Board, 1990; Strobl and others, 1986) have been compiled from mines, measured sections, and shallow drill holes. Estimates for the deeper resources (those usually considered too deep for economic

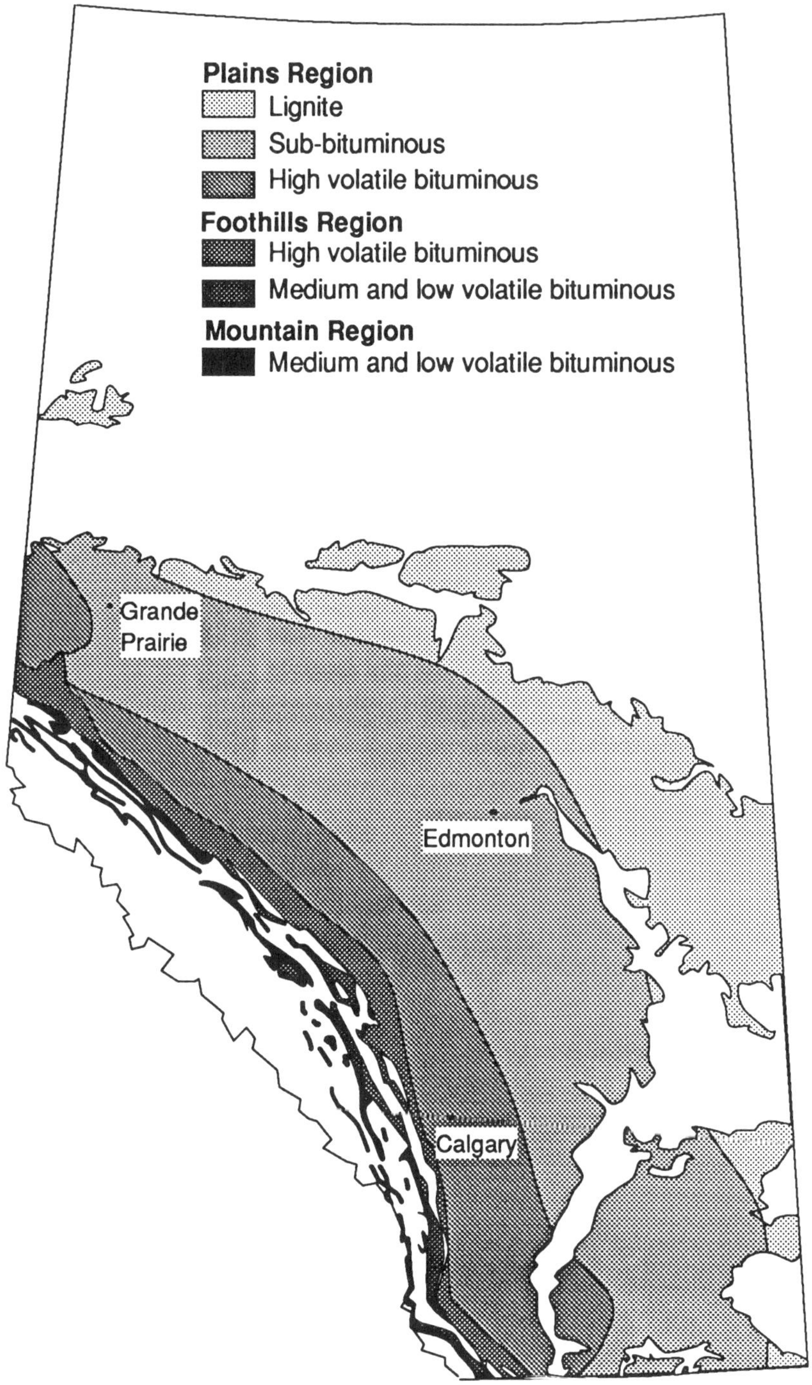

Figure 1. Coal distribution in Alberta by physiographic province and rank. After Alberta Economic Development and Trade (1984).

mining) are based on downhole geophysical logs and limited sampling. Table 1 summarizes the resource estimation criteria used by the various sources cited herein. Because the objective of this paper is to estimate coalbed methane potential in Alberta, only the total size of the coal resource and the probable gas content of the coal (based on coal rank) will be addressed.

It is obvious that the three studies used different methods for different purposes. Williams and Murphy's estimates were selected as those most likely to reflect the overall geologic situation of the region. The other estimates, although of good quality, were too restrictive in their scope for our needs. Table 2 illustrates that two of the estimates are only subsets of the third. The other authors concurred with our approach. R. Venter of the ERCB (1990, pers. comm.) suggested that the ultimate potential coal resource in Alberta would be about three times greater than the ERCB's published figures.

The ERCB's and Strobl and others' studies were focused at depths too shallow for a meaningful total coal resource

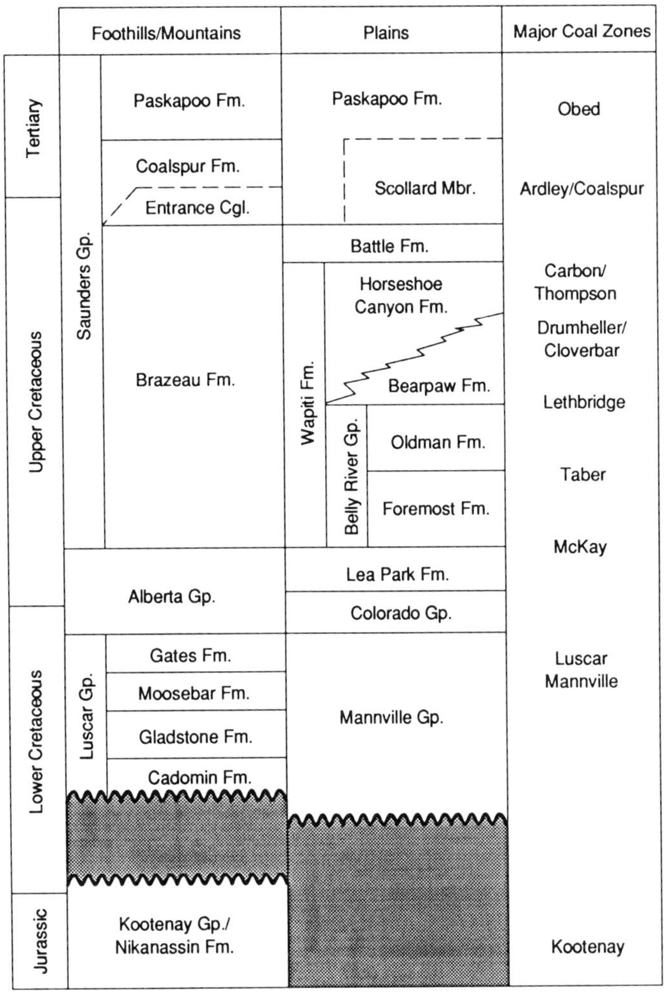

Figure 2. Coal-bearing formations and coal zones in Alberta.

estimate that could be used to estimate coalbed methane potential. Williams and Murphy's estimate appeared to be the best available. Because gas content and recoverability from Alberta's coals are unknown, the total coal resource in place was discounted by 15% and by 30% to represent a range of possible values.

COAL RANK

Actual gas-content measurements on Alberta coals are not available in the public domain at this time; therefore,

Table 1. Criteria used for coal resource estimation in Alberta.

Reference	Minimum Thickness		Depth Limit		Spacing of Observations		Coal Density		Number of Observations
	(ft)	(m)	(ft)	(m)	(no./mi²)	(no./km²)	(T/CF)	(t/m³)	
Williams and Murphy	1	0.3	None	None	1/36	1/93	0.041	1.30	2,448 wells 6,593 points
Strobl and others	1.5	0.5	1,312	400	2–3/36	2–3/93	0.043	1.37	4,737 wells
ERCB	1	0.3	1,969	600	NG	NG	0.041–0.052	1.30–1.68	Not stated

Table 2. Estimates of in-place coal resources, all ranks, in Alberta.

Reference	Resource (short tons)	Resource (t)
ERCB	$2,646 \times 10^9$	$2,400 \times 10^9$
Strobl and others	356×10^9	323×10^9
Williams and Murphy	$>7.7 \times 10^{12}$	$>7 \times 10^{12}$

estimates were made. The first step in this process was the classification of total coal resources by ASTM rank. Data are not as readily available in the literature for such a classification of deep coals as for resource quantity estimation. Consequently, a rather simplistic approach was taken—examining coal resource distribution maps, as well as cross-sections, and making assumptions on the various rank distributions that should be expected.

The rank distribution apparent at the surface has been shown to be misleading as a basis for determining rank at depth (Figure 3) (Nurkowski, 1985). The most expedient approach therefore was to arbitrarily divide the resource into likely rank groupings. (All references here to coal rank will follow ASTM designations.) Because most of the coal resources lay under the plains, we have assigned one-half the total 7.7 trillion short tons (7.0×10^{12} t) to high-volatile bituminous (hvb) and subbituminous (sb) ranks, as shown in Table 3. The remainder was divided equally between medium-volatile (mvb) and low-volatile (lvb) bituminous. Table 3 shows the corresponding discounted resource quantities for each rank category. The quantities of lignite and semianthracite are believed to be too small to consider.

GAS CONTENT

Following a literature review, each rank was assigned a probable gas content. Among the sources consulted were Kim (1976), McFall and others (1987), and King and Ertekin (1989). As actual gas-content measurements for Alberta coals become available, they will be integrated into the geologic framework described above. The following values were selected:

hvb + sb	352 CF/T (11 cc/g)
mvb	416 CF/T (13 cc/g)
lvb	449 CF/T (14 cc/g)

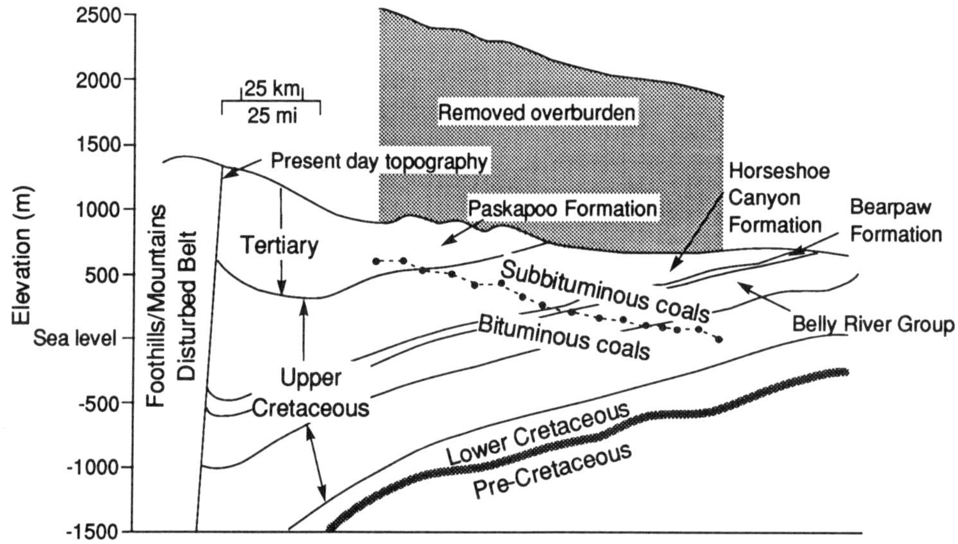

Figure 3. Plains cross-section southwest to northeast showing generalized subsurface stratigraphy, reconstructed overburden, and approximate change in rank of coal-bearing units. After Nurkowski (1985).

Table 3. Coal resource classification in Alberta, by rank. Total resource has been discounted by 15% and 30% to allow for unknown recoverability.

	Total		Discounted 15%		Discounted 30%	
Coal Rank	(10^{12} short tons)	(10^{12} t)	(10^{12} short tons)	(10^{12} t)	(10^{12} short tons)	(10^{12} t)
hvb + sb	3.9	3.50	3.3	3.0	2.70	2.450
mvb	1.9	1.75	1.7	1.5	1.35	1.225
lvb	1.9	1.75	1.7	1.5	1.35	1.225
Total	7.7	7.0	6.6	6.0	5.4	4.9

RESOURCE ESTIMATION

Due largely to the lack of reliable gas-content and rank-distribution data, we are not comfortable in presenting one single value for Alberta's coalbed methane resource. Instead we offer a range of possible values (Table 4) that correspond to the total coal resource in place and to the discounted coal resource values from Table 3. One could lower the gas contents used to achieve the same type of reduction. Figure 4 presents a family of curves that allow readers to make their own adjustments to our calculations based on various gas contents.

CURRENT RESEARCH

The Alberta Geological Survey currently is conducting a basinwide study of all coal resources, regardless of depth, in Alberta. This study involves detailed evaluation of petroleum exploration and development wells and the development of a stratigraphic framework for the coal zones of the basin. An area of more than 4,000 townships is being evaluated. Coals are being identified from geophysical logs, and several hundred cuttings and core samples are being analyzed for vitrinite reflectance. From this extensive data set, a series of planimetric maps will be produced showing structure, thickness, and rank of each coal zone. Results of this work will represent a significant refinement in our understanding not only of the total coal resource but

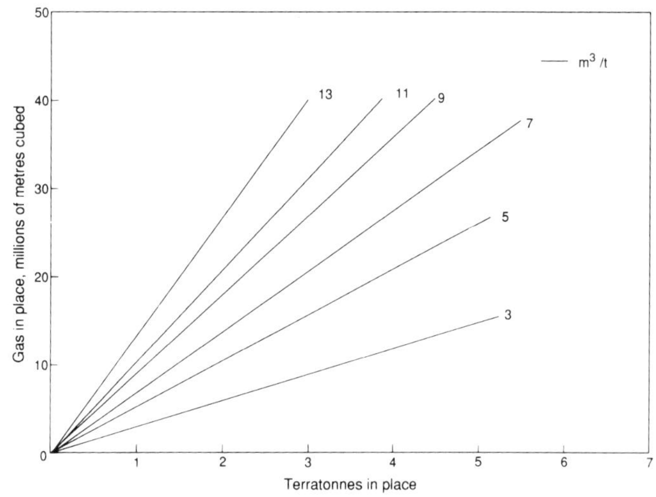

Figure 4. Correlation of coal resource and gas in place for various values of gas content (m^3/t).

Table 4. Coalbed methane potential of Alberta, by coal rank category. Coal resource values from Table 3.

	COAL RESOURCE					
	Total		Discounted 15%		Discounted 30%	
	(10^{12} short tons)	(10^{12} t)	(10^{12} short tons)	(10^{12} t)	(10^{12} short tons)	(10^{12} t)
All Ranks	7.7	7.0	6.6	6.0	5.4	4.9
	METHANE RESOURCE					
Coal Rank	(TCF)	(10^{12} m^3)	(TCF)	(10^{12} m^3)	(TCF)	(10^{12} m^3)
hvb + sb	1,360	38.50	1,165	33	952	26.950
mvb	803	22.75	689	19.5	562	15.925
lvb	865	24.50	742	21	606	17.150
Total	3,028	85.75	2,596	73.5	2,120	60.025

also of rank distribution within the deep coals of the province. All results are being integrated into a sophisticated database. The maps and cross sections will be integrated electronically into the Geoscience Information System, which can generate maps on demand. Information that becomes available can easily be integrated into this new framework.

CONCLUSIONS

The values cited for coalbed methane in place are theoretical and likely to be optimistic under some conditions. A range of possible values could have been used much as a range of coal quantities were used. The question can be resolved only by actual gas-content measurements across the basin, together with a more rigorous survey of coal quantities, coal quality, and geologic settings. Data are now being collected in various geologic settings across Alberta and integrated into our coal database and geologic framework. The resource calculations given here will be adjusted and refined as more data become available.

REFERENCES

Caldwell, W.G.E., 1984, Early Cretaceous transgressions and regressions in the southern Interior Plains, *in* Stott, D.F., and Glass, D.J., eds., The Mesozoic of middle North America: Canadian Society of Petroleum Geologists Memoir 9, p. 173–203.

Energy Resources Conservation Board, 1990, Reserves of coal, Province of Alberta: Energy Resource Conservation Board, ST90–31.

Kim, A.G., 1976, Estimating methane content of bituminous coalbeds from adsorption data: U.S. Bureau of Mines Rept. of Investigations 8245, 22 p.

King, G.R., and Ertekin, T.M., 1989, A survey of mathematical models related to methane production from coal seams, part 1—Empirical and equilibrium sorption models, *in* The 1989 Coalbed Methane Symposium, Proceedings—University of Alabama, Tuscaloosa, April 17–20, 1989: School of Mines and Energy Development, Gas Research Institute, U.S. Mine Safety and Health Administration, and Geological Survey of Alabama, p. 125–138.

McFall, K.S., Wicks, D.E., Kelso, B.S., Sedwick, K., and Brandenburg, C, 1987, An analysis of the coal and coalbed methane resources of the Piceance Basin, Colorado: SPE/DOE 16418 [presented at the SPE/DOE Joint Symposium on Low Permeability Reservoirs, Denver, Colorado, May 18–19].

Mellon, G.B., 1967, Stratigraphy and petrology of the Lower Cretaceous Blairmore and Mannville groups, Alberta Foothills and Plains: Alberta Research Council Bulletin 21, 270 p.

Nurkowski, J.R., 1985, Coal quality and rank variation within Upper Cretaceous and Tertiary sediments, Alberta plains region: Alberta Research Council Earth Science Rept. 85-1, 39 p.

Smith, G.G., 1989, Coal resources of Canada: Geological Survey of Canada Paper 89-4, 127 p.

Strobl, R., McCabe, P.J., Macdonald, D.E., and Richardson, R.J.H., 1986, Estimate of inferred resources of coal in the plains region of Alberta to a depth of 400 metres: Alberta Research Council Open-File Rept. 1987–05, 31 p.

Williams, G.D., and Murphy, M.C., 1981, Deep coal resources of the Interior Plains, estimated from petroleum borehole data: Geological Survey of Canada Paper 81–13, 15 p.

Yurko, J.R., 1976, Deep Cretaceous coal resources of the Alberta Plains: Alberta Research Council Rept. 75–4, 47 p.

MAROON BELLS

Coal Resources and Coalbed Methane Potential of a Major Alberta Coal Zone

R.J.H. Richardson[1]

1. Alberta Research Council, Edmonton, Alberta

ABSTRACT

Total coal resources of Alberta have been broadly estimated by the Alberta Energy Resources Conservation Board at 2.6 trillion short tons (2.4 trillion t). The energy in this coal is 23 times that available in the province's substantial remaining reserves of conventional oil and more than 3.7 times that of its massive oil sands deposits. Although most of this coal resource will never be mined, the prospect of recovering methane from a substantial part of it is promising both because of the availability of coal and favorable geological conditions. Studies to quantify the gas potential of several Alberta coal zones are now under way.

One such coal zone, the Ardley, has calculated resources on the order of 468.5 billion short tons (425 billion t) along a 250-mile-long (400-km) subcrop and dipping, to a depth of about 2,950 ft (900 m), 93 miles (150 km) southwestward into the Alberta syncline. The geologic model developed for the Ardley coal zone involves a continental depositional setting in an alluvial plain environment characterized by widespread peat swamps far removed from marine conditions. Because of extremely low relief, relatively rapid subsidence, and sediment starvation, thick and extensive peats developed. In general the seams thicken, and more seams are present, with increasing depth. These deeper coal resources may contain large volumes of gas.

INTRODUCTION

The total coal resource in Alberta has been broadly estimated by the Alberta Energy Resources Conservation Board (ERCB) at 2.6 trillion short tons (2.4 trillion t). Nikols and Rottenfusser (this volume) have accepted Williams and Murphy's (1981) larger estimate of 7.7 trillion short tons (7 trillion t) based on recent discussions with the ERCB staff and because that estimate was "most likely to reflect the overall geologic situation" (that is, the estimate included deeper coal resources). Even with the lower ERCB estimate, the energy in the coal still is about 23 times the energy available in the province's substantial remaining reserves of conventional oil (estimated 23.7×10^{18} Btu, or 25×10^{18} J; Alberta Energy Resources Conservation Board, 1986) and more than 3.7 times that of Alberta's massive oil sands deposits. The coal resource in place in Alberta is enormous by either estimate. The review by Nikols and Rottenfusser (this volume) addresses coalbed methane potential of the entire Alberta basin and will not be reviewed here. Instead this paper will focus on coalbed methane potential of only one coal zone—the Ardley coal zone of the Paskapoo Formation (Paleocene).

ARDLEY COAL ZONE

Economically the Ardley is one of the most important coal-producing zones in Alberta, where near-surface deposits are mined to fuel mine-mouth power stations. More than 16.5 million short tons (15 million t) annually are mined along the subcrop west of Edmonton (Figure 1).

Previous Work

Research on the Ardley coal zone dates to the 1920s (Allan, 1924). A recent evaluation of the coal resources by Richardson and others (1988) reviews previous work and details the geology of the zone. That study, although focused on coals to a depth of 1,300 ft (400 m), was based on examination of data from 1,408 oil and gas wells to depths of more than 2,950 ft (900 m) and from 98 Alberta Research Council coal-exploration drill holes in the shallow subcrop.

Regional Coal Geology

In south-central Alberta, north of Calgary, from a 250-mile-long (400-km) subcrop, the Ardley coals of the lower Paskapoo Formation (Upper Cretaceous to Paleocene) dip 93 miles (150 km) southwestward into the Alberta syncline (Figures 1 and 2). They attain a maximum depth of about 2,950 ft (900 m) at the edge of the tectonically disturbed area. Ardley coals are contained in a wedge of fine- to coarse-grained continental (fluvial and shallow lacustrine) clastic sedimentary rocks (Figure 3).

Depositional Model

The Ardley coal zone is associated with mainly fine-grained fluvial sediments. Two areas in the west-central

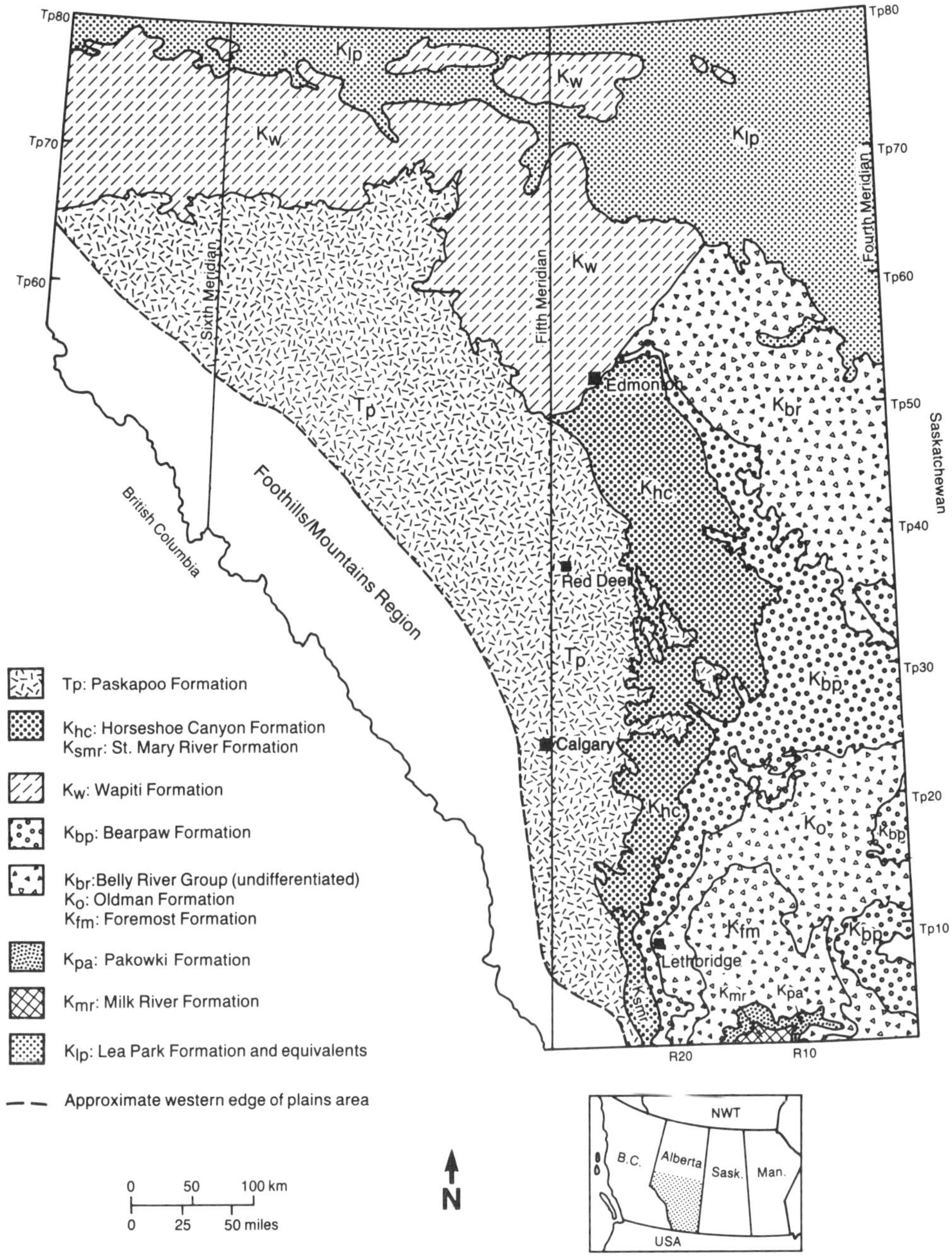

Figure 1. Generalized geology of south Alberta. Ardley coal zone is present north of Calgary near the base of the Paskapoo Formation.

part of the basin have particularly great amounts of coal, with cumulative (net) thicknesses up to 79 ft (24 m) and individual seams to 13 ft (4 m) thick.

A depositional model has been developed to show the relationship between coal accumulation and regional tectonic setting. Major depositional controls appear to relate to variation in subsidence rates and isolation from the effects of clastic deposition in a pervasive peat-forming environment. The coal seams are remarkably persistent, with thick seams correlatable over tens of miles. The peats accumulated in a tectonically active foreland basin, characterized by greater subsidence in the west and southwest due to loading and/or tectonic influences along the western basin margin. Thickness of the Ardley coal zone ranges from 82 ft (25 m) in the east to more than 660 ft (200 m) thick in the west. The total number of seams increases westward in response to greater subsidence. The lateral persistence of many seams suggests development, at times, of extensive peat mires over large areas of the basin.

The development of major river systems parallel to the paleo-mountain front is a possible mechanism for sheltering the eastern peat mires from clastic input (Figure 3) over long periods of time. Areas more proximal to the western basin margin would have experienced greater influx of clastics, thereby preventing the formation of thick coals. Several wells drilled farther westward, in the disturbed

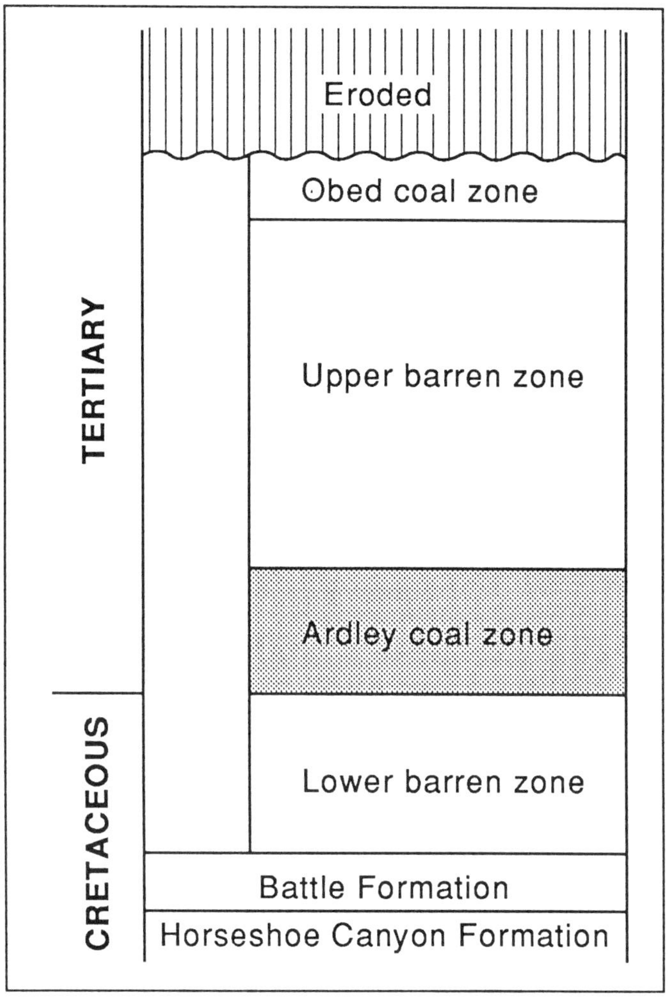

Figure 2. Informal stratigraphic terminology of the Paskapoo Formation, Alberta. Base of the Ardley coal zone lies at the Tertiary/Cretaceous boundary.

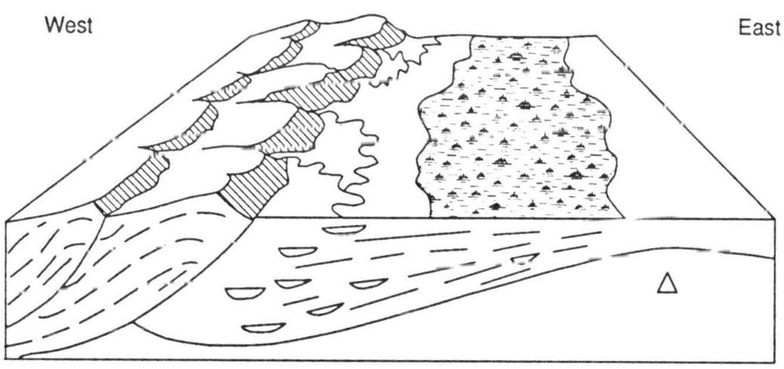

Figure 3. Schematic block diagram showing important tectonic control on peat accumulation during early Tertiary time. Increased subsidence related to tectonic loading in the west and a small foreland bulge (not discussed in this paper) helped confine major river systems to an area parallel to the disturbed belt. Peat swamps, sheltered from clastic input, were extensively developed east of the river systems toward the eastern edge of the basin.

belt, to some extent substantiate this idea in that they penetrate less cumulative coal.

The absence of coal development in the southern part of the basin, toward the United States border, may be related to climatic constraints. Jerzykiewicz and Sweet (1986) suggested that the presence of caliche beds, red beds, and impoverished palynological assemblages in Ardley-equivalent strata in the southern Alberta foothills may be indicative of a semiarid environment unfavorable for peat development.

Coal Resources

Coal Seam Development

Over much of the area at least one seam 6.5 ft (2 m) thick or greater is present. Areas where seams are 13 ft (4 m) or more thick commonly are located within an east-west trend southwest of Edmonton (Figure 4). Generally the Ardley coal zone consists of 3 to 12 seams (Figure 5), but as many as 18 seams are present in the west-central part of the basin. Areas of thickest coal seams and greatest number of seams coincide in two well defined trends of thick cumulative (net) coal development, as shown in Figure 6—a north-south-oriented trend south of the Athabasca River, and an east-west-oriented trend mostly west of the North Saskatchewan River, each containing 40 to 80 ft (12 to 24 m) of cumulative coal.

Depth to "Best 25-m Zone"

A "window" concept was applied in generating a map (Figure 7) of depth to the "best 25-m zone." The term *window* is defined as a zone within a given stratigraphic interval that contains the most net coal. This map is useful in determining the depth of the major pay zone because the total coal zone thickens to more than 660 ft (200 m) westward. Figures 6 and 7 reveal that the greatest cumulative (net) coal and the depth to the best 25-m (82-ft) zone lie between 1,640 ft (500 m) and 2,625 ft (800 m). At depths where the greatest cumulative coal and the depth to the best 25-m zone coincide, formation pressures should be sufficient to contain gas that may be present within the coal.

Coal Resources

Coal resources of the Ardley zone have been calculated at 468.5 billion short tons (425 billion t) (Strobl and others, 1986; this amount does not include equivalent Wapiti and Coalspur coals in the northern tectonically disturbed parts of the basin). New information, based on the same data used in Strobl and others' Ardley resource calculation is presented in Figure 8. This chart shows estimated resources for Ardley coals in 650-ft (200-m) depth intervals to a maximum depth of 3,937 ft (1,200 m), although relatively little coal exists below 2,950 ft (900 m). For each depth interval, resources are shown for three net-coal thickness categories—>1.6, >5 ft, and >8 ft (>0.5, >1.5, and >2.5 m). The >1.6-ft (>0.5 m) thickness bar shows the total amounts of coal, for each depth slice, which when added together approximate the total calculated resource of Ardley coals. The >5-ft (>1.5 m) and >8-ft (>2.5 m) thickness bars at each

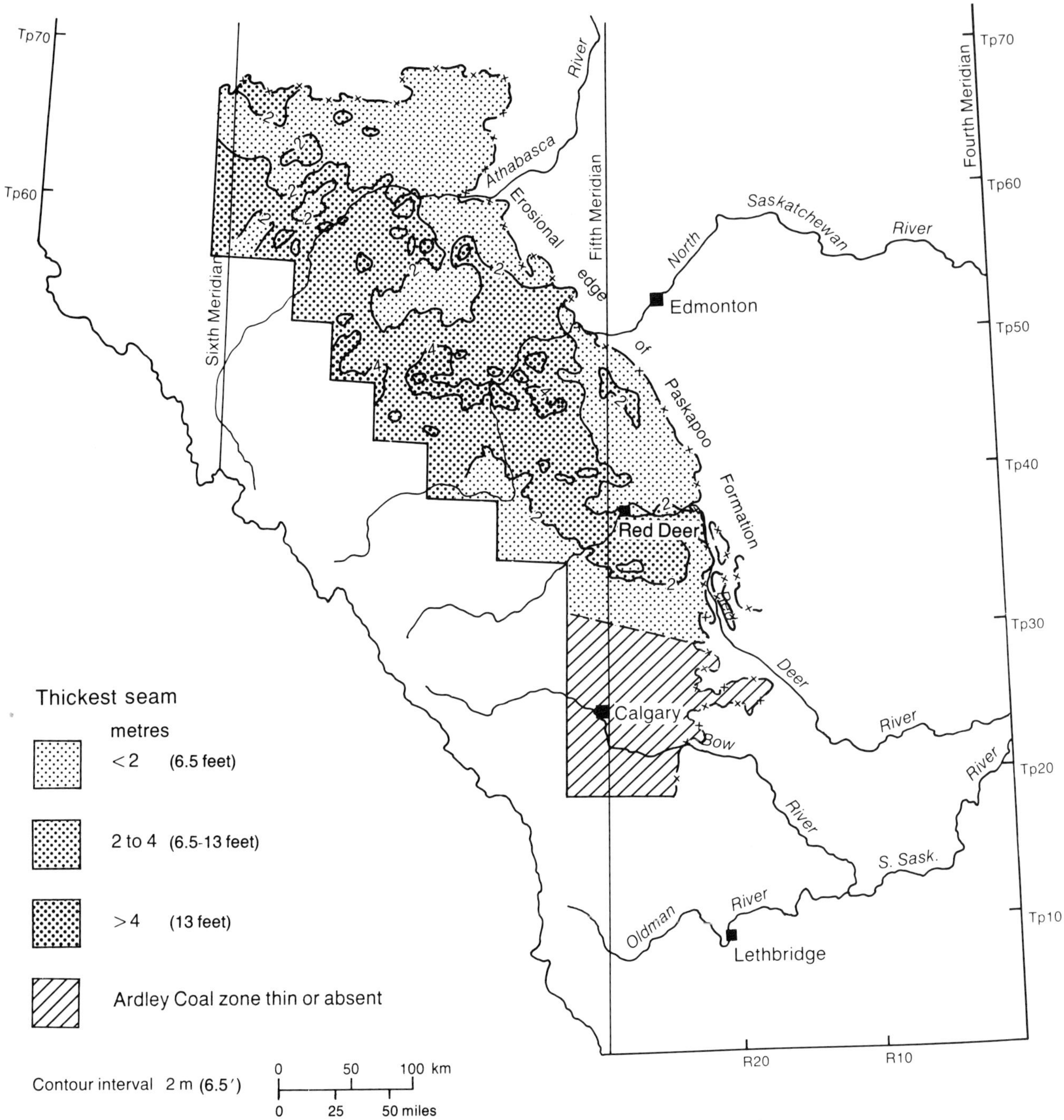

Figure 4. Isopach map of thickest seam in Ardley coal zone, southern Alberta.

depth indicate the resources of thicker coals that would be potential exploration and prime targets, respectively, for coalbed methane. Nearly 55 billion short tons (50 billion t) of coal in seams greater than 8 ft (2.5 m) are present in the depth interval of 1,300 to 1,970 ft (400 to 600 m), and approximately 77 billion short tons (70 billion t) are present between 1,300 and 2,625 ft (400 to 800 m).

COAL RANK

Except for the subcrop area, including the major producing Ardley mines, little is known about the rank of Ardley coals. Historically the Ardley coal has been shown everywhere on maps as subbituminous even at depths west of the subcrop edge where no data existed. In a petrographic study of coals from the Alberta plains, Parkash (1985) examined five Ardley coals from widely separated areas along its subcrop. These coals ranged from subbituminous C to subbituminous A in rank, based on mean vitrinite reflectance. However, Parkash cautioned on using vitrinite reflectance on coals of rank lower than bituminous because of considerable overlap in reflectance values between subbituminous B and A coals and bituminous C

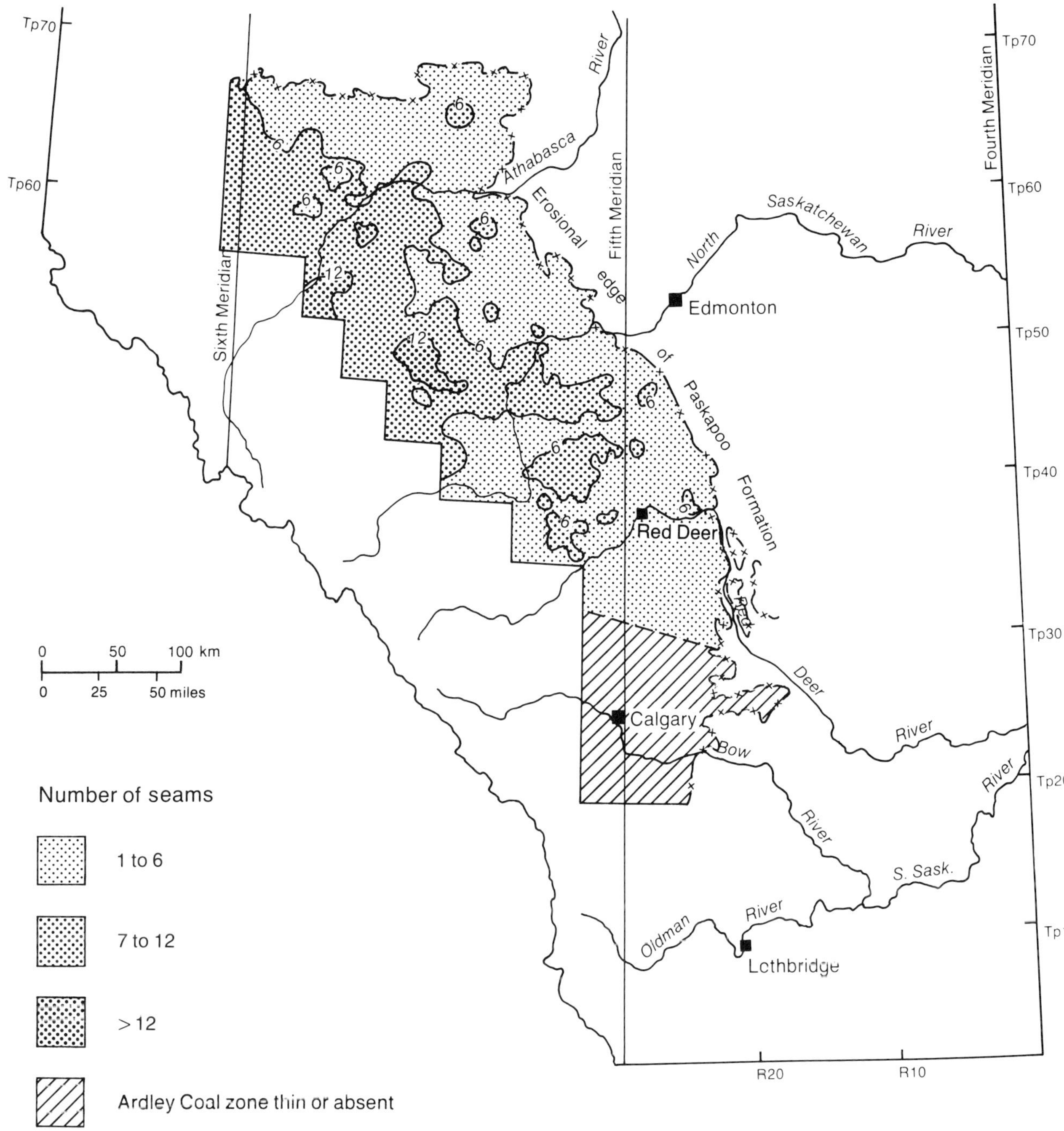

Figure 5. Number of seams in the Ardley coal zone, southern Alberta. Up to 18 seams are present in the west-central part of the basin.

and B coals. Care with the measurement of huminite macerals and a large number of measurements (usually 50) are needed to ensure reliable results. Parkash found a good relationship of carbon content to low-rank coals, supporting the reliability of his rank determinations.

In 1984 Nurkowski reported on 22 Ardley coal samples from four oil and gas wells forming an east-west cross section (Figure 9) near the center of the basin. Samples were of high-volatile C bituminous rank, with several nearing high-volatile B bituminous rank (average maximum vitrinite reflectance of 0.48% to 0.63% R_m). High values in the west are related to greater depth of burial. However, as shown on the vertical lines in Figure 9, the variation in vitrinite reflectance gradient from 0.07% R_m/100 m to 0.02% R_m/100 m from east to west indicates higher paleo-geothermal gradient in the east. From well 46–12W5, for example, six samples having average reflectance of 0.62% R_m were collected from well depths of 1,860 to 2,149 ft (567 to 655 m). From well 46–17W5, located 40 miles (64 km) farther west, six samples having average reflectance of 0.63% R_m were collected from well depths of 2,169 to 2,530 ft (661 to 771 m). Recently Osadetz and others (1990) studied paleogeothermal gradients and changes in the geothermal gradient field of the Alberta plains.

Nurkowski (1985) used the relationship between calorific value (moist mineral-matter free basis) and burial depth to determine the thickness of overburden required to mature various low-rank coals. Based on 685 coal analyses, he determined that 3,000 to 6,200 ft (900 to 1,900 m) of overlying strata in the basin had been removed since coalification during middle Tertiary time. The calculated depth to high-volatile C bituminous coals is shown in Figure 10. A comparison of Figures 7 and 10 reveals that most Ardley coals should fall within the high-volatile C bituminous rank and therefore may be producers of methane.

COALBED METHANE RESOURCE

Without actual gas-content measurements and sufficient rank-distribution data, any estimate of the total methane content of Ardley coals must be considered speculative. However, a large volume of coal within the Ardley zone occurs at a rank at which thermogenic gas normally is produced.

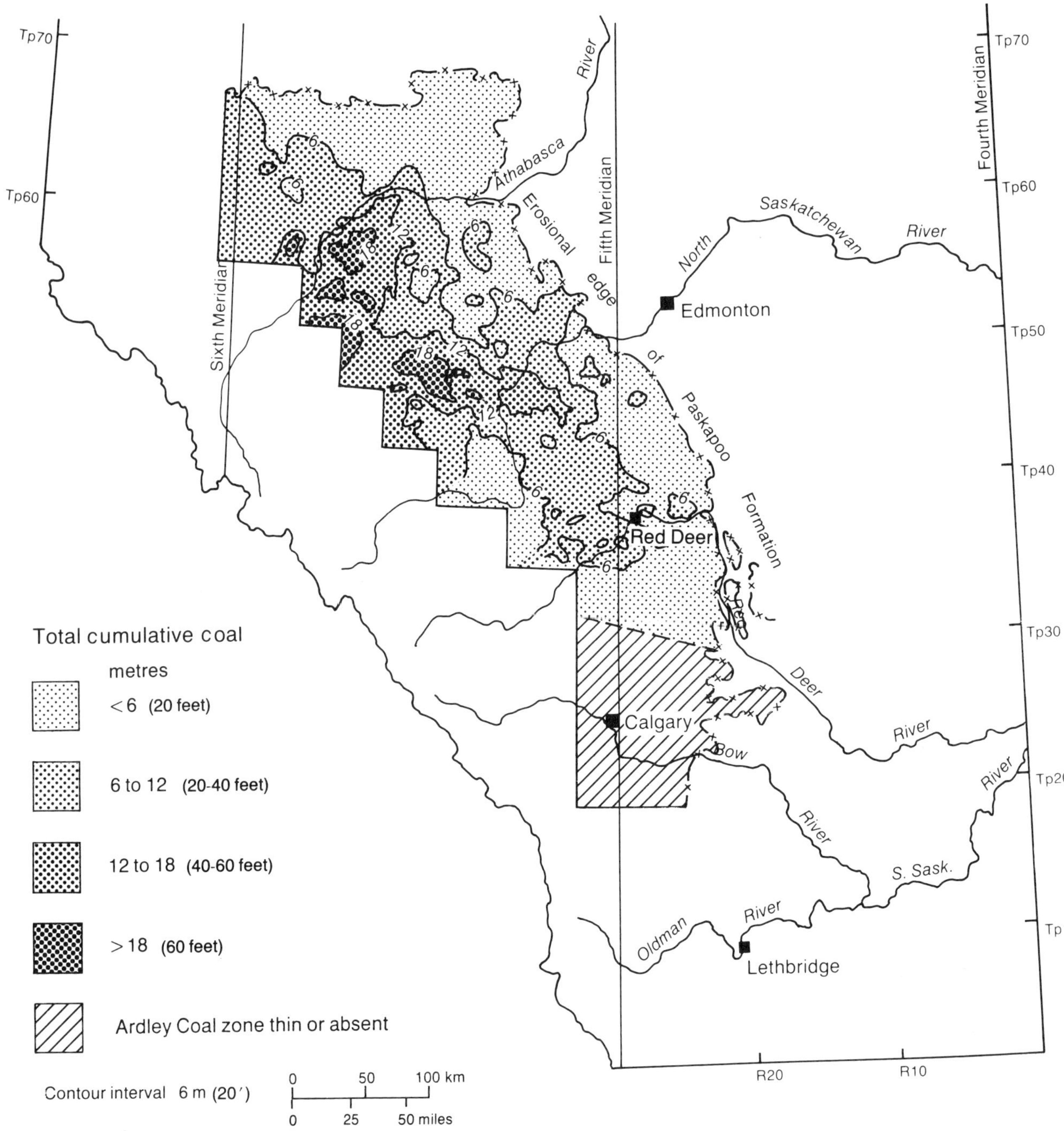

Figure 6. Cumulative (net) coal isopach map of Ardley coal zone, southern Alberta, showing two trends containing 40 to 80 ft (12 to 24 m) of cumulative coal.

Based on Nikols and Rottenfusser's (this volume) estimated gas content of 352 CF/T (11 cc/g) for high-volatile bituminous and subbituminous coals in Alberta, the 358 billion short tons (325 billion t) of Ardley coal deeper than 1,300 ft (400 m) (Figure 8) potentially contain more than 124 TCF (3.5×10^{12} m^3) of gas. About 0.77×10^{12} m^3, or approximately 25 TCF, could be available in seams thicker than 8 ft (2.5 m).

CURRENT RESEARCH

The Alberta Geological Survey is conducting an extensive basinwide study of all coal resources in Alberta (Nikols and Rottenfusser, this volume). Additional data on Ardley coal resources and vitrinite reflectance data for rank determination are being collected as part of the study. In the case of the Ardley coal zone, relatively little change is expected in our knowledge of coal resource distribution and seam characteristics. However, because virtually no hard data on rank are now available for the coal zone, new information critical for a more realistic assessment of coalbed methane potential in Alberta is expected. Additional information being gathered through industry exploration will be released by 1993 as the confidentiality period expires.

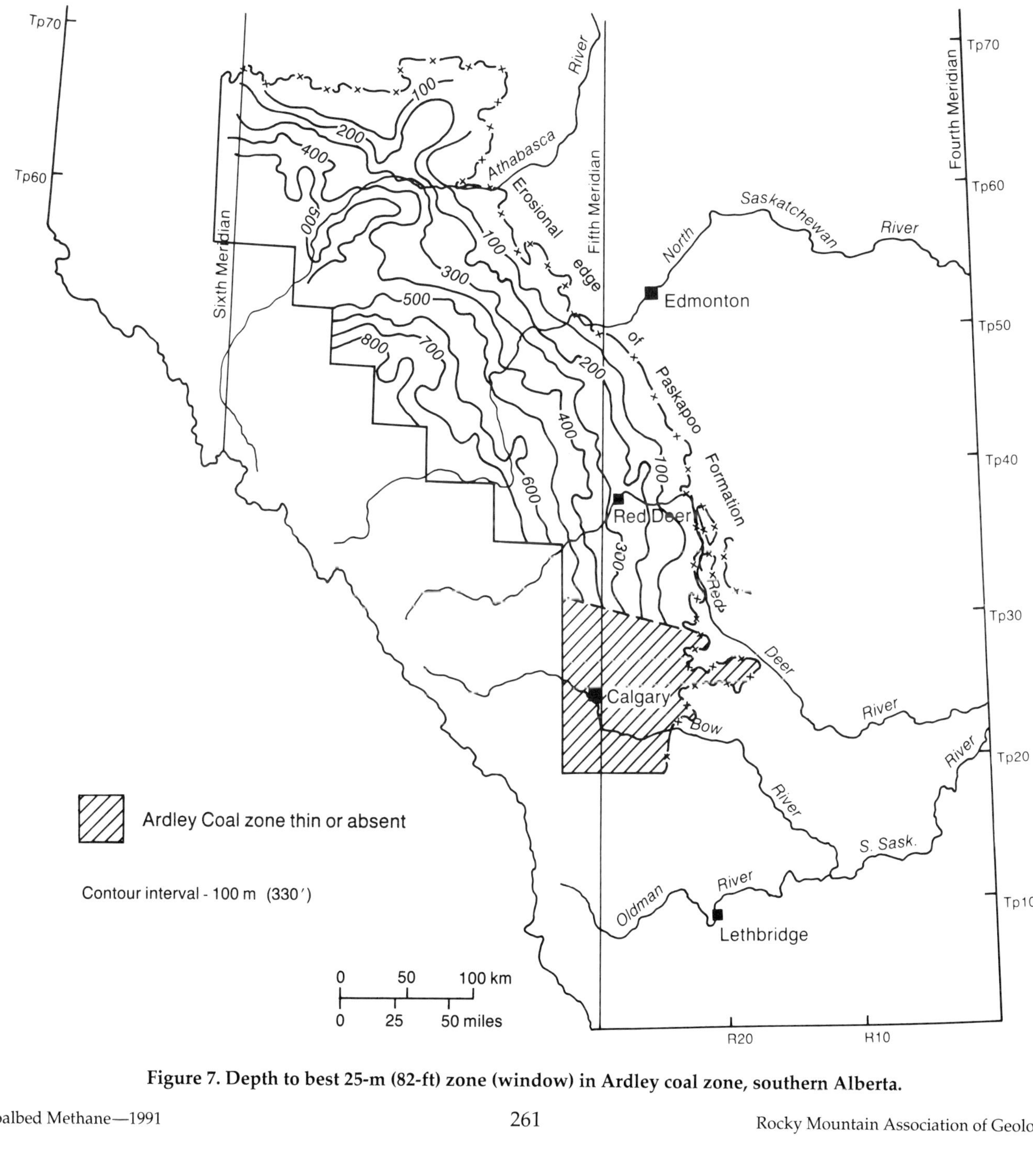

Figure 7. Depth to best 25-m (82-ft) zone (window) in Ardley coal zone, southern Alberta.

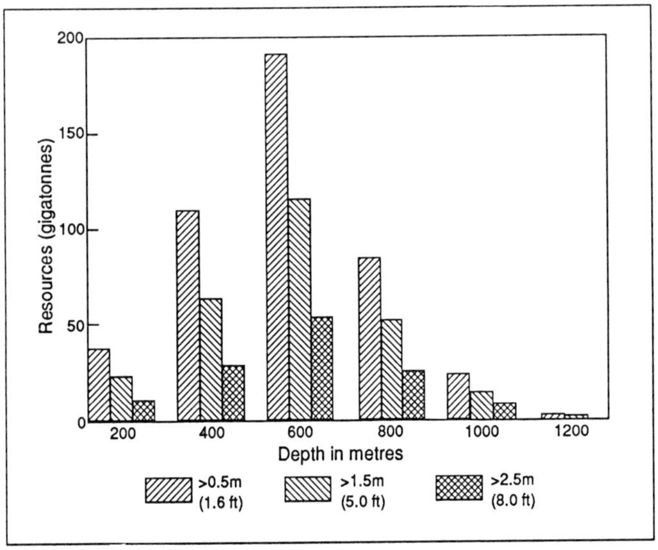

Figure 8. Estimated resources for Ardley coals in 650-ft (200-m) depth intervals to a maximum depth of 3,900 ft (1,200 m) and in seam-thickness categories of >1.6 ft (>0.5 m), >5-ft (>1.5 m), and >8-ft (>2.5 m).

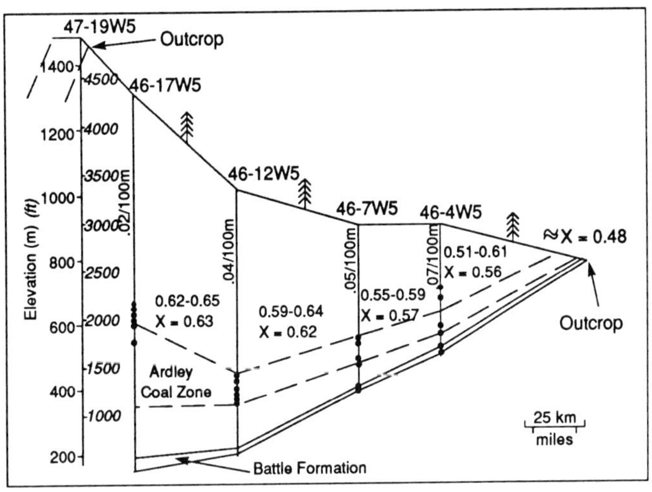

Figure 9. East-west cross-section through four wells near the center of the basin, showing average (x) and range of vitrinite reflectance values of 22 Ardley coal samples of high-volatile C bituminous rank. From Nurkowski (1984).

CONCLUSIONS

With approximately 77 billion short tons (70 billion t) of coal in seams thicker than 8 ft (2.5 m), at depths of 1,300 to 2,625 ft (400 to 800 m), and in areas where as many as 18 seams and seams 13 ft (4 m) thick can be found, the Ardley coal zone contains abundant resources for a coalbed methane exploration prospect. In the western part of the basin, within a large primary prospect area, Ardley coal at depths of 1,640 to 2,625 ft (500 to 800 m) in a major pay zone ("best 25 m") should have sufficient formation pressure to contain coalbed gas.

Most Ardley coals at depth should fall within high-volatile C bituminous rank and, therefore, may be producers of methane. The 358 billion short tons (325 billion t) of Ardley coal deeper than 1,300 ft (400 m) potentially contains more than 124 TCF (3.5×10^{12} m^3) of methane. Of that amount, about 25 TCF (0.77×10^{12} m^3) could be available in seams thicker than 2.5 m (8 ft).

Without actual gas-content measurements and a sufficient number of rank-distribution data, any estimate of the total methane content of Ardley coals must be considered speculative.

ACKNOWLEDGMENTS

Rudy Strobl, a major contributor to the original Ardley coal resource study (Richardson and others, 1988) supervised the production of several maps used in this paper. Discussions with Peter McCabe (U.S. Geological Survey) provided a basis for discussion of the coal depositional model. Brian Rottenfusser is thanked for reviewing the manuscript.

REFERENCES

Alberta Energy Resources Conservation Board, 1989, Reserves of coal, Province of Alberta: Alberta Energy Resources Conservation Board, Rept. ST90–31.

Allan, J.A., 1924, Carbon, Big Valley, and Ardley coal areas survey: Scientific and Industrial Research Council of Alberta, Fifth Annual Rept., p. 37–45.

Jerzykiewicz, T., and Sweet, A.R., 1986, Caliche and associated impoverished palynological assemblages—An innovative line of paleoclimatic research into the uppermost Cretaceous and Paleocene of southern Alberta, *in* Current research, part B: Geological Survey of Canada Paper 86–1B, p. 653–663.

Nurkowski, J.R., 1984, Coal quality, coal rank variation and its relation to reconstructed overburden, Upper Cretaceous and Tertiary plains coals, Alberta: AAPG Bulletin, v. 68, no. 3, p. 285–295.

Nurkowski, J.R., 1985, Coal quality and rank variations within Upper Cretaceous and Tertiary sediments, Alberta plains region: Alberta Research Council Earth Science Rept. 85–1, 39 p.

Osadetz, K.G., Pearson, D.E., and Stasiuk, L.D., 1990, Paleogeothermal gradients and changes in the geothermal gradient field of the Alberta Plains, *in* Current research, part D: Geological Survey of Canada Paper 90–1D, p. 165–178.

Parkash, S., 1985, Petrographic studies of coals from Alberta plains: Alberta Research Council Bulletin 50, 47 p.

Richardson, R.J.H., Strobl, R.S., Macdonald, D.E., Nurkowski, J.R., McCabe, P.J., and Bosman, A., 1988, An evaluation of the coal resources of the Ardley coal zone, to a depth of 400 m, in the Alberta plains area: Alberta Research Council Open-File Rept. 1988–2, 96 p.

Strobl, R., McCabe, P.J., Macdonald, D.E., and Richardson, R.J.H., 1986, Estimate of inferred resources of coal in the Plains region of Alberta to a depth of 400 metres: Alberta Research Council Open-File Rept. 1987–5, 31 p.

Williams, G.D., and Murphy, M.C., 1981, Deep coal resources of the Interior Plains, estimated from petroleum borehole data: Geological Survey of Canada Paper 81–13, 15 p.

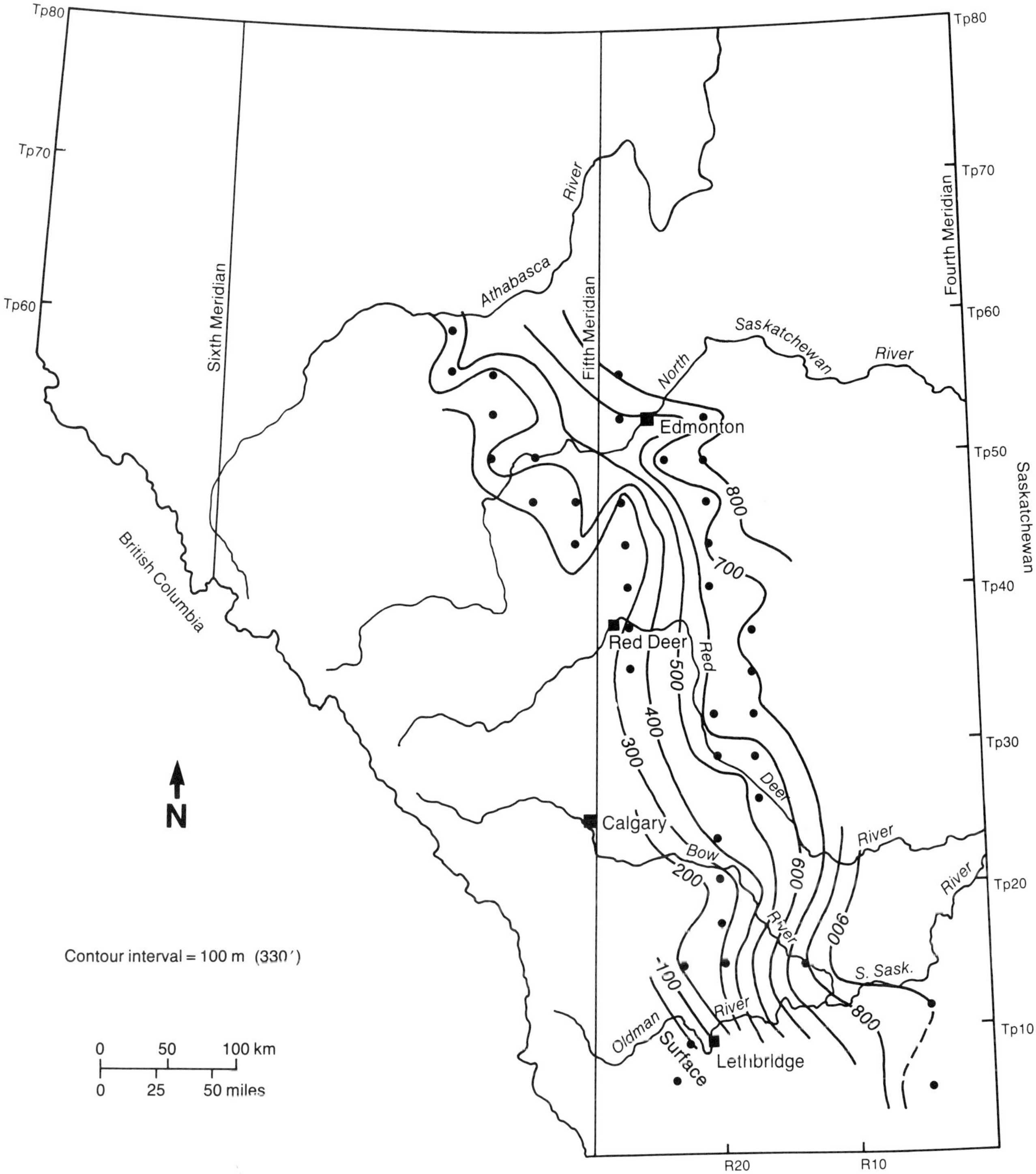

Figure 10. Calculated depth to high-volatile C bituminous coal in the plains area of Alberta. From Nurkowski (1985).

Glenwood Canyon

Coal and Coalbed Methane Resources of Montana

Elizabeth Bartow Campen[1]
James R. Gruber, Jr.[2]

1. Independent Consulting Geologist, Billings, Montana
2. U.S. Bureau of Land Management, Billings, Montana

ABSTRACT

With coal underlying approximately 35% of the state, Montana's coal resources are the second largest in the United States. These resources are estimated at 478 billion short tons (433.6 billion t). Associated in-place coalbed methane resources are estimated to be 14 TCF (401×10^9 m^3).

The coals of Montana range from Jurassic to early Tertiary in age and from lignite to low-volatile bituminous in rank. Thickness, rank, maceral composition, and proximate and ultimate analyses all vary both vertically and laterally. The state contains eight major coal resource areas. A large percentage of Montana's coal consists of Fort Union Formation lignite in eastern Montana, which generally is considered too low in rank to contain significant methane resources. Most of the state's other coal deposits are higher in rank and exhibit many recorded methane shows.

During Cretaceous and Tertiary times, regressive-transgressive cycles resulted in deposition of numerous coal-bearing sequences. Major marine regressions allowed the formation of large peat swamps, followed by transgressions that covered the swamps with impervious marine clay, thus preventing the escape of methane already being generated.

About 75 percent of Montana's coal resources occur at depths less than 1,000 ft (305 m), making it attractive for methane production. In addition, the fact that associated water appears to be fresh will help minimize environmental problems concerning water production and disposal. Pipelines run close to most major coal deposits. Exploration for coalbed methane in Montana still is in its infancy but at this time shows commercial promise.

INTRODUCTION

Coal is one of Montana's most abundant resources. More than 478 billion short tons (433.7 billion t) (Wood and Bour, 1988) underlie approximately 51,000 mi^2 (132,000 km^2) or 35% of the state. This versatile energy resource has been used from the days of the earliest homesteaders for heating and locomotive power generation, through the metal-mining period at the turn of the century when coal was developed for its coking qualities, to the present-day national energy policy efforts, which include coal-fired electrical generation and synthetic fuel possibilities. Oddly enough, throughout this period of history, methane in coal often was noted but regarded as either a hazard in underground mining or a nuisance encountered when drilling exploratory oil and gas wells or water wells.

Only recently has the petroleum industry viewed coal as both a source rock and reservoir rock containing reserves of producible natural gas. Numerous gas shows in coal-bearing Cretaceous sandstone sequences that were judged as "watered out" must be reexamined with coalbed methane technology in mind. Likewise, the anomalously thick and numerous Tertiary-age coals in the Powder River basin are equally as promising as those currently producing south the border in Wyoming. Montana should be viewed as a major player, and the state's coal deserves a close look because of its abundance, diversity, and compatibility with a natural gas industry that is firmly established within the state.

DEPOSITIONAL HISTORY OF MONTANA COAL

Montana's oldest major coalbeds began as peats deposited in the central part of the state in the Great Falls/Lewistown and Lombard coal fields during the last major Jurassic marine regression. Near sea-level Morrison-type deposition was extensive and affected an area of approximately 1,400 miles (2,253 km) north-south and 800 miles (1,287 km) east-west (Moberly, 1960). Petersen (1972) considered the dark shale and coaly deposits of the Lewistown-Great Falls coal fields to have formed in fluvial flood-plain and lacustrine environments. Epeiric seas to the north and southeast contributed to the tropical or subtropical climate in which dinosaurs thrived in swamps and lush forests of cycads, small conifers, mosses, and other warm-weather flora.

Although continental, nearshore, and lacustrine deposition characteristic of Morrison time continued into the Cretaceous, no Lower Cretaceous coalbeds have been recorded in Montana. Not until deposition of the Claggett cyclothem (Campanian) was the environment again conducive to peat accumulation (Kauffman, 1980). Laramide uplift to the west contributed clastic sediments that, together with eastward progradation of the western shoreline of the Cretaceous epicontinental seaway, gave rise to broad, flat, wet coastal lowlands that became ideal sites for peat accumulation during deposition of the Eagle Sandstone facies. South-central Montana must have been such a place, where an enormous swamp of at least 65,000 mi^2 (168,394 km^2) covered the area from Harlowton south to the Wyoming border. Coal deposition was widespread and in places thick; 70 ft (21 m) of total coal has been recorded in the Trail Creek/Livingston area.

Simultaneously coal also was forming in marsh, flood-basin, and bay environments in the northwestern Great Plains area of Montana. A well in sec. 26, T19N R16E recorded three coal zones between the top of the lower member and the base of the nonmarine upper member of the Eagle Sandstone (Rice and Shurr, 1983).

Deposition of the nonmarine facies of the Eagle Sandstone was followed by a widespread transgression (Claggett Shale) and later regression (Judith River Formation). The Great Plains area once again became a humid, maritime setting with swamps which later resulted in discontinuous coal and carbonaceous shale facies. Coal deposits in the Havre area bear testament to temperate climate zones in what is now one of the coldest winter areas in Montana. Apparently the depositional environment of southern Montana differed as no coal seams are recorded in the Judith River Formation in that area.

By Paleocene time the seaway had retreated to the east to become the Cannonball sea. Fort Union coal deposition in eastern Montana indicates that stratigraphic sequences were controlled by local uplifts, the result of Laramide orogeny and basin lineaments (Flores, 1983). Although controversy exists regarding depositional environments of the thick coals in the southern Powder River basin, it is generally agreed that the subbituminous and lignite fields of Montana's Fort Union region (Powder River, Williston, Bull Mountain, and Bighorn basins) are the result of deposition in fluvial and deltaic environments. Flores (1983) attributed the massive coal deposits in the lower Tongue River Member of the Fort Union Formation to river channel-levee facies, flood-plain and lacustrine basin facies, and backswamp facies that covered the entire region. During upper Tongue River deposition these facies were succeeded

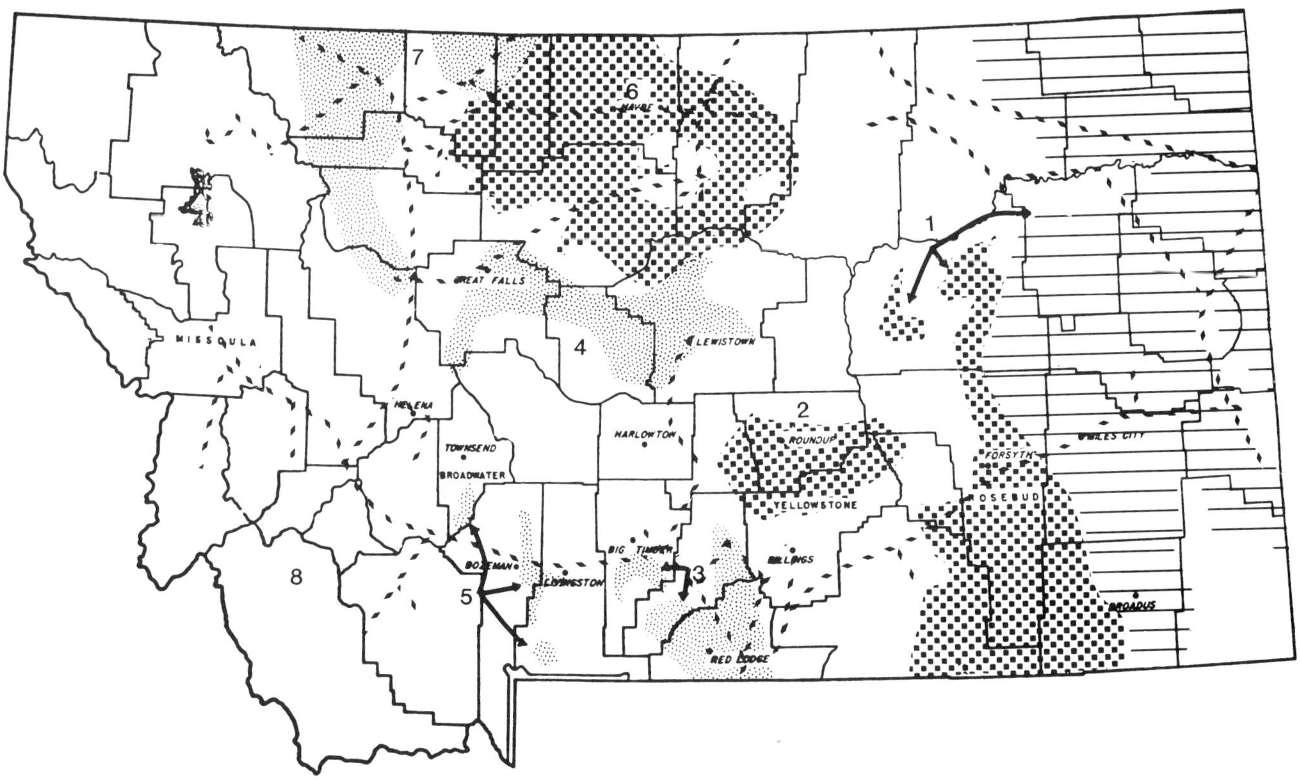

Figure 1. Coal regions of Montana: 1, Eastern (subbituminous and lignite areas); 2, Bull Mountain Basin; 3, Red Lodge; 4, Great Falls-Lewistown; 5, Southwest; 6, Bearpaw Mountains; 7, Blackfoot-Valier; 8, Western. Coal rank patterns—Bituminous (fine dots), subbituminous (coarse dots), lignite (horizontal lines). Gas pipeline (dashed line). After Cole and others (1982).

by those of an anastomosing fluvial system. Individual coalbeds in the Montana part of the Powder River basin reach thicknesses of 70 ft (21.4 m), with composite thicknesses greater than 125 ft (38 m).

COAL-BEARING DEPOSITS

Most of Montana's coal deposits (Figure 1) are found in eight regions (Bateman, 1966) and are the more likely targets for gas exploration. However, geologists should not limit their interest only to these regions. The terrestrial (fresh water) facies of the Eagle Sandstone, in particular, contains significant coal resources outside these delineated areas that should also be prime targets for coal-gas exploration based on lithology, structure, and depth.

Eastern Region (Area 1)

Relatively flat-lying strata of the coal-bearing Paleocene Fort Union Formation underlie most of eastern Montana (Figure 1). To the north, Fort Union beds dip gently eastward into the Williston basin, and in the south dip southward into the Powder River basin. These gently dipping beds are interrupted by several structural features—Cedar Creek anticline, the north end of the Black Hills uplift, Porcupine dome, Bowdoin dome, and Poplar nose. The major coal-bearing member of the Fort Union Formation has been eroded from most of these structures. Faulting is relatively rare, although en echelon normal faults have been mapped along the eastward extensions of the Lake basin and the Nye-Bowler basement wrench zones in the northern Powder River basin.

Coal deposits underlie the entire eastern region, but many seams are discontinuous and vary greatly in thickness. Coal rank increases progressively from lignite at the Alberta and North Dakota borders southwestward to subbituminous B in Powder River, Rosebud, and Big Horn Counties. As rank increases and moisture content decreases, heating value of the coal rises from less than 7,000 Btu (7,392 kJ) to more than 9,000 Btu (9,504 kJ).

Many of the numerous coal fields in the region exhibit special characteristics in terms of lithology, total coal thickness, structure, and hydrology. For simplicity we have divided the region into a lignite area and a subbituminous area, as shown on Figure 1.

Lignite Area

The Tongue River Member of the Fort Union Formation underlies most of the lignite area, but in some localities it has been eroded, exposing the Lebo and Tullock Members. In the northern one-third of the lignite area, little is known of either the stratigraphy or the coalbeds because of the cover of glacial drift. In the south the Tongue River Member contains many thick coal seams that are continuous over large areas and are relatively free from partings, especially in the eastern half. In the western half of the lignite area, the beds generally are discontinuous and thin. Many shallow coalbeds in the southern lignite area have burned to a bright brick-red clinker that is used locally for road surfacing, landscaping, and roofing.

Subbituminous Area

Garfield County coals in the northern subbituminous area occur, for the most part, in the lower members of the Fort Union Formation, where the Tongue River Member has been eroded. The Tullock Member contains 11 coalbeds, and two are found in the Lebo Member.

South of the Yellowstone River, most of the land is underlain by the Tongue River Member, which is 1,500 to 1,700 ft (458 to 519 m) thick. In places the Tongue River contains more than 20 coalbeds, some of which are thicker than 70 ft (21 m). More than half of the coal in Montana can be found in this area of Rosebud, Big Horn, and Powder River Counties.

Bull Mountain Basin (Area 2)

Bull Mountain coal field occupies the central Bull Mountain basin in southern Musselshell County and northeastern Yellowstone County. The basin is a broad, shallow, east-west-trending syncline whose flanks dip at angles of 1° to 5° on the south, east, and west, and as much as 30° on the north.

Stratigraphically, the Tongue River Member of the Fort Union Formation here consists of alternating massive sandstone, claystone, shale, and subbituminous coal in a 1,700-ft (519-m) section. Coal occurs in 26 beds ranging from several inches to 14 ft (4.3 m) thick. Most crop out on the sides and around the bases of mesas.

Ground water is scarce and of poor quality in the southern part of the basin due to a preponderance of alkaline salts in the claystone and shale. The northern basin, however, contains abundant fresh-water-bearing sandstones and coals. Notable springs and artesian wells occur on the north side of the basin.

Red Lodge Region (Area 3)

The Red Lodge coal region of Carbon and Stillwater Counties includes the Red Lodge, Bridger, Silvertip, and Stillwater coal fields in the Montana part of the Bighorn basin as well as lands north of the Nye-Bowler lineament in the Reedpoint syncline. The Montana Bighorn basin is an asymmetrical syncline trending northwest. Coal-bearing rocks of the Upper Cretaceous Eagle Sandstone crop out along the eastern edge of the basin and continue northwestward along strike. The beds dip gently westward and lay beneath Fort Union Formation coal in the western three-fourths of the basin. Coal-bearing rocks are limited to the west partly by the Beartooth fault, along which Precambrian crystalline rocks of the Beartooth Plateau are thrust over younger sedimentary rocks.

Red Lodge Field

The coal-bearing sequence of the Fort Union Formation is 825 ft (251 m) thick and lies under as much as 2,000 ft (610 m) of cover. Eight principal coalbeds range in thickness from 3.5 to 10 ft (1 to 3 m) and dip southwestward 18° at Red Lodge, 9° at Bear Creek, and 3° slightly farther south. Most of the coal is exposed along drainages, and the total coal-bearing sequence has been eroded away just east of the town of Bearcreek.

The coal is of high-volatile C bituminous rank, with heating values between 10,000 and 12,000 Btu (10,560 and 12,672 kJ). Geophysical logs indicate that the regional water table lies directly above the coals in this area. The presence of water and methane was observed and documented at a time of extensive underground mining early in this century.

Bridger Coal Field

Three discontinuous coals occur in shale beds of the Eagle Sandstone. Coal thickness ranges from 2.5 to 6 ft (0.8 to 1.8 m), thinning to the south. The coal is of high-volatile C bituminous rank, with an "as-received" heating value of 10,040 Btu (10,602 kJ).

From literature on the old mine workings, it was observed, for example, that "gas does not occur and open-flame lamps are in use," and that "2,000 gal of water is pumped daily" at the Bridger mine.

Silvertip Coal Field

On the Montana-Wyoming state line, coal in the Eagle Sandstone is exposed on the surface in the northwest-trending Silvertip anticline. This structure, about 8 miles long, has steeply dipping flanks cut by numerous northeast-trending normal faults. The coalbeds are the same three present in Bridger coal field and probably are of similar rank and quality.

Stillwater Coal Field

Bituminous coal of the Eagle Sandstone exists in the upper Stillwater basin near the town of Nye. A 4- to 5-ft-thick (1.2- to 1.5-m) bed exhibiting quality character similar to the Bridger field is exposed in outcrop along the Stillwater River. The areal extent of this deposit is uncertain because of the complex structure in the vicinity and because of glacial cover.

No gas or water was encountered in mines along the outcrop when the coal in this area was worked for local use during the first half of the century.

Great Falls-Lewistown Region (Area 4)

The Great Falls-Lewistown coal region is located in Cascade, Judith Basin, and Fergus Counties along the northern slopes of the Little Belt and Big Snowy Mountains. Most of the coalbeds are depositionally discontinuous, with thicker accumulations occurring in six localized basins between the communities of Great Falls and Lewistown.

The Jurassic Morrison Formation hosts bituminous coal that dips northward as much as 25° along the mountain front but rapidly diminishing basinward to 3° to 5°. The coal throughout most of this region is overlain by a uniform basal sandstone of the Cretaceous Kootenai Formation as thick as 60 ft (18.3 m). Maximum overburden on the coal in this area is approximately 600 ft (183 m).

Southwest Region (Area 5)

Electric Coal Field

Electric coal field covers a small area of Park County in southwestern Montana. Coal-bearing rocks of the Eagle Sandstone occur in highly folded and faulted strata preserved in a fault block downdropped several thousand feet. These beds, exposed in the limbs of two small anticlines, dip steeply and are cut by many small faults. Three coalbeds 3 to 5 ft (1 to 1.5 m) thick contain several clay and sandstone partings. The coal varies in rank from high-volatile A to low-volatile bituminous, depending on the amount of deformation in the enclosing rocks. Heating value is 11,400 Btu (12,038 kJ), and most of the coal formerly mined was used for coke manufacture.

Livingston-Trail Creek Coal Field

Coal-bearing rocks of the Eagle Sandstone lie in a narrow Y-shaped belt in Gallatin and Park Counties on the steeply inclined flanks of a northwest-trending anticline and extend northward along the eastern flank of the Bridger Range. Dips are as high as 65°. The formation crops out along strike of the area's prominent uplifts and extends to depths of thousands of feet into the Crazy Mountain basin.

The area contains numerous thin coals, most of which are 2 to 5 ft (0.6 to 1.5 m) thick in a 600-ft-thick (183-m) section. Coal rank varies with the amount of deformation, ranging from high-volatile A bituminous to high-volatile B and C.

Lombard Coal Field

Approximately 5 mi^2 (13 km^2) of northern Gallatin and Broadwater Counties are underlain by highly sheared Morrison Formation strata containing coal as thick as 6 ft (1.8 m) in lenticular pockets. Coal rank ranges from high-volatile A bituminous to medium-volatile B bituminous and locally is graphitic.

Bearpaw Mountains Region (Area 6)

An area of flat-lying rocks covering more than 10,000 mi^2 (25,907 km^2) surrounds the Bearpaw Mountains and includes the Sweetgrass Hills in northern and northwest Montana. Coal is present throughout most of this area, but beds tend to be thin and lenticular. Little is known of the areal extent or true thickness of beds because of the thick cover of glacial drift.

Coal has been mined from the Eagle Sandstone at a number of localities, primarily along the banks of the Missouri River, where three or four lenticular seams occur in the lower part of the upper member of the Eagle Sandstone. The coal is of subbituminous rank, with a heating value of 9,100 Btu (9,610 kJ).

Coal also is found in the Cretaceous Judith River Formation in localities along the Missouri River and northward. In Milk River coal field near the communities of Havre and Chinook, one bed ranges from 2.5 to 7 ft (0.8 to 2.1 m) thick. An analysis of coal "as received" yielded a heating value of 8,590 Btu (9,071 kJ).

Blackfoot-Valier Region (Area 7)

Coal-bearing rocks occur in a belt extending south from Canada through Glacier, Pondera, and Teton Counties in northwestern Montana (Figure 1). In the eastern two-thirds of this belt, strata dip gently west off the flank of the Sweetgrass arch. The western one-third lies within the Disturbed Belt, in which strata are intricately folded and thrust faulted. Coal in thin sporadic beds is found in the Paleocene St. Mary River Formation, which is equivalent to the Fort Union Formation, and in the lower part of the Cretaceous Two Medicine Formation in strata that are equivalent to the Eagle Sandstone. A thickness of 2 ft (0.6 m) of coal is found in the Two Medicine Formation in Valier coal field, and in the Blackfoot field 3 ft (1 m) of coal occurs at the base of the St. Mary River Formation.

Coal in the Two Medicine Formation is found in small areas preserved by faulting in the Disturbed Belt as far south as the Missouri River.

Western Region (Area 8)

Coal occurs in Tertiary sedimentary rocks that fill many intermontane valleys in western Montana. The rocks are of lacustrine, fluvial, and colluvial origin and Eocene and Oligocene or Miocene in age. The rocks have been deformed slightly into broad, shallow anticlines, synclines, and monoclines, and are cut by minor faults. Coalbeds tend to be thin and lenticular, and mostly lignitic.

Coal has been mined in Flathead coal field on the North Fork Flathead River in northern Flathead County in far northwestern Montana. A composite bed of subbituminous B coal made up of several benches, the thickest of which is about 3 ft (1 m), occurs in the vicinity of Coal Creek. Heating value of this coal is 8,120 Btu (8,575 kJ). In Drummond and Missoula coal fields many thin lignite beds lie within a sequence of bone and shale as much as 25 ft (7.6 m) thick. As-received samples of lignite from these fields indicate heating values of 8,700 and 6,700 Btu (9,187 to 7,075 kJ), respectively.

COALBED METHANE RESOURCES

Methane is a major component of the volatilization of humic matter. Because methane generation is on-going throughout the coalification process, much more gas is generated in coal than is present at any given time. Varying quantities of generated gas are adsorbed in the coal, expelled into other strata (and contained in structural or stratigraphic traps), or lost to the atmosphere.

Adsorbed gas is the focus of current coalbed methane technology. Generally, the amount of methane adsorbed in coal, per unit volume, is a function of coal rank and burial depth (pressure). In Montana, gas-content data (from desorption testing) are very limited; consequently, estimates of methane resources (Table 1) are derived from Kim's (1977) equation for bituminous and subbituminous coal. Lignite, which contains gas in "reported" quantities throughout Montana, does not appear to be a commercially significant methane resource and is not defined in this paper.

Coal resource estimates are derived from Wood and Bour (1988). Subtotals of these estimates are modified to

Table 1. Estimated coal and coalbed resources of Montana. Coal resources modified from Wood and Bour (1988).

Coal Region	Coal Rank*	Coal Resource (billion short tons)	Coal Resource (billion t)	Coal Depth (ft)	Coal Depth (m)	Methane Content (CF/T)	Methane Content (g/cc)	Methane Resource (TCF)	Methane Resource (10^9 m^3)
1 Eastern									
subbituminous	sbC-sbB	260.35	236.19	500	152	35	1.1	9.11	257.81
lignite	lig	173.56	157.45	250	76	–	–	–	–
2 Bull Mountain Basin	sbB-sbA	6.34	5.75	500	152	50	1.6	0.32	9.06
3 Red Lodge	hvCb	4.48	4.06	1,000	305	300	9.4	1.34	37.92
4 Great Falls-Lewistown	hvCb-hvBb	0.82	0.74	1,000	305	350	10.9	0.29	8.21
5 Southwest	hvAb-lvBb	0.33	0.30	1,000	305	450	14.1	0.15	4.24
6 Bearpaw Mountains	sbB	10.50	9.53	500	152	40	1.2	0.42	11.89
7 Blackfoot-Valier	hvCb-hvBb	7.28	6.60	1,000	305	350	10.9	2.55	72.16
8 Western	lig	0.55	0.50	250	76	–	–	–	–
– Undefined	–	13.89	12.60	–	–	–	–	–	–
Total		478.10	433.72					14.18	401.29

* Coal rank abbreviations: lig—lignite; sb(A, B, C)—subbituminous (A, B, C); hv(A, B, C)b—high-volatile (A, B, C) bituminous; lv(B)b—low-volatile (B) bituminous.

ERA AND PERIOD		SERIES	FORMATION AND MEMBER		Missoula (1)	Lombard (2)	Belt (3)	Electric (4)	Livingston (5)	Stillwater (6)	Milk River (7)	Red Lodge (8)	Roundup (9)	Colstrip (10)	McCone Co (11)	Broadus (12)	Sheridan Co (13)	Girard (14)
CENOZOIC	Tertiary	Miocene or Oligocene	Kishenehn(?) Formation of Daly (1913)		X													
		Eocene	Wasatch Formation															
		Paleo-cene	Fort Union Formation	Tongue River Member								X	X	X	X	X		X
				Lebo Shale Member														
				Tullock Member														
MESOZOIC	Cretaceous	Upper	Hell Creek Formation (Lance Fm)														X	
			Lennep Sandstone (Fox Hills Sandstone)															
			Bearpaw Shale															
			Judith River Formation								X							
			Claggett Shale															
			Eagle Sandstone					X	X	X								
			Telegraph Creek Formation															
			Colorado Shale															
		Lower	Kootenai Formation															
	Jurassic	Upper	Morrison Formation			X	X											

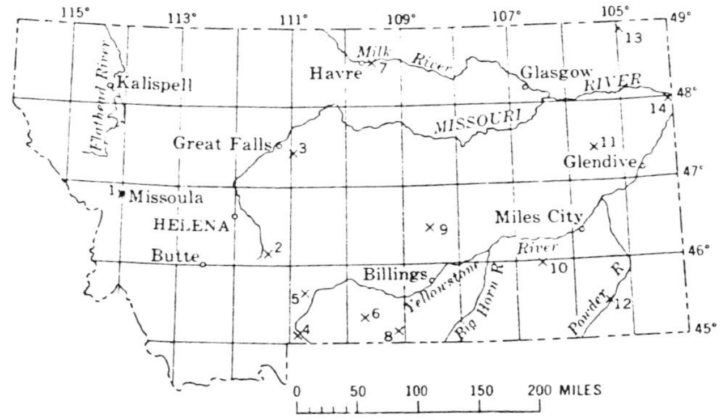

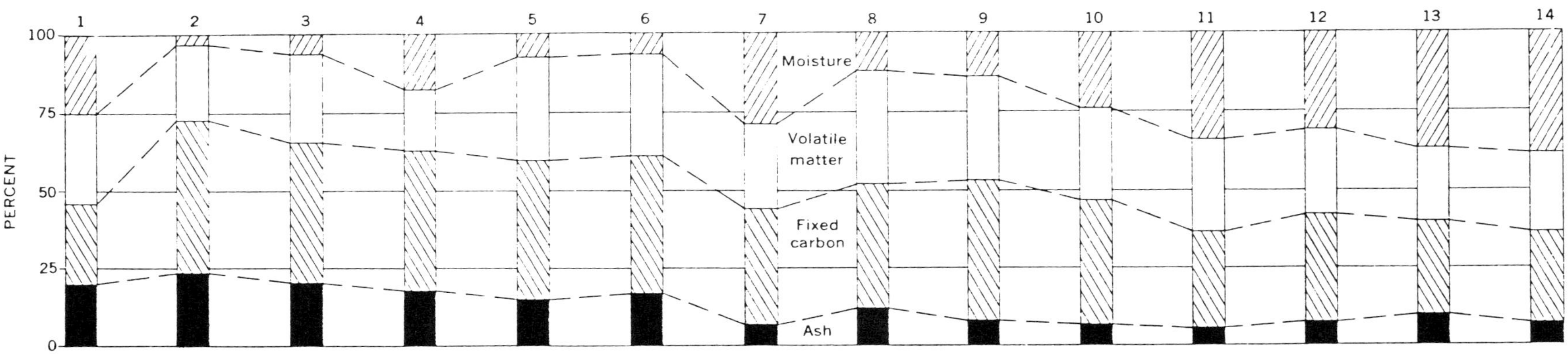

Figure 2. Generalized Upper Cretaceous and Tertiary stratigraphic section of Montana, showing position of significant coal-bearing formations by coal field or area. Average proximate analyses from selected Montana coal fields. After Roberts (1966). Numbers on index map correspond to numbered field names in stratigraphic chart and numbers on coal-analysis chart.

correspond with the coal regions (Table 1) as defined in Bateman (1966). Coal resources are based on the occurrence of one or more coalbeds greater than 14 in. (36 cm) thick at maximum depths of 6,000 ft (1,830 m). The general quality of coal in various fields throughout Montana is compared in Figure 2. Data from the literature, drill holes, and mining records indicate that tectonics played a significant role in shaping the present-day quality of Montana's coal, with the exception of the Eastern region (Table 1). These data also indicate the approximate average depths of bituminous coal, subbituminous coal, and lignite coal to be 1,000, 500, and 250 ft (305, 152, and 76 m), respectively.

Using Kim's equation, methane resources have been estimated for each of the coal regions containing definable bituminous or subbituminous coal. Within these coal regions, methane resources have been calculated according to coal quantity, coal rank, and average depth.

NATURAL GAS IN MONTANA

Production of natural gas in the United States totaled approximately 747 TCF (21.14×10^{12} m^3) through the end of 1989. At that time proven reserves were estimated by the Energy Information Administration of the U.S. Department of Energy at approximately 167 TCF (4.73×10^{12} m^3), excluding 24.6 TCF (696×10^9 m^3) of North Slope reserves in Alaska. An additional 866 TCF (24.51×10^{12} m^3) of probable, possible, and speculative resources have been calculated by the Potential Gas Committee through 1990. This figure includes 145 TCF (4.1×10^{12} m^3) of coalbed methane "most likely recoverable." Thus, total estimated gas resources amount to 1,033 TCF (29.23×10^{12} m^3) with coalbed methane accounting for 14% of the total (Potential Gas Committee, 1991). Various estimates place the total in-place coalbed methane resource of the United States at about 400 TCF (11.32×10^{12} m^3).

Total cumulative gas production in Montana through 1989 was 2.5 TCF (70.8×10^9 m^3). Tonnsen (1989) estimated remaining gas reserves, excluding coalbed methane, at 2.9 TCF (82×10^9 m^3). Major gas-producing companies in Montana in 1989 were Norfolk Energy (22%) and Montana Power (15%). The remaining gas producers are independents who produce less than 7% each. To date no coalbed methane has been produced in Montana, but several independents currently are involved in extensive exploration projects.

Seventy-eight percent of the gas produced in Montana in 1989 came from Cretaceous formations. Other gas-producing horizons include, in descending order of production—Silurian/Ordovician, Mississippian, Devonian, Jurassic, and Pennsylvanian. Gas from Tertiary horizons has been tested, but none has yet been produced.

Gas Research Institute predicts that the share of gas supplied to the north-central states from the Rocky Mountain region will increase from 20% to 40% between 1987 and 2000 (Tonnsen, 1989; Wood, 1987). Montana currently has over 3,000 miles (4,827 km) of natural gas pipelines, part of the "central system" that extends from the Wyoming Overthrust Belt to the Great Lakes and eastern states (Tonnsen,1989).

CONCLUSIONS

The majority of Montana's coalbed methane resources of 9.11 TCF (257.8×10^9 m^3) lies in the Eastern coal region, predominantly in the Powder River basin. Despite relatively low gas contents, 35 CF/T (1 cc/g), large methane resources exist because of the enormous amount of coal present.

Other regions containing coal of lesser quantity carry significantly larger volumes of methane, up to 450 CF/T (14 cc/g), and are of particular interest because of the shallow depths, associated geologic structure, reservoir quality of bounding strata, and proximity to natural gas pipelines. Of particular interest is the Cretaceous Eagle Sandstone, a proven conventional gas reservoir. Coalbed methane contained within the Eagle Sandstone is undoubtedly a principal source of gas in the fresh-water facies of the formation.

The massive volume of coalbed methane in the Powder River basin and the geologic setting associated with Montana's higher quality bituminous coal, together with the existing natural gas infrastructure, present a new frontier promising great rewards for the petroleum industry.

REFERENCES

Bateman, A.F., Jr.,1966, Montana's coal resources, *in* Proceedings of the First Montana Coal Resources Symposium: Montana Bureau of Mines and Geology Special Publication 36, p. 6–11.

Cole, G.A., Berg, R.B., Cromwell, Vern, and Sonderegger, J.L., compilers, 1982, Energy resources of Montana: Montana Bureau of Mines and Geology Geologic Map 28, scale 1:500,000.

Flores, R.M., 1983, Cretaceous and Tertiary fluvial deposits in Rocky Mountain coal basins, AAPG field seminar fluvial systems—Their economic and field applications: AAPG Education Department.

Kauffman, E.G., 1980, Major factors influencing the distribution of Cretaceous coal in the Western Interior United States, *in* Carter, L.M., ed., Proceedings of the Fourth Symposium of the Geology of Rocky Mountain Coal—1980: Colorado Geological Survey Resource Series 10, p. 1–3.

Kent, H.C., 1990, Future supply of natural gas in the United States: Paper presented to Research Commission of Interstate Oil Compact Committee, December 4, 1990, Phoenix, Ariz.

Kim, A.G., 1977, Estimating methane content of bituminous coalbeds from adsorption data: U.S. Bureau of Mines Rept. of Investigations 8245, 22 p.

Moberly, Ralph, 1960, Morrison, Cloverly, and Sykes Mountain Formations, Northern Bighorn basin, Wyoming and Montana: GSA Bulletin, v. 71, no. 8, p. 1137–1176.

Murray, D.K., 1990, Future supply of natural gas in the United States: Paper presented to Research Commission of Interstate Oil Compact Committee, December 4, 1990, Phoenix, Ariz.

Petersen, J.A., 1972, Jurassic System, *in* Mallory, W.W., ed., Geologic atlas of the Rocky Mountain region: Rocky Mountain Association of Geologists, p. 177–190.

Potential Gas Committee, 1991, Potential supply of natural gas in the United States (December 31, 1990): Golden, Colo., Potential Gas Agency, 169 p.

Rice, D.D., and Shurr, G.W., 1983, Patterns of sedimentation and paleogeography across the western interior seaway during time of deposition of upper Cretaceous Eagle sandstone and equivalent rocks, Northern Great Plains, *in* Reynolds, M.W., and Dolly, E.D., eds., Mesozoic paleography of the west-central United States: SEPM Rocky Mountain Section, Rocky Mountain Paleogeography Symposium 2, p. 337–358.

Roberts, A.E.,1966, Geology and coal resources of the Livingston coal field, Gallatin and Park Counties, Montana: USGS Professional Paper 526–A, 56 p.

Sholes, Mark, 1985, Coal resources, in Ryan, Janie, compiler, Proceedings, Montana Coal Forum—Sept. 20–22, 1984: Montana Bureau of Mines and Geology Special Publication 93, p. 2–19.

Tonnsen, John, 1989, Natural gas pipelines and regulations in Wyoming and adjacent areas, *in* Eisert, J., ed., Gas resources of Wyoming: Wyoming Geological Association, 40th Field Conference, p. 31–39.

Wood, G.H., Jr., and Bour, W.V., III, 1988, Coal map of North America: USGS Coal Map, 44 p., scale 1:500,000.

Preliminary Results of a Coalbed Methane Assessment of the Wind River Indian Reservation, Wyoming

Ronald C. Johnson[1]
Charles E. Barker[1]
Mark J. Pawlewicz[1]
Bonnie L. Crysdale[1]
Arthur C. Clark[1]
Dudley D. Rice[1]

1. U.S. Geological Survey, Denver, Colorado

ABSTRACT

The Wind River Indian Reservation, which occupies much of the western half of the Wind River basin, Wyoming, contains significant coal deposits in the Upper Cretaceous Mesaverde and Meeteetse Formations. Preliminary results of a coalbed methane assessment of the Reservation indicate that significant amounts of coalbed methane occur at shallow depths in low-rank Mesaverde coals (vitrinite reflectance 0.40% to 0.50%) near Hudson, in the southern part of the Reservation. Total desorbed gas plus calculated lost gas for cores taken from depths of 470 to 590 ft (143.4 to 180 m) range from 49 to 72 CF/T (1.54 to 2.24 cc/g).

Desorbed gases are chemically dry (97.3% to 99.7% methane) and contain minor amounts of carbon dioxide (0.61% to 0.83%). From isotopic analysis of the C_1 fraction, $\delta^{13}C_1$ values of –47.0‰ to –53.0‰ suggest that the gas could be either in situ generated biogenic or migrated thermogenic in origin. Attempts to desorb methane from Mesaverde coal cored near Alkali Butte, in the southeastern corner of the Reservation, and from Meeteetse coal cored near the Welton coal mine in the north-central part of the Reservation were not as successful.

The vitrinite reflectance method was used to obtain subsurface rank information from nine wells on the Reservation. Mean random vitrinite reflectance of Mesaverde and Meeteetse coals ranged from 0.5% at a depth of 3,770 to 3,790 ft (1,150 to 1,156 m) in the south-central part of the Reservation to 2.1% at a depth of 18,380 to 18,410 ft (5,606 to 5,615 m) along the deep trough of the basin in the northeastern part of the Reservation.

INTRODUCTION

In April 1990 the U S Geological Survey began a three-year assessment of coalbed methane and undiscovered conventional oil and gas resources of the Wind River Indian Reservation, Wyoming. This report presents the preliminary results of the assessment for coalbed methane, which, when completed, will include maps of coal thickness, depth, and rank of important coal-bearing horizons; vitrinite reflectance profiles to determine the rate of increase in coal rank with depth; burial reconstructions; and results of desorption experiments on coals cored at depths of less than 1,000 ft (305 m). Presented here are vitrinite reflectance analyses obtained from nine drill holes on the Reservation and desorption data obtained during the first year of a two-year coring program. Three published desorption measurements for shallow coal on or near the Reservation suggested little or no methane in the coal at this depth (Rieke and Kirr, 1984). However, significant amounts of methane were desorbed from Mesaverde Formation coal recovered from one of three coreholes drilled during the first year of this study.

GEOLOGIC FRAMEWORK

The Wind River Indian Reservation in central Wyoming occupies much of the western half of the Wind River basin (Figure 1), a structural and sedimentary basin formed during the Laramide orogeny in Late Cretaceous through early Eocene time (Keefer, 1970). The basin is bounded on the north by the Washakie Range, Owl Creek Mountains, and Bighorn Mountains; on the south by the Wind River Range and Granite Mountains; and on the east by the Casper arch. The Reservation covers parts of the Owl

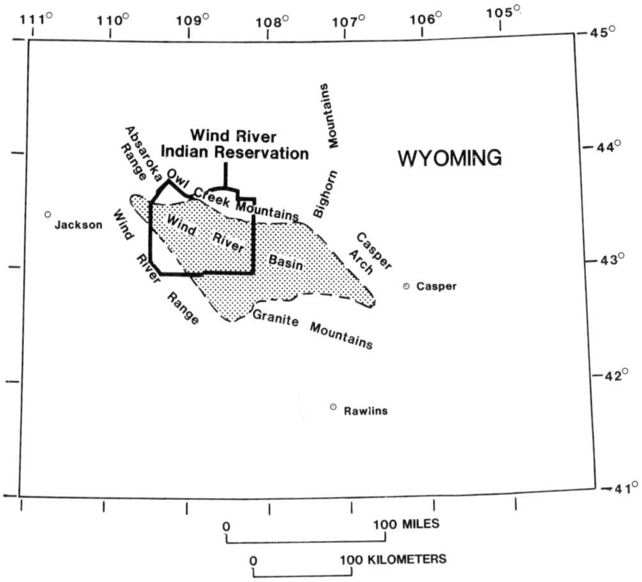

Figure 1. Location of Wind River Indian Reservation, Wyoming, with respect to Wind River basin and surrounding uplifts.

Creek Mountains, Washakie Range, and Wind River Range. Keefer (1970) has detailed the development of this structurally complex basin. The deep trough of the basin lies adjacent to the Washakie Range, Owl Creek Mountains, and Casper arch, which were uplifted and thrust southwestward and southward along a series northwest- and west-trending reverse faults, which in turn created a series of en echelon northwest-trending, southeast-plunging anticlinal structures in the northwestern part of the basin. These structures are highly asymmetrical, with steeply dipping to overturned southwestern limbs and more gently dipping (less than 25°) northeastern limbs. Most of these structures occur within the boundaries of the Reservation. A similar series of northwest-trending, northwest-plunging anticlines formed along the south-central basin margin along southwest-thrusting reverse faults related to uplift of the Granite Mountains. Many of these anticlines also occur within the Reservation. The southwestern flank of the basin dips gently northeastward off the Wind River Range.

Significant coal deposits occur within the boundaries of the Reservation in the Upper Cretaceous Mesaverde and Meeteetse Formations (Figure 2). Coals also are present in the Paleocene Fort Union Formation, but they have not been studied in detail. Windolf and others (1982) estimated total coal resources within the Reservation in beds thicker than 30 in. (76 cm) and at depths less than 6,000 ft (1,830 m) to be nearly 8.5 billion short tons (7.7 billion t); nearly 14.5 billion short tons (13.2 billion t) of coal are contained in beds 30 in. (76 cm) or more thick at depths to 12,000 ft (3,660 m). Mesaverde coals crop out in the north-central, central, and southeastern parts of the Reservation; outcrops of Meeteetse coals are confined to the north-central part of the Reservation (Figure 3). Coal in these areas has been mined in the past, but no mines presently are active.

The Mesaverde and Meeteetse Formations are exposed mainly along the flanks of northwest-trending anticlines, which were actively rising during latest Cretaceous, Paleocene, and early Eocene times. As a result of the uplift and subsequent erosion, the older strata were partly removed before deposition of the lower Eocene Indian Meadows Formation. On many anticlines, such as the Sage Creek and Winkleman anticlines, the Mesaverde and Meeteetse Formations were completely eroded for a considerable distance down the flanks of the structures. The Mesaverde and Meeteetse Formations also were eroded from the southwestern flank of the basin adjacent to the Wind River Range during this time.

COAL-GAS GENERATION AND STORAGE

It is generally believed that the gas found in coalbeds was generated largely in situ rather than having migrated into the coal from other sources (Patteisky, 1951; Venter and Stasson, 1953; Kim, 1978). Methane in coalbeds can be either biogenic or thermogenic, or a combination of both. Distinguishing between biogenic and thermogenic methane is important in assessing the coalbed methane potential of an area. Biogenic methane is generated at temperatures less than 167°F (75°C) through degradation of organic matter by anaerobic microorganisms (Kim and Douglas, 1972; Rice and Claypool, 1981). Thermogenic methane is generated at higher temperatures by progressive devola-

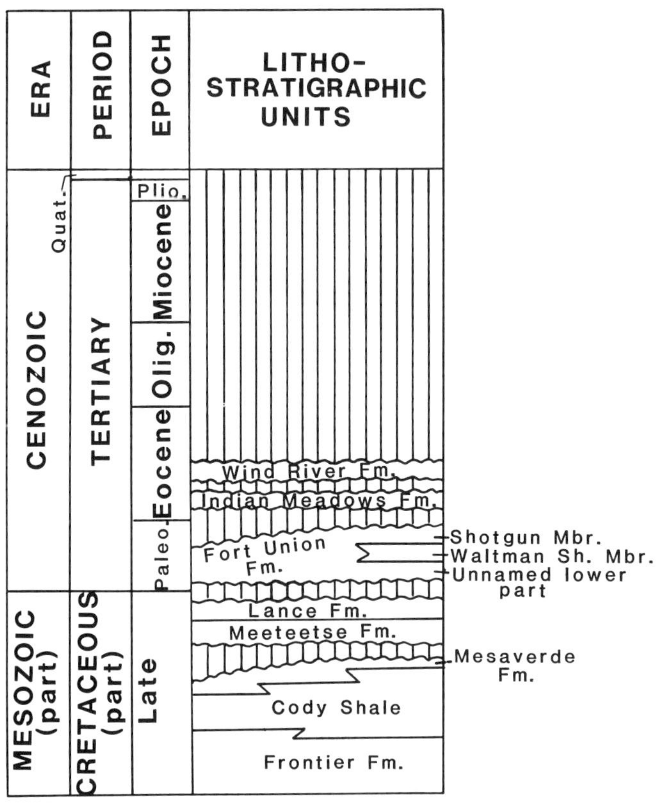

Figure 2. Generalized stratigraphic section of Upper Cretaceous and Tertiary rocks on the Wind River Indian Reservation.

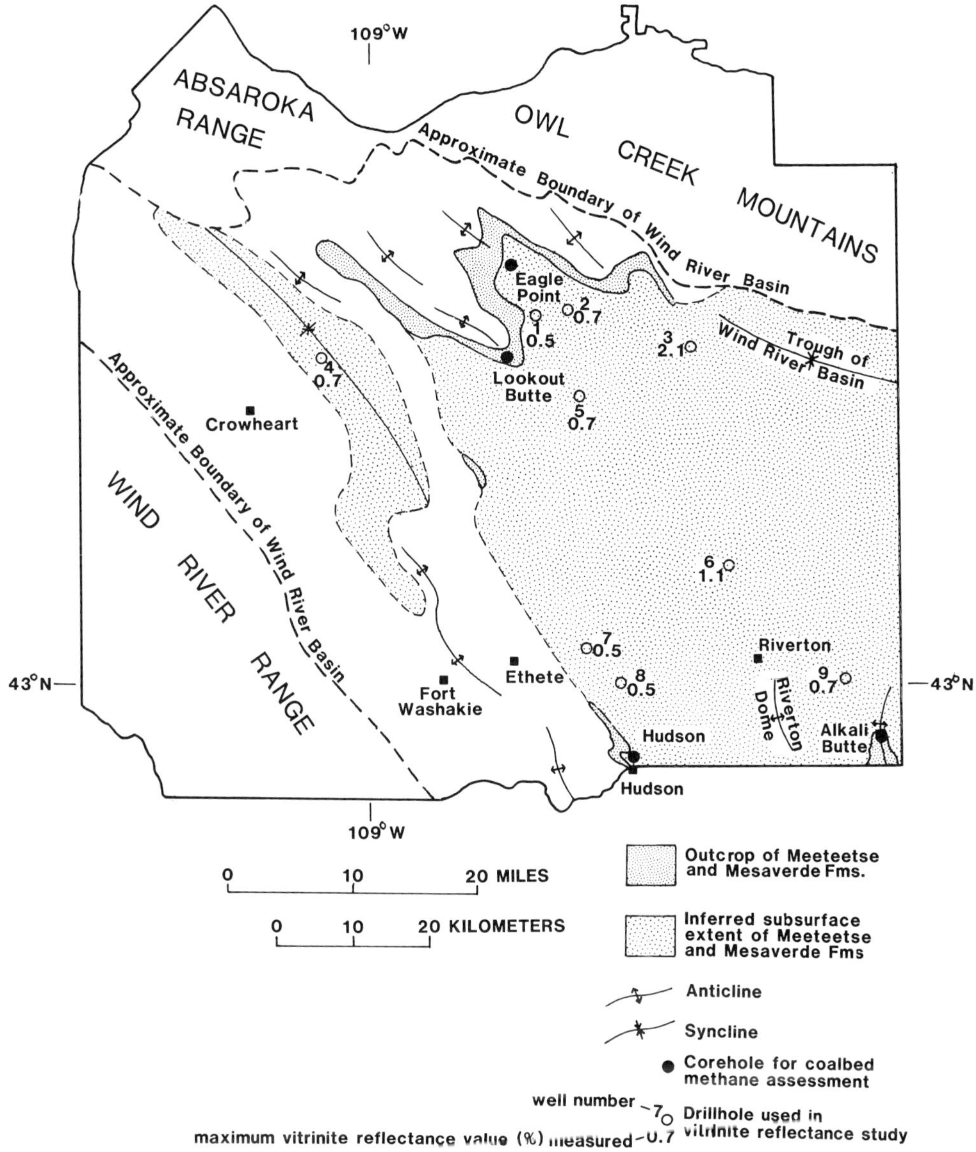

Figure 3. Generalized map of the Wind River Indian Reservation, showing structure and exposures and inferred subsurface extent of Mesaverde and Meeteetse Formations, locations of drill holes cored for coalbed methane, and highest vitrinite reflectance values measured in drill holes (hole numbers are keyed to Table 1).

tilization of coal as it becomes buried and heated within a subsiding sedimentary basin.

The total volume of thermogenic methane generated by coal increases with increasing coal rank (Jüntgen and Karweil, 1966a, 1966b; Karweil, 1969; Jüntgen and Klein, 1975). The onset of significant thermogenic gas generation occurs at about the rank of high-volatile B bituminous (37% volatile matter), or equivalently a mean random vitrinite reflectance value of about 0.7%. Methane found in coals with reflectance values less than 0.7% should, therefore, be largely of biogenic origin, unless the gas has migrated into the coal from more mature sources. The amount of biogenic gas that coals can produce is much more difficult to predict because the decomposition of organic matter by microorganism can be affected by many factors.

Recent studies also have shown that the amount of methane found in coal increases with increasing rank (Ulery, 1987, 1988; Ulery and Puglio, 1984). Coal rank is, therefore, the most important factor to determine when assessing the coalbed methane potential of a region, particularly an area such as the Wind River Indian Reservation, where little coal desorption information is available.

As for composition, biogenic gas usually is isotopically light ($\delta^{13}C_1$ values lighter than –60‰) and consists mainly of methane ($C_1/C_{1-5} > 0.99$; Rice and Claypool, 1981; Schoell, 1983). Thermogenic gases generally exhibit $\delta^{13}C_1$ values heavier than –50‰ and contain significant amounts of heavier hydrocarbons ($C_1/C_{1-5} < 0.95$). Gases with $\delta^{13}C_1$ values in the range of –60‰ to –50‰ usually are interpreted to be a mixture of biogenic gas and early thermogenic gas (Schoell, 1983). Exceptions are biogenic gases generated by fermentation instead of the more common carbon dioxide reduction (Whiticar and others, 1986), and biogenic gases that have undergone partial microbial oxidation in near-surface environments (Rice and Claypool, 1981; Whiticar and Faber, 1986). In both cases, $\delta^{13}C_1$ values can fall within the range of mixed or early thermogenic gases.

Most methane-rich gas is stored in coalbeds either by molecular attraction onto large internal surface areas (adsorption), or by absorption within the microstructure (aromatic rings) of the coal (Oberlin and others, 1980; Van Krevelen, 1981). Both the amount of surface area and development of the microstructure increase with increasing coal rank; thus the ability to store methane also increases with increasing rank. Lesser amounts of methane are contained in the natural fracture systems of the coal, both as free gas and as gas adsorbed onto the walls of the fractures and held there by micropore capillary pressure. Desorption of this gas can occur if the capillary pressure decreases, for example, as a result of uplift and erosion (Close and Erwin, 1989). Because free gas in the fracture system is believed to be in equilibrium with the gas adsorbed within the microstructure, the loss of free gas from the coalbed can induce desorption. The seam can lose part or all of its methane if this process continues long enough. Some of the gas can migrate into adjacent permeable rocks (Ulery, 1988). Since middle or late Pliocene time, several thousand feet of strata have been eroded in the area of the Wind River Reservation (Keefer, 1970), reducing capillary pressures in the coals and undoubtedly releasing some methane. The circulation of near-surface ground water through porous sandstones adjacent to the coalbeds may have caused further loss of methane, particularly at shallow depths.

COAL RANK INVESTIGATION

A considerable amount of coal-rank information was available from shallow coreholes drilled for Windolf and others' (1982) recent investigation of coal resources on the Wind River Reservation. They found that ranks of shallow coals generally were low, ranging from subbituminous C to high-volatile C bituminous; coal ranks in the north-central part of the Reservation are somewhat higher than those that occur in the southeast. However, little coal rank information is available for coals deeper than 1,000 ft (305 m) on the Reservation. To fill this gap, coal chips were collected for R_m analysis from nine wells on the Reservation (Figure 3, Table 1). The samples were mounted in epoxy on microscope slides, planed, and polished. Mean random vitrinite reflectance (R_m) was measured from randomly oriented grains in plane-polarized, incident white light, through a 546-nm monochromatic filter, in immersion oil on a Zeiss Universal microscope with a nonrotating stage. A detailed description of this technique is given in Bustin and others (1983) . Generally 35 to 50 readings per sample were taken.

R_m values for Mesaverde and Meeteetse coals ranged from 0.5% at a depth of 3,770 to 3,790 ft (1,150 to 1,156 m) in the Exxon 1–534 Shoshone-Arapaho well (sec. 2, T1S R2E) to 2.1% at a depth of 18,380 to 18,410 ft (5,606 to 5,615 m) in the Coastal Owl Creek 1 well (sec. 26, T5N R3E) (Figure 3, Table 1). Vitrinite reflectance profiles were constructed for the Owl Creek 1 and the Bridger Petroleum 1–20 Tribal wells (Figures 4 and 5), and "best-fit" straight lines were constructed using a computer program. Figure 6 is a plot of R_m vs depth for all subsurface data collected to date. A vitrinite reflectance value of 0.7%, believed to represent the onset of significant thermogenic gas generation, occurs at depths of about 10,500 ft (3,202 m) in the Coastal well and 8,000 ft (2,440 m) in the Bridger Petroleum well.

COALBED DRILLING PROGRAM

A small-scale drilling program was carried out during October and November 1990 to evaluate coalbed methane resources in the Mesaverde and Meeteetse Formations on the Reservation. Because of budgetary constraints, drilling depths were limited to 1,000 ft (305 m) or less. Such shallow depths are less than ideal for evaluating coalbed methane potential, and it must be emphasized that failing to find coalbed methane at these depths does not preclude its occurrence in the area. As discussed above, much of the methane that was originally present in shallow coals possibly may have been lost to the atmosphere. The three published desorption measurements from shallow Cretaceous coals on and near the Reservation indicated little or no methane (Rieke and Kirr, 1984).

Numerous shallow holes have been drilled in the Mesaverde and Meeteetse outcrop areas to evaluate coal resources (Windolf and others, 1982). This information was useful in selecting sites for the drilling program, although it was difficult to predict the depth at which the coals would be found because of steep structural dips. Four sites (Figure 3) were, however, selected—(a) 2.5 miles (4 km) north of Alkali Butte (SE/4 sec. 29, T1S R6E); (b) one mile (1.6 km) north of Hudson (NE/4 sec. 1, T2S R2E); (c) one mile (1.6 km) north of Lookout Butte near the Barquin coal mine (NE/4 SE/4 sec. 32, T5N R1E); and (d) 4 miles (6.4 km) east of Eagle Point near the Welton coal mine (C sec. 30, T6N R2E). The Alkali Butte hole was located only 75 ft (23 m) northwest of two wells drilled to evaluate coal resources. Because each of the other three sites was located away from any other wells, two closely spaced wells were planned to be drilled at each site—a pilot well to locate the coals, and a second well to core the coals. Only 1 ft (31 cm) of coal was encountered in the pilot well at the Barquin site, so the corehole was not drilled. An old well 1,500 ft (458 m) northwest of the Barquin site had encountered coalbeds with combined thickness of 14 ft

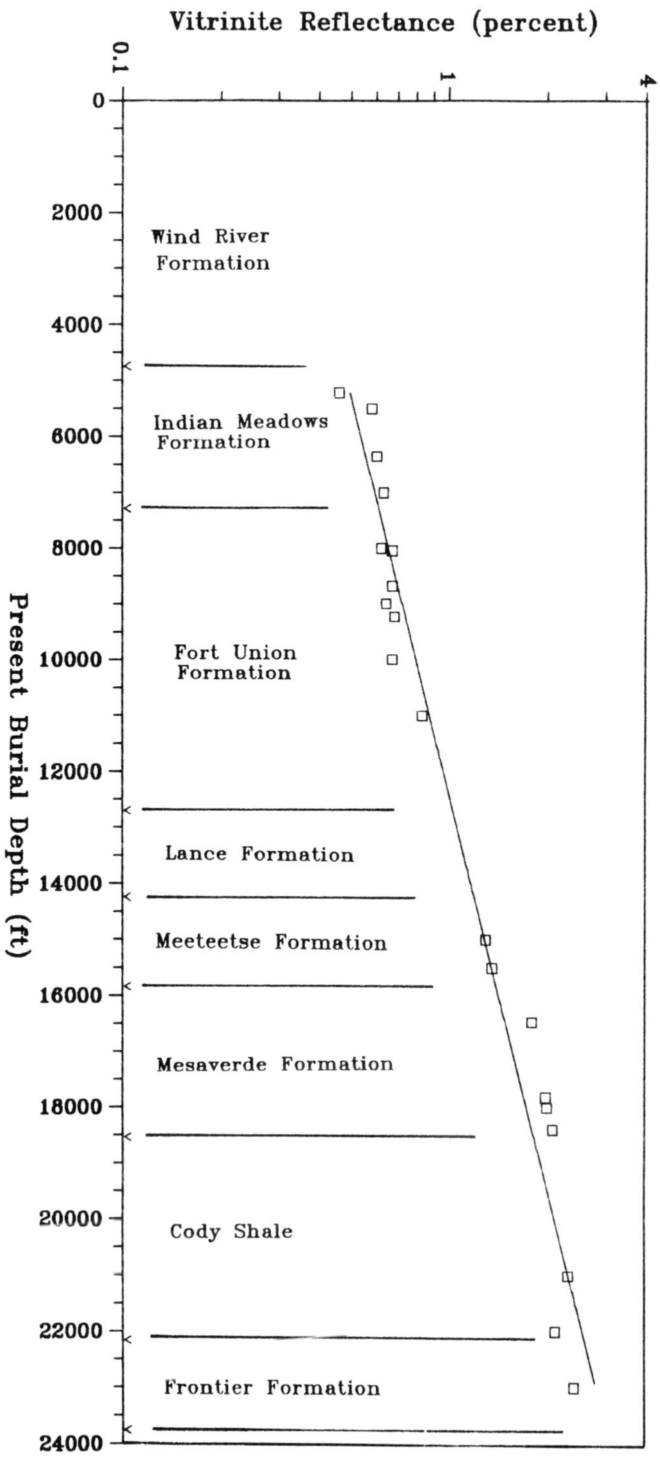

Figure 4. Vitrinite reflectance profile for the Coastal Owl Creek No. 1 well, sec. 26, T5N R3E.

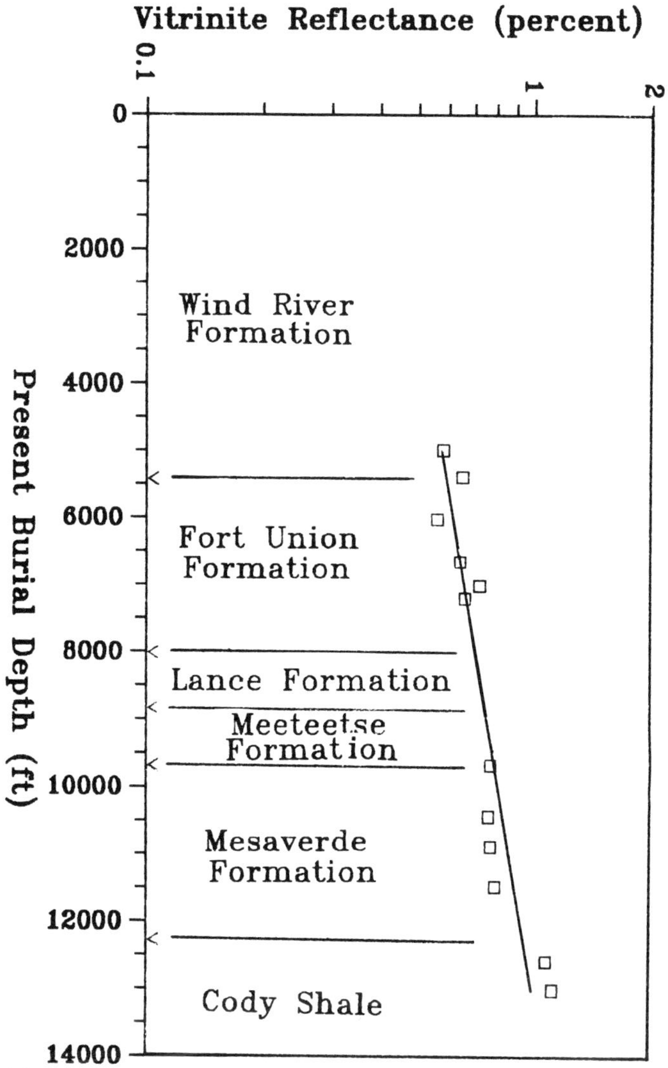

Figure 5. Vitrinite reflectance profile for the Bridger Petroleum 1–20 well, sec. 20, T2N R4E.

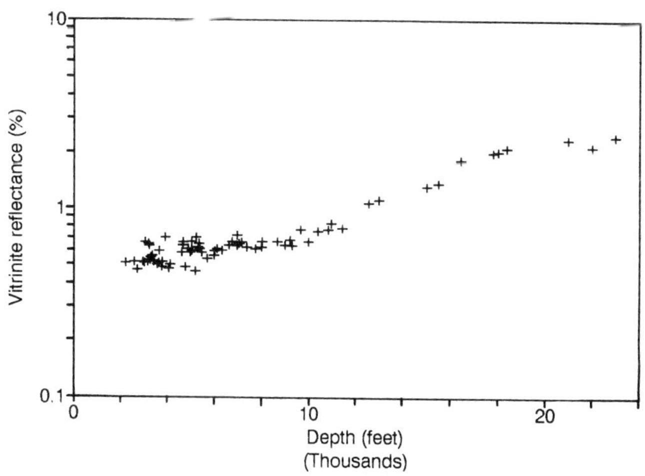

Figure 6. Cumulative profile of all subsurface vitrinite reflectance information collected on the Wind River Reservation.

(4.3 m). Significant coal was encountered in the pilot wells at the other three sites, all of which subsequently were cored.

DESORPTION METHODOLOGY

Coal cores were desorbed using the method of Close and Erwin (1989). For correction of volume measurements of desorbed gas, ambient pressure was recorded digitally with a Solomat model 500E meter equipped with a 515BPX barometric pressure probe. Plastic desorption canisters with aluminum covers were modified so that core tem-

Table 1. Results of vitrinite reflectance analysis from nine oil and gas wells and from three wells drilled for coalbed methane on the Wind River Indian Reservation. See Figure 3 for drill hole locations.

Well	Location sec	Location Twp	Location Rge	Sample Depth (ft)	Sample Depth (m)	Mean Random Vitrinite Reflectance (%)	Formation
American Quasar 1 Shoshone-Fed.	10	5N	1E	3,740–3,830	1,141–1,168	0.52	Mesaverde
				4,100–4,180	1,251–1,275	0.50	Mesaverde
Emanuel 1 Tribal A	7	5N	2E	3,240–3,260	988–994	0.63	Lance–Meeteetse
				3,340–3,370	1,019–1,028	0.56	Lance–Meeteetse
				4,650–4,680	1,418–1,427	0.64	Mesaverde
				5,450–5,470	1,662–1,668	0.59	Mesaverde
				5,630–5,660	1,717–1,726	0.67	Mesaverde
Coastal 1 Owl Creek	26	5N	3E	5,200–5,240	1,586–1,598	0.46	Indian Meadows
				5,500	1,678	0.58	Indian Meadows
				6,300–6,400	1,922–1,952	0.60	Indian Meadows
				7,000	2,135	0.62	Indian Meadows
				8,000	2,440	0.62	Fort Union
				8,030–8,060	2,449–2,458	0.67	Fort Union
				8,670–8,690	2,644–2,650	0.67	Fort Union
				9,000	2,745	0.64	Fort Union
				9,220–9,240	2,812–2,818	0.68	Fort Union
				10,000	3,050	0.67	Fort Union
				11,000	3,355	0.83	Fort Union
				15,000	4,575	1.30	Meeteetse
				15,490–15,510	4,724–4,731	1.36	Meeteetse
				16,450–16,470	5,017–5,023	1.80	Mesaverde
				17,800–17,830	5,429–5,438	1.98	Mesaverde
				18,000	5,490	2.00	Mesaverde
				18,380–18,410	5,606–5,615	2.08	Mesaverde
				21,000	6,405	2.31	Cody
				22,000	6,710	2.12	Frontier
				23,000	7,015	2.42	Frontier
Gulf 1-35–1A Crowheart	35	5N	3W	4,050–4,100	1,235–1,251	0.48	Fort Union
				4,600–4,650	1,403–1,418	0.58	Fort Union
				4,670–4,700	1,424–1,434	0.66	Fort Union
				5,050–5,100	1,540–1,556	0.67	Mesaverde
				5,330–5,340	1,626–1,629	0.62	Mesaverde
				5,370–5,400	1,638–1,647	0.61	Mesaverde
Dome 1–17 Tribal-Hunt	17	4N	2E	4,760–4,780	1,452–1,458	0.49	Shotgun Mbr., Fort Union
				4,910–4,930	1,498–1,504	0.61	Shotgun Mbr., Fort Union
				6,030–6,050	1,839–1,845	0.59	Waltman Sh. Mbr., Ft. Union
				6,140–6,160	1,873–1,879	0.61	Lower Mbr., Fort Union
				6,760–6,780	2,062–2,068	0.67	Lower Mbr., Fort Union
				7,080–7,100	2.159–2,166	0.65	Lower Mbr., Fort Union

perature inside the canisters also could be measured. This was accomplished by connecting the meter to a miniature type K thermocouple inserted into a thin copper tubing sheath that was closed at one end and threaded into the canister cover at the other end. The tubing protruded about 5 in. (13 cm) into the canister and was in contact with the core sample. Before beginning a test, each canister was sealed, pressurized with 10 to 15 psi (69 to 103 kPa) of air, and immersed in water to check for leaks. Each unit was reimmersed in water to check for leaks immediately after filling with coal.

Monitoring core temperature allowed us to correct for any changes in core temperature occurring during the critical first hour or so of the desorption process. Initial core temperature generally was only about 55°F (13°C), the temperature of the circulation water when it reached the surface. Canisters and manometers were kept inside insulated boxes at an ambient temperature of approximately

Table 1, continued.

Well	Location sec	Twp	Rge	Sample Depth (ft)	(m)	Mean Random Vitrinite Reflectance (%)	Formation
Dome 1–17 Tribal-Hunt, continued	17	4N	2E	7,380–7,390	2,251–2,254	0.62	Lower Mbr., Fort Union
				7,760–7,790	2,367–2,376	0.61	Lance
				9,300–9,310	2,837–2,840	0.63	Meeteetse
Bridger 1–20	20	2N	4E	4,960–5,010	1,513–1,528	0.58	Wind River
				5,330–5,430	1,626–1,656	0.65	Wind River
				6,000–6,030	1,830–1,839	0.56	Fort Union
				6,630–6,670	2,022–2,034	0.64	Fort Union
				6,990–7,020	2,132–2,141	0.72	Fort Union
				7,180–7,210	2,190–2,199	0.66	Fort Union
				9,650–9,680	2,943–2,952	0.77	Meeteetse
				10,400–10,450	3,172–3,187	0.76	Mesaverde
				10,860–10,900	3,312–3,324	0.77	Mesaverde
				11,450–11,490	3,492–3,504	0.79	Mesaverde
				12,540–12,590	3,825–3,840	1.07	Cody
				12,970–13,020	3,956–3,971	1.11	Cody
Exxon 1–731 Shoshone-Arapaho	20	1N	2E	2,200–2,290	671–698	0.51	Fort Union
				2,950–2,980	900–909	0.52	Fort Union
				3,140–3,200	958–976	0.52	Fort Union
				3,400–3,480	1,037–1,061	0.52	Fort Union
				3,570–3,620	1,089–1,104	0.50	Mesaverde
Exxon 1–534 Shoshone-Arapaho	2	1S	2E	3,270–3,280	997–1,000	0.52	Fort Union
				3,650–3,660	1,113–1,116	0.51	Mesaverde
				3,770–3,790	1,150–1,156	0.49	Mesaverde
Continental A–1 Tribal	1	1S	5E	5,070–5,080	1,546–1,549	0.59	Lance–Meeteetse–Mesaverde
				5,250–5,260	1,601–1,604	0.70	Mesaverde
				5,690–6,000	1,735–1,830	0.54	Mesaverde
USGS CBM–1 Alkali Butte	29	1S	6E	35–40	10.7–12.2	0.33	Mesaverde
				586.2	178.8	0.48	Mesaverde
				658.3	200.8	0.47	Mesaverde
				975.3–975.8	297.5–297.6	0.44	Mesaverde
USGS CBM–2 Hudson	1	2S	2E	483	147	0.40	Mesaverde
				582	178	0.47	Mesaverde
				591	180	0.50	Mesaverde
USGS CBM–4 Welton	30	6N	2E	100–105	31–32	0.39	Meeteetse
				670–675	204–206	0.49	Meeteetse

70°F (21°C) in a heated trailer on the drillsite. In this way, temperature differences between each canister and manometer could be minimized. In addition, the length of the gas line connecting the canister and manometer was kept at a minimum to reduce temperature differences in the measurements. A propylene glycol and water mix was used as the manometer fluid to prevent freezing in the event the trailer's heater failed. The canisters later were transferred to a laboratory at the Denver Federal Center. The complete stepwise procedure used is outlined in Barker and others (in press).

DRILLING SUMMARY AND DESORPTION RESULTS

Alkali Butte Site

The Alkali Butte site was drilled from October 11 to 16, 1990. A total of 17 ft (5.2 m) of coal in four beds at depths of 577 to 587 ft (176 to 179 m), 654 to 658 ft (199 to 201 m), and 963 to 967 ft (294 to 295 m) were cored and desorbed. Because the corehole (USGS CBM–1) deviated considerably from the older hole drilled nearby for coal evaluation, it was difficult to predict where coalbeds would be encountered. As a result, approximately 8 ft (2.4 m) coal, at a depth of 842 to 850 ft (256.8 to 259.2 m), was missed. No detectable methane was desorbed from any of the coals, nor was any gas observed bubbling from the surface casing. A previous attempt to desorb methane from the coal here also was unsuccessful (Rieke and Kirr, 1984).

Hudson Site

The Hudson site was drilled between October 22 and 25. A total of 11 ft (3.4 m) of coal was recovered from two intervals—478 ft to 482 ft (145.8 to 147 m) and 577 to 587 ft (176 to 179 m). The only other drill hole in the immediate area is a coal test hole located 3,000 ft (915 m) to the east. Because of the lack of subsurface information, precise coal depths could not be predicted; however, surface information suggested that the coals would be penetrated at depths between 700 and 800 ft (214 to 244 m). Despite the shallower than expected drilling depth, gas was noticeably bubbling from the core when it reached the surface. Gas bubbled vigorously from the surface casing, and minor water flowed from both drill holes immediately after they were completed. Bubbling had nearly ceased and water levels had declined by the time the holes were cemented on November 27. The coals continued to desorb methane for more than two months. Desorbed methane contents (uncorrected for ash content) ranged from 45.5 to 67.3 CF/T (1.42 to 2.10 cc/g); estimated lost-gas values ranged from 2.2 to 4.5 CF/T (0.07 to 0.14 cc/g) (Figure 7, Table 2). Gassy coal mines have been reported in the Hudson area (DeBruin and Jones, 1989).

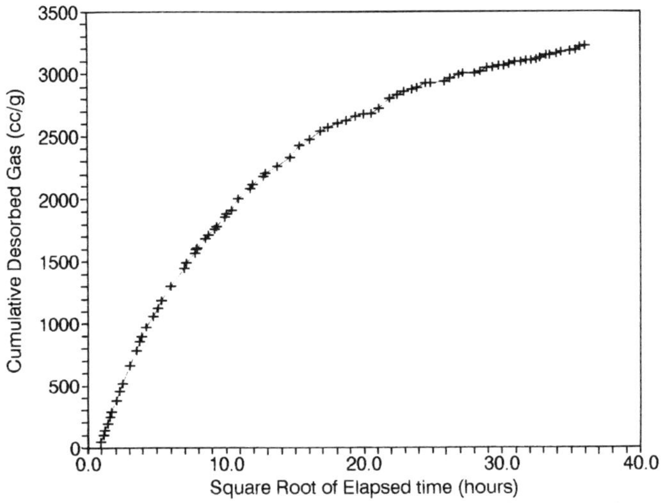

Figure 7. Methane desorption curve for coal from depth of 582.25 to 583.33 ft (177.6 to 177.9 m) at the Hudson site (USGS CBM–2).

Welton Mine Site

The Welton mine site at Eagle Point was drilled from November 5 to 13, 1990. From these drill holes (USGS CBM–4) 11 ft (3.4 m) of the Welton coalbed was cored from 745 ft to 756 ft (227.2 to 230.6 m). No significant amounts of gas were desorbed, and no gas was observed bubbling from the surface casing of either hole. An attempt was made to desorb coal cuttings collected during air drilling of the pilot hole. Because the cuttings reached the surface in only about five minutes, they should have retained most of their methane. Enough coal cuttings were collected on two large screens (about 2 ft square) to fill two 12-in. (31-cm) canisters. No detectable methane was desorbed. The field technique of desorbing methane from pilot-hole cuttings may be used in deciding whether or not to drill and core a second hole at a given site during the drilling program scheduled for 1991.

Table 2. Results of desorption experiments on coal from the Hudson site (USGS CBM–2).

Sample Depth (ft)	Sample Depth (m)	Desorbed Gas (cc/g)	Desorbed Gas (CF/T)	Lost Gas Estimate (cc/g)	Lost Gas Estimate (CF/T)	Desorbed Gas + Lost Gas (cc/g)	Desorbed Gas + Lost Gas (CF/T)
483.83–484.75	147.6–147.8	1.63	52	0.09	3	1.72	55
484.75–485.67	147.8–148.1	1.61	52	0.08	3	1.69	55
485.67–486.58	148.1–148.4	1.54	49	0.08	3	1.62	52
486.58–487.42	148.4–148.7	1.44	46	0.10	3	1.54	49
582.25–583.25	177.6–177.9	2.10	67	0.14	4	2.24	72
583.25–584.25	177.9–178.2	1.86	59	0.11	4	1.97	63
584.25–585.25	178.2–178.5	1.90	61	0.12	4	2.02	65
585.25–586.25	178.5–178.8	1.96	63	0.10	3	2.06	66
586.25–587.25	178.8–179.1	1.42	45	0.07	2	1.49	48
587.25–588.25	179.1–179.4	1.76	56	0.09	3	1.85	59
590.67–591.67	180.2–180.5	1.90	61	0.13	4	2.03	65

Barquin Mine Site

The Barquin mine site near Lookout Butte was drilled from October 27 to 31, 1990. The pilot hole (USGS CBM–3) was drilled to a depth of 910 ft (277.6 m), and only 1 ft (31 cm) of coal was encountered; consequently, a second hole was not drilled. Some gas bubbled from the surface casing during and immediately after drilling.

Discussion

Several geologic and hydrologic factors at the four sites drilled were evaluated in an attempt to understand why gas was present at Hudson but not at the Alkali Butte and Welton sites. Because the Barquin site was not cored, it remains uncertain whether coal in that area contains methane. The possibility of encountering low water tables was a major concern prior to drilling because the coals could have lost methane if the hydraulic head had been decreased. Few water wells have been completed in the Mesaverde and Meeteetse Formations; consequently, little information is available about the near-surface hydrology of these formations. We feared that low water tables would indeed be encountered because of the semiarid conditions on the Reservation. However, water tables were very high—near the tops of the wellbores at all four sites. All the sites produced considerable quantities of water during drilling, and the Hudson and Barquin wells flowed minor quantities of water for several days. Therefore, it appears that sufficient present-day hydraulic head exists for the coals to retain methane.

Another concern was that water tables may have been low at times during the geologic past, specifically during hiatuses represented by multiple unconformities above the coaly intervals at all four sites. However, a major unconformity 100 ft (31 m) above the coaly interval at Hudson suggests that such features cannot explain why gas is found only at Hudson. Furthermore, coal rank is low at all three localities (0.33% to 0.50% R_m; Table 1). The only major difference identified so far is that the coaly interval is less sandy at Hudson than at the other sites. If the sandstones at the other sites are highly porous, then much of the methane could have desorbed out of the coals and have been carried away by percolating ground water, as suggested by Ulery (1988) for Pennsylvania coals. Further study is necessary to understand the spotty distribution of methane in the shallow coals on the Reservation.

CHEMICAL AND ISOTOPIC COMPOSITION OF GASES

Three desorbed gas samples were collected from the Hudson core and analyzed for chemical and isotopic composition (Table 3). Gas samples were collected from the canisters nine days after desorption had begun. No samples were collected from the hydrometers used in the desorption experiments because they fill partially with atmospheric gas when disconnected from the cannisters. Instead, sampling was accomplished by connecting a short tube directly from the canisters to a graduated cylinder inverted in a pail of water that had been disinfected with zephiran chloride. This procedure also enabled measurement of the desorbed gas at the same time. The gas was then transferred to a glass bottle and sealed.

One gas sample also was collected from the casing of one of the twinned holes, from which gas had bubbled vigorously prior to penetrating the coal. Gas volume did not noticeably increase after penetrating the coal, however, and most likely this gas originated largely from sandstone and shale above the coal rather than from the coal itself. If so, then a comparison of chemical and isotopic compositions of coal gas and sandstone gas at Hudson could be made. In the San Juan basin the chemical and isotopic compositions of coal gas and gas from overlying sandstone differ significantly (Rice and others, 1988). This sample was collected by inverting a water-filled bottle, fitted with a funnel, in the casing. Enough gas to completely evacuate the water from the bottle was collected in about five minutes. A small quantity of disinfected water then was leaked into the bottle to lessen the possibility of microbial contamination.

The three desorbed gas samples from the low-rank coals (0.40% to 0.50% R_m) at Hudson site were chemically dry (97.3% to 99.7% methane), isotopically light ($\delta^{13}C_1$ values of –47.0‰ to –53.0‰), and contained varying amounts of atmospheric gas (29% to 45%) but only minor carbon dioxide (0.61% to 0.83%) (Table 3). Atmospheric contamination appears to have resulted from a combination of air contained in the short tube prior to collecting the sample and air remaining in the canisters when they were sealed. The casing-gas sample had a $\delta^{13}C_1$ value of –52.1‰ and consisted of 98.6% methane and about 1.0% carbon dioxide. This is virtually indistinguishable from the desorbed gas.

Three possibilities exist for the origin of these gases. First, they may be of biogenic origin, generated by the *fermentation* pathway instead of by carbon dioxide reduction, which is the dominant pathway for ancient biogenic gas accumulations. Methane generated by the fermentation pathway usually has heavier carbon-isotope values than

Table 3. Chemical and isotopic compositions of three desorbed gases and one casing gas from the Hudson site (USGS CBM–2). Location shown on Figure 3.

Sample Depth (ft)	(m)	Composition (%) N_2 and (or) air	C_1	CO_2	C_2	C_3	iC_4	nC_4	C_1/C_{1-5}	$\delta^{13}C^1$ (‰)
483	147.3	44.58	53.30	0.61	0.63	0.58	0.10	0.20	0.9725	–53.0
583	177.8	50.38	48.90	0.57	0.14	–	–	–	0.9971	–47.0
590	179.9	28.66	70.30	0.83	0.21	–	–	–	0.9970	-51.3
Casing	–	8.64	90.11	1.00	0.24	–	–	–	0.9864	–52.1

those generated by carbon dioxide reduction and may overlap those of thermogenic methane. However, they can be distinguished by deuterium isotope (δD_1) analysis (Whiticar and others, 1986). Gases desorbed and produced from shallow, low-rank (0.3% to 0.5% R_m) coal of the Fort Union Formation in the Powder River basin have similar $\delta^{13}C_1$ isotopic values (–54‰ to –61‰) and, based on deuterium isotopic measurements, are interpreted to be biogenic and generated by the fermentation pathway. Second, the gas may be biogenic in origin but subsequently altered by *microbial oxidation* in the near-surface sediment/water environment. Oxidation results in the enrichment of heavy isotope ^{13}C in methane (Whiticar and Faber, 1986). Third, the gas may be *thermogenic*, having migrated from deeper, more mature source rocks. In this case, the coalbeds acted as a seal. Detailed investigation of the gases will help resolve their origin and facilitate resource assessment.

TOTAL COAL THICKNESS IN THE HUDSON AREA

The thickness of coal in the Hudson area is evaluated here briefly. This information may be useful in assessing whether coalbed methane can be economically produced in the area. The number of coalbeds and total thickness of coal in several Hudson area wells are shown in Table 4. Drill-hole data are sparse. Many of the holes from which data are available were drilled more than twenty years ago, and only limited suites of logs were run. Estimating thicknesses of coalbeds thinner than 4 ft (1.2 m) using geophysical logs is imprecise; however, because completions in thin coalbeds have been successful in other areas (Reinecke and others, this volume), it was decided to include in the study all beds thicker than 2 ft (0.6 m).

Based on limited well control available, total thickness of coalbeds thicker than 2 ft (0.6 m) in the Hudson area appears to range from 27 to 32 ft (8.2 to 9.8 m) in six to eleven beds (Table 4). These are distributed through a zone 550 to 1,400 ft (167.8 to 427 m) thick. Maximum thickness of an individual seam is 6 to 7 ft (1.8 to 2.1 m).

The total thickness of coal appears to increase markedly toward Riverton dome, located about 12 miles (19.3 km) east of Hudson. Here, total thickness of coalbeds thicker than 2 ft (0.6 m) ranges from 90 to 110 ft (27 to 34 m) in 22 to 29 beds at depths of 3,150 to 4,260 ft (961 to 1,299 m) (Table 4). Maximum thickness of an individual seam is 10 to 11 ft (3.1 to 3.4 m), and the coals are distributed through a zone 860 to 1,080 ft (262 to 329 m) thick. The reason for this increase in thickness is not entirely clear because no drill-hole information is available between Hudson and Riverton dome. However, Paape (1968) showed a stacking of near-shore deposits in the Mesaverde Formation just east of the Riverton dome area, suggesting that lower coastal plain and deltaic conditions favorable for peat accumulation may have persisted for an unusually long time in the Riverton dome area.

SUMMARY

Preliminary evaluation of coalbed methane resources of the Wind River Indian Reservation indicates that significant coalbed methane occurs at shallow depths near Hudson in the southern part of the Reservation. Total desorbed gas plus lost-gas estimates of 49 to 72 CF/T (1.54 to 2.24 cc/g) are favorable for coal of this rank and depth. These should be considered minimum values inasmuch as they are uncorrected for ash content, which has not yet been determined for these coals.

Both the desorbed gases and casing gas from the Hudson site are chemically dry (97.3% to 99.7% methane), isotopically light ($\delta^{13}C_1$ values of –47‰ to –53‰) and contain minor amounts of carbon dioxide. The low coal rank (0.40% to 0.50% R_m) together with isotopic analysis of the C_1 fraction indicate either biogenic gas generated by fermentation and/or altered by microbial oxidation, or thermogenic gas that migrated from more mature source rocks. Further study is needed to define the origin of these gases.

Attempts to desorb methane from coals at Alkali Butte and the Welton mine area were less successful. However, because coals at shallow depths commonly have lost methane as a result of various geologic phenomena, the deeper coals in these areas still likely contain some methane. To date, the only major factor that explains the difference in gas occurrences at the Hudson site and the Alkali Butte and Welton sites is the scarcity of sandstone in the coaly

Table 4. Estimates of total coal thickness for seven wells in the Hudson and Riverton dome areas, southern Wind River Indian Reservation.

Well	Location			Depth of Coaly Interval		Total Thickness of Coal >2 ft		Number of Coalbeds
	sec	Twp	Rge	(ft)	(m)	(ft)	(m)	
Exxon 1–731 Shoshone-Arapaho	20	1N	2E	2,200–3,600	671–1,098	23	7.0	7
Exxon 1–534	2	1S	2E	3,090–3,800	942–1,159	27	8.2	8
Pan American K–1 Shoshone-Arapaho	24	1N	2E	6,050–6,700	1,845–2,044	32	9.8	11
Pan American L–1 Shoshone-Arapaho	34	1N	2E	3,980–4,530	1,214–1,382	27	8.2	6
Atlantic 10 Tribal	25	1S	4E	3,290–4,150	1,003–1,266	99	30.2	22
Arco 30 Tribal	36	1S	4E	3,150–4,230	961–1,290	90	27.4	24
Arco 40 Tribal	25	1S	4E	3,310–4,260	1,010–1,299	112	34.2	29

interval at Hudson. If the sandstones interbedded with the coals at Alkali Butte and Welton are highly porous, then methane could have desorbed from the coals and migrated through these aquifers. Additional work is needed to better understand the spotty distribution of coalbed methane at shallow depths on the Reservation.

Although coalbeds in the Hudson area tend to be relatively thin, they are not necessarily less favorable to coalbed methane development than are thicker coals. In the Piceance basin of western Colorado, thin coals in the upper Cameo coal zone in the Williams Fork Formation have proven to be more productive than thicker coals in the lower part of the zone, possibly because the thinner coals are more highly fractured (Reinecke and others, this volume). Total coal thickness in the Hudson area is about 27 to 32 ft (8.2 to 9.8 m). Total coal thickness is considerably greater east of the Hudson drill site at Riverton dome, where about 100 ft (31 m) of coal in beds thicker than 2 ft (0.6 m) appears to be present.

Mean random vitrinite reflectance for Mesaverde coal from the three shallow coreholes drilled on the Reservation range from 0.33% to 0.50%. For the Mesaverde and Meeteetse Formations, R_m values determined from oil and gas tests range from 0.5% at a depth of 3,770 to 3790 ft (1,150 to 1,156 m) to 2.1% at a depth of 18,380 to 18,410 ft (5,606 to 5,615 m).

REFERENCES

Barker, C.E., Johnson, R.C., Clark, A.C., and Crysdale, B.L., (in press), A stepwise procedure for desorbing coal gases: USGS Open-File Rept.

Bustin, R.M., Cameron, A.R., Grieve, D.A., and Kalkreuth, W.D., 1983, Coal petrology, its principles, methods, and applications: Geological Association of Canada Short Course Notes, v. 3, 230 p.

Close, J.L., and Erwin, T.M., 1989, Significance and determination of gas content data as related to coalbed methane reservoir evaluation and production implications, *in* The 1989 Coalbed Methane Symposium, Proceedings—University of Alabama, Tuscaloosa, April 17–20, 1989: School of Mines and Energy Development, Gas Research Institute, U.S. Mine Safety and Health Administration, and Geological Survey of Alabama, p. 37–55.

DeBruin, R.H., and Jones, R.W., 1989, Coalbed methane in Wyoming, *in* Eisert, J., ed., Gas resources of Wyoming: Wyoming Geological Association, 40th Field Conference Guidebook, p. 97–103.

Johnson, R.C., and Nuccio, V.F., 1986, Structural and thermal history of the Piceance basin, western Colorado, in relation to hydrocarbon occurrence in the Mesaverde Group, *in* Spencer, C.W., and Mast, R.F., eds., Geology of tight gas reservoirs: AAPG Studies in Geology, Series 24, p. 165–206.

Johnson, R.C., and Rice, D.D., 1990, Occurrence and geochemistry of natural gases, Piceance basin, northwest Colorado: AAPG Bulletin, v. 74, no. 6, p. 805–829.

Jüntgen, H., and Karweil, Joachim, 1966a, Gasbildung und Gasspeicherung in Steinkohlenflözen, part I—Gasbildung: Erdol und Kohle, Erdgas, Petrochemie, v. 19, p. 251–258.

Jüntgen, H., and Karweil, Joachim, 1966b, Gasbildung und Gasspeicherung in Steinkohlenflözen, part II—Gasspeicherung: Erdol und Kohle, Erdgas, Petrochemie, v. 19, p. 339–344.

Jüntgen, H., and Klein, J., 1975, Entstehung von Erdgas aus Kohligen Sedimenten: Erdol and Kohle, Erdgas, Petrochemie, v. 28, no.2, p. 65–73.

Karweil, J., 1969, Aktuelle Probleme der Geochemie der Kohle, in Schenk P.A., and Havernaar, I., eds., Advances in organic geochemistry, 1968: New York, Pergamon Press, p. 59–84.

Keefer, W.R., 1970, Structural geology of the Wind River basin, Wyoming: USGS Professional Paper 495–D, 35 p.

Kim, A.G., 1978, Experimental studies on the origin and accumulation of coalbed gas: U.S. Bureau of Mines Rept. of Investigations 8317, 18 p.

Kim, A.G., and Douglas, L.J., 1972, Hydrocarbon gases produced in a simulated swamp environment: U.S. Bureau of Mines Rept. of Investigations 7690, 15 p.

Lebedev, V.S., Ovsyannikov, V.M., and Mogilevskiy, G., 1969, Separation of carbon isotopes by microbiological processes in the biochemical zone: Geochemistry International, v. 6, no. 5, p. 971–976.

Oberlin, A, Boulmier, J.L., and Villey, M., 1980, Electron microscopy of kerogen microstructure—Selected criteria for determining the evolutionary path and evolution stage of kerogen, *in* Durand, B., ed., Kerogen: Paris, Éditions Technip, p. 191–242.

Paape, D.W., 1968, Geology of Wind River Basin of Wyoming and its relationship to natural gas accumulation, *in* Beebe, B.W., and Curtis, B.F., eds., Natural gases of North America: AAPG Memoir 9, v. 1, p. 760–779.

Patteisky, K., 1951, The occurrence and emission of methane in Carboniferous rocks: Bergbau Arch., v. 12, no. 2, p. 29–61.

Rice, D.D., 1983, Relation of natural gas composition to thermal maturity and source rock type in San Juan basin, northwestern New Mexico and southwestern Colorado: AAPG Bulletin, v. 67, no. 8, p. 1199–1218.

Rice, D.D., and Claypool, G.E., 1981, Generation, accumulation, and resource potential of biogenic gas: AAPG Bulletin, v. 65, no. 1, p. 5–25.

Rice, D.D., and Flores, R.M., 1990, Coalbed methane potential of Tertiary coal beds and adjacent sandstone deposits, Powder River basin, Wyoming and Montana (abs.): AAPG Bulletin, v. 74, no. 8, p. 1343.

Rice, D.D., Threlkeld, C.N, Vuletich, A.K., and Pawlewicz, M.J., 1988, Identification and significance of coal-bed gas, San Juan basin, northwestern New Mexico and southwestern Colorado, *in* Fassett, J.E., ed., Geology and coal-bed methane resources of the northern San Juan basin, Colorado and New Mexico. RMAG Guidebook, p. 51–59.

Rieke, H.H., and Kirr, J.N., 1984, Geologic overview, coal, and coalbed methane resources of the Wind River basin—Wyoming, *in* Rightmire, C.T., Eddy, G.E., and Kirr, J.N., eds., Coalbed methane resources of the United States: AAPG Studies in Geology, Series 17, p. 295–334.

Schoell, Martin, 1983b, Genetic characterization of natural gases: AAPG Bulletin, v. 67, no. 12, p. 2225–2238.

Ulery, J.P., 1987, Vitrinite reflectance measurements as an indicator of coalbed gas contents: Quarterly Review of Methane From Coal Seams Technology, v. 5, no. 1, p. 38–39.

Ulery, J.P., 1988, Geologic factors influencing the gas content of coalbeds in southwestern Pennsylvania: U.S. Bureau of Mines Rept. of Investigations 9195, 26 p.

Ulery, J.P., and Puglio, D.G., 1984, Application of vitrinite reflectance to interpret gas content, maximum depth of burial, and paleogeothermal gradient of coal beds in Dunkard Basin (abs.): AAPG Bulletin, v. 68, no. 12, p. 1929–1930.

Van Krevelen, D.W., 1981, Coal—Typology, chemistry, physics, constitution, 2d printing: Amsterdam, Elsevier, 514 p.

Venter, J., and Stasson, P., 1953, Drainage and utilization of firedamp: U.S. Bureau of Mines Information Circular 7670, 22 p.

Whiticar, M.J., and Faber, Eckhard, 1986, Methane oxidation in sediment and water column environments—Isotope evidence: Organic Geochemistry, v. 10, p. 759–768.

Whiticar, M.J., Faber, E., and Schoell, M., 1986, Biogenic methane formation in marine and freshwater environments—CO_2 reduction *vs.* acetate fermentation, isotope evidence: Geochimica et Cosmochimica Acta, v. 50, no. 5, p. 693–709.

Windolf, J.W., Jr., Hickling, N.L., and Warlow, R.C., 1982, Coal resource assessment of the Wind River Indian Reservation, Fremont and Hot Springs Counties, Wyoming: USGS Administrative Report to the Bureau of Indian Affairs, 186 p.

Coalbed Methane in Wyoming

Richard W. Jones[1]
Rodney H. DeBruin[1]

1. Geological Survey of Wyoming, Laramie, Wyoming

ABSTRACT

Biogenic or thermogenic coalbed methane may exist in all of Wyoming's coal fields. Biogenic methane is derived from microbial action on low-rank coals that has occurred at relatively shallow depths and at low temperatures prior to thermogenic methane generation by devolatilization. Thermogenic methane is derived from higher temperature coalification and is associated with higher rank coals that have been or are presently more deeply buried—and (or) have been more thermally metamorphosed—than low-rank coals.

Biogenic methane is found at relatively shallow depths both in coalbeds and in sandstone reservoirs associated with coalbeds. Although much of this methane occurs in thermally immature, low-rank coals with low gas yields—usually less than 100 CF/T (3.1 cc/g)—the great number of coals and the exceptional thicknesses of some in Wyoming's coal-bearing areas indicate a large in-place methane resource. The most significant quantity of biogenic methane occurs in shallow Tertiary coalbeds (and associated sandstones) in the Powder River Coal Field, where production of the methane is now occurring.

Thermogenic coalbed methane probably exists in many Wyoming coal fields, where Cretaceous and some Tertiary coalbeds buried in the structurally deep parts of sedimentary basins have become more thermally mature (higher in rank) and have generated large volumes of methane. This methane has become trapped either in the coal or in reservoir rocks associated with the coal. Although these higher rank coals usually are not as thick as low-rank coals, their methane content probably is higher, exceeding 500 CF/T (15.6 cc/g) in western Wyoming. Exploration for thermogenic methane is occurring in many Wyoming coal fields. and production of methane from coal has been reported on the northern flank of Rock Springs uplift and in southeastern Washakie basin.

Our estimate of in-place coalbed methane resources in Wyoming ranges from 7.25 to 145.0 TCF (0.205×10^{12} to 4.1×10^{12} m^3). The wide range of this estimate reflects very limited data on the gas content of coalbeds and assumes that the statewide gas content of all coal is between 5 and 100 CF/T (0.2 and 3.1 cc/g). If only a small part of this resource can ultimately be recovered, the volume of methane in coalbeds is still quite large. This coalbed methane resource has the potential of becoming a significant contribution to Wyoming's energy industry.

Evaluation of Fruitland Coal Properties and Development Economics, San Juan Basin, Colorado and New Mexico

Rahul Dhir[1]
Matthew J. Mavor[1]
Jay C. Close[1]

1. Resource Enterprises Inc., Salt Lake City, Utah

ABSTRACT

This paper documents the reservoir properties, deliverability, and development economics of four Fruitland Formation coal-gas wells in the San Juan basin of Colorado and New Mexico. The impact of critical reservoir properties on development economics is demonstrated, and ranges of reservoir properties required for commercial development are outlined. Guidelines for rapid estimation of economic feasibility of Fruitland coal-gas development projects in the San Juan basin also are established.

Detailed estimates of reservoir properties, including pressure, thickness, ash content, gas content, gas storage capacity, diffusion coefficients, permeability, cleat frequency and orientation, coal rank, and coal composition have been obtained by analysis of core, log, and production test data. Well deliverability has been projected based on these reservoir properties and on matching of well performance with a coalbed reservoir simulator. The economics of development for each location are included.

INTRODUCTION

The purpose of this paper is to present the reservoir properties, deliverability, and development economics of Fruitland Formation coal-gas wells located in the San Juan basin of Colorado and New Mexico as illustrated by case studies of four wells. The San Juan basin has been the focus of significant activity for the development of natural gas resources from the Fruitland coal seams. Extensive formation evaluation research has been conducted to characterize coals of the Fruitland Formation. Documentation of some of these efforts can be found in Mavor and Close (1989a, 1989b, 1989c, 1989d). The impact of critical reservoir properties on development economics is demonstrated. The range of critical reservoir properties required for commercial coal-gas development in the San Juan basin is outlined, and guidelines for rapid, accurate estimation of economic feasibility of coalbed methane development projects in the San Juan Basin also are established.

FORMATION EVALUATION METHODOLOGY

Accurate and reliable formation evaluation data are essential for proper reservoir characterization and for proper assessment of deliverability and project economics. To predict the producibility of coalbed methane reservoirs, estimates of critical reservoir properties are required. For San Juan basin coal reservoirs, the critical properties and sources of their estimates are as follows:

Critical Property	Estimate Source
Thickness	Open-hole density log
Gas content	Onsite desorption tests
Sorptive capacity	Sorption isotherm
Diffusion coefficient	Onsite desorption tests
Cleat permeability	Open-hole drillstem test
Pressure	Open-hole drillstem test

Because most coalbed methane wells are stimulated with hydraulically induced fractures to increase production rates, it is necessary to quantify the effects of induced-fracture properties on production behavior. Quantification requires estimates of the tip-to-tip length and the conductivity (permeability-width product) of the induced fracture.

To evaluate these reservoir properties, extensive formation evaluation efforts on four case-study wells were conducted, when possible, using methods discussed in Mavor and others (1990a). The evaluations included 3.5-in.-diameter (8.9-cm) oriented cores taken through the coal seams of interest. Coal samples from the core and from drill cuttings were placed in canisters for onsite desorption experiments. Open-hole drillstem tests of the coal seams were conducted following coring. Upon reaching total depth of the well, open-hole logging suites were run. Upon completion of the desorption experiments, sorption isotherms were measured in the laboratory. Numerous proximate analyses were performed to determine the rank and composition of the coal. Vitrinite reflectance was measured to confirm coal rank. Detailed descriptions of the cleat systems were performed to assist in quantifying diffusivity and permeability characteristics. Post-stimulation production tests were conducted in two wells.

Estimates of the reservoir properties obtained from analysis of these data, together with well locations, are summarized in Table 1. The reservoir properties listed generally were obtained from direct analysis of coal samples, wireline log data, and well test data. However, in some cases, estimates have been obtained by history matching gas and water production data with a coalbed reservoir simulator. The simulator used for this purpose is the COMETPC–3D simulator (Sawyer and others, 1990). Fracture system porosity and total compressibility estimates were derived in this manner. In some cases estimates of certain properties were not possible for the well in question, and values were assumed based on data collected from other wells. Estimates of fracture length most commonly were derived in this manner, as we had few reliable sets of post-stimulation production-test data. Fracture conductivity was assumed to be infinite for all wells, as all the test data to which we had access has resulted in this conductivity estimate. Assumptions for unknown properties were required for this presentation, as we wish to present guidelines for development economics that may be achieved for each of these wells.

It should be noted that not all the properties listed in Table 1 are required for production-rate estimates. Some nonessential data, such as pure coal gas-content estimates, also have been included for a more complete reservoir description.

Table 1. Fruitland coal reservoir properties and locations of four coalbed methane wells, San Juan basin.

Reservoir Property	Hamilton 3	NEBU 403	S. Ute-Mobil 36-1	Colorado 32-7 No. 9
Gas gravity, °API	0.6	0.6	0.6	0.6
Average in situ gas content, SCF/T (cc/g)	352.6 (11.0)	386.8 (12.1)	420.4 (13.1)	344.3 (10.7)
* Average ash-free gas content, SCF/T (cc/g)	526.7 (16.4)	560.7 (17.5)	601.4 (18.8)	526.0 (16.4)
Average sorption time, hours	9.8	6.4	22.6	10.1
* Ash-free Langmuir volume, SCF/T (cc/g)	685.5 (21.4)	756.4 (23.6)	710.8 (22.2)	671.9 (21.0)
Langmuir volume (in situ), SCF/T (cc/g)	426.0 (13.3)	524.2 (16.4)	494.7 (15.4)	439.4 (13.7)
Langmuir pressure, psia (kPa)	388.5 (2,679)	309.0 (2,131)	239.0 (1,648)	287.6 (1,983)
Completed coal thickness, ft (m)	33.0 (10)	44.4 (13.5)	42.7 (13.0)	27.4 (8.4)
* Total coal thickness, ft (m)	45.0 (13.7)	50.5 (15.4)	66.6 (20.3)	41.3 (12.6)
* Average ash content, %	32.6	30.7	30.4	34.6
* Total gas in place, 320 acres, BSCF (10^6 m^3)	10.4 (294.3)	13.0 (367.9)	16.8 (475.4)	8.3 (234.9)
Completed gas in place, 320 acres, BSCF (10^6 m^3)	7.6 (215.1)	11.4 (322.6)	10.8 (305.6)	5.5 (155.6)
Absolute permeability, md	6.7	5.0	39.3	1.7
Fracture spacing, number/in. (number/cm)	10 (4)	25 (10)	10 (4)	10 (4)
Initial reservoir pressure, psia (kPa)	1,273 (8,777)	1,726 (11,900)	1,366 (9,419)	1,620 (11,170)
Pressure datum, ft (m)	2,842 (867)	2,949 (899)	2,345 (715)	2,830 (863)
Initial hydrostatic gradient, psi/ft (kPa/m)	0.448 (10.1)	0.585 (13.2)	0.583 (13.2)	0.57 (12.9)
Cleat porosity, %	0.8	0.8	0.8	0.8
Cleat water saturation, %	100	100	100	98
Pore-volume compressibility, 10^{-6} psi^{-1} (10^{-6} kPa^{-1})	100 (14.5)	100 (14.5)	100 (14.5)	100 (14.5)
Temperature, °F (°C)	115 (46)	116 (47)	125 (52)	115 (46)
* Vitrinite reflectance, %	0.78	0.88	1.34	1.27
* Coal rank	hvAb	hvAb	mvb	mvb
Induced fracture tip-to-tip length, ft (m)	350 (107)	350 (107)	350 (107)	368 (112)

* Not required for estimation of production rate.

Well Locations

Well Name	Operator	State	County	Sec.	Twp	Rge
Hamilton 3	Mesa Limited Partnership	New Mexico	San Juan	30	32N	10W
Northeast Blanco Unit 403	Blackwood and Nichols Co.	New Mexico	Rio Arriba	9	30N	7W
Southern Ute-Mobil 36–1	McKenzie Methane Corp.	Colorado	La Plata	36	34N	10W
Colorado 32–7 No. 9	Mobil Oil Corp.	Colorado	La Plata	4	32N	7W

WELL DELIVERABILITY

The reservoir and formation properties listed in Table 1 were used to generate forecasts of well deliverability, using COMETPC–3D. For the purpose of this analysis, it was assumed that well spacing was 320 acres (130 ha) and that flowing bottom-hole pressure would be reduced to 100 psig (690 kPa) over a three-month period and would remain at that level for the duration of the project life. Because of negligible present worth of production at later times, a project life of 20 years was considered adequate.

Results

Results of the reservoir simulations are summarized in Table 2. Based on the analysis of these results, we conclude that gas in place alone is not a definitive indicator of cumulative recovery and gas production rates from coal reservoirs. Gas production and recovery also are a strong function of the magnitude of fracture permeability and the desorption characteristics of the coal. The application of more effective completion technology that would enable placement of longer conductive fractures, facilitating better access to the reservoir, should result in enhanced recoveries. variations can be attributed primarily to differences in absolute permeability of the fracture systems with secondary influence from differences in gas content and desorption characteristics of the coal.

Colorado 32–7 No. 9

Average gas production over a 20-year period is predicted to be 59 MSCFD (1,670 m^3/day), with peak production of 68.5 MSCFD (1,938.6 m^3/day) after 21 months. Analysis of gas-content and isotherm data indicate that the coal is at an "undersaturated" condition; that is, the estimated critical desorption pressure (reservoir pressure required to initiate desorption) is lower than the initial reservoir pressure. Significant depressurization via dewatering would be required to achieve sustained gas-production rates. The relatively low fracture-system permeability in this well results in lower depressurization and dewatering rates and thus low cumulative gas recovery and recovery efficiency.

Hamilton 3 and Northeast Blanco Unit 403

Average 20-year gas production rates for the Hamilton 3 and Northeast Blanco Unit 403 wells are estimated to be 256 MSCFD (7,245 m^3/day) and 196 MSCFD (5,547 m^3/day), respectively. Gas production is estimated to peak initially at 590 MSCFD (16,697 m^3/day) from Hamilton 3 and at 274 MSCFD (7,754 m^3/day) from Northeast Blanco Unit 403 after 9 months. Gas recovery from these wells is anticipated to be higher than that from the Colorado 32–7 No. 9 primarily because of higher fracture-system permeability.

Table 2. Results of reservoir simulations for four San Juan basin coalbed methane wells.

Result	Colorado 32–7 No. 9	Colorado 32–7 No. 9 (enhanced)	Northeast Blanco Unit 403	Hamilton 3	Southern Ute-Mobil 36–1
Net recovery, BSCF (10^6 m^3)	0.34 (9.6)	0.53 (15.0)	1.13 (32.0)	1.49 (42.2)	4.15 (117.4)
Total gas in place, BSCF (10^6 m^3)	8.30 (234.9)	8.30 (234.9)	13.00 (367.9)	10.40 (294.3)	16.80 (475.4)
Completed gas in place, BSCF (10^6 m^3)	5.50 (155.6)	6.91 (195.6)	11.40 (322.6)	7.63 (215.9)	10.80 (305.6)
Recovery factor, %	6.13	7.63	9.91	19.53	38.46
Average gas rate, MSCFD (m^3/day)	58.8 (1,664)	91.4 (2,586.6)	196.0 (5,547)	255.8 (7,239)	702.5 (19,881)
Peak gas rate, MSCFD (m^3/day)	68.5 (1,938.6)	115.9 (3,280)	273.8 (7,748)	589.7 (16,688)	3,091.9 (87,500)
Time to peak gas, months	21	16	9	1	1
Average water rate, BPD (m^3/day)	16.3 (2.6)	21.5 (3.4)	33.7 (5.4)	23.8 (3.8)	41.9 (6.7)

Gas-in-place volumes are relatively similar among the northern San Juan basin wells. Initial gas-in-place volumes on 320 acres (130 ha) range from a low of 8.3 BSCF (234.9 × 10^6 m^3) for the Colorado 32–7 No. 9 well to a high of 16.8 BSCF (475.4 × 10^6 m^3) for the Southern Ute 36–1. Typically, the completed interval is less than the gross coal thickness. Gas-in-place volumes based on completed interval thickness on 320 acres (130 ha) range from a low of 5.5 BSCF (155.6 × 10^6 m^3) for the Colorado 32–7 No. 9 well to a high of 11.4 BSCF (322.6 × 10^6 m^3) for the Northeast Blanco Unit 403.

Cumulative recovery and recovery factors are significantly different among the four wells. Recovery factors after 20 years of production range from a low of 6% for the Colorado 32–7 No. 9 well to a high of 38% for the Southern Ute 36–1 well. For the same two wells, cumulative recovery ranges from a low of 0.34 BSCF (9.6 × 10^6 m^3) to a high of 4.15 BSCF (117.4 × 10^6 m^3), respectively. These

The differences in production rate, recovery efficiency, and cumulative recovery between the Hamilton 3 and Northeast Blanco Unit 403 wells can be attributed primarily to variation in desorption characteristics. Analyses of gas-content and isotherm data from these wells indicate that the coal at the Hamilton 3 location is "saturated," whereas coal at the Northeast Blanco Unit 403 location is "undersaturated," even though gas content and gas in place are higher for Northeast Blanco Unit 403. Under these conditions, gas production from Hamilton 3 would be initiated as soon as reservoir pressure falls below the initial reservoir pressure. Depressurization via dewatering would be required to initiate gas production from Northeast Blanco Unit 403, resulting in lower gas recovery.

Desorption characteristics and initial saturation conditions of the coal have a significant influence on gas recovery. Gas storage capacity for a given coal seam is determined as a function of pressure by a sorption isotherm. Based on initial reservoir pressure, gas content, and isotherm characteristics, the pressure required to initiate desorption can be estimated. This pressure is termed the *critical desorption pressure* (Mavor and others, 1990b).

The effect of degree of saturation on development economics underscores the importance of gas content and isotherm data for determining the coal's desorption characteristics and the relevance of these data to the determination of cumulative recovery.

Southern Ute-Mobil 36–1

Average gas production over a 20-year period is predicted to be about 700 MSCFD (19,810 m^3/day), with a peak production of 3,100 MSCFD (87,730 m^3/day). Cumulative gas recovery and recovery efficiency are anticipated to be the highest for this well even though the completed gas in place is less than that for the Northeast Blanco Unit 403. High fracture permeability and the "saturated" condition of the coal account for the high predicted cumulative gas recovery.

Influence of Advanced Completion Technology

The current focus of research in coalbed methane technology is the development of more effective completion techniques. The ramifications of implementing such a technique for hydraulic fracturing was illustrated by simulating gas and water production from the Colorado 32–7 No. 9 well, completed with a 275-ft (83.9-m) (half-length) hydraulic fracture—compared to the original 168 ft (51.2 m)—and sufficient fracture-height growth to communicate with an additional 7 ft (2.1 m) of coal. Average gas-production rates for this case ("enhanced" case in Table 2) were predicted to be 90 MSCFD (2,547 m^3/day), a gain of more than 50% over the original completion technique. Cumulative recovery and recovery efficiency after 20 years of production were forecasted to be 0.64 BCF (18.1×10^6 m^3) and 7.63%, respectively.

DEVELOPMENT ECONOMICS

Development economics for the four wells were analyzed using a model developed for the analysis of coalbed methane projects. A parametric study also was performed to analyze the effects of critical reservoir properties on development economics of Fruitland coal wells in the San Juan basin and to provide guidelines for determining economic feasibility of coalbed methane projects.

Economic Analysis Results

The following investment criteria were used to evaluate project economics—payout term, discounted cash-flow rate of return, and unit development costs on a \$/MSCF basis. Results of the before-tax and after-tax analyses are presented in Table 3.

Discounted cash-flow rate of return and unit development costs serve as effective measures of the economic viability of coalbed gas projects. As is obvious from the analyses, development economics for coal reservoirs within the same basin can vary significantly with variations in geologic and reservoir properties and completion technology. A proper geologic and formation evaluation program and accurate economic analysis are essential for proper determination of the feasibility of a well or project. The IRC Section 29 nonconventional-fuel tax credit, implementation of more effective completion technology, and reduction in well completion costs are all demonstrated to have a significant positive impacts on project economics.

Payout Term

Payout terms for the four wells (Table 3) were obtained by determining the time required for cumulative revenue and tax credits to equal drilling, completion, and operating costs based on a 100% working interest. Revenue in this case is defined as gross income less royalty and production taxes. Because of high initial capital investment, high operating expenses, and water-disposal costs associated

Table 3. Results of economic analysis for four San Juan basin coalbed methane wells.

Result	Colorado 32–7 No. 9	Colorado 32–7 No. 9 (enhanced)	Northeast Blanco Unit 403	Hamilton 3	Southern Ute-Mobil 36–1
Net recovery, BSCF (10^6 m^3)	0.34 (9.6)	0.53 (15.0)	1.13 (32.0)	1.49 (42.2)	4.15 (117.4)
Development costs					
Total, 1,000 \$	658	548	699	678	727
Unit, \$/MSCF (\$/m^3)	1.95 (0.069)	1.04 (0.037)	0.62 (0.022)	0.46 (0.016)	0.18 (0.006)
Before-tax DCFROR, %	–3.41	6.06	16.15	28.34	173.52
Payout term, years	19.0	8.0	5.5	3.9	2.3
After-tax DCFROR, %	3.19	17.05	32.29	57.23	331.60

with dewatering, coal-gas projects typically are characterized by relatively long payout terms. For the wells studied during this analysis, projected payout terms ranged from 2.3 years for the Southern Ute-Mobil 36–1 well to 19 years for the Colorado 32–7 No. 9.

In general, use of the payout term for project economic evaluation is a less important indicator of project worth than the rate of return and unit development costs, discussed below.

Discounted Cash-Flow Rate of Return

The discounted cash-flow rate of return (DCFROR) is defined as the rate of return at which the present worth of cash flow from an investment equals the present worth of the entire investment (Stermole, 1984). DCFROR measures the efficiency with which a project utilizes investment funds. Before-tax DCFROR values measure the project economics on a "cash-on-cash" basis without benefit of the Section 29 tax credit. For the given wells these values ranged from –3.4% for the Colorado 32–7 No. 9 well to 174% for the Southern Ute-Mobil 36–1 (Table 3). Under the modified cost scenario, the "enhanced technology" production case for the Colorado 32–7 No. 9 had a pretax DCFROR of 6.1%, which reflects significant impact of technology improvements. After-tax DCFROR values, which reflect the added benefit of the tax credit, follow a similar trend and were significantly higher than before-tax values—3.2% to 332%, respectively, for the wells cited above.

Unit Development Costs

Development costs were computed on a \$/MSCF basis. Total development costs for each well were calculated by adding the present value of all expenses (operating, administrative, and water disposal) to the present value of the initial net investment. These development costs then were divided by forecasted net recovery to determine the unit development costs, which ranged from a high of \$1.95/MSCF (\$0.069/m^3) for the Colorado 32–7 No. 9 well to \$0.18/MSCF (\$0.006/m^3) for the Southern Ute-Mobil 36–1 (Table 3). Application of more effective completion technology for Colorado 32–7 No. 9 results in a 46% reduction in development costs to \$1.04/MSCF (\$0.037/m^3). Using a 15% pretax DCFROR as a lower limit, it can be concluded that unit development costs at 40% of current wellhead gas prices or less would be desirable to achieve significant economic returns from a given well.

Parametric Study

The direct impact of geologic variables on the ultimate economics of a given project in the San Juan basin was illustrated by conducting a parametric study, using the base-case cost scenario presented in Table 4.

For a given cleat system absolute permeability value (k, md), before-tax and after-tax DCFROR can be represented as a linear function of a composite geologic variable that incorporates the basic reservoir characteristics. This

Table 4. Assumptions for economic analysis.

Costs	Base Case	Modified Cost Case
Cost and Expenses		
Drilling, completion, facilities	\$500,000	\$400,000
Tangible fraction	0.5	0.5
Intangible fraction	0.5	0.5
Lease operating expenses (\$/well-month)	\$1,000	\$1,000
G&A expenses (\$/well month)	\$500 esc @4%/yr	\$500 esc @ 4%/yr
Water disposal	\$0.50/bbl	\$0.25/bbl
Gas Marketing		
Gas price	\$1.50/MSCF esc @5%/yr	\$1.50/MSCF esc @5%/yr
Taxes		
Severance taxes	6%	6%
Ad valorem taxes	5%	5%
Federal tax rate	36%	36%
Depletion rate	15%	15%
Depreciation model	MACRS, 200% depreciation rate	MACRS, 200% depreciation rate
Tax credits	Initial \$0.89/MSCF esc @4.3%/yr until Jan 01	Initial \$0.89/MSCF esc @4.3 %/yr until Jan 01
Division of Interest		
Working interest	100%	100%
Net revenue interest	82.5%	82.5%

composite variable is simply represented as the product of initial reservoir pressure (p_i, psia), initial gas content (g_c, SCF/T), and completed interval thickness (h, ft) and can be written as $p_i g_c h$ (psia-SCF-ft/T).

The relationship between DCFROR and reservoir properties is illustrated as a direct function of the composite variable and fracture-system permeability on a log-log "iso-econ" plot. These specific "iso-econ" lines serve as a useful guideline in assessing the economic feasibility of a given project in the San Juan basin. Similar guidelines can be established for other geologic and cost scenarios.

To estimate the iso-econ lines, it was assumed that the reservoir was "saturated." It should be recognized that project economics also would vary with variations in nongeologic variables, such as division of interest, drilling and completion costs, and operating expenses. To isolate the effects of reservoir and geologic characteristics on project economics, these nongeologic variables were kept constant at the base-case values shown in Table 4.

A total of 32 reservoir simulation projections of fluid-production rates were made for a wide range of reservoir properties, including three values of fracture-system permeability, as shown in Table 5. Gas and water production forecasts for each set of reservoir properties were used in the economic model to analyze project economics. DCFROR values were calculated both on a before-tax and after-tax basis. The results from the reservoir simulation and economic analyses were divided into three groups, corresponding to each permeability value. Analysis of the variation in economic results with the composite variable, $p_i g_c h$ was used to establish a statistically significant correlation between the two variables using a linear-regression model for each permeability value. Results for the 5-md case are illustrated in Figure 1. Based on this analysis, the economic results (before-tax and after-tax DCFROR) can be represented as a linear function of $p_i g_c h$ in the form of the following relationship:

$$\text{DCFROR} = a_0 + a_1(p_i g_c h) \quad (1)$$

where a_1 and a_2 are constants for a given value of permeability. The relationship between these constants and absolute permeability is as follows:

$$a_0 = -a_{01}(a_{00}{}^k) \quad (2)$$

and:

$$\log a_1 = a_{10} + a_{11} \log k \quad (3)$$

These coefficients for the before-tax and after-tax cases are summarized as follows:

Coefficient	Before-Tax Value	After-Tax Value
a_{00}	0.9706	0.9963
a_{01}	0.1187	0.1010
a_{10}	–8.239	–8.086
a_{11}	0.6689	0.7714

The linear relationships between (after-tax) DCFROR and $p_i g_c h$ for various permeability values are illustrated in Figure 2. Note that, as stated earlier, the above relationships apply only to the cases evaluated in this study. Use of the relationships for other cost and reservoir property scenarios may result in significant errors in the DCFROR value.

Table 5. Range of reservoir properties for parametric analysis.

Reservoir Property	Minimum	Maximum
Permeability, md	1	50
Pressure gradient, psi/ft (kPa/m)	0.35 (7.9)	0.585 (13.2)
Initial pressure, psia (kPa)	350 (2,413)	2,340 (16,134)
Depth, ft (m)	1,000 (305)	4,000 (1,220)
Gas content, SCF/T (cc/g)	278 (8.7)	463 (14.4)
Completed thickness, ft (m)	15 (4.6)	66.6 (20.3)

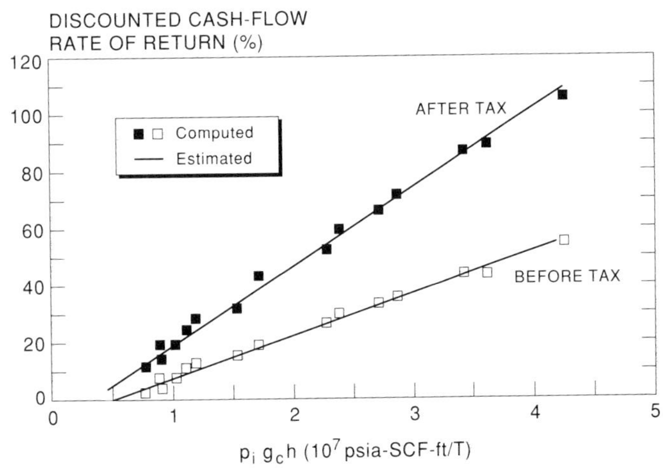

Figure 1. Computed DCFROR vs estimated DCFROR in before-tax and after-tax analyses, assuming 5-md permeability.

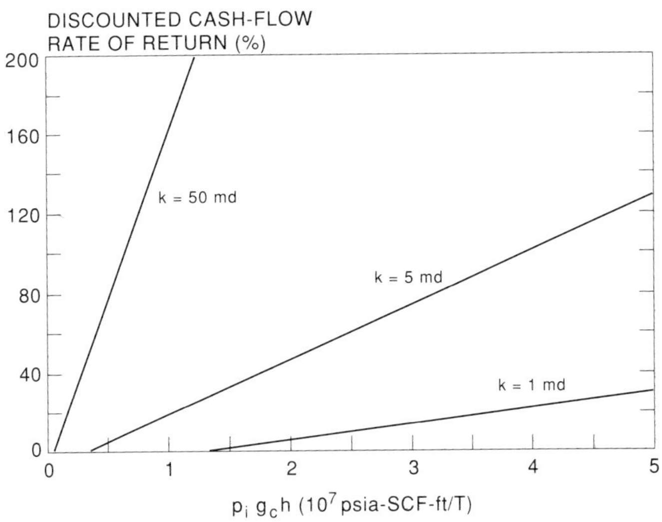

Figure 2. Results of a parametric study of after-tax DCFROR vs critical reservoir parameters for permeabilities of 1, 5, and 50 md.

Using the linear correlation established above, a series of iso-econ lines for various values of DCFROR was generated (Figure 3) that represent DCFROR as a function of $p_i g_c h$ and permeability. For the given cost assumptions, these iso-econ lines represent a framework for the rapid estimation of economic feasibility of Fruitland coal projects in the San Juan basin. The applicability of this framework is demonstrated with data from the Hamilton 3 well plotted on Figure 3. For a $p_i g_c h$ of 1.48×10^7 psia-SCF-ft/T and absolute permeability of 6.7 md, after-tax DCFROR is estimated to be 54%, compared with the after-tax DCFROR of 57.2% calculated from the reservoir simulation and economic analysis.

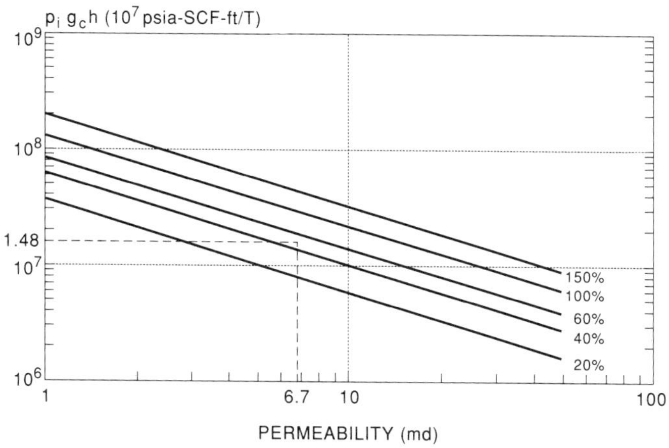

Figure 3. After-tax DCFROR as a function of critical reservoir parameters ($p_i g_c h$ and permeability).

The framework above also is useful for determining the range of minimum reservoir properties for economic development. This application is illustrated assuming an after-tax DCFROR of 20% as a minimum required return on investment. For a permeability value of 1 md, a value for $p_i g_c h$ of 3.7×10^7 psia-SCF-ft/T would be necessary to obtain the minimum required return. Various combinations of reservoir properties could result in the desired value of the composite variable. One possible scenario could be p_i = 1,500 psia, g_c = 435 SCF/T, and h = 57 ft. Similarly, for a permeability of 5 md, a $p_i g_c h$ of 1.04×10^7 psia-SCF-ft/T would be required to obtain a minimum return, and a possible scenario could be p_i = 1,250 psia, g_c = 420 SCF/T, h = 20 ft. For a permeability of 50 md, a value of $p_i g_c h$ of 0.16×10^7 psia-SCF/T-ft would be required to obtain a minimum return, and a possible scenario could be p_i = 1,000 psia, g_c = 400 SCF/T, and h = 4 ft.

Economic Analysis Methodology and Assumptions

The economic analysis and parametric study were performed using results from the reservoir simulation and a specialized economics model that accurately incorporates the effects of the IRC Section 29 nonconventional-fuel tax credit. Gas price forecasts, drilling and completion costs, water-disposal costs, and regular operating and administrative expenses were included in the analysis. Production forecasts from the reservoir simulation results were used as a basis for the revenue stream and water-disposal costs in the model.

A base-case cost scenario, outlined in Table 4, was used to analyze the development economics for all four wells. Under this scenario a "generic" coalbed methane well would cost \$500,000 to drill, complete, and equip with production facilities. Water-disposal costs were incorporated at a unit cost of \$0.50/bbl (\$3.14/m^3), and water-production forecasts from the simulation were used to estimate cumulative water-disposal charges. Monthly lease operating expenses were assumed to be \$1,000/well, and monthly administrative expenses were assumed to be \$500/well initially, both escalating annually at 4%. It should be noted that well costs and operating expenses for specific projects in the San Juan basin may vary (often significantly) from these base costs. However these cost estimates should serve as a useful reference for analyzing the development economics of Fruitland coal wells.

To evaluate the potential benefits of advanced technology, a modified cost scenario was used to analyze the economics of production from the Colorado 32–7 No. 9 well. As outlined in Table 4, this scenario is based on the assumption that the application of advanced drilling and completion technologies, which are unavailable at this time, would result in 20% reduction of total well costs. Possible advances that could significantly reduce well costs include reductions in the costs of tubulars, stimulation, formation evaluation, water lifting, and surface treating facilities. Under this scenario we further assumed that water-disposal costs would be reduced to \$0.25/bbl (\$1.57/m^3).

A conservative gas price forecast was used for this analysis. Average wellhead gas prices for 1990 were assumed to be \$1.50/MSCF (\$0.05/m^3) with a real escalation of 5% in each succeeding year.

It was assumed that gas production from all the wells would be eligible for nonconventional-fuel tax credits established under Section 29 of the Internal Revenue Code of 1986 (Nemirow, 1988). This credit would be allocated on the basis of the net revenue interest. IRC Section 29 creates a tax credit for qualified fuels sold by a taxpayer to an unrelated party during the taxable year, provided that (1) production of the fuel is attributable to the taxpayer; (2) the fuel is produced in the United States; (3) the gas is sold at a lawful price; and (4) the fuel is produced from "wells" drilled after December 31, 1979, and before January 1, 1991. Production from qualified "wells" is subject to the credit until January 1, 2001. The value of this credit is anticipated to be approximately \$0.89/MSCF (\$0.03/m^3) in 1990, and it was assumed that it would escalate at 4.3% per year until its expiration on January 1, 2001. This tax benefit scenario was discussed by Sööt (1988). Implicit in this treatment is the assumption that inflation-adjusted oil prices, on which the credit is based, would not rise to levels high enough to effect either a partial or complete phaseout of the credit. For the purpose of this analysis, the tax credit was treated as an add-back item and included in the after-tax cash flow, which was used to calculate the after-tax discounted cash-flow rate of return.

Under a revision of the tax code in 1990, the last dates for having drilled or placed in service a qualified well and for production to qualify for the tax credit were extended by two years to January 1, 1993, and January 1, 2003, respectively. The results and conclusions of this study are not, however, expected to change substantially with these extended deadline dates; therefore, the economic analyses were conducted based on the original dates specified in the 1986 tax code.

The Modified Accelerated Cost Recovery System (MACRS) was used to calculate the depreciation allowance for tangible project investments, based on a 7-year recovery period and a 200% declining balance (Prentice Hall Information Services, 1990; Russell and Bowhay, 1989). MACRS is a modification of the original ACRS that applies generally to property placed in service after 1986. Under this system no distinction is made between new and used property, and salvage value is disregarded. The depreciation allowance is calculated by using declining-balance depreciation and switching to straight-line depreciation in the year when the straight-line rate exceeds the declining-balance rate.

Depletion allowance enables a mineral interest owner to recover costs through federal tax deductions for depletion over the life of the property (Prentice Hall Information Services, 1990; Russell and Bowhay, 1989). For this analysis the depletion allowance was calculated using both cost and percentage depletion methods, and the method giving the greater tax deduction was used.

SUMMARY

Detailed deliverability and project economics analyses were performed for four Fruitland Formation coalbed gas wells in the San Juan basin of Colorado and New Mexico. These analyses demonstrate that gas recovery from coalbed reservoirs is a strong function of the following reservoir properties—initial gas in place, initial reservoir pressure, magnitude of fracture-system permeability, and desorption characteristics of the coal. Discounted cash-flow rate of return and unit development costs were demonstrated to be accurate measures of the economic viability of Fruitland coal development projects. Project economics, as measured by DCFROR, also were determined to be a function of the above specified reservoir properties. For a given value of fracture-system permeability, DCFROR was shown to be a linear function of a composite geologic variable ($p_i g_c h$). A graphical method was presented for the rapid estimation of economic feasibility of Fruitland coal development projects in the San Juan basin.

CONCLUSIONS

The primary conclusions from the analysis done for this study are as follows:

Gas production and recovery from coalbeds are most affected by the volume of gas initially in place, magnitude of fracture system permeability, and desorption characteristics of the coal.

Development economics for coalbed resource development projects within the same basin vary significantly because of variations in geologic and reservoir properties and completion technology.

Because of high initial capital investment, high operating expenses, and water-disposal costs associated with the dewatering phase, coalbed methane projects typically are characterized by relatively long payout terms.

Unit development costs at 40% of current wellhead gas prices or less are desirable to achieve significant economic returns from a "typical" Fruitland coal well in the San Juan basin.

For a given permeability value, before-tax and after-tax discounted cash-flow rates of return can be represented as linear functions of a composite geologic variable that includes the basic reservoir characteristics of initial reservoir pressure, initial gas content, and completed interval thickness.

Economic returns from a coal-gas development project can be graphed as a series of "iso-econ" lines that establish a basic screening criterion for economic assessment of Fruitland coal-gas development projects in the San Juan basin.

The Section 29 nonconventional-fuel tax credit, implementation of more effective completion technology, and reduction in well completion costs have significant positive impacts on project economics.

The application of more effective completion technology, which would enable placement of longer conductive fractures, facilitating better access to the reservoir, would result in significantly enhanced recoveries of gas from Fruitland Formation coal.

ACKNOWLEDGMENTS

The authors are grateful to the Gas Research Institute for funding our efforts and for permission to publish this work. Special thanks are due to Robert Dern, CPA, Resource Enterprises Inc., for his assistance with the development of the economic analysis model.

REFERENCES

Dougherty, E.L., 1985, Guidelines for proper application of four commonly used investment criteria: SPE Paper 13770 [presented at the SPE 1985 Hydrocarbon Economics and Evaluation Symposium in Dallas, Texas, March 14–15].

Mavor, M.J., and Close, J.C., 1989a, Western Cretaceous Coal Seam Project, evaluation of the cooperative research well Hamilton #3 operated by Mesa Operating Limited Partnership: Gas Research Institute, Topical Rept. GRI–90/0040, 40 p.

Mavor, M.J., and Close, J.C., 1989b, Western Cretaceous Coal Seam Project, evaluation of the cooperative research well Northeast Blanco Unit operated by Blackwood & Nichols Co. Ltd.: Gas Research Institute, Topical Rept. GRI–90/0041, 46 p.

Mavor, M.J., and Close, J.C., 1989c, Western Cretaceous Coal Seam Project, evaluation of the cooperative research well Southern Ute-Mobil 36–1 operated by McKenzie Methane Corporation: Gas Research Institute, Topical Rept. GRI–90/0042, 18 p.

Mavor, M.J., and Close, J.C., 1989d, Western Cretaceous Coal Seam Project, evaluation of the cooperative research well Colorado 32–7 #9 operated by Mobil Oil Corporation: Gas Research Institute, Topical Rept. GRI–90/0043, 21 p.

Mavor, M.J., Close, J.C., and McBane, R.A., 1990, Formation evaluation of exploration coalbed methane wells: CIM/SPE Paper 90–101 [presented at International Technical Meeting of the Petroleum Society of CIM and Society of Petroleum Engineers, Calgary, Alberta, June 10–13].

Mavor, M.J., Owen, L.B., and Pratt, T.J., 1990, Measurement and evaluation of coal sorption isotherm data: SPE Paper 20728 [presented at the 65th Annual Technical Conference and Exhibition, New Orleans, Louisiana, September 23–26].

Nemirow, L., 1988, Structuring coal seam methane transactions to obtain the Section 29 credit: Eastern Mineral Law Foundation, Coalbed Methane Special Institute, Nashville, Tennessee.

Russell, C.W., and Bowhay, R.W., 1989, Income taxation of natural resources: Paramus, N.J., Prentice Hall Inc., 4529 p.

Sawyer, W.K., Paul, G.W., and Schraufnagel, R.A., 1990, Development and application of a 3D coalbed simulator: CIM/SPE Paper 90–119 [presented at International Technical Meeting of the Petroleum Society of CIM and Society of Petroleum Engineers, Calgary, Alberta, June 10–13].

Sööt, P.M., 1988, Non-conventional fuel tax credit, *in* Fassett, J.E., ed., Geology and coal-bed methane resources of the northern San Juan basin, Colorado and New Mexico: Rocky Mountain Association of Geologists Guidebook, p. 253–255.

Stermole, F.J., 1984, Economic evaluation and investment decision methods, 5th ed.: Golden, Colorado, Investment Evaluations Corp., 469 p.

Prentice Hall Information Services, 1990, Federal tax handbook: Paramus, N.J., Prentice Hall Information Services, 738 p.

GOLDEN GATE

Coalbed Gas Development in the Western United States—An Update

Kurt M. Petersen[1]

1. Davis, Graham & Stubbs, Denver, Colorado

ABSTRACT

The recovery of coalbed gas as an energy resource has become increasingly attractive due to recent technological advances and changing market conditions. The western United States in general, with the San Juan, Piceance, Raton, Uinta, Green River, Wind River and Powder River basins in particular, have very large coal reserves and have been and will continue to be the focus of coalbed gas development. This increased interest has brought two legal issues to the fore—first, the issue of ownership of the coalbed gas and the rights to develop that gas; and second, the issue of competing rights when both the coal and the gas are to be developed. These issues are far from being resolved, but from a legal standpoint, the risks are ascertainable and generally manageable.

For the first issue, the ownership of coalbed gas must be analyzed separately for each type of land ownership—fee, federal, Indian and state. Although this has not been conclusively resolved, virtually all coalbed gas development to date has been done pursuant to the terms of oil and gas leases or instruments specifically covering coalbed gas. It appears to be an acceptable business risk at this time for the owner of the oil and gas on fee, federal, Indian, or state lands, or its lessee, to continue to develop coalbed gas in this manner, provided there are no specific contrary grants in the chain of title.

As to the second issue concerning the potential conflict if coal and coalbed gas are to be concurrently developed, the major areas of concern are access to the target coal seam through shallower coal seams and possible damage to the coal seam by the development of the gas. The legal risk here also seems manageable, especially if mineral developers formally address the issue by entering into an agreement covering concurrent development.

The key to eliminating the legal risks associated with a potential coalbed-gas ownership dispute or concurrent development conflict lies with the early identification of possible conflicting claimants to those rights. The magnitude of the legal risk is directly related to the minability of the targeted coal seam—the greater the minability, the greater the chance for conflict. Careful and extensive analysis of land ownership records, with an eye toward any special interests in the coal, should identify the legal risks and the likelihood of disputes. If such disputes seem likely, advance negotiation is the key to minimizing the legal risks associated with coalbed gas development.

Editors' Note: The style of headings and references in this paper differs from the traditional style of the RMAG guidebook. Because the paper was written primarily for lawyers and landmen, we decided to leave the paper in the standard legal-paper format, as was done for a similar paper in the 1988 guidebook for the San Juan basin. Nonetheless, this chapter is most informative and is recommended for reading by all.

CONTENTS

I. INTRODUCTION

This paper is a general discussion of the legal issues and risks associated with coalbed gas[1] development in the western United States, and an update of *Coal-Bed Gas Development in the San Juan Basin—A Primer for Lawyer and Landman*, which was presented at the 1988 symposium, *Geology and Coal-Bed Methane Resources of the Northern San Juan Basin, Colorado and New Mexico*, sponsored by the Southern Ute Indian Tribe in cooperation with the U.S. Geological Survey, U.S. Bureau of Indian Affairs, Rocky Mountain Association of Geologists, and Gas Research Institute, in Durango and Ignacio, Colorado.[2] The issues addressed by this paper have received substantial attention from legal commentators.[3] This article does not reiterate the legal debate among such commentators. Rather, it summarizes and updates the current state of the law regarding coalbed gas development and outlines a course of action by which coalbed gas developers may minimize disputes or costly litigation in this area.

This paper is organized as follows—Part II focuses on legal issues of ownership of and access to coalbed gas on federal lands. Part III considers these issues as they relate to Indian lands, and Part IV examines these issues as they relate to state- owned minerals. Part V discusses such issues as they relate to fee lands; particular attention is paid to the holding of the Pennsylvania Supreme Court in *United States Steel Corp. v. Hoge, et al.*[4] and its impact on coalbed gas development activities on fee lands in the West. Finally, Part VI offers some conclusory thoughts and summarizes the steps necessary to minimize and mitigate the legal risks associated with coalbed gas development.

II. COALBED GAS ACTIVITIES ON INDIAN LANDS

This section focuses on the special problems the coalbed gas developer may encounter when developing coalbed gas on federal lands. Because Congress has never specifically addressed the question of whether the right to extract coalbed gas is part of a Federal coal estate or part of the Federal oil and gas estate, the developer cannot proceed entirely risk free. The ownership issue must be resolved by construing existing Federal statutes which provide for the reservation of mineral resources to the United States and authorize the leasing of those mineral resources.

A. The 1981 Solicitor's Opinion

A thorough analysis of the issues surrounding the ownership of the coalbed gas in federally owned coal is contained in the Opinion rendered on May 12, 1981, by the Solicitor of the Department of the Interior.[5] After examining the relevant federal statutes, the Solicitor concluded that, first, a reservation of coal to the United States does not include coalbed gas and, second, a federal reservation of gas does encompass gas found in deposits of coal and, therefore, such gas is disposable as a gas under Section 17 of the Mineral Leasing Act. Thus, according to the Solicitor, where the oil and gas and coal are federally owned, a federal oil and gas lease covers gas found in coal seams, and development and production of coalbed gas by the federal oil and gas lessee is subject to the regulations governing traditional oil and gas drilling and production operations.[6] Where the oil and gas is owned in fee and the coal is federally owned, the oil and gas owner or lessee owns or has the right to develop the coalbed gas found in the federally owned coal, subject to the government's right to protect its coal resource from damage.

In reaching these conclusions, the Solicitor first considered the laws surrounding federal reservations of coal in patents. Patents issued pursuant to the Act of March 3, 1909,[7] or the Act of June 22, 1910,[8] reserved to the United States "all coal" and the right to prospect for, mine, and remove the "coal deposits" underlying the patented lands. In examining the legislation and its history, the Solicitor found that, in the 1909 and 1910 Acts, Congress intended to, and did, reserve only the coal and not other minerals found in association with coal. The Solicitor then concluded that, except for coal, all minerals, including coalbed gas, passed to the surface owner at the time a patent was issued pursuant to those two Acts.[9]

Next, the Solicitor considered whether federal reservations of oil and gas included coalbed gas. After re-

1. For purposes of uniformity, the term coalbed gas is defined to mean all natural gas, including methane, which is or can be produced from coal formations or seams and all zones in communication therewith and all associated hydrocarbons contained therein.

2. Kemp and Petersen, Coalbed Gas Development in the San Juan Basin: A Primer for Lawyer and Landman; Geology and Coal-Bed Methane Resources of the Northern San Juan Basin, Colorado and New Mexico, Rocky Mountain Association of Geologists (1988) (also reprinted in 25 Pub. Land and Resources L. Dig. 231 (1989)) (hereinafter, Kemp and Petersen at __).

3. In addition to those legal periodicals cited by Kemp and Petersen, the following recent articles have addressed the legal issues associated with coalbed gas development: The Eastern Mineral Law Foundation has presented Special Institutes on Coalbed Methane Development on November 3 and 4, 1988, November 9 and 10, 1989, and November 5 and 6, 1990. The papers presented at these institutes contain detailed discussions on various legal aspects associated with coalbed gas development. See also, Counts, Coal-Bed Methane—Ownership Questions Can Stymie Development, Oil and Gas Journal, January 1, 1990; Brody, Coalbed Methane Development and Related Issues; The Landman, May/June 1991.

4. 468 A.2d 1380 (Pa. 1983) (hereinafter referred to as *Hoge*).

5. Ownership of and Right to Extract Coalbed Gas in Federal Coal Deposits (M-36935) (Memorandum of the Solicitor to the Secretary of the Department of the Interior) (May 12, 1981) 88 I.D. 538 (1981) (hereinafter, generally referred to as 1981 Solicitor's Opinion).

6. 43 C.F.R. parts 3000 and 3100 (1990).

7. 30 U.S.C. § 81 (1990).

8. 30 U.S.C. §§ 83–85 (1990).

9. 88 I.D. at 542.

viewing the statutes and the legislative history, the Solicitor concluded that where coalbed gas occurs in lands on which oil and gas were reserved to the United States in agricultural patents issued under the Act of July 17, 1914,[10] coalbed gas was included in the term "gas" and was reserved to the United States.

After reviewing the Mineral Leasing Act of 1920[11] and its legislative history, the Solicitor found that the Mineral Leasing Act refers only to gas or natural gas, with no qualifying adjectives, thus supporting a nonrestrictive reading of those terms to include coalbed gas.[12] The Solicitor noted that the standard federal oil and gas lease grants to the lessee the right to drill for, extract, and dispose of all oil and gas, except helium, and that current federal oil and gas lease forms do not except coalbed gas from the granting clause. The Solicitor concluded that coalbed gas is leasable under the oil and gas leasing provisions of the Mineral Leasing Act and, thus, drilling for and production of federal coalbed gas are governed by traditional oil and gas operating regulations.[13]

Conversely, the Solicitor found that the coal-leasing provisions of the Mineral Leasing Act did not provide for a coal lessee's extraction of minerals related to or associated with coal.[14] The Solicitor concluded that those statutory provisions do not authorize a coal lessee's extraction of coalbed gas, other than the ventilation of the gas required by mine safety and health laws. However, the Solicitor made it clear that nothing in his Opinion warrants title to any coalbed gas deposit nor detracts from any coal lessee's obligation to comply with the methane ventilation provisions of the federal and State laws relating to the safety and health of the nation's coal miners. The Solicitor also noted that the oil and gas lessee does not have a license to develop the coalbed gas in a manner that would damage the coal or create hazards for subsequent mining operations. In addition, the Solicitor stated that the rights of an oil and gas lessee would be restricted to the rights not previously granted to the coal lessee.[15]

The author is not aware of any recent or impending change in or challenge to the position set forth in the 1981 Solicitor's Opinion. The formal Opinion, however, of the Solicitor is not binding upon the courts and can be revoked by a subsequent Solicitor. Although a court will not be bound by the Solicitor's interpretation of the federal statutes in question, the current interpretation should continue to be persuasive in a judicial setting, since courts tend to defer to the judgment of the agency charged with implementing the laws they are interpreting, and since the current interpretation does not favor the interest of the United States. Until the issue is settled by litigation, reliance on the Solicitor's Opinion involves some risk. This risk, however, seems minimal at this time because states such as New Mexico, Utah, and Wyoming have informally followed the federal position and because the Solicitor's Opinion has gone unchallenged for more than ten years. The longer the 1981 Solicitor's Opinion goes unchallenged, and the more development that occurs pursuant to its ownership model, the less is the risk that it will be overturned. Thus, the 1981 Solicitor's Opinion is still good law within the federal agencies having jurisdiction over mineral leasing activities on federal lands.

B. Legal Risks Associated with Coalbed Gas Development on Federal Lands

Coalbed gas found in federal coal deposits may be explored for and produced under either a fee or federal oil and gas lease, depending on who owns the noncoal minerals. The legal risks associated with developing the gas in the federal coal deposit are fairly clear. The first is the above-referenced risk that the 1981 Solicitor's Opinion will somehow be overturned. This risk seems remote at the present time and appears to be an acceptable business risk. Second, even if no coal lease has been issued, there is a risk that the U.S. Bureau of Land Management (BLM) will nonetheless move to protect its coal resource and inhibit the coalbed gas developer's access to the resource. This risk, too, seems manageable. Third is a legal risk that the federal government has issued a coal lease and that either the BLM or the federal coal lessee will oppose the oil and gas lessee's attempts to gain access to the coal to develop the resource. Here the risk in proceeding with the development of coalbed gas must be assessed case by case, depending on the possibility of development under the coal lease.

C. Structuring Coalbed Gas Development on Federal Lands

It is the author's experience that the BLM is very interested in cooperating with private companies in the development of coalbed gas on federal lands. As industry interest in coalbed gas increases, several companies have formed federal units to assist them in developing their coalbed gas prospects and to protect correlative rights.[16] In addition, a prudent coalbed gas developer should explore other development avenues, including forming working interest units or entering into development contracts. A prudent coalbed gas

10. 30 U.S.C. § 121 (1990).

11. 30 U.S.C. § 181, *et seq.* (1990).

12. 88 I.D. at 544–549.

13. 43 C.F.R. Parts 3000 and 3100 (1990).

14. 88 *Id.* at 547.

15. *Id.* at 549; *see also* Lear, Multiple Mineral Development Conflicts: An Armageddon in Simultaneous Mineral Operations, 28 Rocky Mtn. Min. L. Inst. 79, 147 (1983) (hereinafter, Lear at __).

16. An example of such a unit was the Cactus Valley Unit, BLM Contract No. 14–08–0001–20420, which unitized approximately 30,000 acres for the development of coalbed gas in Pitkin and Garfield Counties, Colorado.

developer should explore all available options before structuring development.

One additional consideration when developing coalbed gas on federal lands should be mentioned. Although it heretofore has not been a major issue, the environmental impacts of coalbed gas development are beginning to receive increased scrutiny. At this time, the focus of concern seems to be the effect of dewatering coalbed gas wells on aquifers and the method selected to dispose of the water. If the coalbed gas project is a "major action affecting the human environment," the requirements of the National Environmental Policy Act of 1969 (NEPA) may be triggered.[17] If NEPA requirements are triggered, the developer would be required to prepare an environmental assessment or an environmental impact statement examining the proposed project. A developer pursuing coalbed gas on federal lands should thoroughly evaluate the proposed project in light of NEPA requirements and work with the BLM to ensure that all such requirements are met.[18]

III. COALBED GAS ACTIVITIES ON INDIAN LANDS

The legal risks associated with coalbed gas development on Indian lands are not as straight forward as those on federal lands. Indian lands have been developed for coalbed gas on the Southern Ute Indian Reservation in southwestern Colorado and on the Jicarilla Apache Reservation in northwestern New Mexico, and will likely be involved in development of coalbed gas reserves in Utah, Wyoming, and Montana. Because the nature of Indian ownership necessarily will vary from reservation to reservation, or allottee to allottee, this paper will address such ownership generally, leaving the specific analysis to each particular title situation.[19]

This section discusses two issues—(1) ownership of minerals on an Indian reservation, and (2) the right to develop coalbed gas under those lands. The ownership of Indian land, and hence the ownership of the minerals beneath the land, differs from other types of ownership within the United States. This fact affects all operational phases of mineral resource development on Indian land. Historically, mineral developers operated on Indian land in much the same manner as on federal land, recognizing that title matters and document execution required more attention.[20] In recent years, as tribes have moved toward control of their own destiny vis-a-vis mineral development, these historical patterns are changing. Moreover, with interest in coalbed gas escalating, the issue of rights to develop coalbed gas under a lease issued pursuant to the Indian Mineral Leasing Act of 1938 recently was addressed by a Solicitor's Opinion issued in 1990.

A. Mineral Ownership on Indian Lands

A coalbed gas developer interested in drilling for coalbed gas on an Indian reservation and in assessing the legal risks associated with such operations must understand the ownership of the minerals on the reservation. Although an Indian tribe or an individual Indian owns the reservation lands, this ownership, whether by the tribe or by an individual, differs from other types of land ownership in the United States in two significant respects. First, no sale or other disposition (e.g., lease) of land by any Indian or Indian tribe is valid unless it complies with certain statutory restrictions and is approved by the federal government.[21] Second, Indians are considered wards of the United States government and, as such, they are to be protected.[22] This trust relationship is most clearly seen in the way title to much of the Indian lands is held—that is, by the United States, in trust for the particular Indian tribe.

As can be gleaned from the history of the formation for any Indian reservation, two major types of Indian mineral ownership exist—(1) ownership by the tribe, and (2) ownership by individual Indian allottees or their successors in interest.

1. Tribal Lands

Land ownership by an Indian tribe is ownership in common. Equitable ownership is held by the tribe for the benefit of all members of the tribe; individual tribal members have no alienable or inheritable interest in the tribal lands. Tribal governments decide how to use the land, and individual tribal members can control this use only to the extent of their participation in tribal government.[23]

Interests in real property have been acquired by Indian tribes in at least four ways—(a) by treaty, (b) by act of Congress, (c) by Executive Order, and (d) by purchase.[24] For most purposes, tribal rights created by these four

---

17. National Environmental Policy Act of 1969, 42 U.S.C. 4321–4370 (1990).

18. Triton Oil & Gas Corporation is conducting an environmental assessment in connection with its proposed 96-well project in the Green River basin in Wyoming. See, "Plains elk herd, groundwater questions hold up Triton Oil & Gas Corp.'s coalbed methane project," Western Oil World, May 1991.

19. For a more thorough discussion and analysis of the ownership under the Southern Ute Reservation, *see* Kemp and Petersen at 263–268.

20. Moore, Mineral Development on Indian Lands—Cooperation and Conflict, 28 Rocky Mtn. Min. L. Inst. 1 (1983) (hereinafter, Moore at __).

21. This restraint upon alienation has been imposed through Congressional adoption of a series of Trade and Intercourse Acts enacted between 1790 and 1834. The sections of these acts prohibiting unapproved land acquisitions from Indians are generally called Nonintercourse Acts by the Supreme Court. Oneida Indian Nation v. Oneida, 414 U.S. 661 (1974). *See generally*, F. Cohen, Handbook of Federal Indian Law 510–522, 618–628 (1982 ed.) (hereinafter, cited as Cohen at __).

22. This trust relationship was premised on the constitutional provisions granting to Congress the right to regulate commerce with the Indians, found in Article I, section 8, clause 3 of the U.S. Constitution, and on the treaty-making powers granted by Article VI.

23. Cohen at 475–476.

24. *Id.* at 472.

mechanisms are "as sacred as the fee-simple title of whites."[25] Because the nature of the property interest granted by each method has been subject to some controversy in the past, the actual property rights conveyed as granted by each method must be analyzed on a case-by-case basis. A general discussion of these methods, however, follows.

a. Treaty and Congressional Action. In general, Indian tribes were initially granted the entire surface and mineral interest in the lands of the reservation by the documents establishing the reservation. The tribe often ceded portions of their land back to the United States, with the tribe retaining both the surface and mineral estate in the lands not ceded.

b. Executive Order. Portions of certain reservations were set aside by the Executive Orders.[26] Courts have sustained the validity of reservations established in this manner and the power of the President to create them, based largely upon the acquiescence of Congress in such practices.[27] In addition, Congress has specifically authorized by statute the leasing of oil and gas on Executive Order reservations in the same manner as on congressionally created reservations.[28] By implication, then, the Executive Order reservations conveyed both the surface and the mineral interest under the subject lands.

c. Purchased Lands. Indian tribes have the capacity to enter into contracts for the acquisition of land, and the validity of such acquisitions has been recognized by both Congress and the courts.[29] The purchase of such land vests the tribe with an enforceable property interest, irrespective of the form in which title to the property is held. But, as with all transfers of property, the transferee can only obtain what the transferor owned, and nothing more.

d. Regrant of Minerals Pursuant to the Indian Reorganization Act of 1934. In the mid 1930s, pursuant to the Indian Reorganization Act,[30] the federal government granted Indian tribes the minerals (coal and/or oil and gas) previously reserved in patents issued to non-Indians that covered lands formerly held by the tribe and subsequently ceded to the federal government. This unique chain of title typically is as follows. First, the tribe cedes to the federal government reservation land not allotted to individual Indians nor retained by the tribe. Second, the federal government issues patents to non-Indian third parties, reserving coal and/or oil and gas. Third, the federal government, pursuant to the Indian Reorganization Act, regrants the tribe the mineral interest reserved by the federal government in the patent. It is this author's opinion that the right to develop coalbed gas in this situation depends on the precise mineral rights reserved by the federal government in the patents. The majority of federal reservations involved coal under patented lands reserved pursuant to the Act of March 3, 1909, or the Act of June 22, 1910, and oil and gas reserved pursuant to the Act of July 17, 1914.[31] If the analysis of the 1981 Solicitor's Opinion is followed, when the government reserved coal, it did not reserve coalbed gas; when the government reserved oil and gas, it did reserve coalbed gas. These reserved mineral rights then were transferred to the tribes.[32] If the Solicitor's Opinion is relied upon, the oil and gas owner of the tract or its lessee, not the tribe, has the right to develop the coalbed gas found in the coal regranted to the tribes pursuant to Indian Reorganization Act. (Hereafter, coal with this particular chain of title will be referred to as Severed Indian Coal.)

e. Summary of Tribal Ownership. In general, an Indian tribe owns the minerals and hence the coalbed gas under the tribal lands acquired by treaty, congressional action, and Executive Order. The tribe also owns whatever property rights it has purchased or was granted, which may or may not include coalbed gas depending on the nature of the conveyance. The tribe or one of its lessees has the right to develop the coalbed gas under such lands.

---

25. Mitchel v. United States, 34 U.S. (9 Pet.) 711, 746 (1835). The exceptions to this rule are that: (1) aboriginal title may be extinguished by the United States without paying compensation as required by the fifth amendment, Tee-Hit-Ton Indians v. United States, 348 U.S. 272 (1955), and (2) title held pursuant to an unratified executive order may be revoked without requiring the payment of compensation under the fifth amendment, Sioux Tribe v. United States, 316 U.S. 331 (1942).

26. For example, the tract of land covered by the Oil and Gas Lease at issue in the 1990 Solicitor's Opinion was established as part of the Jicarilla Apache Indian Reservation by an 1887 Executive Order. 1 C. Kappler, Indian Affairs, Laws and Treaties, 875 (1904).

27. Arizona v. California, 373 U.S. 546, 598 (1963), decree entered, 376 U.S. 340 (1964). In addition to judicial recognition of the validity of creating reservations by Executive Order, Congress has acted in a number of ways to assure that title to many Executive Order reservations stands on the same footing as title to reservations created by statute. One such method has been congressional acknowledgement of this presidential power by enacting legislation that recognized and confirmed tribal title to a number of reservations created by Executive Order. *See generally* Cohen at 495.

28. Act of March 3, 1927, 25 U.S.C. §§ 398a–398e (1990).

29. Cohen at 482.

30. Indian Reorganization Act, 48 Stat. 984, ch. 576 (June 18, 1934) codified at 25 U.S.C. §§ 401–479 (1990).

31. Minerals were reserved pursuant to many laws, some of which are the Acts of March 3, 1909 (30 U.S.C. § 81), June 22, 1910 (30 U.S.C. §§ 83–85), July 17, 1914 (30 U.S.C. §§ 121–123), February 27, 1917 (30 U.S.C. §§ 86–89) and December 29, 1916 (43 U.S.C. §§ 291–301). The precise nature of each reservation must be examined to ascertain the nature of the mineral reservation.

32. Mineral reservations pursuant to other Acts of Congress should be analyzed in the same manner as did the Solicitor. The author has not undertaken an analysis of those other Acts because the majority of the federal reservations involved the three Acts analyzed in the 1981 Solicitor's Opinion.

2. Allotted Lands

Indian lands can also be owned by individual Indians pursuant to federal statute or treaty providing for the distribution of tribal property in severalty or pursuant to the General Allotment Act of 1887.[33] This new form of Indian ownership was designed to assimilate Indian tribal members into the American system by transferring parcels of land to them individually.[34] An allotted parcel of land may be owned by the United States in trust for an individual Indian (trust allotment) or owned by the Indian subject to a restriction on alienation in favor of the United States (restricted fee allotment).[35] Either type of allotment may be owned beneficially by more than one Indian or as cotenant by devise, descent, or conveyance. Because all such allotments came from tribal lands, such allotments conveyed the entire fee interest in the land, including the minerals. Lands can lose their allotted status and hence the governmental trust supervision and restriction on alienation in various ways, including patenting, competency determinations, sales to nontrust status, death and inheritance by non-Indians, mortgage foreclosures, condemnations, leasing, or easements.[36] If the restrictions on alienation are removed, the land and minerals under the lands may be conveyed as if the land were fee land.

B. Coalbed Gas Development on Indian Lands

The following two sections discuss the legal risks associated with development of coalbed gas found on tribal lands and allotted lands.

1. Tribal Lands

The developer's first step in evaluating possible legal risks associated with coalbed gas development on tribal lands is to thoroughly examine the manner in which the tribe acquired its mineral interest. The legal risk in this situation will be affected by the development patterns used by the individual tribe. Mineral deposits on tribal lands are leased under the authority of the Indian Leasing Act of 1938[37] and the Indian Mineral Development Act of 1982.[38] To date, the majority, if not all, coalbed gas has been developed pursuant to the terms of tribal oil and gas leases[39] or minerals agreements[40] specifically covering coalbed gas. Within that development scenario, the first legal risk associated with coalbed gas development on tribal lands concerns these leases or agreements and their execution—the risk that the lease or minerals agreement was not executed in accordance with federal[41] and tribal rules and regulations. This risk is straightforward and can be eliminated by careful examination of the existing lease and leasing procedure or by careful compliance with applicable rules and regulations when taking a new lease or minerals agreement.

The second legal risk involves the typical question faced by coalbed gas developers—whether the oil and gas lessee or the coal lessee has the right to develop the coalbed gas. This issue has not been specifically addressed either by legislation or by the courts as it pertains to tribal lands. The specific legal risk here is that this de facto grant of the right to develop coalbed gas pursuant to the tribal oil and gas lease or minerals agreement will be challenged. In fact, one such challenge already has been made, and another challenge may be likely in the near future. The Jicarilla Apache Tribe made the first such challenge by questioning whether the terms of an older Indian oil and gas lease covered the development of coalbed gas. This challenge became the subject of the 1990 Solicitor's Opinion. The second likely challenge is that the tribes will question a non-Indian third party's rights to develop coalbed gas from the Severed Indian Coal. Each particular challenge is addressed in the following paragraphs.

a. 1990 Solicitor's Opinion. On October 16, 1990, the Solicitor of the Department of the Interior issued an opinion on the "Rights to Coalbed Methane Under an Oil and Gas

33. 24 Stat. 388 (1887) codified as amended at 25 U.S.C. §§ 331–355 (1990). The General Allotment Act is commonly referred to as the Dawes Act; it established the legal framework for most present-day allotment law. Before 1887, allotment of Indian lands required the consent of the Tribe, and provisions to that effect were included in many treaties. After 1887, allotment of Reservation lands was mandatory.

34. As a companion policy to allotment, the Federal government sought the cession of Tribal Lands after each Tribal member had received an allotment. These surplus lands were sold to or homesteaded by non-Indians. By 1920, it was generally recognized that the allotment policy had failed, and in 1934 Congress passed the Indian Reorganization Act (IRA), 48 Stat. 984, ch. 576, codified at 25 U.S.C. § 461 *et seq.*, repudiating the policy of allotment of tribal lands. Application of the IRA depended on tribal consent and many tribes rejected it. Despite these rejections, the Interior Department has not initiated any new allotment programs since 1934, and has routinely extended the trust or restricted period on allotments to non-IRA tribes. The current extension is until January 1, 1989, for all trust and restricted patents that would have expired prior to 1989. 48 Fed. Reg. 34,026 (1983).

35. Cohen at 615–616.

36. Moore at § 26.01 [3].

37. 25 U.S.C. §§ 396a–396g (1990).

38. 25 U.S.C. §§ 2101–2108 (1990).

39. The form commonly used in the past was United States Department of the Interior, Bureau of Indian Affairs, Oil and Gas Mining Lease, Form S–157 (July 1964).

40. Such minerals agreements are executed pursuant to the terms of the Indian Mineral Development Act of 1982, 25 U.S.C. § 2101 *et seq.* (1990).

41. 25 C.F.R. § 211 *et seq.* (1990). 25 C.F.R. §§ 211, 212, and 225 have been the subject of recent Department of the Interior action. New Final Rules, promulgated in August 1987 to be effective on October 23, 1987 (52 Fed. Reg. 31916, August 24, 1987), were withdrawn and reproposed with a 60-day comment period ending December 21, 1987. As of July 1, 1990, the New Final Rules had not been promulgated.

Lease for Lands in the Jicarilla Apache Reservation."[42] The Solicitor examined the terms of a 1952 Oil and Gas Lease between the Jicarilla Apache Tribe and Magnolia Petroleum Company (the "Lease"), which defined gas to include "all other natural gases." The Solicitor first analyzed whether "coalbed methane" was a natural gas found in oil and gas deposits within the meaning of the Lease. After analyzing the *Hoge* and *Rayburn* cases, the physical state of the coalbed methane before it is produced, and whether the term "deposit" was intended to exclude coalbed methane, the Solicitor concluded that the terms of the Lease unambiguously granted the oil and gas lessee the right to produce coalbed methane.[43] Next, the Solicitor examined whether the absence of specific intent to include coalbed methane as a gas under the terms of the Lease, in light of the unambiguous general intent to lease all gas, created a reservation of that gas. After analyzing the *Rayburn* and *Hoge* decisions and cases addressing whether various gases are "gas," "natural gas," or "gas deposits,"[44] the Solicitor concluded that the rights to develop coalbed gas were not reserved under the Lease, and were granted to the lessee. The Solicitor also found that conclusion consistent with terms of the Indian Mineral Leasing Act of 1938, the purpose of which was to provide the tribes with much needed revenue.[45] The Solicitor concluded his analysis by examining the limitation of the rights granted in the Lease. The lessee has the right to extract the coalbed gas but does not have the right to extract coal or cause unauthorized damage to the coal resource without compensating the tribe. Thus, the Solicitor found that the terms of the Lease unambiguously granted the lessee the right to produce coalbed gas.

As with the 1981 Solicitor's Opinion, reliance on the 1990 Solicitor's Opinion involves a certain degree of risk. Give the brief time that has elapsed since the 1990 Solicitor's Opinion was issued, it is too early to tell whether it will have as great an impact on the development of coalbed gas on Indian lands as the 1981 Solicitor's Opinion has had on federal lands. Nevertheless, the mere existence of the 1990 Solicitor's Opinion should be of some assurance to the lessee's holding oil and gas leases issued under the Indian Mineral Development Act of 1938.

b. Coalbed Gas Development From Severed Indian Coal. The coalbed gas developer should be aware of the potential challenge by a tribe of a non-Indian third party's rights to develop coalbed gas found in the Severed Indian Coal. The two specific legal risks associated with this type of development are fairly clear. The first is the theoretical risk that the tribe may dispute the 1981 Solicitor's Opinion analysis and claim the rights to develop the coalbed gas by virtue of their ownership of the coal. A court analyzing this mineral reservation likely will follow the Solicitor's position, based on the deference given to agency determinations and the fact that the Solicitor's interpretation does not favor the United States. This analysis might be different if the tribe had reserved the minerals, based on the general rule of construction liberally construing statutes and conveyances in favor of Indians. However, this rule is not applicable in this situation because the United States, not the Indians, reserved the minerals. Moreover, to take the tack that the coal owner has the rights to develop the coalbed gas would be a complete reversal of prior leasing policy and, accordingly, would cause serious problems.

The second legal risk is that the tribe will move to protect its coal resource from damage or require the oil and gas lessee to compensate the tribe for any such damage. This risk, which is discussed above in the context of coalbed gas development on tribal lands, appears manageable at present.

2. Allotted Lands

Allotted Indian lands (which include the entire interest in the land, surface, and minerals) may be leased for the development of oil and gas[46] and other minerals[47] pursuant to the Indian Leasing Act of 1909[48] or the Indian Mineral Development Act of 1982.[49] The definition of gas in such leases is broad enough to include coalbed gas, but does not specifically so state. As with tribal lands, these rules and regulations do not specifically address the rights of the oil and gas lessee and (or) the coal lessee to develop the coalbed gas. The rules and regulations allow for the issuance of oil and gas leases and leases for other minerals under allotted lands, but the priority or conflict between lessees is not addressed.

The legal risks associated with developing coalbed gas on Indian allotted lands are similar to those on Indian tribal lands—the risk that the specified leasing procedures were not correctly followed; the risk that a coal lessee might challenge the oil and gas lessee's right to develop the

42. Rights to Coalbed Methane Under an Oil and Gas Lease for Lands in the Jicarilla Apache Reservation (M–36970) (Memorandum of the Solicitor to the Secretary of the Department of the Interior) (October 16, 1990) 98 I.D. 59 (1990) (hereinafter generally referred to as the 1990 Solicitor's Opinion).

43. *Id.* at 62.

44. *See, e.g.*, Navajo Tribe of Indians v. United States, 364 F. 2d 320, 327 (Ct. Cl. 1966) (as a matter of law, the term gas in an Indian oil and gas lease includes helium); Northern Natural Gas Co. v. Grounds, 441 F. 2d 704, 714–715 (10th Cir.), *cert. denied* 404 U.S. 951 (1971) (under State law, court looks to "general intent of the parties, grant of gas in the lease includes helium").

45. Cotton Petroleum Corp. v. New Mexico, 109 S. Ct. 1698, 1709 (1989).

46. 25 C.F.R. § 212.4 (1990).

47. 25 C.F.R. § 212.6 (1990).

48. 25 U.S.C. § 396 (1990).

49. 25 U.S.C. § 2101 *et seq.* (1990). Section 2102(b) specifically allows an individual Indian owning a beneficial or restricted interest in a mineral resource to include such resource in tribal minerals agreements.

coalbed gas; and the risk that the allottee, through the BLM, the U.S. Bureau of Indian Affairs (BIA), or tribal action, may move to protect the coal resource, affecting the oil and gas lessee's access to the coal.

If lands have passed out of allotted status, federal rules and regulations are no longer applicable, and the lands may be transferred in the same manner as fee lands. The instruments transferring title to the lands should be carefully examined to determine the actual property rights transferred and, if it can be discerned, the intent of the parties regarding the right to develop coalbed gas.

IV. COALBED GAS ACTIVITIES ON STATE LANDS

We turn now to the situation where the coalbed gas developer has targeted a coalbed gas prospect involving state lands. Very few western states have faced the issue of whether coalbed gas is included in a state coal leasehold estate or, on the other hand, whether the resource is covered by a state oil and gas lease. Most states in the West recognize that coalbed gas is covered by a state oil and gas lease, subject to the state's right to protect its coal resources and, possibly, subject also to the coal lessee's right to ventilate the gas during mining. Colorado has specifically addressed this issue and has taken this approach. Assuming that other states formally follow Colorado's lead, the legal risks in developing coalbed gas from state lands will not involve the ownership question, but rather the question of the state oil and gas lessee's ability to have access to the gas found in the state's coal deposit. This risk will surface in situations where the state has leased, or where there is a potential that it will lease, the coal.

However, even with respect to the access issue, there are good legal guideposts to follow. Commentators on this subject indicate that the western states are recognizing that the swing of western jurisprudence, consistent with the nation's energy needs and public policy regarding development of energy resources, is slowly moving toward the adoption of multiple-use concepts and away from the traditional "first in time, first in right" theory of access to different mineral deposits.[50] However, with respect to state-owned minerals, the current status of this balancing act is varied, depending on the particular state involved.

A. Colorado

Colorado is the only western state that has faced the issue of coalbed gas development under state lands head-on. Colorado sought to avoid disputes when in the early 1980s relatively aggressive oil and gas development occurred on the western slope of Colorado, which development was in the middle of long-time producing coal fields. The problem was addressed by amending the terms of the state coal mining and oil and gas leases. Paragraph 22 of the current coal mining lease issued by the State Board of Land Commissioners[51] allows the coal lessee to vent coalbed gas as a mining safety procedure but specifies that coalbed gas should be developed under the terms of an oil and gas lease. Similarly, the current Colorado oil and gas lease issued by the State Board of Land Commissioners contains a provision specifically covering coalbed gas development.[52]

Although the language in both the coal and oil and gas leases addresses the ownership issue, it does not address the questions of access and/or protection of the coal. Lease provisions do not define the rights and obligations of the parties in sufficient detail to enable an oil and gas lessee to quantify the legal risks associated with developing the coalbed gas where a prior coal lease has been issued. These issues would have to be faced on a case-by-case basis and may, in a contentious situation, finally be resolved by order of the Board of Land Commissioners.

B. Other Western States

No other western state has addressed the issues of ownership and access to coalbed gas as comprehensively as Colorado. After discussions with the agencies administering state lands in, and a review of oil and gas leases forms of New Mexico, Montana, Utah, Oregon, Arizona, California, Nebraska and North Dakota, the general consensus is that the oil and gas lessee has the right to develop the coalbed gas. The provisions of the respective

---

50. *Id.*

51. State of Colorado, State Board of Land Commissioners, Coal Mining Lease, form SLB 237–(8/72) (rev. 3/77, 3/81, 2/85). Paragraph 22 provides as follows:

> COAL MINE GAS—Methane Gas or other volatile gases produced, saved, and/or sold by the coal mining Lessee from minable coal measures shall be the property of that Lessee provided that the gas is removed as a mining safety procedure prior to mining and that a royalty be paid to Lessor per the terms set forth in Lessor's then current oil and gas leases. Gas that is uneconomical to produce may be vented or flared provided that the venting or flaring complies with all Federal, State and Local requirements. Methane gas or other volatile gases produced by the oil and gas Lessees from the minable coal measures and from roofs and floors of minable coal measures prior to mining shall be the property of the oil and gas Lessee under the terms of the oil and gas lease, so long as the terms in this paragraph are consistent within the law.

52. State of Colorado Board of Land Commissioners, Oil and Gas Lease, form 193–5 (Rev. 4/84). The lease specifically provides as follows:

> Coal Mine Gas—Coal Mine Gas is included herein; however, volatile coal mine gas may be produced, saved and/or sold by a coal mining Lessee from minable coal measures and from roofs and floors of minable coal measures and the gas shall be the property of that Lessee, provided that the gas is removed only as a mining safety procedure prior to or during mining and that a royalty be paid to Lessor per the terms set forth in Lessor's then current oil and gas leases. Gas that is uneconomical to produce may be vented or flared, provided that such venting or flaring complies with all Federal and State requirements. Gases produced by the Oil and gas Lessee from the minable coal measures and from roofs and floors of minable coal measures prior to, during or after mining shall be the property of the oil and gas Lessee under the terms of the lease. Oil and gas operations shall not render coal seams unminable.

state oil and gas leases and coal leases do not, however, specifically address the issue of which lessee has the right to develop coalbed gas.

V. COALBED GAS ACTIVITIES ON FEE LANDS

A question common to nearly all jurisdictions concerns ownership of coalbed gas under fee lands is, "Is the substance owned by the coal owner or by the oil and gas owner?"[53] Because of the lack of judicial precedent in the West, the answer is not readily apparent.

Any judicial examination of the issue of ownership of coalbed gas will necessarily turn on the resolution of two issues. First, is coalbed gas a gas, or is it an integral and intrinsic part of the coal? The answer turns on the legal definition of coalbed gas and which party is deemed to own it. Eventually, in every jurisdiction where coalbed gas is found, that definition will have to be enunciated by judicial decree and (or) legislation before coalbed gas development can proceed entirely risk-free. However, such judicial or legislative edicts are relatively unlikely in many states, at least in the near future.[54] Until then, the coalbed gas developer can consider the following legal analysis.

The second issue that may be the focus of any litigation relating to ownership of coalbed gas is the intent of the parties to the particular grants, reservations, or conveyances of the coal rights on the one hand, and the gas rights on the other. The precise issue here is whether the parties to a particular mineral conveyance (be it a coal grant or an oil and gas lease) intended coverage of coalbed gas. Resolution of this issue will be on a case-by-case basis.[55] Again, analysis of existing precedent can serve to illuminate the true level of the risk.

Specifically, decisions of the Pennsylvania Supreme Court and a federal district court in Alabama, both discussed below, present an opportunity to view the way in which courts may approach and decide these questions of defining coalbed gas and construing the intent of the parties to a particular mineral conveyance.

A. The Hoge Case

In a case of first impression, the Pennsylvania Supreme Court rendered the first decision regarding ownership of methane found in coal seams. In *U.S. Steel v. Hoge*, the Pennsylvania Supreme court held that coalbed gas belonged to the owner of the coal vein, which had previously been severed from other mineral strata. Although the decision was based on construction of a mineral severance deed with unique language, the court appears to adopt a broad rule that ownership of gas in general, follows ownership of the stratum in which the gas is found.

The facts of this case are as follows. The plaintiff, United States Steel Corporation ("U.S. Steel"), owned a seam of the Pittsburgh coal under a tract of land pursuant to a 1920 coal severance deed to its predecessor. The defendants, owners of other rights in the tract of land, leased all oil and gas rights to defendant Mary Cunningham in 1976. U.S. Steel operated a coal mine near the defendant's tract. In 1978 Cunningham drilled into the plaintiff's coal seam to extract the entrapped coalbed gas. In response, U.S. Steel sought an injunction to prevent both the surface rights owners and the oil and gas lessees from drilling into and removing gas from the coal seam. In addition, U.S. Steel sought to quiet its title to the ownership of the gas located within the coal seam.

U.S. Steel argued that the 1920 conveyance of the coal seam included title to the gas in the seam because the parties were aware that the gas existed when the transaction was made, and that the right to ventilate the coalbed gas, expressly granted in the deed, necessarily implied such a degree of control as to be tantamount to ownership. Coalbed gas, the plaintiff argued, traditionally has not been viewed as a natural gas. U.S. Steel also contended that use of artificial stimulation methods to remove coalbed gas could make it more hazardous and expensive to later mine the coal.

Defendant Cunningham argued that the right to extract all gas underlying the tract she had leased included the gas contained in coal seams, that coalbed gas is a natural gas, and that its removal, with or without hydrofracturing, does not harm the structural integrity of the coal seam. She also argued that the plaintiff's failure to object to her application for State drilling permits estopped U.S. Steel from subsequently judicially attacking the defendant's drilling operations.

Citing early Pennsylvania precedent, the majority opinion of the Pennsylvania Supreme Court first determined that, under Pennsylvania law, gas necessarily belongs to the owner in fee, so long as it remains part of the property and

---

53. *See* McGinley, Legal Problems Relating to Ownership of Gas Found in Coal Deposits 80 W. Va. L. Rev. 369, 370 (1978) (hereinafter, McGinley at ___). Mr. McGinley also considers the possibility that the coalbed resource may belong to the surface owner.

54. Legislative edicts concerning coalbed gas ownership have been unsuccessful to date. Rather than stimulating development by clarifying the ownership question, such legislation has had the opposite effect by clouding title. Virginia passed and has since repealed the Migratory Gas Statute, Va. Code Ann. § 55–154.1 (1986) because it clouded title to all migratory gases by creating a presumption of ownership in the surface owner. In addition, any transfer of title pursuant to the act may have been considered a taking and thus unconstitutional. One further attempted legislative edict was House Bill 1078 (1991) proposed by Congressman Rayhall. This bill proposed that in states where the ownership of coalbed gas was not settled as a matter of state law, the ownership of that gas should be split equally between the coal owner, the oil and gas owner, and the surface owner. This bill met still opposition in committee and never made it to the House floor.

55. For discussions of the general rules relating to the construction of legal instruments and the principles governing the interpretation of mineral conveyances, see Craig and Meyers, Ownership of Methane Gas in Coal Beds, 24 Rocky Mtn. Min. L. Inst. 767, 776–91 (1978) (hereinafter Craig and Meyers at _); McGinley at 387–392; Olson, Coal-Bed Gas: Legal Considerations Affecting Its Development as an Energy Resource, 13 Tulsa L. J. 377, 392–397 (1978) (hereinafter, Olson at ___).

does not migrate from the property.[56] Since the gas was part of the fee, the court reasoned, a conveyance of that part of the fee in which the gas was situated necessarily included the gas as well. Therefore, the gas present in the coal must necessarily belong to the coal owner.[57] The court also recognized that the owner of the coal could allow others certain rights respecting the gas.

Second, the court analyzed the provisions of the severance deed to determine the effect of the reservation, by the grantor, of the right to drill through the granted coal stratum in order to produce oil and gas. The court held that, although the unrestricted term "gas" was used in the reservation clause of the severance deed, when viewed in light of the conditions existing at the time of its execution, it was inconceivable that the parties intended a reservation of ownership of the right to extract a "valueless waste product with the attendant potential for damages resulting from its dangerous nature."[58]

It followed then, that the grantor of the severance deed reserved the right to drill through the severed coal seam for oil and gas implied that the parties intended the gas to which the grantor retained the fee simple title to be "that which was generally known to be commercially exploitable."[59]

In sum, the Pennsylvania Supreme Court's majority decision in *Hoge* appears to be founded on the broad rule that ownership of gas, in general, follows ownership of the stratum in which the gas is found. Although nothing in the opinion states that a grantor of a coal severance deed could not maneuver around that rule by granting a fee interest in the coal seam and expressly reserving the gas found therein, the court found that such was not the specific intent of the parties nor the effect of the language of the severance deed in question.

Two justices of the Pennsylvania Supreme Court dissented from the majority. Significantly, however, the dissenting justices also embraced the general rule adopted by the majority that "subterranean gas is owned by whoever has title to the property in which the gas is resting."[60] The dissent disagreed only with the majority's interpretation of the parties' intent with respect to the reservation of the right to drill through the coal to produce gas, believing that the effect of this language was to reserve the gas present in the coal seam.[61]

56. 468 A.2d at 1383, *citing* Hamilton v. Foster, 272 Pa. 95, 102, 116 A.50, 52–53 (1922), and Westmoreland & Cambria Natural Gas Co. v. DeWitt, 130 Pa. 235, 249, 18 A. 724, 725 (1889).

57. 468 A. 2d at 1383.

58. *Id.* at 1384–1385.

59. *Id.* at 1385.

60. *Id.* at 1388.

61. *Id.* at 1389.

1. Impact of the *Hoge* Decision on Western Coalbed Gas Operations

The *Hoge* decision, because of its timing and notoriety, has had an impact on coalbed gas operations in the West. From a purely legal perspective, however, the *Hoge* case will have a diminished impact in the West for several reasons. First, the decision, although it represents statewide judicial authority in Pennsylvania, does not represent binding judicial precedent for other states. Its legal significance in the West will depend entirely on its persuasiveness.

Second, most fee oil and gas leases in the West cover lands in which ownership of the coal has not been severed from the oil and gas. In such cases, unless a coal lease exists that predates the coalbed gas developer's oil and gas lease and that arguably includes the right to extract coalbed gas, the only risk is that the oil and gas lessor may assert that he did not intend to include coalbed gas in the oil and gas lease. This risk predated the *Hoge* decision; thus, that decision should have little, if any, relevance to the resolution of this issue, the question being one of intent of the parties to the lease as reflected by the language of the lease.

Third, in those situations where the coal has been severed prior to granting the coalbed gas developer's oil and gas lease, the *Hoge* decision may well be rejected by western courts because Pennsylvania apparently has adopted the rule that ownership of gas follows ownership of the stratum in which it is found.[62] The majority of the coal- and oil and gas-producing jurisdictions in the West have not followed the theory of ownership of strata. Rather, such jurisdictions have adopted either the (a) theory of ownership in place, which holds that oil and gas are owned by the surface owner and are capable of being sold or leased in place; or (b) the nonownership theory, which holds that the oil and gas are owned when captured at the surface.[63] Accordingly, western courts would continue to treat the severance of coal or oil or gas or any other specifically named mineral resource as creating a separate corporeal estate in fee simple in that resource. Under this rule, a showing that coalbed gas is a gas and that, as such, it is mineralogically different from coal should result in a holding that the owner of the coal does not own the gas trapped therein purely by virtue of his ownership of the coal.

Fourth, *Hoge* turns on the majority's interpretation of the intent of the parties to a severance deed involving the carving out of one specifically defined seam of coal. That situation is unlikely to be found in a severance deed in the West; rather, the usual severance deed in the West severs the coal by grant or reservation from the surface and (or) other minerals.

Fifth, it is doubtful that coalbed gas litigation in the Rocky Mountain region would involve the complication of

62. *Id.* at 1383, *citing* Kier v. Peterson, 41 Pa. 357 (1862) (owner of subterranean salt entitled to oil commingled with it).

63. 1 Williams & Meyers, Oil and Gas Law § 203.3 (1983 ed.).

the coal owner being granted the express right to ventilate the gas found in the coal seams. This seems particularly true in light of the relative depth and unminability of many deeper western coalbeds that are now the target of coalbed gas developers.

Sixth, western courts might well reject the application of the specific intent test employed by the court in *Hoge* and use a general intent test instead. Some commentators have criticized the use of the specific intent test to determine how parties to a conveyance intended to dispose of substances unknown (or substances with little commercial value) at the time of the conveyance.[64] Under the application of a test of general intent, construction of a severance deed would focus on the proper definition of gas and coal, that is, on whether the parties to the instrument used language that naturally and inevitably encompassed the mineral in question, and not on what the parties to the instrument thought they were doing. Application of a general intent test would likely impel a western court to find that a gas grant (or reservation) encompasses coalbed gas, subject perhaps to the gas grantee's duty to protect the structural integrity of the coal.

Several western courts have applied the general intent test to find that a lease of gas includes all kinds of natural gases, whether hydrocarbon or nonhydrocarbon in nature, and whether or not commercially exploitable at the time of lease execution.[65] Under this reasoning, a holding that the coalbed gas developer's oil and gas lease covers coalbed gas would not be defeated by a showing that the lessor did not know about either the developer's intent to extract the gas found in coalbeds or the commercial exploitability of coalbed gas.

The prediction that western courts would not follow the *Hoge* precedent should include a word of caution. The *Hoge* decision impacts only the issue of ownership of gas found in coal deposits.[66] Other issues, such as access to the coal, still may come into play even if it is certain that the lessor of an oil and gas lease owns the gas in the coal seam. This means that, even if the legal risks relative to ownership of the coalbed gas are acceptable to the developer, the developer should remember that, under certain circumstances, he still may have to face considerable legal risks relative to the extraction of the gas.[67] Mitigation of these risks is detailed below in Part VI.

Finally, a word should be said about the possible impact of the *Hoge* decision on the federal situation, discussed in Part II. The author believes there is no significant risk that the *Hoge* decision will cause the federal government to reconsider its view that its patent reservations of coal did not include coalbed gas because the principles of state law that governed the Pennsylvania Supreme Court's decision are not generally applicable to federal statutory construction issues governing reservations in federal patents.

B. The Rayburn Case

Although *Hoge* remains the leading case on the issue of coalbed gas ownership, this issue has been before a federal district court in Alabama and the Eleventh Circuit Court of Appeals. The dispute in *Rayburn et al. v. USX Corporation*[68] involved a deed from the plaintiffs' predecessor in interest to USX Corporation (USX) conveying the minerals in approximately 2,874 acres (1,163 ha) of land in Tuscaloosa County, Alabama, reserving the oil and gas. The plaintiffs claimed that coalbed gas fell within the definition of gas excepted from the deed to USX. Conversely, USX claimed that coalbed gas is inherent to the coalbed and, as such, it too passed with the mineral rights. Despite urgings by the parties to decide the broader issue of coalbed gas ownership as a matter of law, the court specifically confined its decision to the language of the deed in question.

To interpret the deed, the court first attempted to determine the intent of the parties from the circumstances surrounding the conveyance. The court examined all information available to the parties on the date the deed was executed (1960) and found the record bare of any indicia of common knowledge of the extraction of coalbed gas for commercial purposes, or for future possibilities. The court then examined the language of the deed in question, which clearly stated that the grantors reserving the oil and gas rights were "subject to the requirement that all coal seams located in said land penetrated in such exploration or drilling operations shall be encased or grouted off. ..."[69] The court found this language determinative; it clearly expressed the intention that coalbed gas not be available to any oil and gas well drilled by the grantors or their lessees. Relying on this language, together with the fact that coalbed gas was not considered to be commercially recoverable at the time of the reservation, the court concluded that it

64. *See* Comment, New Values Under Old Oil and Gas Leases: Helium, Who Owns It? 61 Mich. L. Rev. 1168 (1964); *see also* Craig and Meyers, *supra* note 55.

65. *See, e.g.*, Northern Natural Gas Co. v. Grounds, 441 F. 2d 704 (10th Cir. 1971), *affirming* 292 F. Supp. 619 (D. Kan. 1968); Navajo Tribe of Indians v. U.S., 364 F. 2d 320, 176 Ct. Cl. 502 (1966). It is also the policy of the State of Colorado and the BLM to permit helium to be developed pursuant to the terms of an oil and gas lease. In addition, the Regional Solicitor in Denver, Colorado, has concluded that carbon dioxide is included in a reservation of oil and gas in a land patent under the Act of July 17, 1914 (30 U.S.C. § 121). Reservation of Carbon Dioxide Gas in Land Patent, Memorandum to the Colorado State Director, BLM, from Regional Solicitor, Denver, Colorado (July 12, 1979).

66. The lower court decisions that were reversed, because they awarded the coalbed gas to the gas owner, concerned themselves with not only the ownership question but also access issues and fashioned a holding that protected the integrity of the coal. When the Pennsylvania Supreme Court awarded the gas to the coal owner, the access issues became irrelevant.

67. For a discussion of cases involving these issues, *see* Lear at 198–209.

68. Rayburn, et al. v. USX Corporation, Civ. Act. No. 85–G–2661–W (N.D. Ala. 1987) aff'd 844 F. 2d 796 (11th Cir. 1988).

69. *Id.* at 2.

was the intention of the parties that coalbed gas not be included in the reservation. This case was affirmed by the Eleventh Circuit Court of Appeals without opinion.[70]

Unlike the *Hoge* decision, the *Rayburn* court confined its decision to the language of the instruments in question and specifically declined to answer the broader question of whether coalbed gas is or is not included in the term gas, and is or is not severed with the mineral coal. This type of decision, based solely on a deed executed before the commercial potential of coalbed gas generally was known, sheds little light on the issue of coalbed gas ownership in general but may be indicative of how courts handle such disputes.

C. The Pinnacle Petroleum Case

One other court, the Circuit Court of Mobile County, Alabama, currently is considering the issue of coalbed ownership in *Pinnacle Petroleum Company vs. Jim Walter Resources, Inc.*[71] This case involves the ownership of coalbed gas, specifically gob gas, produced from wells drilled into areas that had been mined by the longwall method and that subsequently collapsed. The issue in this case is whether the gob gas well is a post-mining method of extraction that communicates only with the area that has been mined, and thus, the ownership of the gas in that area no longer belongs to the coal owner, but instead has reverted to the original grantor or his lessee. The court in a recent summary judgment motion, held that the the owner of the coal was deemed to own the coalbed gas under the particular title documents in question. This summary judgment decision is, however, of limited precedential importance in the West because it is limited to the terms of specific conveyancing documents, it is specific to gob gas wells, which are rare in the West, and it is from a lower state court. The coalbed gas developer, however, should keep apprised of the status of this case as it moves through the Alabama judicial system.

VI. CONCLUSION—MITIGATING THE RISKS

Assuming that coalbed gas development in the West continues to attract government and industry interest as an energy resource, the complex legal issues surrounding ownership of and access to coalbed gas will become of increasing concern. Although those issues are far from being resolved, the risks from a legal standpoint are ascertainable and, in most cases, manageable.

In the area of fee coal and oil and gas, the ownership question has been decided judicially only by the Pennsylvania courts and, arguably, by the Alabama courts, in favor of the owner of the coal seam. In the West the issue has not been litigated. Even if a dispute as to ownership were to be settled judicially in favor of the owner of the gas rights, many issues undoubtedly would remain relating to access to and methods of extracting the coalbed gas, and the corresponding rights of the coal owner.

As to federal minerals, the Solicitor of the Department of the Interior has determined that coalbed gas is leasable under the oil and gas provisions of the Mineral Leasing Act, and this interpretation has gone unchallenged for more than a decade. Moreover, for access to the coalbed gas, the BLM requires the coalbed gas developer to comply with fairly significant procedures before the coalbed gas can be removed.

As to Indian-owned minerals, current BIA and BLM policy and the 1990 Solicitor's Opinion give the allotted or tribal oil and gas lessee the right to develop the coalbed gas. This policy has not been litigated to date.

As to State-owned minerals, Colorado has declared that its oil and gas leases cover coalbed gas, but the coal lessee has the right to the gas during mining. Furthermore, the oil and gas lessee may not render the coal unminable. Although not enunciated as a formal policy, coalbed gas development of state-owned minerals in other western states is conducted pursuant to an oil and gas lease rather than a coal lease.

Thus, with respect to federal, Indian, fee, and state domains of coal and oil and gas in the West, while ownership of coalbed gas may be and now is (at least in practice) awarded to the holder of the oil and gas rights, a significant amount of law remains to be settled. Assuming continued development of the resource, the developer should keep abreast of new developments by paying close attention to the courts, state legislatures and conservation agencies, tribal councils, Department of the Interior, and Congress.

Taking all this into account, the coalbed gas developer should proceed by identifying and mitigating the special legal and business risks associated with exploration and production of coalbed gas by anticipating disputes and resolving them in advance of exploration and production. This is discussed in more detail below.

A. Tract Identification—The Key to Determining the Risks

The key to anticipating disputes regarding coal and coalbed gas rights lies with the early identification of possible conflicting claimants to those rights. To identify the various owners of and conflicts involving the target prospect, the coalbed gas developer's land personnel should go beyond the usual search of county real estate records and (or) federal, Indian, and state oil and gas leasing records. Additional steps should be taken to determine not only whom to lease from, but also who may, under certain circumstances, have conflicting claims to the coal deposit. In areas where the federal and (or) state governments or Indian tribes own the coal deposits, a detailed surface inspection to identify the presence of any coal mining activity, as well as a search of the respective coal-leasing records, should be made.

70. 844 F. 2d 796 (11th Cir. 1988).

71. Pinnacle Petroleum Company v. Jim Walter Resources, Inc., Circuit Court of Mobile County, Alabama, Case No. CV87–3012.

If no special coal interests can be identified—such as fee simple conveyances to known coal operators, conveyances, or reservations of coal rights that severed the coal and oil and gas rights, coal leases, or surface disturbances indicating coal mining activity—then no further title search is necessary. On the other hand, if some evidence of a special interest in the coal is found, the landman's work should be broadened to determine the intention of the coal owner or lessee. Such work might include, depending on the ownership of the subject lands, examination of state mining and reclamation agency records, Indian tribal records, county mining records, county and U.S. Forest Service water-use records, or county zoning records, as appropriate. In addition, the coal owner or lessee can be approached directly about its intentions to develop the coal. Where the coal is determined to be minable, as a basic minimum it must be determined whether (1) coal mining operations are under way or planned; (2) sufficient time remains for coalbed gas recovery before the reservoir is to be mined; and (3) the coal operator currently has an economic coalbed gas-recovery operation planned or in effect.

B. Low-risk Situations—Nonsegregated Ownership

The simplest and lowest-risk route from tract identification through recovery of coalbed gas occurs where the coalbed gas developer leases a tract (federal, Indian, state, or fee) having all the oil, gas, and coal rights vested in the lessor(s), and none of the mineral owners are associated with development of coal.

This situation may be the result of either an unminable coal seam or a mineral owner uninterested in coal development. Even in this low-risk situation, however, one cannot unequivocally conclude that the granting clause of a typical oil and gas lease or minerals agreement includes coverage of coalbed gas. Although it may be an acceptable business risk to proceed with coalbed gas development under a standard-form oil and gas lease, inclusion of certain language in the oil and gas lease may reduce the risks to near zero, depending on certain factors. If it is determined that the mineral owner is uninterested in coal development, or takes a position consistent with that of the coalbed gas developer and supports production of coalbed gas under the lease, any such risk may be almost totally eliminated by the inclusion of lease language to that effect. Such language should specifically grant the right to explore for and produce coalbed gas and specifically subordinate any subsequent grant of coal rights to the rights of the oil and gas lessee. In most instances, inclusion of these two special provisions should not materially affect or alter the progress or expense of lease negotiations.

Coal seams at depths below 3,000 ft (915 m) or that dip steeply may not be economical to develop. In these situations the coal owner or lessee may even encourage coalbed gas development. Certain problems may, however, occur in the situation where a coalbed gas developer targets a deep, unminable coal seam by drilling through shallower, minable coal zones to reach the objective horizon. The access question presents the legal risks here, with the ultimate risk being that the non-oil and gas minerals owner will not execute a lease or will exact too high a price for limited access. Risks can be mitigated in this situation if the coalbed gas developer finds a mineral owner willing to negotiate a reasonable oil and gas lease. The oil and gas lease in this situation should include language addressing (1) timing of and operations the coal rights owner is permitted to undertake; (2) what coal operations are prohibited during the life of the oil and gas lease; (3) exact location of the wells to optimize coalbed gas production and conform with the design of the coal mine; and (4) extent of and limits on liability for coal seam damage.[72]

The notable exception to the above generalizations occurs where a coal operator or railroad owns all the coal rights under the target prospect. In this case the prospective coalbed gas developer may confront risks associated with the unsettled legal questions and issues surrounding access to and extraction of coalbed gas. In this situation the coalbed gas developer faces an uphill struggle when attempting to lease and drill for coalbed gas. Negotiations with the mineral owner may be very difficult; as a general rule, most coal companies or coal developers (1) view coalbed gas development as a needless and possibly costly headache; (2) will go to great lengths to avoid having wellbores drilled into or through their target coal seams; and (3) feel (probably with some justification) that standard ventilation technology is sufficient to alleviate all but the most severe methane problems. The benefits of possible royalty income and degasification of the coal seam before mining offer the coal interest owner little incentive to cooperate.

C. Higher Risk Situations—Segregated Ownership

The coalbed gas developer enters into an arena of higher risk where the tract-identification process reveals that the ownership of or rights to develop the coal have been severed from the ownership of or rights to develop the oil and gas. If certain factors are present, the risk may be minimal. Such factors might include a target coal seam that is unminable by current technology, or the absence of any intention on the part of the coal owner to develop the coal. The tract-identification process setting forth the exact nature of the conveyance or reservation, as well as the nature of the coal resource, should indicated the likelihood of any dispute.

D. Highest-Risk Situation—The Presence of Coal Development

In the event that the tract-identification process indicates that coal operations are pending or under way or that the coal or coal rights are owned by a person or entity intending

72. For a discussion of possible lease provisions see Lowe, Coalbed Methane Leasing—Covering all the Bases, E. Min. L. Fdn. Special Inst., Paper 7 (1990).

or likely to mine the coal, the coalbed gas developer faces a potentially difficult situation. Certain measures, however, can be taken to mitigate the likelihood of a dispute that would interrupt the planned coalbed gas operations.

Where a dispute is likely because of the nature and scope of existing or anticipated coal mining operations, the coalbed gas developer should attempt negotiations with the coal owner to provide for the orderly and equitable development of both resources. Such negotiations should address, at a minimum, the timing and priority of respective operations, permitted activities and locations, and liability and damages. Bearing this in mind, once a lease has been obtained from the oil and gas owner, the coalbed gas developer should attempt to secure a subordination agreement from any prior segregated coal interest owner. Ideally, such a subordination should provide for a disclaimer of ownership of the coalbed gas, prohibition of mining operations during the life of the lease, and payment of any damages associated with coal left in place. It should also recite a specific grant of the right to stimulate any wells drilled on the leased lands.

It is entirely possible, if not probable, that a coal interest owner may elect not to sign a subordination agreement (perhaps arguing that the *Hoge* precedent should be followed in the particular jurisdiction involved). That risk is inherent in targeting a coalbed gas prospect containing segregated ownership of coal and oil and gas where coal development is possible. Unless the nonsubordinating parties own horizons that may be isolated from the wellbore without destroying the economic viability of the prospect, the coalbed gas developer may face such a risk of litigation that it will not make sense to proceed without careful consideration of the situation, including the advice of counsel.

E. Access to and Protection of Coalbeds

It is clear from the preceding narrative that certain legal risks may be anticipated, and that certain procedural hurdles must be cleared by a coalbed gas developer in the early stages of planning a coalbed gas prospect. These legal risks and procedural hurdles relate to the concept of multiple-use or multiple-mineral development and pertain to the ways in which the two resources, coal and coalbed gas, should be developed. Determining who owns and has the right to develop coalbed gas is not the end of the story—questions of access will move to the fore. How to develop both the coal and coalbed gas has been dealt with by the federal government where the coal is federally owned. The development situation where an Indian tribe, a state, or an individual owns the coal is less well defined.

Multiple-mineral development issues, where gas or oil is one mineral and coal is the other, typically relate to the issue of access through the coal to reach the gas or oil. The wellbore is the bane of the coal operator and usually the focus of his opposition. These issues typically have been resolved in favor of such access, provided, however, that the oil and gas developer compensates the coal owner for damage to his resource. In the area of coalbed gas, however, more than access through the coal is involved. We are talking about access into the coal and the extraction techniques employed, which may or may not render the coal unminable or more expensive to mine.

As is discussed below in greater detail, the federal government seems to afford great protection to its coal resource. The Indian and state protections are not as clearly delineated.

1. Federal Coal

Where coal is federally owned and a federal coal lease is outstanding, the coalbed gas developer can expect significant problems in gaining unfettered access to the resource.[73] These problems will be greater if the coal lease was issued prior to the federal oil and gas lease under which the coalbed gas developer desires to drill. However, even where no federal coal lease is in effect covering the target coal seam, our experience indicates that the BLM may take specific actions to protect the federal coal resource from possible damage resulting from drilling and completion operations, notwithstanding that the target coal seam is below the currently accepted minable depth of 3,000 ft (915 m).

Such protective actions normally are accomplished through restrictive stipulations attached to an Application for Permit to Drill as a condition of approval. Those stipulations can take the form of requiring submission of well logs to the BLM District Mining Supervisor, approval prior to hydrofracture stimulation of the well, and casing and cementing the wellbore through any coal seams encountered. In any event, case-through stipulations and restrictions against hydraulic fracturing will, in all likelihood, have to be removed before the developer commences drilling operations. If this cannot be accomplished by negotiation and (or) submission of fracture-treatment design information, an administrative appeal may be necessary to protect the coalbed gas developer's rights as lessee.

Where the coal is federally owned but the oil and gas is owned in fee, and the coalbed gas developer contemplates drilling a well into a federal coal seam as lessee under a fee oil and gas lease, a search of BLM records should be made to determine whether a federal coal lease has been issued. The author has cautioned clients (in Drilling Title Opinions, for example) to notify the BLM of their drilling plans even if no coal lease has been issued. In an abundance of caution, it also has been suggested that they obtain BLM approval of extraction techniques that arguably might damage the coal, particularly where the coal seam is above 3,000 ft (although the author, in practice, has not been involved with any western coalbed gas projects where the target seam is so shallow). A more conservative approach would

73. *See* Lear at 89–95.

be to attempt to obtain a subordination agreement from the BLM. Such an agreement should, ideally, provide for the BLM's disclaimer of any ownership interest in the coalbed gas. The agreement might also preclude the commencement of mining operations during the life of the oil and gas lease, provide for payment of any damages associated with well stimulation, and recite a specific grant of the right to stimulate any wells drilled on the leased lands. Significant delays might arise if such an avenue is taken because no codified procedures exist to guide the BLM in negotiating such an agreement.

Where the developer's search of BLM coal-leasing records reveals that a coal lease has been issued, prior to drilling under a fee oil and gas lease, the BLM likely will urge that any and all access and extraction issues be resolved by negotiation between the oil and gas lessee and the federal coal lessee.

2. Indian Coal

Where the coal is owned by an Indian tribe, the problems a coalbed gas developer may face are not as clearly defined. There are no general regulatory provisions either authorizing or prohibiting multiple-mineral leasing on either allotted or tribal lands, but any Indian mineral leases are made expressly subject to such restrictions or stipulations of the Secretary of the Interior as may be necessary for the protection of natural resources. In addition, individual tribes may have implemented specific regulations to protect the tribe's coal resources. The regulations applicable to oil and gas operations on Indian lands require the operations to be conducted in a "good and workmanlike manner";[74] however, they do not specifically address multiple-mineral development, in particular, the development of coal and coalbed gas.

The BIA does have authority to create and enforce stipulations pertaining to multiple-mineral development conflicts on Indian lands, but no standard stipulations were found. In a situation where all the minerals are Indian-owned, it would be prudent to negotiate leases or minerals agreements with language addressing multiple-mineral development. In situations where the Indian tribe or allottee owns only the coal, and coal development is likely, prudence would dictate that the fee oil and gas lessee approach the Indian tribe, BIA, and BLM and inform them of its proposed plan of development. If it is likely that no coal development will occur on the tract, it may be an acceptable business risk to ignore multiple-mineral development issues and proceed with development. If negotiations between the parties take place, they should focus on the multiple-mineral development issues of access and extraction as outlined above.

3. State-Owned Coal

Individual statutes of each state should be examined to determine if and how state-owned coal is protected. States such as Colorado provide for such protection,[75] but other states such as New Mexico do not. If a state does move to protect its coal resource, the oil and gas lessee may be forced to negotiate with appropriate state agencies to address the multiple-mineral development questions of access and extraction. The deeper and more unminable the coal seam, the more unlikely any significant restriction will be imposed upon the oil and gas lessee.

F. Conclusion

Although many legal questions remain to be settled in the West, the law has developed sufficiently to properly analyze the issues and foresee with relative accuracy the degree of risk and even the outcome of disputes between owners of oil and gas rights and owners of coal rights regarding the ownership of the coalbed gas, or the right to drill into the coal seam and extract such gas. Where the western coalbed gas developer has targeted a prospect, it must determine such risks and the likelihood of such disputes and, where they exist to a sufficient degree, negotiate solutions to them in advance of coalbed gas exploration and production. Aside from certain procedural differences, the same strategy, combining vigilance and, where the likelihood of a dispute arises, advance negotiation, applies equally to mineral resources under federal, Indian, state, or private ownership.

74. 25 C.F.R. § 211.19 (1990).

75. C.R.S. §§ 34–61–101 through 108.

Road Log Through Southeastern Piceance Basin

Bradford C. Boyce[1]
Clarence L. Harr[2]
Bruce S. Kelso[3]

1. Bowen/Edwards Associates Inc., Duramgo, Colorado
2. Consulting Petroleum Geologist, Grand Junction, Colorado
3. ICF Resources Inc., Lakewood, Colorado

Focusing on the Basin's Coalbed Gas Resources
Trip Begins and Ends in Historic Glenwood Springs, Colorado

Thursday, September 19, 1991

Assembly point:	Ramada Inn, Glenwood Springs, Colorado
Departure time:	7:30 a.m.
Distance:	220 miles
Number of stops:	3
Return time:	Approximately 6:30 p.m.

INTRODUCTION

This bidirectional field trip passes through some of Colorado's most scenic and geologically interesting countryside. Starting in Glenwood Springs, home of the famous and therapeutic hot springs, the divided trip will proceed in opposite directions. Both groups will travel along the same route and rendezvous for lunch at a spectacular viewpoint on Grand Mesa. Trip length exceeds 220 miles, and elevation ranges from over 10,500 ft to less than 4,500 ft. During any time of the year in Colorado, one can expect changeable weather, especially when traveling such a distance through these elevation changes. Please prepare yourselves accordingly.

From historic Glenwood Springs, built around the "Yampa Hot Springs" and adjacent Leadville Formation vapor caves in the late 1800s, the field trip traverses through rocks from the Mississippian to Holocene pediment and basaltic flows. In one direction, the tour proceeds south up the Roaring Fork River and its tributary, the Crystal River, then over McClure Pass at 8,755 ft. In the opposite direction, the trip heads west from Glenwood Springs and down the Colorado River, up and over the west end of the basalt-capped Grand Mesa, then up the North Fork Valley and over McClure Pass to return to Glenwood Springs.

The town of Glenwood Springs is well known for its hot springs mineral baths and the naturally heated swimming pool. These springs issue mainly from the cavernous Mississippian Leadville Limestone. The Leadville crops out at both ends of Glenwood Canyon (at Dotsero on the east and Glenwood Springs on the west). Springs at these locations bring strongly mineralized waters directly into the Colorado River between Gypsum and New Castle, Colorado. The main sources of the relatively high salt content of the springs are evaporite minerals in the Eagle Valley Evaporite (Pennsylvanian) and the alkali-rich Mancos Shale (Upper Cretaceous).

In the mid-1980s, concern over the excessive salinity in the Lower Colorado River Basin (Arizona, Nevada, southern California, and Mexico) spawned a major U.S. Bureau of Reclamation study of the salinity problem. In one local attempt to attack this problem, a company designed a cogeneration project to be located in Glenwood Springs (close to a major source of salinity). This project would have generated electricity by using heat from the hot springs and by burning abundant Piceance basin natural gas. At the same time, some of the energy would be used to treat the highly saline waters, creating essentially pure water and by-product salts. Income would have been derived from the sale of power, salts, and large quantities of potable water, with the added benefit of removing much of the salinity problem at its source. Because of residents' concerns over visual and other surface impacts, the community has not entirely supported the project. The company expects to try again in the near future.

Energy resources of the Piceance basin are varied and numerous, including coalbed gas, conventional gas and oil, coal, oil from oil shale, and geothermal springs, as found in Glenwood Springs. The sedimentary section in the basin is more than 25,000 ft thick, and to date only several wells have penetrated the full section. As this field conference focuses on western coalbed gas potential, discussions for this field trip will address Piceance basin coal and coal-gas resources more than other energy resources. However, this is not to say other resources will be excluded. The stop at Barrett Resources' facility in the Grand Valley is adjacent to the Unocal oil shale project, and a presentation is planned regarding this tremendous resource.

Piceance basin coals are found in the Dakota Sandstone, Mancos Shale, and Mesaverde Group rocks, all of Cretaceous age. The Mesaverde Group contains by far the most economically important coal and is the principal target for coalbed gas exploration. The same coals are exposed along the basin margins, where mines and outcrops have provided a wealth of data. Mesaverde coal is mainly high-volatile (±30% to 35%) bituminous in rank, with site-specific areas of higher maturity. Some subsurface coal mines have experienced severe problems with high concentrations of methane, an obvious indicator of coalbed gas potential. However, the mine gas problem is so severe in some areas that it has taken human lives and contributed to the shutdown of numerous mining operations, the most recent of which is the Redstone Mine.

During the 1985 AAPG-RMAG Rocky Mountain Section meeting, the Energy Minerals Division (EMD) field trip focused on coalbed gas reserves of the southeastern Piceance basin. The road log traversed the same terrain as this field trip and was used as a basis for this trip. We gratefully acknowledge the Energy Minerals Division of AAPG and the authors for permission to use their log. Additional contributions from Bruce Collins, Dan Powell, and Jack Rathbone (photos and comments) provided valuable data to the log, especially to those areas with recent activity related to coalbed gas. Arco Coal, Cyprus Minerals, Barrett Resources Corporation, Unocal, and Oryx Energy Company also contributed data and comments. Bowen/Edwards Associates Inc. was extremely generous in providing technical support through their office personnel.

The 220-mile trip is lengthy, especially considering the mountainous terrain through which it passes. However, the exceptional geology and scenery make the trip educational and enjoyable. The road-log narrative is structured for either direction, and one only need begin on the first or last page of the route description. When following the road log on your own, remember that not all odometers measure distance equally; therefore, you may need to reorient yourself along the route with recognizable features.

Enjoy your trip!

In the immediate vicinity of Glenwood Springs, from the mouth of Glenwood Canyon westward along Interstate 70 and to the south along State Highway 82 toward Aspen, rocks ranging in age from Mississippian to Late Cretaceous are well exposed. A generalized stratigraphic section is provided at the right, with local descriptions at appropriate points along the road log.

Late Cretaceous

Mesaverde Group

Upper paralic/fluvial sandstone and shale with coal; lower marine shale and thin sandstone; 3,500 ft thick.

Mancos Shale

Mainly dark-gray marine shale; up to 4,000 ft thick.

Early Cretaceous

Dakota Sandstone

Gray to tan, fluviodeltaic sandstone and shale; 60–200 ft thick.

Jurassic

Morrison Formation

Gray, green, and red silty shale; about 100 ft thick.

Triassic

Entrada Sandstone

Tan to pale grayish-orange eolian sandstone; 20–120 ft thick.

Triassic to Late Permian

State Bridge Formation

Reddish brown, laterally persistent sandstone and interbedded siltstone; well exposed only in steeper canyon walls. Similar to, but more easily eroded than the underlying Maroon Formation. Thickness ranges from 200 ft or less west of Glenwood Springs to more than 1,000 ft near Carbondale.

Permian to Pennsylvanian

Maroon Formation

Thick (3,000–12,000 ft) red-bed sequence consisting of micaceous, conglomeratic, arkosic sandstone and mudstone with local calcareous units. Colors range from bright brick-red to reddish-gray and reddish-brown. Base of formation is drawn at the color change from red to gray and yellowish-gray below. West of Glenwood Springs a thin, light-colored sandstone sequence at the top of the Maroon has been referred to the "Schoolhouse tongue of the Belden Formation."

Pennsylvanian

Eagle Valley Evaporite

This thick unit is variously referred to as a member of the Minturn Formation in the Eagle Basin or as a member of the Gothic Formation farther west and southwest. The Eagle Valley consists of a thick sequence of gypsum, halite, and interbedded dark-gray shale with thin, calcareous sandstone and shale near the top. The unit typically is diapiric adjacent to the Grand Hogback in the Roaring Fork valley south of Glenwood Springs and to the east in the Eagle Basin.

Belden Formation

Thick, fossiliferous, gray marine shale and limestone. It is not well exposed in the Roaring Fork valley, where it probably pinches out.

Mississippian

Leadville Limestone

Gray, nearly pure limestone with abundant solution caverns near mouth of Glenwood Canyon. Hot springs issue from many of these solution features along the Colorado River. To the southeast, near Marble, the Leadville has been metamorphosed into a fine white marble. The Leadville typically is less than 100 ft thick in the Glenwood Springs-Aspen area.

ROAD LOG

Mileage, South Route
(Mileage, West Route)
Mileage Increment — Description

0
(221.2) Road log starts and ends at the three-way intersection of Grand Avenue and U.S. Highways 6 and 24 in Glenwood Springs. Proceed either south toward Aspen on Grand Avenue or west on Interstate 70. When heading west, descriptions and mileage (numbers in parentheses) begin at the end of the log.
1.5

1.5
(219.7) Traffic signal; Colorado Highway 82 at Grand Avenue and 23d Street. Grand Avenue leads south and west across the Roaring Fork River to Sunlight Ski basin. Stay on four-lane Highway 82, heading toward Aspen. East of highway, excellent exposure of red State Bridge and Maroon Formations; westward, across the river, note color change marking the contact between Eagle Valley Evaporite below and Maroon Formation above.
0.8

2.3
(218.9) Excellent view of Mt. Sopris, elevation 12,959 ft. This Oligocene granodioritic intrusive of the Elk Mountains complex rises approximately 6,000 ft above the valley floor.
0.2

2.5
(218.7) Maroon Formation along road on the east, with poorly exposed State Bridge Formation above.
0.2

2.7
(218.5) Roaring Fork River to the west.
0.2

2.9
(218.3) Across the Roaring Fork River is a battery of abandoned coke ovens. These ovens coked the coal mined from the Mesaverde Group southwest of here.

Generally, a "coking coal" is a coal that, when heated to high temperatures, (570°–750°F) in the absence of air, becomes soft and cellular, not unlike a sponge. The cellular material, when cooled, becomes a hard, porous,

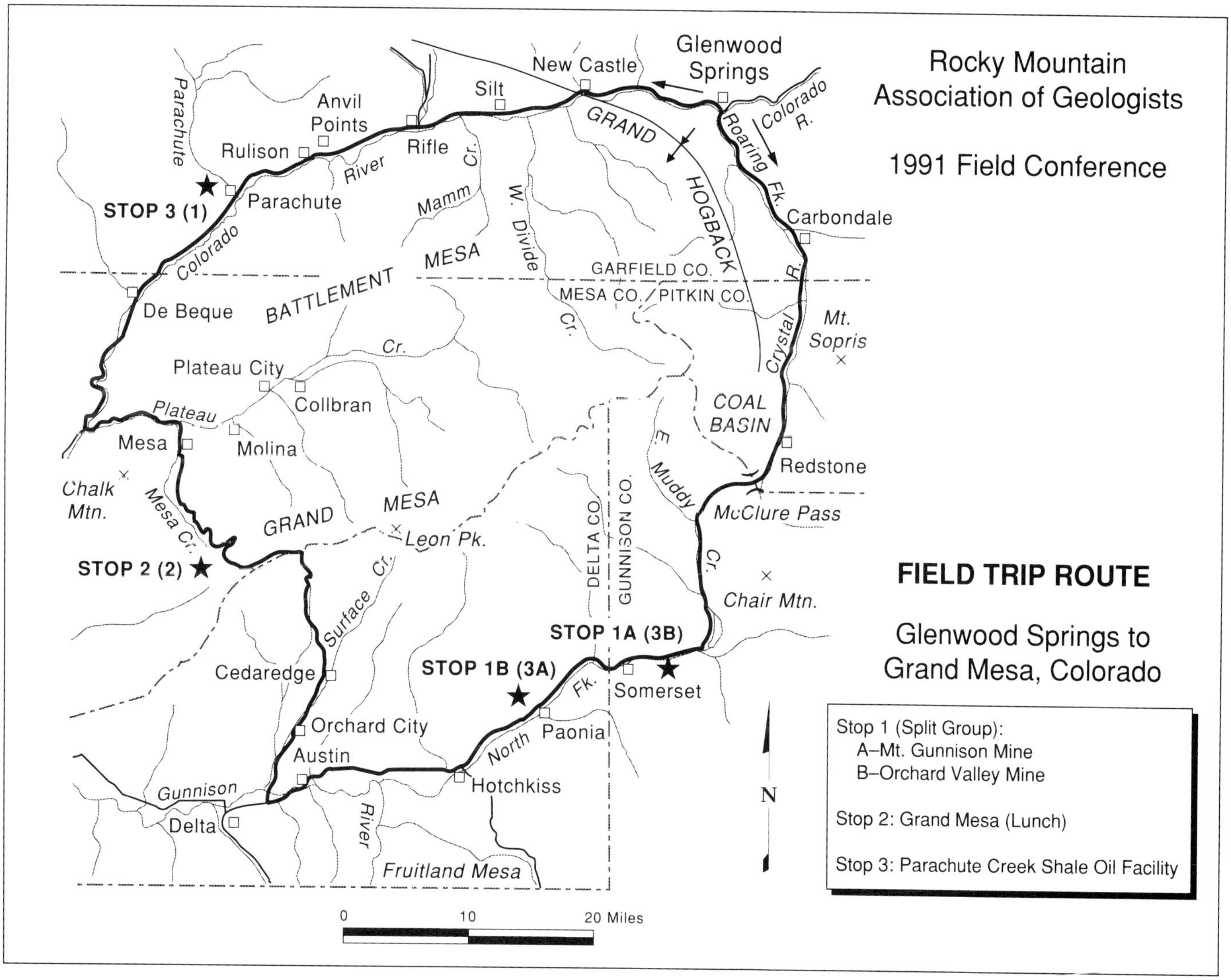

carbonaceous material with virtually no volatiles and a high fixed-carbon content. A premium coking coal generally is low in ash, sulfur, and volatiles and is primarily made of woody plant material. Coked coal, because of its unique physical properties and absence of impurities, is useful in steel-making and chemical processing in which high temperatures and additional carbon molecules are desired.

Economics of shipping coal vs coke created the business of coking coal here. However, as coking coal technology improved, transportation economics became secondary to the coke quality required by the steel industry. The brick beehive-type ovens could not be modified to keep pace with new steel technology. The last operating ovens here were abandoned more than 30 years ago.

0.9

3.8 (217.4) At this location to the east, not visible from the road, are Tertiary basalt flows overlying the Weber/Maroon Formation, which continue for several miles.

2.0

5.8 (215.4) Again, an outstanding view of Mt. Sopris, the most northwesterly pluton in the Elk Mountains. Adjacent upturned rocks range from Upper Pennsylvanian Gothic Formation to Cretaceous Mancos Shale.

1.1

6.7 (214.5) To the east is the entrance to Colorado Mountain College. The contact between Maroon Formation and Eagle Valley Evaporite can be seen on the hillside to the east. Upon examination of these rocks at less than 60 mph, one will notice dramatic structural features, primarily within the evaporite. Its physical characteristics indicate formation flowage induced by diapiric structures, which were influenced regionally by basin-margin tectonics.

0.5

7.2 (214.0) Paved road #113 to the east crosses Cottonwood Pass to Gypsum, 27 miles.

1.0

Exceptional fall view of Mt. Sopris rising over 6,000 ft above the valley to 12,959 ft. Photo by Jack Rathbone.

8.2 (213.0) Bluffs east of the road show chaotic deformation attributable to diapiric deformation in thin sandstone, shale, and gypsum of the Eagle Valley Evaporite.

0.5

8.7 (212.5) To the east, excellent exposure of the contact between Eagle Valley Evaporite and Maroon Formation.

1.6

10.3 (210.9) Scenic overlook to the south. Mancos Shale forms the slopes against the Elk Mountains.

0.7

11.0 (210.2) Junction of Colorado Highways 82 and 133. Turn south onto Highway 133 toward Marble or west onto Highway 82 toward Glenwood Springs.

0.1

11.1 (210.1) Cross Roaring Fork River.

0.4

11.5 (209.7) Railroad crossing.

0.5

12.0 (209.2) Traffic light. Downtown Carbondale to the east. To the west and southwest, Eagle Valley Evaporite and overlying Quaternary terrace gravels are exposed in stream-cut banks. The intermediate ridge is capped by Dakota Sandstone; the ridge on the skyline is the Grand Hogback, consisting of Mesaverde Group sandstone, shale, and coalbeds.

1.8

13.8 (207.4) Colorado Division of Wildlife fish hatchery on right.

0.2

14.0 (207.2) Bridge over Crystal River. Four Quaternary terrace levels are visible here. No detailed study of the terraces and gravels along the lower Crystal River has been published, but preliminary work has been conducted by the Colorado Geological Survey (Stover and Soule, unpublished).

0.1

14.1 (207.1) Mancos Shale in hills ahead marks the west end of the Roaring Fork syncline.

0.6

14.7 (206.5) Entering/leaving Pitkin County.

0.2

14.9 (206.3) If the light is right, one can see the rock glaciers on the northern and northwestern flanks of Mt. Sopris. Birkeland (1973) concluded that some of the rock glaciers are as old as Bull Lake age and that some, which initially formed in late Pinedale time, are presently active.

1.6

16.5 (204.7) To the south-southwest, the red exposure is Maroon Formation. Near Mt. Sopris, the beds below the Maroon Formation have been considered by some workers to be the Gothic Formation, which contains gypsum beds, or Eagle Valley Evaporite by others. The Gothic-Maroon contact is placed at the color change from mostly brown, gray, and green rocks below to mostly reddish-brown rocks above.

0.1

16.6 (204.6) Five miles west are the abandoned Spring Gulch and Thompson Creek coal mines, in the southern Grand Hogback, which can be seen through the notch of Thompson Creek Canyon to the west. The hogback here consists of lower Mesaverde Group coals. A good high-volatile coking coal was produced from the area by Colorado Fuel & Iron from 1887 to 1916 (and coked in the ovens seen earlier just south of Glenwood Springs), by Thompson Creek Coal & Coke from 1951 to 1966, and by Anschutz/Snowmass Coal from 1976 to 1986. Total production was about 7,000,000 short tons. The coal in these mines contained some methane, but gas was not considered a major mining problem.

0.3

16.9 (204.3) Bridge over Thompson Creek. Mancos Shale on both sides of the valley, in the Roaring Fork syncline. Shale is capped by river gravels. Wolf Creek anticline is about 10 miles to the west, at the head of Thompson Creek. This faulted gas field produced until 1976 and currently is used for gas storage by Rocky Mountain Natural Gas Company.

0.4

17.3 (203.9) Maroon Formation outcrops of conglomeratic, coarse-grained sandstone.

0.1

17.4 (203.8) Across the river is a poorly exposed section of Mancos Shale, Dakota Sandstone, Morrison Formation, Entrada Sandstone, and Maroon Formation, dipping into the Roaring Fork syncline.

0.5

17.9 (203.3) Road east leads to Potato Bill Creek.

0.4

18.3 (202.9) Outcrop of conglomeratic Maroon Formation.

0.9

19.2 (202.0) Maroon Formation outcrop.

1.5

20.7 (200.5) Maroon Formation to the east.

0.2

20.9 (200.3) To the east, marble slabs on river bank were used as riprap along the old railroad grade.

1.6

22.5 (198.7) Entering/leaving White River National Forest.

0.5

23.0 (198.2) Avalanche Creek road to the east.

0.6

23.6 (197.6) Gothic Formation (Eagle Valley?) is on left.

0.2

23.8 (197.4) Outcrops of granodiorite. At this exposure the stock is separate from the Sopris pluton but may be connected at depth.

0.7

24.5 (196.7) South edge of granodiorite stock.

0.4

24.9 (196.3) To the east, below the road along the river level, is Penny Hot Springs. Gothic and Maroon Formations are upwarped against granodiorite to the east.

0.1

25.0 (196.2) Exposures in high cliffs to the west are southward-dipping Maroon Formation.
0.2

25.2 (196.0) Peak to the south is Chair Mountain, a glacier-carved portion of the Ragged Mountain laccolith, a quartz-monzonite porphyry.
0.3

25.5 (195.7) On both sides of the road in the high cliffs are outcrops of Maroon Formation that consist of sandstone, conglomeratic sandstone, and conglomerate with little, if any, siltstone or claystone. This part of the Maroon is believed to have been deposited in the headward part of a fluvial system, perhaps as alluvial fans.
0.9

26.4 (194.8) Very large boulders such as those seen at creek level are common along this stretch of the Crystal River. They result from rock falls by oversteepening of the canyon walls during the Pleistocene.
0.7

27.1 (194.1) Outcrops of coarse-grained sandstone and conglomerate of the Maroon Formation.
0.4

27.5 (193.7) Note the drag fold within beds of the Entrada Sandstone. The classic overlying Morrison Formation is on the downthrown (east) side of a north-trending fault. Maroon Formation is on the west side of the fault.
0.6

28.1 (193.1) Road across Coal Creek.
0.1

28.2 (193.0) Beehive coke ovens to the west. Two hundred ovens were in operation here about 1900-1909, supplied by Mesaverde coal mines to the west.
0.1

28.3 (192.9) Road to the east crosses the Crystal River and enters the town of Redstone.
0.1

28.4 (192.8) Road to the west leads to Coal Basin and Mid-Continent Resources mining operations, approximately 5 miles. The coal mines were idled early in 1991, putting more than 400 people out of work. A contributing factor was the very gassy condition of the mine. Coal Basin is a deeply eroded elliptical anticline superimposed on the Grand Hogback. The anticline was formed as a result of an ill-defined igneous intrusion. Rank of Mesaverde coals in the Coal Basin area ranges from the low end of the medium-volatile range for the lowest beds (A and B, 22%-24% maf volatile matter),

View south toward Chair Mountain with remnants of brick beehive coke ovens in the foreground. The ovens coked coals from the mines west in Coal Basin. Photo by Brad Boyce.

Beautiful drag fold west of the Crystal River just north of the town of Redstone. See mile 27.5 for description. Photo by Jack Rathbone.

through the middle and upper end for the Dutch Creek bed (also known as M bed, 27%-30% maf), to high-volatile A for the Placita, the uppermost coal of mineable thickness. The total coal-bearing interval varies but averages about 700 ft thick. Collins (1977) believes the higher rank of the coals in Coal Basin was caused by the underlying igneous heat source, the decreasing rank up the section being the result of increasing distance from this source. For details on the geology and coal resources of Coal Basin, see Collins (1977).

Coal mining began here in 1900, after the construction of two towns (Redstone and Coal Basin), a standard-gauge railroad from Carbondale and narrow-gauge tracks from Redstone to Coal Basin (12 miles of twisting rails that gained 2,300 ft in elevation), the Redstone coke ovens, and the mine itself, all of which began in the early 1890s. Colorado Fuel & Iron extracted just over one million tons from the mine before it was closed early in 1909. Mid-Continent Resources opened its first mine, Dutch Creek No. 1, in a different location in 1956. Approximately 22.5 million tons of premium-grade metallurgical (coking) coal had been produced when the mine was idled in January 1991. The original Coal Basin Mine was reopened in 1973 and produced another million tons before it closed in 1982.

0.1

28.5 (192.7) Exposures to the west reveal the top of the Maroon Formation, Entrada Sandstone, and Morrison Formation.

0.6

29.1 (192.1) At the top of the hill to the west is ledge-forming Dakota Sandstone.

0.2

29.3 (191.9) Conglomerate in Maroon Formation.

0.2

29.5 (191.7) Light greenish-gray bed in Maroon Formation.

0.4

29.9 (191.3) Maroon Formation on both sides of road.

0.1

30.0 (191.2) Hayes Creek Falls to the west in Maroon Formation outcrop.

0.1

30.1 (191.1) Roadcuts through light-colored Maroon sandstones.

0.4

30.5 (190.7) Bridge across Crystal River.

0.7

31.2 (190.0) In this area is a poorly exposed section that includes upper Maroon Formation, Entrada Sandstone, Morrison Formation, Dakota Sandstone, and a small section of Mancos Shale. A thrust fault, probably the northwestern extension of the Elk Mountain thrust, trends southward and separates upper Mancos from lower Mancos, Dakota, Morrison, Entrada, and upper Maroon Formations. At this point, the Maroon immediately underlies the Entrada, and throw on the fault is about 3,000 ft.

0.1

31.3 (189.9) Bridge across the Crystal river. Upper Mancos Shale is exposed in stream cut to the west.

0.1

31.4 (189.8) Lower Mesaverde Group fine-grained sandstone in roadcut.

0.2

31.6 (189.6) Exposure of gray shale above basal Mesaverde sandstone.

0.2

31.8 (189.4) Holgate Mine and dump at 11:00. The Placita mine to the west. Both produced some coking coal, with most production consumed by the railroad and marble quarrying operations.

0.2

32.0 (189.2) Placita townsite.

0.1

32.1 (189.1) Holgate Mine at 9:00.

1.0

33.1 (188.1) Gunnison/Pitkin County line.

0.1

33.2 (188.1) Gunnison County Road 3 to the southeast goes to Marble. The field trip stays on Colorado Highway 133. Marble obtained its name from the fine quality white marble formed by metamorphism of Leadville Limestone. Marble quarried along Yule Creek was used in the Lincoln Memorial. The town expanded to approximately 3,000 people in the early 1900s due to its beautiful stone. In 1941 a flood destroyed the marble works, which remained out of service until 1990, when the "revitalized" Yule Marble Company reopened the quarries to begin production.

1.0

34.2 (187.0) Excellent Upper Mesaverde Group exposures in the roadcuts for the next 1.2 miles.

0.9

35.1 (186.1) Thin coalbeds are exposed in the roadcuts.

0.2

35.3 (185.9) Pitkin/Gunnison County line. Excellent Mesaverde exposures along roadcuts.

0.8

36.1 (185.1) The surrounding views here are of the Elk Mountains. On the south side of the river valley is Treasure Mountain (elevation 13,530 ft), the rounded peak, and on its northern shoulder is Whitehouse Mountain (elevation 11,974 ft). These two peaks are expressions of the Treasure Mountain dome, which was formed about 12 million years ago (Miocene) by intrusion of the granite of Treasure Mountain. Precambrian rocks are exposed in the core of the dome, and Paleozoic and Mesozoic sedimentary rocks dip west and northwest off the dome. The mountains west of Treasure Mountain are formed by granodiorite porphyry of the Oligocene Raspberry Creek laccolith and by the eastern part of the Ragged Mountain laccolith.

0.1

Scattered slabs of quarried marble along the Crystal river near Marble, Colorado. The pillars are the remains of the dressing room. Mancos Shale along river in left lower quarter of photo. Photo by Jack Rathbone.

36.2 (185.0) Roadcut exposes sandstone, siltstone, and carbonaceous shale in the Barren Member (or undifferentiated member) of the Mesaverde Group.
0.4

36.6 (184.6) Roadcut through the Barren Member of the Mesaverde Group.
0.1

36.7 (184.5) McClure Pass summit (elevation 8,755 ft) named for Thomas McClure, who achieved fame for developing the Red McClure potato. Entering/leaving Gunnison County and Gunnison National Forest. For the next 1.9 miles, the road crosses glacial deposits.
0.2

36.9 (184.3) The ridge to the southeast is formed by sandstone of the Ohio Creek Member of the Mesaverde Group. The Ohio Creek was redefined by Johnson and May (1980) as the uppermost member of the Mesaverde and represents a deeply weathered zone below the unconformity separating the Mesaverde from the Wasatch Formation. Breakdown of feldspars and accumulation of kaolinite in the weathered zone impart a white color to the sandstones. The base of the Ohio Creek Member is then the base of the lowest white sandstone.
1.5

38.4 (182.8) McClure Pass Campground is on the southeast. A concealed normal fault that trends east-northeast is about 0.2 mile to the south. Displacement along the fault is about 1,500 ft, and the north side is down, placing the Wasatch against the Mesaverde to the south. Mt. Gunnison is the prominent mountain south.
0.3

38.7 (182.5) View to the southeast is down the valley of Lee Creek.
0.1

38.8 (182.5) Road crosses Lee Creek. Outcrops on west along the creek are Wasatch Formation, which dips about 15° S.
0.2

39.0 Road patching is a constant struggle on this active

(182.2) landslide in the Wasatch Formation and will be a never-ending battle until the road is relocated, if ever.

0.6

39.6 (181.6) Road crosses a landslide deposit for another 0.3 mile.

0.5

40.1 (181.1) Roadcut to the west exposes sandstone and shale of the Wasatch Formation, which dips 10°–15° to the south.

0.2

40.3 (180.9) Cross axis of an east-northeast-trending syncline.

0.6

40.9 (180.3) Leaving/entering Gunnison National Forest. Chair Creek joins Lee Creek from the east.

0.8

41.7 (179.5) Lateral and terminal moraines from glaciation (probable Bull Lake age) in Lee Creek Valley are at about road level to the right and left.

0.5

42.2 (179.0) View of Chair Mountain and The Raggeds to the east. Chair Mountain is to the north of the Raggeds and is separated from them by Buck Creek Basin, a deep glacial valley. Chair Mountain does look like a chair when viewed from north of Redstone and is part of the Ragged Mountain laccolith. The rolling slopes up to the laccolith consist of Wasatch Formation, glacial deposits, and landslide deposits.

0.5

42.7 (178.5) Outcrops along the road and on the hillside to the west are Wasatch Formation, which consists mainly of varicolored mudstone, siltstone, and fine- to coarse-grained, lenticular feldspathic and arkosic sandstone. Wasatch Formation in this area also contains occasional thin, erratic, lignitic coals.

0.2

42.9 (178.3) Gravel road on west side winds northwest to Plateau Creek Valley and town of Collbran. County Road 265 travels across the east end of Grand Mesa.

0.3

43.2 (178.0) Cross East Muddy Creek.

0.8

Roadcut heading up to McClure Pass with excellent exposures of Mesaverde along the cuts. Photo by Jack Rathbone.

44.0 Reddish-tan sandstone bed to the left across the creek is
(177.2) the uppermost sandstone of the Ohio Creek Member of the Mesaverde Group.

0.8

44.8 Bridge over East Muddy Creek.
(176.4)

0.2

45.0 Scenic overlook turnout to the west. Across East
(176.2) Muddy Creek is an excellent exposure of thick white sandstone and thinner reddish-tan sandstone, both in upper Ohio Creek Member of the Mesaverde Group. Ragged Mountains are to the east.

0.3

45.3 Cross Buck Creek.
(175.9)

0.5

45.8 On the hillside across the creek to the right is a reddish-
(175.4) tan sandstone bed above a thick white sandstone. Unconformity at the base of the Wasatch Formation is just above the reddish-tan sandstone, as determined by Johnson and May (1980).

0.8

46.6 Wasatch Formation, whose lower part is of Paleocene
(174.6) age, unconformably overlies Ohio Creek Member of the Mesaverde Group. Near the base of the Wasatch is a thin, discontinuous, bony coalbed. The coal is up to 2.5 ft thick and has been prospected near the top of the hillside to the west across East Muddy Creek (Godwin, 1968).

0.4

47.0 Bridge over East Muddy Creek.
(174.2)

0.2

47.2 Overlook on southeast side of road. Roadcut in Ohio
(174.0) Creek Member of the MesaverdeGroup to the west exposes a thick channel sandstone.

0.7

47.9 Roadcut along this part of the new highway exposes
(173.3) the Ohio Creek Member of theMesaverde. The highway was relocated to its present position to avoid the huge landslide that occurred in 1986. The toe of the landslide is visible at creek level to the east. Lenticular sandstone and interfingering shale in the roadcut on the right are evidence that the Ohio Creek was deposited in a system of meandering streams.

0.9

48.8 Cross West Muddy Creek. On the east side of Muddy
(172.4) Creek is the Amoco Hotchkiss Ranch No. 2 well drilled in the spring of 1982. This was an unsuccessful coalbed gas attempt, and Amoco has since abandoned its efforts on the 100,000-acre Megas Federal unit.

Public data is sparse regarding Amoco's Mesaverde coal-gas tests in southeastern Piceance basin. Amoco drilled five wells in the Megas Federal unit, all in section 20, T12S R89W, Gunnison County. The initial test, Hotchkiss Ranch No. 1 well, was drilled in 1978 and reported as a Rollins Sandstone test. However, Amoco cut coal cores and almost immediately abandoned the well. The Hotchkiss Ranch No. 2 well was drilled to a depth of 2,252 ft, only 400 ft north of the No. 1 well in 1982 and successfully recovered coal from coring. Desorption of the high-volatile B bituminous coal indicated gas contents in excess of 450 SCF/T. Amoco continued production testing through 1989 and then decided to abandon the well primarily because of mechanical and completion difficulties.

In 1984 Amoco drilled the Powers Federal well, which when production tested, appeared to have the greatest potential of the wells drilled to date. As noted above, this area is prone to landslides, and at this well Amoco experienced the effects of one landslide that sheared off the well casing near the surface. Amoco successfully plugged the well, which may have been its best coalbed gas well in the basin. The two Megas wells in the west half of section 20 also encountered gassy coals; however, numerous hole problems resulted in their abandonment.

Compliments to Amoco for its attempt at coal-gas production in an extremely difficult environment.

0.9

49.7 Road to the east crosses Muddy Creek for access to the
(171.5) Paonia Reservoir State Recreation Area. The 3-mile-long reservoir ensures adequate irrigation water through the summer for the North Fork Valley. At times of low water levels, silt deposits in the reservoir are visible, and one may wonder how long it will be before the reservoir completely fills with sediment from Muddy Creek.

0.4

50.1 Roadcut to the west through white, friable sandstone of
(171.1) the Ohio Creek Member of the Mesaverde Group.

0.5

50.6 Light tan to white sandstone to the east is near the base
(170.6) of the Ohio Creek Member ofthe Mesaverde Group.

0.4

51.0 Williams Creek flows into Paonia Reservoir from the
(170.2) opposite side.

0.4

51.4 White kaolinitic sandstone in roadcut on west is near
(169.8) the base of the Ohio Creek Member of the Mesaverde Group.

0.6

52.0 Deep Creek enters the reservoir from the east. Marcel-
(169.2) lina Mountain (elevation 11,348 ft) is the sharp peak in the background (south) and is another Oligocene granodiorite porphyry laccolith. Good Mesaverde exposures on the west for next 1.3 miles.

1.3

53.3 The earth-filled dam of Paonia Reservoir (elevation
(167.9) 6,450 ft).

0.1

53.4 Excellent exposures of the Barren member or undiffer-
(167.8) entiated Mesaverde for the next 0.5 mile.

0.5

53.9 Gunnison County Road 12 to the east goes southeast
(167.3) over Kebler Pass to Crested Butte. The road follows Anthracite Creek up toward West Beckwith Peak. Anthracite and Muddy Creeks join here to form the North Fork of the Gunnison River, which flows generally west and southwest. Paonia Reservoir dam to the north. Anthracite Creek is aptly named, as its head-

waters are located in Crested Butte coal field in which anthracite coal was common. Local intrusives and not burial depth generated the high rank of coals in the Crested Butte field.

0.8

54.7 (166.5) Gravel pits to the south are used by the State Highway Department for road work on Colorado Highway 133.

0.4

55.1 (166.1) Cross Thompson Creek.

1.1

56.2 (165.0) Raven Gulch to the south.

0.1

56.3 (164.9) Drive onto new section of highway.

0.2

56.5 (164.7) Cross Hawks Nests Creek. The old Hawks Nest Mine, which produced from the F coal zone, is below the highway to the north.

0.3

56.8 (164.4) To west-northwest is the portal of the Hawks Nest East Mine, which produced steam coal from the F zone. The East and West Mines closed in January 1983 when Williams Corporation bought Western Slope Carbon (Northwest Company).

0.5

57.3 (163.9) Clinker and coal exposed in the roadcut to the north are in the F coal zone.

0.4

57.7 (163.5) The Hawks Nest West Mine is above the Western Slope Carbon offices to the north. It produced coal from the F coal zone as did the old Black Beauty Mine in the same vicinity. Black Beauty Mine closed in 1970.

0.3

58.0 (163.2) The storage silo and loadout facility to the north were part of the Hawks Nest operation. Farther north is the site of the Oliver No. 1 Mine, which produced coal from the D and E coal zones until 1944. Coal also was produced from the D and E coal zones south of the river from the Oliver No. 2 Mine.

0.1

58.1 (163.1) Coal exposed in roadcut to the east is in the D coal zone of the upper coal member (Paonia Shale Member of Collins, 1976).

0.5

58.6 (162.6)

STOP 1A (or 3B)—Mt. Gunnison No. 1 Mine*

The Mesaverde Formation in this part of the basin was deposited as part of the Piceance delta system. For general discussion purposes, the Mesaverde Formation has been divided into five members—Rollins Sandstone, lower coal member, upper coal member, barren member, and Ohio Creek Member (Wellborn, 1982; Johnson and May, 1980). The major coals on the Mt. Gunnison property are designated A through F, in ascending order, with the first three coals in the lower coal member and the second three in the upper coal member. The mine, operated by West Elk Coal Inc., currently is working the F coal zone and produces approximately 500,000 short tons annually. Most of the produced coal is sold to the Labadie power plant of Union Electric in Missouri. The F coal contains approximately 35% volatile matter with minimal sulfur and a heating value of 11,500 Btu.

This mine is interesting in that Arco representatives claim it is one of only several recently started successful underground mines. Part of its success is attributed to the gentle dip of the coalbed, with mining moving updip, thereby allowing gravity to assist in the operations. Secondly, minimal methane occurs in the mine. The F coal is exposed in outcrop almost entirely around the mountain from which it is mined. Knowing that pressure is required to retain coalbed gas, it is reasonable to believe that these particular coal seams have not retained pressure head because of their outcrop exposure, thus allowing gas to desorb and escape.

0.2

58.8 (162.4) Across the North Fork to the south is the main portal to the old Bear Mine, which produced coal from 1934 to 1982, in the C coal zone. The cliff east of the Bear Mine portal is formed by sandstone in the upper part of the lower coal member.

0.6

59.4 (161.8) To the north along the road are thick sandstones in the lower coal member.

0.2

59.6 (161.6) On the south side of North Fork River is the Bear No. 3 Mine, which produces coal from the C zone.

0.2

59.8 (161.4) Entering/leaving the community of Somerset, home of several hundred people. The town was established in 1902 by the D&RGW railway and named after a coal mining town in Pennsylvania that in turn was named for Somersetshire in England. Somerset is built in a hazardous location on a recent debris flow from Elk Creek (Junge, 1978).

0.2

60.0 (161.2) On the north side is the storage silo and loadout facilities for the Somerset Mine, which closed in 1985. This mine produced coal from the B and C zones of the lower coal member. The marginal coking coal mined here was classified as high-volatile B bituminous and was shipped to U.S. Steel's now dismantled Wellington coal preparation plant in the Sunnyside district of eastern Utah. When blended with Coal Basin coal, a high-quality coke can be produced. This plant provided coke-oven feed stock for the U.S. Steel Geneva steel mill at Orem, Utah, which is now owned by Basic Manufacturing and Technologies of Utah Inc. Elk Creek is to the north.

0.2

60.2 (161.0) Entering/leaving Somerset.

0.4

60.6 (160.6) To the south is the intake on the North Fork River for the Fire Mountain Canal. Outcrop on the north is the

* For logistical reasons, part of the field trip group will visit Mt. Gunnison No. 1 Mine (mile 58.6 or 162.6), and the other part will visit Orchard Valley Coal Mine (mile 68.7 or 152.5).

massive lower bench of Rollins Sandstone. Wellborn (1982) divided the Rollins into three benches according to their weathering characteristics—the lower bench is massive and blocky; the middle bench weathers to a honeycomb pattern; and the upper bench has a tabular appearance.

0.7

61.3 (159.9) Cross Bear Creek. The mountain to the north was named Fire Mountain by the Indians, apparently from the hot steam and gases that vented through fractures from burning coalbeds.

0.8

62.1 (159.1) Delta/Gunnison County line.

0.3

62.4 (158.8) Cross D&RGW railway tracks. On the hillside to the north is the gradational contact between offshore marine sediments of the upper Mancos Shale and delta-front deposits of the Rollins Sandstone.

0.4

62.8 (158.4) Cross D&RGW railway tracks.

0.2

63.0 (158.2) Cross Hubbard Creek. A basaltic plug of Miocene age is exposed about 4 miles to the north at Iron Point. This intrusion, a basaltic dike 2 to 3 miles north of Somerset, and nearby intrusions in the West Elk Mountains, created high heat flow that raised the rank (and gas content) of much of the coal in the Somerset area to low-grade bituminous coking coal.

0.5

63.5 (157.7) Road to the west leads to the site of Blue Ribbon Coal Mine, about 2 miles north up Hubbard Creek. Coal was produced from the F seam in the upper coal member. The mine closed in 1983.

0.2

63.7 (157.5) Historical marker for the Dominguez-Escalante expedition is on the south. The expedition traveled up the North Fork of the Gunnison Valley in August 1776, in search of a new route from Santa Fe, New Mexico, to Monterey on the California coast. At this location the expedition turned northward up Hubbard Creek, then over Grand Mesa. The expedition failed, being turned back because of mishaps and the approach of winter. Cliff-forming sandstone to the north (about 150 ft thick) is Rollins Sandstone, a delta-front and/or barrier-beach sandstone.

0.4

64.1 (157.1) King Mine (known locally as the Bowie Mine after the family that mined it) was on the mountainside to the north. A steep cableway connected the mine entry to the crushing and railway facilities. Coal was produced from the B coal seam in the lower coal member.

0.2

64.3 (156.9) Entering the old townsite of Bowie. The town was first called Reading but was changed to Bowie in the 1900s to honor the man who managed the Juanita Coal and Coke Company, which operated the mines. The mines stayed under the Bowie family management until 1974, when Coors Energy took over. The town is built on recent debris flows (Junge, 1978) and is abandoned.

0.6

64.9 (156.3) Cross Fire Mountain Canal. Across the river to the south is Lennox Mesa, composed mainly of landslide deposits near the top of the Mancos Shale. Above Lennox Mesa on the northwest slope of Jumbo Mountain are several recent landslide scarps in the lower Mesaverde.

2.3

67.2 (154.0) Redtop Peak is south across the river. It is Mancos Shale capped by landslide material composed partly of red clinker from the lower Mesaverde.

0.4

67.6 (153.6) To the south are the Cyprus Minerals coal storage silos and loadout facilities for the company's Orchard Valley Mine. Trucks haul coal about 3 miles down to this facility from the Orchard Valley Mine. Coal is produced from the D seam in the upper coal member. See description at mile 68.7.

1.1

68.7 (152.5)

STOP 1B (or 3A)—Orchard Valley Coal Mine*

Intersection of Colorado Highway 187. Road south leads to Paonia. The road north heads up Stevens Gulch about 2 miles to Orchard Valley Coal Mine, which is owned by Cyprus Minerals and employs more than 70 people. The mine opened in 1976 and reached full production in 1978. The mine produces D seam coal, which contains approximately 35% volatiles and 11,400 Btu. This mine differs from the Mt. Gunnison No. 1 in that at Orchard Valley mining is downdip. However, gas and water are only minor difficulties, and Cyprus claims economic success.

The mine always suffered from "heating pillars" and in the spring of 1986 experienced a brief but spectacular fire. Injuries were minimal. The mine was sealed after a three-day attempt to extinguish the fire. A new mine was opened several months later up Roatcap Creek, the next drainage to the west, which the mine produces from today.

1.3

70.0 (151.2) Lower mesa to the north is Pitkin Mesa, a Pinedale outwash surface.

0.6

70.6 (150.6) Looking south-southeast, the westernmost peaks of the West Elk Mountains are quite prominent. Lands End Peak is aptly named. Elevations range from 10,800 ft to over 12,000 ft.

1.1

71.7 (149.5) To the north is Stucker Mesa, a Bull Lake outwash surface.

1.8

73.5 (147.7) Sunshine Mesa to the north is a Pinedale outwash surface.

0.2

73.7 (147.5) Dominguez-Escalante historical marker to the south. There are two distinct levels of riverine terrace across

* For logistical reasons, part of the field trip group will visit Mt. Gunnison No. 1 Mine (mile 58.6 or 162.6), and the other part will visit Orchard Valley Coal Mine (mile 68.7 or 152.5).

the North Fork. The lower terrace of Pinedale age is Stewart Mesa, and the upper terrace is unnamed and probably is of Bull Lake age. Black Canyon is the higher terrain to the south.

2.1

75.8 (145.4) Cross Jay Creek. The highway here is on Hanson Mesa, a Pinedale riverine terrace. Next highest mesa level to the north is Powell Mesa, a Pinedale outwash surface.

1.7

77.5 (143.7) Entering the town of Hotchkiss, a community of about 1,000 people. The town is mainly a center for fruit growing and cattle and sheep operations.

0.2

77.7 (143.5) Junction of Colorado Highways 133 and 92. Continue either west on Highway 92 or east on Highway 133, if running the trip in reverse.

0.7

78.4 (142.8) Cross D&RGW railway tracks.

0.2

78.6 (142.6) Cross Leroux Creek. The road ascends/descends the east side of Rogers Mesa, a Pinedale outwash surface. The distal parts of this mesa and that of Surface Creek Mesa project to about 220 ft above present base level. By comparison, the Bull Lake outwash surface on Redlands Mesa to the north projects to about 1,050 ft above base level (Cole and Sexton, 1981).

0.4

79.0 (172.2) Dominguez-Escalante historical marker on south side.

1.0

80.0 (171.2) To the south are Grand View and Fruitland Mesa, which are formed on Dakota Sandstone that dips gently north-northeast on the flank of the Gunnison uplift. The Gunnison uplift is dissected along its axis by the Gunnison river, which formed the imposing Black Canyon. It is interesting of why the Gunnison River maintained this axial course through extremely resilient rock when only a mile south the rocks are much less resistant.

No other canyon provides such depth and narrowness in the country, over 2,400 ft deep and in places at river level, only 40 ft wide. The Painted Wall, aptly named because of its highly distorted rocks interlaced with pink granites and pegmatites against a black matrix, is approximately 2,300 ft high. Excellent views south, and on clear days, the northernmost peaks of the San Juan Mountains are quite scenic.

1.5

81.5 (139.7) Rogers Mesa Community Center on south. Road to south goes to Lazear and Hotchkiss National Fish Hatchery (2.5 mi) along the North Fork of the Gunnison. Nearby Doughty Spring is a highly radioactive spring in the North Fork area.

0.5

82.0 (139.2) We are passing through one of the principal orchard areas of Delta County.

1.1

83.1 (138.1) Descend/ascend the west side of Rogers Mesa.

0.2

83.3 (137.9) Enter the Mancos Shale desert.

0.5

83.8 (137.4) Cross D&RGW railway tracks. Juana Lopez Member of the Mancos Shale is exposed near the top of the bluff to the north.

0.2

84.0 (137.2) Road to the south goes about 1 mile to the confluence of the North Fork and Main Gunnison Rivers. South and upstream of the confluence, the Gunnison River flows through the Black Canyon. The lower stretches of the Black Canyon are raftable and contain some of the finest fishing in Colorado. This river access is where most rafters disembark.

0.5

84.5 (136.7) Road to the south follows Sulfur Gulch for about 0.5 mile to an abandoned sulfur mine. Sulfur-impregnated rocks in the upper Dakota Sandstone and basal Mancos Shale have been extracted for use as soil conditioners. The sulfur is the surface manifestation of hydrogen sulfide and sulfur dioxide gas seeps (Cadigan and others, 1971).

0.7

85.2 (136.0) A small gravel-mining operation south on top of the bluff is on a riverine surface, probably of Pinedale age.

1.0

86.2 (135.0) Crossing a broad alkali flat in Mancos Shale in the Lawhead Gulch area.

1.3

87.5 (133.7) Road north leads to Redlands Mesa, an extensive Bull Lake outwash surface. The lower mesa is Shamrock Mesa, a Pinedale riverine surface (Cole and Sexton, 1981).

1.2

88.7 (132.5) Roadcut exposes Juana Lopez Member of the Mancos Shale. The member consists mainly of sandy calcareous shale and is situated about 300 to 350 ft above the base of the Mancos. A zone of large, rounded, limonite-stained calcareous concretions occurs just below the Juana Lopez in some areas. Some of these concretions can be seen in the roadcut.

1.1

89.8 (131.4) Roadcut exposes Dakota Sandstone and basal Mancos Shale. Upper Dakota Sandstone here consists of interbedded sandstone and shale deposited in a shoreline setting related to the transgressing Western Interior seaway. Sandstones are cross stratified to ripple bedded. The Dakota grades into thin interbedded siltstone and shale of the overlying Mancos. This is the anticlinal nose of the Laramide Gunnison uplift. The steep west flank of the uplift is a faulted monocline.

0.2

90.0 (131.2) Bridge over Currant Creek.

0.3

90.3 (130.9) Near the top of the hillside north of the Gunnison River are several short adits into thin coal seams of the Dakota Sandstone. The canyon cut by the Gunnison River exposes Dakota Sandstone, Burro Canyon Formation, and Morrison Formation.

0.7

91.0 Village of Austin to the north.

(130.2)

0.5

91.5 Bridge over Gunnison River.
(129.7)

0.5

92.0 Dominguez-Escalante historical marker to the north.
(129.2)

0.4

92.4 Small community of Read. Road south goes to
(128.8) Montrose via Peach Valley, formed in Mancos Shale just west of the Gunnison uplift.

1.0

93.4 Surface Creek Mesa, now to north, is a Pinedale out-
(127.8) wash surface. Antelope Mesa, the higher mesa (Antelope Hill) covered by orchards is a Bull Lake outwash surface (Cole and Sexton, 1981). Surfaces to the south are younger (Holocene) alluvial terrace gravels of the Gunnison River.

0.3

93.7 Junction of Colorado Highways 92 and 65. Turn north
(127.5) onto Highway 65 (reverse trip turns east onto Highway 92). Distinct pediments are developed on Mancos Shale at the base of Grand Mesa. Several levels can be identified, especially to the northwest toward Grand Junction. Outcrops of Mesaverde are visible midway up Grand Mesa, but most exposures of Mesaverde, Wasatch, and Green River Formations are covered with soil, vegetation, and slide or glacial debris.

0.7

94.4 Bridge over Gunnison River.
(126.8)

0.5

94.9 Traveling along Surface Creek Mesa. Surface Creek
(126.3) west of the road, collects Tongue Creek and many smaller streams on this part of the mesa. Storage reservoirs abound on the mesa, providing late season irrigation water to the orchards and croplands on these slopes and in the Grand Valley. Roadcut for the next 0.8 mile is in Pleistocene outwash gravels of Surface Creek Mesa. These gravels contain mostly basalt fragments from Grand Mesa (in coarse sizes—pebbles to boulders). Pockets of brown and tan sand in this exposure are from sandy layers in the Mancos Shale and Mesaverde sandstones. The mesa rests on Mancos Shale here and on the Mesaverde north of Cedaredge.

0.7

95.6 Enter Orchard City.
(125.6)

0.2

95.8 Good view of clay "badlands" in Mancos Shale to the
(125.4) west.

0.9

96.7 Badland topography in the Mancos Shale extends sev-
(124.5) eral miles to the west along the base of Grand Mesa. Vegetation has a hard time getting established on Mancos Shale because moisture can not penetrate the clays of the soil, which also leads to serious erosion problems.

0.7

97.4 Good view to the north and northwest of Grand Mesa.
(123.8)

1.0

98.4 Good view of Mancos Shale desert to the west. Yellow
(122.8) weathering is common, but fresh exposures are olive gray. Occasional tan sandy and silty stringers are found in the Mancos, as well as pelecypods *Ostrea*, *Inoceramus*, and others, with some *Inoceramus* larger than 3 ft across. Preservation usually is poor.

0.3

98.7 Antelope Hill, 2 miles to the east, has orchards on top
(122.5) in a small veneer of well-drained pediment gravel overlying Mancos Shale. Occasional large basalt boulders have been transported to this area along the pediment surface. Some of these boulders probably are debris from prominences that once extended beyond the present basalt cap on Grand Mesa, 1.5 miles to the north. Jointing and subsequent caving of large wedges of the cap rock continually reduce the size of Grand Mesa.

1.5

100.2 Old Eckert townsite (Eckert is now part of Orchard
(121.0) City). Note the beautiful small church built from Grand Mesa basalt fragments. Many stones actually were cut for the building of the church.

2.3

102.5 Entering/leaving Orchard City. Pediment remnants to
(118.7) the east are developed on Mancos Shale and covered by up to 20 ft of basalt cobbles and boulders.

1.4

103.9 Entering/leaving Cedaredge. Roadcut is in Mancos
(117.3) Shale.

1.8

105.7 Entering/Leaving Cedaredge. Many small coal mines
(115.5) operated in the vicinity, mostly in the area of Tongue Creek, Ward Creek, and their tributaries west of Cedaredge. The largest mines were the Tomahawk, a surface mine that produced nearly 425,000 short tons in the late 1970s and early 1980s, and the Red Canon No.1, a deep operation that produced from the 1920s through 1983, with a total production of nearly 370,000 short tons, mostly between 1980 and 1983. In spite of very low sulfur and ash contents, coal in this area is on the borderline between high-volatile C and subbituminous A, and its lower Btu content has made it difficult to compete with higher rank coals farther east in the Paonia-Somerset area. Methane was never a problem in the underground mines in this area, although a small oil seep was encountered in the Red Canon No. 1 Mine.

0.3

106.0 Distinctive juniper-covered high surface to the east is
(115.2) Cedar Mesa, a Bull Lake(?) pediment.

0.4

106.4 Occasional large basalt boulders from Grand Mesa are
(114.8) seen on this higher part of Surface Creek Mesa.

4.1

110.5 Road junction with County Road 2225 to the west.
(110.7) Views south are of the San Juan Mountains, Uncompahgre Plateau to the southwest, and West Elk Mountains to the east. West, the Abajo and La Sal Mountains may be visible near Moab, Utah.

1.0

111.5 Road cut exposes slide material (or till) on uphill side
(109.7) of road.

3.7

115.2 Cross old Grand Mesa road and Kiser Creek. Mesa-

(106.0) verde is buried beneath these gravels.

1.6

116.8 Probably the highest exposure of Tertiary Green River
(104.4) Formation along the road on the south side of Grand Mesa.

0.5

117.3 Enter/exit Grand Mesa National Forest. The south side
(103.9) of the Mesa has fewer conifers, more aspen, and the scrub oak zone is higher than on the moister, north-facing slope.

2.2

119.5 Large unstable slump blocks and creep make the road
(101.7) uneven, requiring constant repairs.

0.5

120.0 Road crosses the base of joint-induced slump blocks,
(101.2) which are common all along the Mesa rim. Many of the lakes along the edge of the mesa, like Island Lake, have formed in detachment chasms produced by this slumping. The basalt cap on the mesa covers river gravels that are rich in quartzite cobbles, suggesting that the flows occupied pre-existing river valleys. The cap consists of three to five separate flow sequences, with a total thickness of about 250 ft.

1.4

121.4 **STOP 2 (2)—Grand Mesa (Lunch)**
(99.8)

Ward Lake-Carp Lake turnoff. This road leads to numerous lakes with a large community of summer homes and continues across the mesa to Collbran. The view from our lunch stop is south toward the San Juan Mountains on the horizon. West is the Uncompahgre uplift, which redirected the Gunnison and Colorado Rivers, forcing the river around the north end of the structure. A prominent deep valley, Unaweep Canyon, containing only minor drainage, cuts the uplift today, and provides evidence for this river piracy. Beyond the Uncompaghre uplift the La Sal Mountains and Abajo

The beautiful basalt stone church in Orchard City, Colorado. See description at mile 100.2. Photo by Jack Rathbone.

Mountains of Utah's Canyon Lands may be visible. They are located in the Paradox basin. Black Canyon of the Gunnison is south-southeast, and West Elk Mountains are southeast.

0.2

121.6 (99.6) Island Lake turnoff.

1.0

122.6 (98.6) Resort and store on south overlooking Island Lake. This lake is a favorite for ice fishers because the hike from the road to the lake is only several hundred yards. With 6 ft or more of snow common in winter, most lakes are accessible only by snowmobile, skis, or snowshoes.

0.5

123.1 (98.1) Leon Peak is visible to the east along the crest of the mesa and is probably part of the original vent-filling conduit fissures.

1.4

124.5 (96.7) Highest point on Colorado Highway 65 (elev. 10,850 ft) the and Mesa/Delta County line. Lakeview parking area and scenic trail to the east. From the southwest end of the loop trail is a geologically and scenically spectacular view of the south landslide bench, numerous sag lakes, valley erosion of 5,000 ft into Mancos Shale, and arch of the Uncompahgre Plateau. The basalt on the rim shows glacier shaping, striations, and polish produced by rocks embedded in the ice that was spilling from the ice cap over the rim at this location.

1.6

126.1 (95.1) Junction with Lands End road. Lands End overlooks Grand Junction and west with superb scenery. It is well worth the drive.

1.0

127.1 (94.1) Glaciated terrain is clearly visible across the top of the mesa. Ten meters east of the road, as it reaches the crest of the hill, are several roche moutonnée of basalt with striations and streamlined forms that indicate ice movement from the central ice cap of Grand Mesa across the rim and down the drainage to the north and west. Basalt erratics sit on the whalebacks. Marginal meltwater channels are right and left of the road as it crosses the causeway of fill dirt over the marsh land. Channels that originally formed against receding ice now are the headwaters of Kannah Creek. Terminal moraines of the last (Grand Mesa: Pinedale: Wisconsin) ice cap are 2.5 miles to the west.

0.2

127.3 (93.9) Turnout on west side of the road with scenic overlooks. Good view of joint-induced "grabenlike" topography to the west that clearly shows how the "World's Largest Flattop Mountain" is being reduced in size by slope failure. Roadcut to the east displays three separate lava flow units separated by baked soils from which "cooked" snails and plant remains have been collected. The road descends off the basalt cap for 350 vertical feet. This grade in winter has horribly gusty winds, rockslides, avalanches, and roadbed failures, but despite these conditions, the road is rarely closed over the mesa.

2.7

130.0 (91.2) A 300-ft cliff of basalt on small isolated point to the south. Less than a mile to the northeast are several lakes that lie in sags created by slump-block rotation.

1.0

131.0 (90.2) Exit for Mesa Lake Resort, which has cabins for rent and a summer restaurant. The nearest lakes are always stocked with rainbow and brook trout, and an occasional cutthroat native is taken.

0.3

131.3 (89.9) Parking for Jumbo Reservoir.

0.6

131.9 (89.3) Road swings westward. This relatively flat bench contains many sags and undrained depressions produced by large-scale slumping of the north rim of the mesa. Failure is deep seated and probably carries to mudstones of the Wasatch Formation. The basalt cap has been broken, and blocks several hundreds of feet wide and thousands of feet long have rotated and subsided from the rim along progressive failure planes.

0.3

132.2 (89.0) Turnout. Good view of Plateau Creek drainage and Roan Cliffs to the north. Landslide scars are visible below the road.

0.1

132.3 (88.9) Area of numerous landslides. The road traverses unstable Green River Formation, and this area is prone to landslides and creep. The road for the next mile was badly damaged by major landsliding in 1983.

1.2

133.5 (87.7) Turnout for small recreation area. This is the site of the old Grand Mesa Ski Area, run by the Grand Mesa Ski Club, a group of local volunteers, until Powderhorn Ski Area opened in 1966. Two large Poma lifts and several rope tows and associated ski runs were used, but now only brush-filled liftlines and ski runs are visible.

0.7

134.2 (87.0) Leaving/entering Grand Mesa National Forest.

0.7

134.9 (86.3) Roadcut on south side is in Green River Formation. Soil and basalt talus are the typical surface materials, but the highway cut has exposed bedrock.

0.8

135.7 (85.5) Road to Powderhorn Ski Area to the south. Base of the ski runs is at the base of the steeper slope where Green River Formation is protected by the basalt cap of Grand Mesa. Lifts climb to the top of the cap rock. The break in slope at the base of the steeper slopes also marks landslide and mudflow deposition of material from above. Upper terminals of the chair lifts end on West Bench, a large-scale slump feature of the basalt cap.

0.7

136.4 (84.8) Mesa Creek and Coon Creek (about 1 mile to the east) drain a broad amphitheater from Chalk Mountain on the west to "The Breaks" on the skyline to the east. This gently sloping shoulder of Grand Mesa is underlain by Wasatch and Green River Formations, but it is covered by soil and debris from slides and glacial material.

1.6

138.0 (83.2) Broad road loop near the Half Circle C Ranch is to establish a good road grade. The new road to the top of the mesa was constructed by the U.S. Bureau of Public Roads in the 1950s.

1.7

139.7 (81.5) Glimpse of Jerry Creek Reservoir on the north side of Plateau Creek. Jerry Creek Reservoirs are two storage facilities for the Ute Water District, which provides domestic water for areas outside the cities in the Grand Valley.

0.5

140.2 (81.0) Roadcut mostly in colluvium on east with Wasatch Formation.

0.7

140.9 (80.3) Chalk Mountain, 4 miles to the west, is "white" because of exposures of Green River Formation. About 0.5 mile to the east are variegated clays and gravels in the Wasatch Formation.

1.0

141.9 (79.3) Scattered large rounded basalt boulders from the cap rock of Grand Mesa, directly south. Grand Mesa, the "World's Largest Flattop Mountain."

1.3

143.2 (78.0) Enter the town of Mesa. Note dip of the rocks to the north and into the Piceance basin. View north is across the Colorado River and De Beque Canyon. Skinner Ridge and South Shale Ridge are visible where Chevron and Conquest Oil are attempting production from Cameo coal seams. Coal depths range from 3,500 to 5,500 ft. Roan Cliffs, containing the oil shale, are slightly to the east.

0.7

143.9 (77.3) The "Beehive", an appropriately named promontory, is visible 2 miles east. The hill is a 500-ft exposure of upper Mesaverde Group rocks.

0.5

144.4 (76.8) Roadcut for the next 0.3 mile is through a Pleistocene terrace cut on upper Mesaverde.

0.1

144.5 (76.7) Recent slides across Mesa Creek to the west.

0.2

144.7 (76.5) Roadcut exposes Mesaverde on east side of the road.

0.1

144.8 (76.4) Buff sandstone ledges of the Mesaverde across the creek to the west.

0.1

144.9 (76.3) Intersection of Colorado Highways 65 and 330. Remain on Highway 65. Highway 330 proceeds east through Plateau gas field (Mesaverde production) to Collbran. At Red Mountain, near Collbran, Resource Enterprises Inc., with funding from the Gas Research Institute (GRI), drilled three deep coal-seam research wells to test Cameo coal-gas potential. The third hole was a short-radius lateral drainhole drilled from a vertical wellbore. The coals were found to be extremely low in permeability; however, coal-gas contents suggested potential if higher permeability coals could be located. Numerous reports are available describing the Deep Coal Seam Project.

0.1

145.0 (76.2) Cross bridge over Plateau Creek.

0.6

145.6 (75.6) Double N Ranch to the west. This is an old stage coach station.

0.5

146.1 (75.1) Large alluvial fan fills the south side of the valley.

0.3

146.4 (74.8) Wellhead west on bluff. This is still considered the Plateau field.

0.1

146.5 (74.7) Outwash gravels form surfaces on south side of valley.

0.4

146.9 (74.3) Cross bridge over Shire Gulch, type locality of the Shire Member of the Paleocene-Eocene Wasatch Formation. South edge of Shire Gulch gas field, which is centered about 1 mile to the north. Production is from Morrison, Cedar Mountain, Dakota, Mancos, and Mesaverde rocks.

0.1

147.0 (74.2) Valley of Plateau Creek widens ahead. This is the possible westward terminus of ice tongue(s) during the late Pleistocene.

0.5

147.5 (73.7) Dike construction on the left protects Ute Water Pipeline and highway from flooding.

0.3

147.8 (73.4) Excellent view of slump features in the upper Mesaverde across the creek to the west.

0.9

148.7 (72.5) Outwash gravels (Bull Lake?) in cut on right. Note landslide topography across valley.

0.5

149.2 (72.0) Massive channel-fill sandstone forms cliff to the east.

0.3

149.5 (71.7) Large alluvial fan from wash on south side of Plateau Creek. Boulders are Grand Mesa basalt (Pliocene).

2.8

152.3 (68.9) High cliff to the east formed by thick fluvial channel-fill sandstone in the middle Mesaverde.

0.5

152.8 (68.4) Trailer and ranch house to west. Note intricate festoon cross-bedding in fluvial channel-fill sandstone in cliff to the east. Distance to the Colorado River from the meander is only 0.2 mile.

0.5

153.3 (67.9) Ranch buildings to east and west. This meander has cut northward to within 0.15 mile laterally of the Colorado River on Beavertail Meander. The bed of Plateau Creek is about 75 ft higher than that of the Colorado River at this point.

0.8

154.1 (67.1) Junction with north branch of Colorado Highway 65.

0.1

154.2 (67.0) Cross Plateau Creek.

0.5

154.7 (66.5) Parking area and information sign on west side.

0.1

154.8 Mesaverde sandstone "toreva" block to east is called
(66.4) "Praying Bear."

0.1

154.9 Grand Mesa Exit. Turn north onto Interstate 70, (re-
(66.3) verse trip turns east onto Colorado Highway 65). Bridge over Plateau Creek. Upriver in the Colorado is the diversion dam of the U.S. Bureau of Reclamation's Highline Project. Completed in 1915, it annually diverts 730,000 acre-ft of water for power and for irrigation of 35,000 acres in Grand Valley. An additional 35,000 acres are irrigated from other diversions downstream.

0.5

155.4 U.S. Bureau of Reclamation diversion dam. See notes at
(65.8) mile 154.9 (66.3).

0.6

156.0 Beavertail Mesa and meander loop of the Colorado
(65.2) River are to the south. (See note for mile 153.3.) Cross river.

0.3

156.3 Highway tunnel through Mesaverde sandstone.
(64.9) Colorado River has cut De Beque Canyon into Mesaverde rocks. The west end of the canyon also marks the edge of the Piceance basin. Mesaverde coals (Cameo group) crop out and are mined along the basin margin. Coal seams up to 40 ft thick generally are low in sulfur and high in volatiles. These same rocks, dipping east into the basin, produce gas at depth. Note the lenticular nature of the sands, which assists in trapping gas. Generally though, structural features are required for commercial production from lenticular Mesaverde sands.

0.3

156.6 Cross Colorado River.
(63.9)

5.8

162.4 East end of De Beque Canyon. Cliffs to the north are
(58.8) the Roan Cliffs, composed of Eocene Green River Formation. Below the Green River is Wasatch Formation. At the top of the cliffs is basal Uinta Formation. See mile 156.3 for De Beque Canyon discussion and mile 180.3 (40.9) for oil shale discussion.

1.5

163.9 Heron colony in trees by river.
(57.3)

2.4

166.3 De Beque anticline to the west.
(54.9)

1.1

167.4 De Beque Exit. A gentle arch that can be seen in the
(53.8) Paleocene-Eocene Wasatch Formation is known as De Beque anticline. Chevron and Conquest Oil and Gas are attempting coalbed gas completions just northwest of here. Conquest is on South Shale Ridge, 1 mile northwest of De Beque, and Chevron is farther north on Skinner Ridge. Target Cameo coals lie at depths between 3,500 ft and 5,500 ft. To date, reported

De Beque Canyon, cut by the Colorado River, exposes along its length the entire Mesaverde section. Photo by Jack Rathbone.

Mt. Callahan, north of highway near Parachute Colorado, is capped by basalt flows, probably of the same origin as those on the Battlement Mesa and Grand Mesa. Photo by Jack Rathbone.

production is limited; however, neither company has pulled out from their projects. Published production statistics show gas with minimal associated water production.

1.6

169.0 (52.2) Bridge over Colorado River. Along this stretch of the Colorado River, waterfowl and birds of prey, including bald eagles, are common. On the skyline directly east is Mt. Logan (elevation 8,413 ft) and the flare stack of the Occidental Oil Shale Inc. underground in situ retort on its Logan Wash project. In 1982 two retorts were burned, producing 196,000 bbl of oil. The project was completed with these burns.

2.5

171.5 (49.7) Garfield County line. Good exposures of Green River and Wasatch Formations to the west and north. The contact between formations generally is marked by a color change from pinkish-buff Wasatch to pale greenish-gray Green River. Southeast is Battlement Mesa, a basalt-capped mesa similar to Grand Mesa but smaller.

8.8

180.3 (40.9)

STOP 3 (or 1)—Parachute Creek Shale Oil Facility

Grand Valley (Parachute) exit. Exit Interstate 70, turn north. The road log continues in either direction exclusive of the short drive up Parachute Creek. Unocal's Parachute Creek shale oil plant and mine is the only commercial-scale facility ever operated in the United States. During eight years of startup and full operation, 4.5 million barrels of 34° gravity API syn-crude was produced.

Construction of the 10,000-BOPD design-capacity retort plant was completed in August 1983 at a cost of $654 million. Trial operations and modifications of the retort scraper and shale cooling system continued for three years. The first sustained syncrude production of eight days was achieved in 1986. Peak production from the retort system of 7,000 BOPD was reached in 1989. During 1990, Unocal's best year, the operation averaged 6,000 BOPD, producing 1.5 million barrels of oil.

The 4.5 million barrels of syncrude output makes the Parachute Creek operation the largest shale-oil effort in U.S. history. However, sustained production rates high enough to achieve a positive cash flow were never reached. Unocal suspended operations in June 1991 to study alternative technologies to improve the economics.

Unocal received over $114 million of a $400 million federal price support for developing the technology and producing the 4.5 million barrels of oil. In 1990 syncrude price averaged $24/bbl. The support price payment of $25/bbl allowed a total $49/bbl sold. Primary mining operations were directed toward the Long Ridge shale zone in the Parachute Creek Member of the Green River Formation. The Long Ridge zone contains 34 gal of recoverable oil per ton of shale, which equates to a project reserve of 1.6 billion barrels of crude oil. An additional 2 billion barrels of oil are recoverable in lower quality zones.

2.5

182.8 (38.4) Barrett Resources gas-compression station north of the highway.

4.4

187.2 (34.0) Rulison Exit. Several miles south of here was the site of Project Rulison, a 40-kiloton underground nuclear test designed to stimulate natural gas production from tight Mesaverde sands. The test occurred in September 1969. Reports indicated that the natural gases contained a small amount of radiation.

2.3

189.5 (31.7) Overpass of Anvil Points Road. Anvil Points oil shale mine and retorting facility to the north. Entrance to the underground shale mine is barely visible at the top of the road winding up the face of the Roan Cliffs. Surface retort facilities are located at the base of the cliff about 0.5 mile north of here. U.S. Bureau of Mines operated the Anvil Points research center here from 1944 to 1953. Paraho began operations in 1974 and has retorted shale from Israel and Morocco at this plant along with Green River shale. Paraho is now processing oil shale into road-surfacing material and claims it is successful.

0.9

190.4 (30.8) On the bluff overlooking the southward meander in the Colorado River is the U.S. Department of Energy's (DOE) Multiwell Experiment (MWX) site operated by CER Corporation.

The purpose of the experiment was to develop technology for improving economic recovery of natural gas from low-permeability sandstone reservoirs. MWX consists of three closely spaced wells drilled to total depths below 8,000 ft. The wells serve as a subsurface laboratory for examining interwell geologic and reservoir properties and for testing stimulation effectiveness. MWX features a comprehensive formation evaluation program with more than 4,000 ft of core recovered and numerous high-quality conventional and experimental logs tested. The MWX wells have been subjected to eight major hydraulic fracturing treatments and extensive well testing.

In 1990 and 1991, CER drilled a slant/horizontal well through the Mesaverde Group at the site for DOE. The SHCT No. 1 was designed as a completion technology comparison with that of the MWX. Lenticular, fluvial, and paludal sandstones and coals were penetrated at 60°, and the Cozzette Sandstone (marine) was drilled nearly horizontal. A pressure core taken in one of the thinner coals indicated a gas content of 765 SCF/T. The well is being tested through 1991.

2.4

192.8 (28.4) West Rifle Exit.

1.8

194.6 (26.6) To the north are remnants of Union Carbide Corporation's Rifle plant, which processed uranium-vanadium ore during the period 1959-1972. Sources of the ore were mines in the Rifle Creek, Meeker, and Maybell, Colorado, areas. When ore processing ceased, the plant continued to treat vanadium-bearing solutions from the Union Carbide plant at Uravan, Colorado. Various vanadium products made were used principally by the steel industry. Note the "stabilized" tailings pile. Taxpayers' dollars are and will be spent hauling away not only the scrap from this plant, which can be seen in various piles around it, but the tailings themselves as part of DOE's tailings removal program.

1.8

196.4 (24.8) Rifle Exit. Grand Hogback to the north and east.

3.4

199.8 (21.4) Mamm Creek Exit. Southeast of this point is Mamm Creek gas field, discovered in 1965, which produces from three completed zones in Williams Fork Formation sandstone (Mesaverde Group). The field is now shut in, having produced slightly over 0.88 BCF.

During 1986 and 1987, GRI and Resources Enterprises Inc. participated with Koch Exploration to evaluate Piceance basin deep coals at Mamm Creek. Koch Exploration's Mobil 11–20 well was plugged back to approximately 7,700 ft, about 700 ft below the Cameo coals. Measured permeability for the thick (>40 ft) Cameo "A" coal seam was less than 1 md; therefore, a large stimulation would be required for production. Under extensive and detailed monitoring, a large cross-linked gel and sand frac was performed on the coal. After treatment, production was less than 5 MCFD with very little water. This was another indication that deeply buried coals with minimal permeability would be a difficult challenge in unlocking their tremendous gas reserves.

3.6

203.4 (17.8) Town of Silt is north of Interstate 70. Ridges to the north on the skyline are composed of steeply dipping

Wasatch Formation and Mesaverde strata in the Grand Hogback. South of Interstate 70, the distant skyline is formed by nearly horizontal Wasatch and Green River Formations, locally capped by Tertiary basalts, like those on Grand Mesa. The valley here is occupied by low hills of Wasatch Formation and Ohio Creek Member of the Mesaverde Group.

0.5

203.9 (17.3) Divide Creek Road heads south from the south access road of Interstate 70 and up Divide Creek about 15 miles to Uncle Bob Mountain, which is visible on the skyline to the south-southeast. Uncle Bob Mountain is the high area of Divide Creek anticline, a large gas-producing structure in eastern Piceance basin.

In the 1950s, Sun Oil Company developed large gas reserves in fractured, lower (marine) sandstone of the Mesaverde Group on this structure. Oryx Energy Company (a derivative of Sun) currently is developing the coalbed gas potential in coal zones above the marine sandstones in the Mesaverde at depths of 3,000 to 4,250 ft. Among the wells drilled recently (since 1988), two or three are each producing an average 450 MCFD from the Cameo coal zone (lower Williams Fork Formation sandstone). These wells are still dewatering.

The U.S. Forest Service (USFS) has surface management responsibilities for a large part (18,893 acres or 73 percent) of lands in the 25,746-acre Divide Creek unit. The Bureau of Land Management (BLM) has responsibility for managing all Federal minerals and manages part of the surface in the unit (2,768 acres or 11 percent).

An Environmental Impact Statement was prepared in 1991 by the USFS in cooperation with BLM to evaluate and disclose potential environmental impacts from continued development of coalbed methane in the Divide Creek Unit. If approved, during 1991 and 1992, Oryx plans to drill 20 additional coalbed methane wells.

1.0

204.9 (16.3) Mile marker 99 on Interstate 70. View east is across the Grand Hogback to the high area of the White River uplift. View west is toward Anvil Points with bluffs of Green River Formation. Outcrops along road, north side, are Wasatch Formation sandstone and shale.

1.5

206.4 (14.8) Steeply dipping strata to the north are in the Grand Hogback, a monocline that bounds the western and southwestern flanks of the White River uplift and forms the eastern boundary of Piceance basin. Wasatch Formation and Mesaverde Group are exposed here.

2.0

208.4 (12.8) Ohio Creek Member of the Mesaverde Group is the white sandstone forming the dip slope to the north. Above the Ohio Creek the basal Wasatch Formation is exposed.

1.8

210.2 (11.0) Parking area along the south side of Interstate 70. Across river greenish-gray shale and siltstone of Jurassic Morrison Formation are overlain by Dakota Sandstone. Continuing upsection to the south, Mancos Shale sequence includes beds comparable to eastern Colorado units—Graneros Shale; a light-gray calcareous zone representing the Greenhorn interval; dark shale of the Carlile interval; and a second light-gray calcareous zone representing the Niobrara interval. In the foothills of the hogback these beds are overlain by over 2,000 ft of upper Mancos Shale, with thin, poorly exposed marine sandstone of the lower Mesaverde Group forming indistinct ledges in the trees near the top of the hogback.

North of Interstate 70, along old U.S. Highway 6, the stratigraphic interval just above the Dakota Sandstone includes black shale and thin beds of slabby, siliceous shale that probably represent the southwesternmost evidence of the Mowry Shale facies.

1.0

211.2 (10.0) New Castle Exit 105. Between the 1880s and World War I, New Castle was the center of an important coal mining area. The mines declined rapidly with the closure of the Colorado Midland Railroad in 1918. A Midland Route stone bridge pier can be seen in the middle of the river opposite the center of town. John Lawson, who would later become a well-known union organizer and leader of the strike that culminated in the Ludlow Massacre near Trinidad in 1914, was shot here in an argument with a local coal operator in 1903. Although mines in this district were not particularly gassy, they were prone to spontaneous combustion, and the dust was particularly explosive. A stone marker along U.S. Highway 6 at Exit 105 commemorates the miners lost in several violent dust explosions. Both these factors led to persistent mine fires that produced massive clinker in Mesaverde strata to the northwest and odd spots in the hogback south of the river. Steam and smoke occasionally can be seen in several locations, clear indicators that coal adjacent to the long-abandoned mines is still burning. The principal coalbed mined here was the Wheeler, a Cameo (Fairfield) group coal, which reached thicknesses in excess of 40 ft. Massive sands north of the town are the lowermost sands in the Mesaverde Group. Numerous faults occur along the hogback.

1.2

212.4 (8.8) Good Mancos Shale exposure to the north. Red horizon south of highway in Grand Hogback is a clinker bed in the Mesaverde. Portals and water tank low on the hogback slope to the south are tunnel openings to the Storm King Mine. These tunnels are to encounter steeply southwest-dipping coals of the Williams Fork Formation (Mesaverde Group) from below and provide a uniform downgrade to the surface for hydraulic mining and sluicing of the coal. This project has undergone several reorganizations since it was initiated in the mid-1970s and once again appears to be on track.

0.8

213.2 (8.0) Dakota Sandstone caps the hill on south side of the highway.

1.2

214.4 (6.8) Jurassic Morrison Formation overlain by Lower Cretaceous Dakota Sandstone in the exposure to the south.

Steep dips are part of the Grand Hogback structure.
0.3

214.7 (6.5) Outcrops south of the Colorado River are Jurassic Morrison Formation overlying a prominent, thin (20 ft), yellowish-gray interval of Jurassic Entrada Sandstone, in turn overlying brick-red Triassic Chinle Formation.
0.3

215.0 (6.2) Canyon Creek Exit.
0.7

215.7 (5.5) Brick-red Triassic strata form the slopes south of the Colorado River. Permian South Canyon Creek Member of the State Bridge Formation is the thin (dolomitic) layer in the red beds just above the light-colored Pennsylvanian and Permian Weber Sandstone. The highway parallels the Grand Hogback.
0.7

216.4 (4.8) Red beds in slopes on both sides of the highway are Pennsylvanian-Permian Maroon Formation, coarse-grained detritus of which was derived mostly from the ancestral Uncompahgre highland to the southwest. Ripple marks are common and here can be seen on the north side.
0.9

217.3 (3.9) South Canyon Exit. A hot spring issues from the top of the Dakota Sandstone about a mile up the canyon. Several coal mines produced from the Mesaverde Group 2 to 3 miles farther. As in the New Castle area, underground fires are burning in these old workings.
1.1

218.4 (2.8) Red-bed sequence consists of Pennsylvanian and Permian Weber Sandstone at the base, overlain by Lower Triassic and Permian State Bridge Formation, Upper Triassic Chinle Formation, and Middle Jurassic Entrada Sandstone.
0.8

219.2 (2.0) West Glenwood Springs Exit 114. To the north are south-dipping beds of Mississippian Leadville

The steeply dipping eastern monocline exposing an excellent section of the Mesaverde Formation. Note vehicle in foreground and absence of interstate highway. Photo by Jack Rathbone.

Limestone and Pennsylvanian Belden Formation.

1.4

220.6 (0.6) Outcrops north of Interstate 70 are dark-gray calcareous shale and nodular limestone of Pennsylvanian Belden Formation.

0.5

221.1 (0.1) Glenwood Springs Interstate 70 exit and entrance. Quaternary river gravels exposed in roadcut to southwest consist of buff gypsiferous beds of the Eagle Valley evaporitic units of Pennsylvanian-Permian Maroon Formation. To the southeast is Mt. Sopris (elevation 12,959 ft).

Quarries high on slope north of town are Mississippian Leadville Limestone. Crushed, nearly pure calcitic limestone was used in processing of sugar beets at Delta, Colorado.

0.1

221.2 (0) Intersection of Grand Avenue and U.S. Highways 6 and 24, Glenwood Springs.

End/Beginning of trip.

REFERENCES

Birkeland, P.W., 1973, Use of relative age-dating methods in a stratigraphic study of rock glacier deposits, Mt. Sopris, Colorado: Arctic and Alpine Research, v. 5, no. 3, p. 401–416.

Cadigan, R.R., Felmlee, J.K., and Rosholt, J.N., 1976, Radioactive mineral springs in Delta County, Colorado: USGS Open-File Rept. 76–223, 39 p.

Cole, R.D., and Sexton, J.L., 1981, Pleistocene surficial deposits of the Grand Mesa area, Colorado, *in* Epis, R.C., and Callender, J.F., eds., Western Slope Colorado: New Mexico Geological Society 32d Field Conference Guidebook, p. 121–126.

Collins, B.A., 1977, Geology of the Coal Basin area, Pitkin County, Colorado, *in* Veal, H.K., ed., Exploration frontiers of the Central and Southern Rockies: Rocky Mountain Association of Geologists Symposium Guidebook, p. 363–377.

Freeman, V.L., and Bryant, B., 1977, Road logs for Rocky Mountain Association of Geologist Snowmass Conference, September 29-30, 1977, *in* Veal, H.K., ed., Exploration frontiers of the Central and Southern Rockies: Rocky Mountain Association of Geologists Symposium Guidebook, p. 469–497.

Godwin, L.H., 1968, Geologic map of the Chair Mountain quadrangle, Gunnison and Pitkin Counties, Colorado: U.S. Geological Survey Geologic Quadrangle Map GQ-704, scale 1:24,000.

Johnson, R.C., and May, F., 1980, A study of the Cretaceous-Tertiary unconformity in the Piceance Creek basin, Colorado—The underlying Ohio Creek Formation (Upper Cretaceous) redefined as a member of the Hunter Canyon or Mesaverde Formation: USGS Bulletin 1482–B, 27 p.

Junge, W.R., 1978, Geologic hazards, Hotchkiss-Paonia Reservoir area, Delta and Gunnison Counties, Colorado: Colorado Geological Survey Open-File Rept. 78–12.

Murray, D.K., Fender, H.B., and Jones, D.C., 1977, Coal and methane gas in the southeastern part of the Piceance Creek Basin, Colorado, *in* Veal, H.K., ed., Exploration frontiers of the Central and Southern Rockies: Rocky Mountain Association of Geologists Symposium Guidebook, p. 379–405.

Stover, B.K., and Soule, J.M., Surficial geologic map of the 7-1/2' Carbondale quadrangle, Garfield County, Colorado: Colorado Geological Survey unpublished map.

Wellborn, J.F.F., 1982, Stratigraphy of the Mesaverde Formation on the Mt. Gunnison coal property, Gunnison County, Colorado, *in* Southeastern Piceance basin, western Colorado: Grand Junction Geological Society, p. 65–67.